Joanna Bourke is Professor of History at Birkbeck College, University of London. Her books include *Dismembering the Male: Men's Bodies, Britain and the Great War*, *Husbandry to Housewifery: Women, Economic Change and Housework in Ireland 1890–1914* and *Working Class Cultures in Britain 1890–1960: Gender, Class and Ethnicity*. *An Intimate History of Killing* was awarded the 1998 Fraenkel Prize in Contemporary History.

AN INTIMATE HISTORY OF
KILLING

FACE-TO-FACE KILLING IN TWENTIETH CENTURY WARFARE

JOANNA BOURKE

Granta Books
London

Granta Publications, 2/3 Hanover Yard, London N1 8BE

First published in Great Britain by Granta Books 1999
This edition published by Granta Books 2000

Copyright © 1999 by Joanna Bourke

A CIP catalogue record for this book
is available from the British Library.

3 5 7 9 10 8 6 4 2

Typeset by M Rules
Printed and bound in Great Britain
by Mackays of Chatham plc

Contents

List of Illustrations

All illustrations, with the exceptions of numbers 5 and 13, are reproduced with the kind permission of the Imperial War Museum. (While every effort has been made to trace copyright holders of illustrations 5 and 13, this has proved impossible. The publisher is prepared to make suitable acknowledgement in any reissue or reprint of the work.)

Acknowledgements

It would be hopeless to attempt a comprehensive list of the individuals who have helped me write this book. In particular, though, I have amassed a debt to a number of scholars and friends. Some of these people who have shared their ideas with me will wish that I had been less eclectic in my choice of theories and brave enough to develop their insights more fully: I thank them for their generosity. My warmest thanks go to Richard Evans, George Mosse, Avner Offer, and Jay M. Winter. In addition, I am immensely grateful to Jad Adams, John Barnes, Vincent Brome, Nicholas Brown, Roger Cooter, Marianne Elliott, David Feldman, David Fitzpatrick, Roy Foster, Robin Haines, Ken Inglis, Eric Leed, Iain McCalman, Oliver MacDonagh, Catherine Merridale, Margaret Mitchell, Bernard Porter, Dorothy Porter, Robin Prior, John Remy, Sue Rickard, Hilary Sapire, Naoko Shimazu, F. B. Smith, Kosmas Tsokhas, and my extended family. This book could not have been written without the generous financial help of the Australian National University, the Australian War Memorial, Birkbeck College, and the British Academy. Special thanks, however, must go to my colleagues at the Research School of Social Sciences and the Humanities Research Centre (both at the Australian National University) and to my colleagues and students in the History Department at Birkbeck College. Neil Belton of Granta provided valuable support from the book's early days and Jane Robertson was an admirable copy editor. Peter Robinson was my literary agent: he always believed in the project. Finally, a special thanks to Costas Douzinas, without whom the traumatic character of much of the research for this book would have been unbearable.

List of Abbreviations

AA: Anti-Aircraft Artillery
AIF: Australian Imperial Force
ARVN: Army of the Republic of Vietnam
ATS: Auxiliary Territorial Service
AWAS: Australian Women's Army Service
BEF: British Expeditionary Force
FA: Field Artillery
LDV: Local Defence Volunteers
MACV: (US) Military Assistance Command in Vietnam
MIGs: Mikhail Ivanovich Glinka (Soviet jet fighter plane)
NCO: Non-Commissioned Officer
NVA: North Vietnamese Army
PBI: Poor Bloody Infantry
PFC: Private First Class
PTSD: Post-Traumatic Stress Disorder
QARANC: Queen Alexandra's Royal Army Nursing Corps
RADC: Royal Army Dental Corps
RAF: Royal Air Force
RAMC: Royal Army Medical Corps
RUSI: Royal United Services Institute
SAS: Special Agents
SEAL: Sea, Air and Land Capacity
SOE: Special Operations Executive
SPCA: Society for the Protection of Children and Animals
TA: Territorial Army
VC: Victoria Cross (also Viet Cong)
WAAC: Women's Auxiliary Army Corps
WAC: Women's Army Corps
WRAC: Women's Royal Army Corps
WRAF: Women's Royal Air Force
WRNS: Women's Royal Navy Service

INTRODUCTION

The characteristic act of men at war is not dying, it is killing. For politicians, military strategists and many historians, war may be about the conquest of territory or the struggle to recover a sense of national honour but for the man on active service warfare is concerned with the lawful killing of other people. Its peculiar importance derives from the fact that it is not murder, but sanctioned blood-letting, legislated for by the highest civil authorities and obtaining the consent of the vast majority of the population. In the twentieth century, the two world wars and the Vietnam War bloodied the hands and consciences of thousands of British, American and Australian men and women. In this book, the combatants share their fantasies and experiences of intimate killing and, in the process, reveal themselves as individuals transformed by a range of conflicting emotions – fear as well as empathy, rage as well as exhilaration. These men surrender to irrational although sincere moral outrage, embrace the

idea of agency, find relief in agonizing guilt, and attempt to nego-
tiate pleasure within a landscape of extreme violence.

I have chosen to focus on the First World War, the Second
World War and the Vietnam War as three of the most influential
conflicts this century. Other wars, of course, have destroyed millions of lives but none were as terrible and as decisive as these
three wars for British, American and Australian servicemen and
civilians alike.

With bitter humour, a military padre in France during the
First World War stated the most obvious fact of all: 'The soldier's
business is to kill the enemy . . .' he preached, 'and he only tries
to avoid *being* killed for the sake of being efficient.'[1] The nature of
the combatant's 'business' was common knowledge. What is
striking, therefore, is the lengths some commentators will go to
deny the centrality of killing in modern battle. Of course, the
taking of human lives is not a *necessary* component of warfare:
buildings, military installations and farmland are equally valid
targets and there are always ways to injure human beings which
do not involve extermination. But, in the three conflicts examined in this book, human slaughter was at the heart of military
strategy and practice. This fact is glossed over by most military
commentators and denied by others. Accounts of the 'experience' of war prefer to stress the satisfaction of male bonding, the
discomforts of the frontlines, and the unspeakable terror of
dying. Readers of military history books might be excused for
believing that combatants found in war zones were really there to
be killed, rather than to kill.

This book aims to put killing back into military history. After
all, during each war British, American and Australian combatants
themselves had no illusions about the nature of their employment; there was no 'silencing' their savagery. Instead, there was
often a monstrous and multifarious celebration of violence. The
1918 Handbook for the 42nd East Lancashire Division exhorted
officers to 'be bloodthirsty, and never cease to think how you can
best kill the enemy or help your men to do so'.[2] Soldiers were

regaled with gory lectures given by a lecturer dubbed the 'King of No Man's Land' and instructors harangued timid students, reminding them that if they did not enjoy killing they should not be in the infantry.[3] In 1955 two senior American officers directed that 'the killing of an individual enemy with a rifle, grenade, bayonet – yes, even the bare hands – is the mission of the Army . . . This mission has no civilian counterpart.'[4]

In the frontlines, combatants strove to find a language to express their own experiences and this book draws primarily on the narratives of such combatants – men as diverse as Arthur Hubbard ('it makes my head jump to think about it', he stumblingly admitted to his mother after slaughtering three prisoners), Richard Hillary ('I had a feeling of the essential rightness of it all,' he explained when he shot a German pilot dead), and William Broyles (butchery was a 'turn on', he confessed). Men invited their girlfriends to share in the experience of killing, boasting that they were 'fiteing' every day with bayonets, killing 'too [sic] most days, never less than one' and that 'every one I gets under the ribs I thinks of you, mi [sic] dear, and it gives strength to my arm'.[5] They deceived neither themselves nor their families that their actions and fantasies were concerned with killing.[6] Whether killing was symbolic or tangible, the expectation of blood-letting was part of the experience of all soldiers on active service and much was made of this in their letters and journals. Neither combatants nor civilians could be unaware of the mass carnage of war nor of the part they were expected to play in this process of decimation.

Even combatants in those forms of military service which are generally regarded as unsoiled by the proximity of 'real' blood and guts we find emphasizing the rare moments of bloodiness. Artillerymen – usually miles away from their foe – are found describing the 'ghastly' sight of

dead bodies everywhere, some half propped against the trench wall, some lying on their backs, some with their

> faces in the mud . . . the shock was beyond belief. I stared
> at these poor German soldiers. Never have I forgotten the
> look on their faces.[7]

Aircraft artillery is regarded as even more bloodless (so much so that women were allowed to direct fire), yet it was occasionally possible to witness enemy crew jumping out of their aeroplanes and, to the cheering of people on the ground, falling to their deaths ('we didn't know whether to scrape them off or just paint over them', commented one subaltern in 1943).[8] The heavy, slow planes of the First World War meant that gunners were typically close to the enemy when firing.[9] Even during the Second World War, fighter pilots' 'Ten Commandments for Air Fighting' included never firing until they could see the whites of the foe's eyes.[10] On other occasions, fighter pilots could be required to fly low over destroyed U-boats, training their machine-guns on any surviving crew.[11] These excited stories of murder were not necessarily typical. As we shall see later, in particular theatres of war (the Pacific during the Second World War and some aspects of the war in Vietnam) and amongst certain branches of the armed forces (most obviously in elite units such as commandos and rangers, but more generally amongst infantrymen wielding bayonets and fire-throwers) killing was often intimate, though a large proportion of stories were fantastical combat narratives related for their cathartic and consolatory function rather than as an objective recital of 'experience'. Despite the statistical insignificance of face-to-face killing, such stories were of immense personal importance with combatants constantly emphasizing (and exaggerating) any rare moments of intimate killing. Such attention to the details of blood-letting is a feature throughout the informal (and formal) writing of war.

Although the act of taking another person's life can never be banal, time and time again in this book we will hear from men like John Slomm Riddell Hodgson whose letters seamlessly weave together domestic trivia with a narrative of murder, as when he wrote to his parents on 28 March 1915 and said, after thanking them for sending some socks, that: 'you can hardly have too many [socks] in service, especially if you have much murdering to do'. He continued in this vein throughout his wartime correspondence, writing a year later that he had been 'murdering all the time' since his last letter, 'except for one day when it poured with rain the whole day so we stopped in camp'.[12]

In this book, each chapter introduces an 'ordinary' combatant like Hodgson, usually men and women who fought in one of the two world wars or in Vietnam, drawn from a range of economic backgrounds and nationalities. They include members of the Regular forces as well as volunteers and conscripts; they all claim to be 'just like us'. Indeed, it is not necessary to look for extraordinary personality traits or even extraordinary times to explain human viciousness. Numerous studies of cruelty all show how men and women 'like us' are capable of grotesque acts of violence against fellow human beings.[13]

The two processes most commonly regarded as enabling violent acts in modern warfare are 'numbed consciences' and 'agentic modes'. Combatants were able to maintain an emotional distance from their victims largely through the application of (and almost exclusive focus upon) technology. During the Second World War, for instance, a counsellor at Fort Leavenworth observed the enthusiasm with which combatants discussed the 'relative merits of particular weapons'. He noted that physics and ballistics so preoccupied these men that they forgot the 'morbid' and murderous function of their metallic 'arms'.[14] The psychiatrist, Robert Jay Lifton, made a similar observation during the Vietnam War. In his book entitled *Home From the War* (1974), he argued that technology enabled 'numbed killing'. Skill in the performance of the 'job', internal competition for funds, and a 'technological imperative' (that is, the urge to make full use of any equipment) came to dominate the whole process.[15]

Undeniably, warfare has been transformed by the mechanization of the battlefield in the twentieth century. Technological warfare meant that progressively fewer men were required to actually engage in killing. During the 1914–18 conflict, for every soldier in combat there were eight additional soldiers in combat support roles. The ratio of rear-area support troops to combatants was twelve to one by the 1939–45 conflict and by the time of the Vietnam War, of the 2.8 million men who saw service in Vietnam less than 0.3 million faced battle.[16] The ratio of combatants to service personnel was even lower in particular branches of the armed forces, such as the air force. Technology was a factor in rendering the killing process more mechanical. During the First World War, it took (on average) 1,400 shells to kill one soldier and during the Battle of the Somme, a British gunner would have to fire thirty shells to hit one German. Opponents rarely saw each other. By the end of the Second World War, the ratio of soldiers to space was 1 man to every 27,500 metres (compared with 1 man to every 257 square metres during the American Civil War).[17] The long-distant and indirect character of 'area attacking weapons' such as shrapnel, bombs and gas meant that while people could regularly be seen dying, it was rarer to actually see them being killed. During the two world wars, shells, mortars and aerial bombs were responsible for the deaths of over two thirds of all casualties and (contrary to popular belief) less than one half of 1 per cent of wounds were inflicted by the bayonet.[18] In the words of Sir Winston Churchill, modern warfare consisted of the massing of 'gigantic agencies for the slaughter of men by machinery'. Killing was 'reduced to a business like the stock-yards of Chicago'.[19]

However, as we shall see throughout this book, technology still failed to render the dead completely faceless. Combatants used their imagination to 'see' the impact of their weapons on other men, to construct elaborate, precise and self-conscious fantasies about the effects of their destructive weapons, especially when the impact of their actions was beyond their immediate vision. In the words of William J. Simon's poem on the war in Vietnam, 'My Country':

> *They call me Chopper Jockey:*
> *Below in the jungle, carpeted*
> *with blood of men I have killed, I see the*
> *faces of men I have never seen.*[20]

Personal experiences and fantasies of killing, and the lengths to which servicemen went to 'see' the effects of combat, are also central to this book. In another way, personalizing the foe could be crucial to the moral and emotional well-being of combatants and formed a buffer against numbing brutality.

So while technology was used to facilitate mass human destruction, it did very little to reduce the awareness that dead human beings were the end product. What is striking is the extent to which combatants insisted upon emotional relationships and responsibility, *despite* the distancing effect of much technology. Indeed, one of the most disturbing features in the private letters and diaries of combatants is the extent to which they were not 'numbed'. Even the men interviewed by Lifton were painfully attempting to understand their role as agents of death. Indeed, the 'numbed' consciences may belong more to civilians who dispassionately observed the war taking place and regarded it as irrelevant to their lives than to combatants: men who actually killed were more liable to be forced to deal with their own tortured consciences.

An insistence upon personal moral integrity was typical of those engaged in fighting. The combatant was not an isolated individual: his actions were taken on behalf of the nation, a hierarchical military establishment, and an intimate, interdependent platoon – this was what distinguished martial combat from murder. In civilian as well as in martial contexts, the power of such institutions was frequently used to legitimate brutal behaviour: people slipped into an 'agentic mode' and acted in ways they would otherwise find unacceptable.[21] As we shall see in various chapters, most combatants clung to the rationale that they were 'merely obeying orders'. But the importance of this rationale should not be exaggerated. Obedience to higher authorities was fraught with difficulties: after all, what *was* the 'appropriate authority'? Furthermore, there was no necessary contradiction between following orders and accepting personal responsibility for one's actions.[22] While civilians in the past and present have been anxious to exonerate servicemen from responsibility for their actions in battle, combatants themselves were often anxious to accept their *own* agency and to judge and be judged for their deeds. Historians have been rightly cautious about passing judgement on people in the past,[23] but this must be done if we are to grant to servicemen and -women in the past an active, creative role in the making of their own histories. To a large extent, the men who fought in twentieth-century conflicts were civilians first, and servicemen only by historical mishap. These men were passionately engaged in elaborating ways of justifying killing in wartime, and most were eager to assume moral responsibility for their bloody deeds. As sentient humans, they insisted upon bearing a share of responsibility for their own actions: blaming a higher authority was certainly appealing, but only superficially. As we shall be seeing throughout this book, the purposeful actions of individuals were central to the telling of war stories. This is not to deny that issues of agency and responsibility were often uncertain: war presents individuals with irresolvable conflicts. But the notions of responsibility which combatants applied were no less real for being contradictory, flexible, and complex. Individuals in combat were not excused from making moral judgements and, while historians must hesitate about passing sentence, the moral universe of combatants cannot be ignored.

War stories

In war, there is only one incontrovertible 'truth': that of the man in his final death agony. The stories of all other individuals are necessarily fractured. Throughout this book, men and women speak – some through an apathetic silence, others through a barrage of hostility and denial. Men who 'were there' claim a higher knowledge than other commentators,[24] yet, in the heat of battle, experiences were often confused, indeterminate, and unarticulated. As a soldier in a Manchester Regiment described his recent experience in battle:

> Ay! but that was a fight. If only somebody could describe it. But, there, they can't, for it was dark, with very little room, and nobody could exactly say what he saw.[25]

Equally honest was a diary entry which began: 'One thing that I can remember seeing quite distinctly (although I know I never did, but dreamt it in the hospital – it was a true story nonetheless).'[26] We cannot even trust the statements of men recorded under the influence of so-called truth drugs. When hysterical servicemen were given barbiturates to enable them to recover from amnesia, they would recite emotional accounts of their combat experiences and lunge at psychiatrists in the belief that they were the enemy. Rather than being re-enactments of repressed memories, however, most of the events recollected were fictional or highly distorted versions of battle.[27] In a similar way, during the Vietnam War, combatants suffering from Post-Traumatic Stress Disorder were appalled to find themselves sitting next to veterans who were suffering flashbacks of battle, despite never having served in the frontlines. 'Those guys were having heavy flashbacks,' one combat veteran complained,

> I couldn't understand. I said, 'What y'all talking 'bout? You was in artillery. At the base camp. You fired guns from five miles away and talking 'bout flashbacks?'[28]

Despite such problems, and admitting that the pleasures of

spinning elaborate and completely fanciful yarns were substantial,[29] autobiographers and letter-writers still insisted upon the 'truth' of their stories. The authority of the statement 'I was there' was constantly asserted – even while the stories were being moulded to appeal to the audience. One example of this process can be taken from the correspondence of the Australian soldier, William K. Willis. In a letter to a young woman in 1917, he gave a long and extremely improbable description of his recent experiences wielding a bloody bayonet. He then wrote:

> I would like very much to know just how my letters appeal
> to you . . . I aim to make them interesting yet I fear some-
> times that the constant gossip of war may pall upon you. I
> realize that you people are to be excused if you treat the
> description of one of our 'stunts' as an everyday matter,
> because . . . one has to rely almost solely upon the newspa-
> pers which delight to gorge the ears of their readers with
> all kinds of fabrications. I write of the things which I
> happen to be in and of course they appear to me immense,
> but at times I think they bore instead of interest you. I
> always write the truth and never like anything to creep
> into my screeds that is likely to mislead people.[30]

Yet, as we shall see in Chapter 1, Willis's account of combat was highly fanciful, fitting into the conventions of combat narratives.

We must, however, take the stories told by men like Willis seriously. Many of the texts discussed in this book are the texts of trauma: they are the words of men who have killed (or attempted to do so) in battle. They are the texts of executioners, as much as of victims. Their stories are contradictory, consolatory and often fantastical, but bewilderment, hope and fantasy are the very stuff of human experience. Of course, in historical time, many things actually do (or do not) 'happen', but the very act of narrating changes and shapes the 'experience'. Men really did kill: but

from the moment of killing, the event entered into the imagina-
tion and began to be interpreted, elaborated, restructured. As the
Vietnam story-teller, Tim O'Brien, wrote in *The Things They
Carried* (1990):

> it's difficult to separate what happened from what seemed
> to happen. What seems to happen becomes its own hap-
> pening and has to be told that way. The angles of vision
> are skewed . . . And then afterwards, when you go to tell
> about it, there's always that surreal seemingness, which
> makes the story seem untrue, but which in fact represents
> the hard and exact truth as it seemed.[31]

Martial combat has a 'before' and an 'after'. Prior to battle,
men and women imagined what it would be like to slaughter a
fellow human-being and the armed forces attempted to prepare
their personnel to act aggressively and fearlessly. The first three
chapters of this book are largely concerned with the resultant
tension between imagination and subjective experience. Once
engaged in battle, men responded in a variety of ways. Chapters
4 to 6 compare men who acted heroically with those who acted
atrociously, and question the relative importance of love over
hate in propelling the bayonet thrust. The guilt and psychological pain which combatants sometimes felt are then given their
due weight. Morality both restrained men's aggressive impulses,
and enabled violence. Finally, Chapters 9 to 11 investigate groups
labelled 'non-combatants'. Padres, women and other civilians
are heard expressing their own violent desires. Indeed, when
examining the impact of killing, the martial imagination of non-
combatants was typically more brutal than that of servicemen
struggling with their consciences. Throughout this book, the
narrative of personalized killing remains central to men's experience of war, despite the fact that the face of the enemy was
usually veiled. Language and imagination are crucial. When
combatants did not actually *witness* the effect of their actions,

they imagined it. In this sense, war is not hell and the bayoneting soldier is no crazed beast. Intimate acts of killing in war are committed by historical subjects imbued with language, emotion and desire. Killing in wartime is inseparable from wider social and cultural concerns. Combat does not terminate social relationships: rather, it restructures them. Inevitably, fantasy permeates all the narratives. The purpose of this book is to look at the way men experience killing, how society organizes it, and to investigate the ways in which killing is embedded within human imagination and culture in the twentieth century.

Chapter 1

THE PLEASURES OF WAR

Some day when you are hunting in attic trunks
Or hear your friends boasting of their brave fathers,
I know that all excited you will ask me
To tell war stories. How shall I answer you?

R. L. Barth, 'A Letter to my Infant Son', 1987[1]

Stories of combat provide a way of coping with a fundamental tension of war: although the act of killing another person in battle may invoke a wave of nauseous distress, it may also incite intense feelings of pleasure. William Broyles was one of many combat soldiers who articulated this ambiguity. In 1984, this former Marine and editor of the *Texas Monthly* and *Newsweek* explored some of the contradictions inherent in telling war stories. With the familiar, authoritative voice of 'one-who-has-been-there', Broyles asserted that when combat soldiers were questioned about their war experiences they generally said that they did not want to talk about it, implying that they 'hated it so much, it was so terrible' that they would prefer it to remain 'buried'. Not so, Broyles continued, 'I believe that most men who have been to war would have to admit, if they are honest, that somewhere inside themselves they loved it too.' How could *that* be explained to family and friends, he asked? Even comrades-in-arms were wary among themselves: veterans' reunions were awkward occasions precisely because the joyous aspects of slaughter were difficult to confess in *all* circumstances. To describe combat as enjoyable was like admitting to being a blood-thirsty brute: to acknowledge that the decisive cease-fire caused as much anguish as losing a great lover could only inspire shame.

Yet, Broyles recognized, there were dozens of reasons why combat might be attractive, even pleasurable. Comradeship, with its bittersweet absorption of the self within the group, appealed to some fundamental human urge. And then – in contrast – there was the awesome power conferred upon *individuals* by war. For men, combat was the male equivalent of childbirth: it was the 'initiation into the power of life and death'. Broyles had little to say about the 'life' aspect, but argued that the thrill of destruction was irresistible. A bazooka or an M-60 machine-gun was a 'magic sword' or a 'grunt's Excalibur':

> all you do is move that finger so imperceptibly, just a wish
> flashing across your mind like a shadow, not even a full
> brain synapse, and *poof!* in a blast of sound and energy
> and light a truck or a house or even people disappear,
> everything flying and settling back into dust.

In many ways, war *did* resemble sport – the most exciting game in existence, Broyles believed – which, by pushing men to their physical and emotional limits, could provide deep satisfaction (for the survivors, that is). Broyles likened the happiness gener-ated by the sport of war to the innocent pleasures of children playing cowboys and Indians, chanting the refrain, 'bang bang, you're dead!', or to the seductive suspense adults experience while watching combat movies as geysers of fake blood splatter the screen and actors fall, massacred.

There was more to the pleasures of combat than this, said Broyles. Killing had a spiritual resonance and an aesthetic poignancy. Slaughter was 'an affair of great and seductive beauty'. For combat soldiers, there was as much mechanical elegance in an

M–60 machine-gun as there was for medieval warriors in decorated swords. Aesthetic tastes were often highly personal: some Marines favoured the silent omnipotence of napalm which made houses vanish 'as if by spontaneous combustion' while others (such as Broyles) preferred white phosphorous because it 'exploded with a fulsome elegance, wreathing its target in intense and billowing smoke, throwing out glowing red comets trailing brilliant white plumes'. The experience seemed to resemble spiritual enlightenment or sexual eroticism: indeed, slaughter could be likened to an orgasmic, charismatic experience. However you looked at it, war was a 'turn on'.

Crucially, in the context of this chapter, Broyles noted that men's responses to combat contained an element of the carnivalesque. To illustrate this point, he described what his men had done to a North Vietnamese soldier whom they had recently killed. They had propped the corpse against some C-rations, placed sunglasses across his eyes and a cigarette in his mouth, and balanced a 'large and perfectly formed' piece of shit on his head. Broyles described his reaction thus:

> I pretended to be outraged, since desecrating bodies was
> frowned on as un-American and counterproductive. But it
> wasn't outrage I felt. I kept my officer's face on, but inside I
> was . . . laughing. I laughed – I believe now – in part
> because of some subconscious appreciation of this obscene
> linkage of sex and excrement and death; and in part
> because of the exultant realization that he – whoever he had
> been – was dead and I – special, unique me – was alive.

This joyous celebration of the 'material bodily principle', this carnivalesque inversion of what was sacred, was a most potent combination.[2]

William Broyles was not unique in admitting to feelings of pleasure in combat. The enthusiastic glee expressed by many recruits at the idea of shedding human blood can, to some extent, be understood by looking at the complex ways in which martial combat has become an integral part of the modern imagination. Literature and films provide scripts more exotic and thrilling than everyday scenarios and, although such narratives do not directly stimulate imitation, the excitement they generate creates an imaginary arena packed with murderous potential and provides a linguistic structure within which aggressive behaviour might legitimately be fantasized. Furthermore, the archetypes of combat are seductive precisely because of their unreal quality.

Combat Literature and Films

Long before any prospect of real combat, boys and girls, men and women, created narratives of pleasure around acts of killing. Australian lads imaginatively cleared primitive Aborigines from artificial bushland; American kids fought off wild Indians in suburban backyards; English boys slaughtered beastly blacks on playing fields. Combat literature, martial films and war games attracted men to the killing fields. Of course, there was no direct relationship between immersion in combat stories and the urge to kill. Young girls were obviously also entranced by such fictions, although their pleasure derived as much from the fate of the heroine-in-love as from the competence of the hero-in-battle. Any analysis of the diaries, letters, and autobiographies of combatants will reveal the extent to which literary and cinematic images were adopted (and transformed) by men and women prior to combat. War stories constitute our most democratic 'basic training'.

In the three conflicts to be considered in this book – the two world wars and the conflict in Vietnam – literary and cinematic models for combat spanned a range of diverse styles, from popular comics and imperial adventure tales, to classics such as Bunyan's *The Pilgrim's Progress* and the Victorian romances of Tennyson and William Morris, to the very graphic war films which dominated cinematic screens from 1939. This canon has been analysed by innumerable historians and literary scholars, but most skilfully in the context of both world wars by Paul Fussell in his classic books, *The Great War and Modern Memory* (1975) and *Wartime: Understanding and Behavior in the Second World War* (1989) and by Jay Winter in *Sites of Memory, Sites of Mourning* (1995). Throughout the twentieth century, combat literature remained one of the most buoyant genres. Movie-houses also promoted battle enthusiastically. More than one third of all Hollywood feature films produced between 1942 and 1945 were war movies, largely due to the help of the War Department, Navy Department and the Office of War Information.[3] By the time of the Vietnam War, the Department of Defense had lost its monopoly on wartime images (although they remained successful in circulating propaganda films to independent and non-commercial stations throughout the country) and television had taken combat into everyone's living rooms.

It is not difficult to see the attraction of combat literature and films. Everything, and everyone, appeared as more noble and more exotic in such stories than in commonplace encounters. Despite the understandable emphasis that many historians have placed on the literature of disillusionment arising out of the ashes of war, patriotic and heroic depictions of combat never lost their attraction. Once our gaze is turned from a narrow canon represented by poets such as Wilfred Owen, Siegfried Sassoon and Edmund Blunden, 'high diction' with its stock phrases (baptisms of fire, transfigured youth and gallant warriors) emerges as the dominant language of war.[4] The same is true if we turn to films. Patriotic romances such as John Wayne's *The Green Berets* (1968) comforted viewers with a familiar sequence of mythical codes and archetypal characters who were all the more recognizable because they were overblown. In Francis Ford Coppola's film, *Apocalypse Now* (1979), with its exploration of the dark fantasies which enabled men to take pleasure in cruelty, audiences rejoiced in the invincible Colonel Kilgore filming his own private John Wayne movie in the midst of battle, Marines surfing under fire, cosy barbecues on hostile beaches, and Wagner booming from combat helicopters. So-called 'B-movies' took such heroics to even greater lengths. In Cirio H. Santiago's *Behind Enemy Lines* (1987), audiences were supposed to be charmed by supermen mowing down hundreds of enemy soldiers with one machine-gun and infiltrating enemy camps in daylight. They performed extraordinary feats without pause, and never wearied of killing, killing, killing.

Not all these narratives lauded armed prowess or exalted the pleasures of combat. During and after each war there were strong literary and cinematic movements *against* militarism. The horrors of combat were most dramatically represented in the literature of the 1914–18 conflict and in the cinematic and literary representations of the war in Vietnam. In these fictional accounts, there was often little to 'justify' the slaughter and the theme was that of the disillusionment of individuals, puny in comparison to a technologically driven, military imperative. As Michael Cimino, the director of *The Deer Hunter* (1978) claimed, 'any good picture about war' has to be 'anti-war'.[5] However, just like the anti-war protestors who ended up exhorting battle when they chanted 'Ho, Ho, Ho Chi Minh,/ We shall fight and we shall win', anti-war films simply relocated the conflict and quickly re-entered the romanticized canon of war. Take Philip Caputo during the Vietnam War. While he was in training, he recalled finding classroom work mind-numbing: 'I wanted the romance of war, bayonet charges, and desperate battles against impossible odds', he explained, 'I wanted the sort of thing I had seen in *Guadalcanal Diary* and *Retreat, Hell!* and a score of other movies.'[6] Despite the filthy, anti-heroic battle scenes in the films he mentioned, he was entranced by them. Realistic representations of combat are not necessarily pacifist or even pacificistic.[7] It was precisely the horror which thrilled audiences and readers: gore and abjection *was* the pleasure, subverting any anti-war moral. As the narrator of Larry Heinemann's *Paco's Story* (1987) recognized: 'Most folks will shell out hard-earned, greenback cash, every time, to see artfully performed, urgently fascinating, grisly and gruesome carnage.'[8]

Anti-war representations contained an additional ambivalence: while the horror of bloody combat might be meticulously delineated, the enemy remained individual, deeply flawed yet capable of being understood and thus accorded empathy. Debate was focused around good versus bad combatants, rather than about 'combatants'. Thus, the atrocities in Roland Emmerich's *Universal Soldier* (1992) served merely to differentiate between Sergeant Scott (who collected ears) and Private Devreux (who refused to shoot civilians). Both were extremely effective soldiers. Equally, in John Irvin's *Hamburger Hill* (1987), the poignant opening scenes of the Vietnam Veterans' memorial in Washington were utilized simply to justify the actions of the good servicemen, the dropping of napalm was portrayed as though its function was purely aesthetic, and the real enemies were the effete men at home and anti-war protestors. The Vietnam veteran, author, and anti-war campaigner, Tim O'Brien, summed up the dangers of such representation of war by reminding his readers that, for him,

> Vietnam wasn't an unreal experience; it wasn't absurd. It was a cold-blooded, calculated war. Most of the movies about it have been done with this kind of black humorish, *Apocalypse Now* absurdity: the world's crazy; madman Martin Sheen is out to kill madman Marlon Brando; Robert Duvall is a surfing nut. There's that sense of 'well, we're all innocent by reason of insanity'; 'the war was crazy, and therefore we're innocent'. That doesn't go down too well.[9]

In merging 'the horror, the horror' with a shadowy, oriental or fascist evil, viewers were faced with a familiar motif: the enemy as the ultimate 'other'.

So how did the men who are quoted in this book imagine war

prior to entering the armed forces? Combat stories were most powerful when recounted by fathers, older brothers and other close male friends. And the attraction was not restricted to boys. Young girls also found the war adventures of their fathers enticing and a chief motivation for wanting to become combatants themselves. Vee Robinson, for instance, volunteered to work on an AA (anti-aircraft) gun site, downing enemy aeroplanes during the Second World War. She recalled the battle stories told by fathers (in the interwar years) about their experiences between 1914 and 1918, remembering how her school-friends would boast about the heroic deeds carried out by 'our Dads'.[10] At times, the desire to fire guns could be strangely disassociated from the act of destruction. This was the case with Jean Bethke Elshtain when she recalled her desire around the time of the Korean War to own a gun. She called this her 'Joan of Arc period' and admitted that she had 'got the taste for shooting' at a picnic when she turned out to be a better shot than some of the boys. In her words: 'I didn't want to kill anything, save symbolically. But the idea of being a dead-eye shot and the image of going outside, rifle in hand: that intrigued.'[11]

It is clear that fantasies of combat drawn from books and films also dominated many men's daytime and night-time dreams, encouraging them to volunteer in anticipation of being given an opportunity to emulate heroes that they had read about since infancy.[12] Lieutenant-Colonel H. F. N. Jourdain of the Connaught Rangers in the First World War was one such romantic. In his memoirs (published in 1934), he could still conjure up the thrill he felt while reading the war stories of Sir Walter Scott and Charles Lever. These tales prompted him to join the army, in order to 'take part in such stirring times as were depicted'. He singled out Lever, in particular, for preparing him for the type of combat which involved charges, wild cheers which 'rent the skies', and wholesale bayoneting until the enemy scattered and the Rangers 'halted to recover breath and stayed the slaughter'.[13]

Similar heroics were imagined by men who signed up in 1939.

Eugene B. 'Sledgehammer' Sledge had been a sickly child but had entertained himself by translating Caesar's writings on war-fare and he revered Washington, Audubon, Daniel Boone, and Robert E. Lee (both of his grandfathers had been Confederate officers).[14] Similarly, Audie Murphy had grown up in a poor white family of sharecroppers in Texas. As a twelve-year-old boy weeding and hoeing the infertile land, his mind was always on some imaginary battlefield where 'bugles blew, banners streamed, and men charged gallantly across flaming hills'. This was a fan-tastic battleground where enemy bullets always missed while the bullets from his 'trusty rifle forever hit home'.[15] During each of these conflicts, boys and men devoured the romantic, military lit-erature of the preceding war. Thus, during the First World War, literature about the imperial wars entranced thousands; at the time of the Second World War, boys spent hours reading *The Times History of the Great War*; in the Vietnam era popular books (such as *Fighting the Red Devils*) based on Second World War exploits were much in demand.[16]

By the 1960s, new kinds of media offering narratives of killing had overtaken the more traditional literary ones. Television brought the joy of slaughter into the living room. Allen Hunt (for instance) saw himself as a typical son of rural America. From small-town Maryland, Hunt slipped naturally into the navy and then the army. He explained becoming a soldier in terms of an easeful transition:

> I liked the Army, as I was trained to do things that I felt
> comfortable doing. All through my childhood I had
> roamed the woods, hunting, playing Army and 'hide and
> seek'. From reading and watching Television, I had learnt
> basic theories of combat. Growing up at home, I had often
> dreamt about participating in combat; it was an experience
> that I wanted to acquire.[17]

Ron Kovic's memoir, *Born on the Fourth of July* (1976), also described the excitement of combat films such as *To Hell and Back*, in which Audie Murphy jumped on top of a flaming tank in order to turn a machine-gun on to the Germans: 'He was so brave I had chills running up and down my back, wishing it was me up there,' he recalled.[18] Not many years later, he was.

The poet, William D. Ehrhart, made a similar observation in an article entitled 'Why I Did It' (1980) in which he attempted to explain his reasons for joining the Marines at the age of seventeen. He admitted that he had an unrealistic idea of what war actually entailed. His image was framed by real and imaginary men like John Wayne, Audie Murphy, William Holden, Nathan Hale, Alvin York and Eddie Rickenbacker. His childhood was spent building plastic models of bombers and fighter planes, playing cowboys and Indians, and indulging in other war games, and his two most memorable childhood Christmas presents were a lifesize plastic .30 calibre machine-gun, and a .45 calibre automatic cap pistol in a leather holster with USMC embossed on the military-style cover flap. ('I was so proud of that pistol that I ran up the street first chance I got to show it off to Margie Strawser,' he recounted.) Newsreels of troops returning home in 1945 and his envy of friends whose fathers had 'been heroes' in that war were as fresh in his memory in 1980 as they had been when he was a child.[19]

As Ehrhart's memory attested, the industry devoted to the production of martial toys was immense. This can be illustrated by looking at one of the most aggressive military units, the Special Forces or, as they were popularly known, the Green Berets. By the late 1960s, parents could buy Green Beret dolls, records, comic strips, bubble-gum, puzzles and books for their children (or themselves). For ten dollars, the Sears catalogue offered a Special Forces outpost, complete with machine-gun, rifle, hand grenades, rackets, field telephone, and plastic soldiers. For half this price, Montgomery Ward's Christmas catalogue promised to send a Green Beret uniform and, for an additional six pounds, they would throw in an AR-15 rifle, pistol, flip-top military holster and a green beret. Adults were catered for not only by Green Beret books, but also by a major film, *The Green Berets* (1968).[20] Children like Kovic and Ehrhart played with Matty Mattel machine-guns and grenades, and cherished miniature soldiers holding guns, bazookas, and flamethrowers. Every Saturday afternoon, Kovic and his friends would take their plastic battery-generated machine-guns, cap pistols and sticks down to Sally's Wood where they would 'set ambushes, then lead gallant attacks, storming over the top, bayoneting and shooting anyone who got in our way'. Afterwards, they would walk out of the woods 'like the heroes we knew we would become when we were men'.[21] By 1962, imitation guns were the largest category of toys for boys with American sales exceeded $100 million annually.[22] Thus civilians were provided with their first clumsy introduction to a technology of combat which could be harnessed in the event of war.

When a man was inducted into one of the armed forces, these imaginary structures became more significant. Indeed, the military recognized that it was crucial to foment such fantasies if combat effectiveness was to remain high. They acknowledged that the best combatants were men who were able to visualize killing as pleasurable. Military authorities frequently financed and encouraged the artistic production of combat narratives. The First World War saw the creation of film as a new, 'modern' weapon of war. The most important of these films were *The Battle of the Somme* (1916) and *The Battle of the Ancre* (1916) which claimed to show men dying and killing in defence of England's green fields, but were in fact created for propaganda purposes by writers and artists gathering secretly at Wellington House. Hundreds of thousands of people flocked to see these films in their first week. The propaganda machine was even more central to military regimes during the Second World War. One American example was the popular *Why We Fight* (1942), a series of seven films ordered by the US Army Chief of Staff and directed by Frank Capra. This series combined footage from newsreels and films captured from the enemy, in combination with narration and animation. Like other films in this genre, it drew rigid distinctions between good and evil and was shown not only to troops but in cinemas throughout America. During the Vietnam War, the Department of Defense produced other troop indoctrination films, including *Why Vietnam?* (1965), which played on the need to rid Vietnam of the Communists. Again, it was shown not only to troops but also in high schools and colleges. From this point on, the Department of Defense's films substituted a more subtle ethnographic approach for the Second World War's more dramatic Hollywood-style films. Attention shifted to the Vietnamese reliance upon American aid, both medically, educationally and technologically. In films like *The Unique War* (1966) and *Vietnamese Village Reborn* (1967), 'ordinary' scenes from Vietnamese villages were presented to viewers. Films aimed at troops about to be posted to Vietnam were more hard-hitting. In *Your Tour in Vietnam* (1970), the narrator provided his listeners with tough advice and conjured up images of male comradeship, adventure and the thrill of combat, including the dropping of bombs from 'the big B-52s' (the bombs exploded in rhythm with jazz music).[23]

These propaganda films were also part of training regimes. In much the same way as Second World War films such as *Batton* (1943) and *Guadalcanal Diary* (1943) are used by the military establishment today, in the period between 1914 and the end of the Vietnam War, combat films were used as much to excite men's imaginations as to reassure them. One recruit who watched *The Battle of the Somme* just before going into battle told his comrade that seeing the film had made him aware of what they were going to face. 'If it were left to the imagination you might think all sorts of silly b—— things,' he admitted.[24] During the Second World War, the film *The Battle of Britain* (1944) was also said to be effective in making the men who watched it 'feel like killing a bunch of those sons-of-bitches'.[25] Opinion polls conducted during the Second World War showed that troops who had watched *Why We Fight* and the bi-weekly newsreels issued by the War Department were more aggressively pro-war than those who had not seen these films.[26]

Warfare was also reflected through the lens of earlier wars, particularly that of the American frontier. In Britain and Australia (as well as in America), the motif of men envisaging themselves as heroic warriors linked modern warfare with historical conflicts in which it was almost a duty to conquer another race, all for the sake of 'civilization'. During the First World War, trench raiders were described as slipping over the parapet with the stealth of Red Indians: 'No Sioux or Blackfoot Indian straight from the pages of Fenimore Cooper ever did it more skilfully,' Robert William MacKenna recounted.[27] In the Second World War, the most virile fighters were frequently slotted into mythical stories of cowboys and Indians. Men like Captain Arthur Wermuth (known as the 'one-man-army' after having singlehandedly killed more than one hundred Japanese soldiers) identified strongly with the cowhands on his father's range in South Dakota. He insisted on jumping into trenches 'with a cowboy yell'.[28] The souvenirs which were collected from enemy corpses were described as being so intimate that they would have 'curled the scalplock of Pontiac'.[29] One catalogue of over 600 Vietnam war films provides innumerable examples of the importance of 'cowboys and Indians', ranging from *Nam Angels* (1988) in which the hero wore a cowboy hat and swung a lasso, Montagnards hooted like Indians, and bikers charged as though they were on horses not motorbikes, to John Wayne's *The Green Berets*, released twenty years earlier. This was really a western in disguise, with the Viet Cong playing the Indians and the quip 'Due process is a bullet' being substituted for 'the only good Vietcong is a dead one'. The film *Little Big Man* (1970) even explicitly linked the genocide of the Indians in the American West with the war in Vietnam.

In the 'Indian country' of Vietnam, 'The Duke' or John Wayne was the hero most emulated.[30] Indeed, in July 1971, the Marine

Corps League named him the man 'who best exemplifie[d] the word "American" '.[31] He had become the most popular actor on television and men like the ace pilot, Lieutenant Randy Cunningham (who confessed to enjoying downing North Vietnamese MIGs) was proud that his tactical sign was 'Duke', after John Wayne, because (he said) 'I respect his American Ideals'.[32] Philip Caputo imagined himself charging up a beach-head, like John Wayne in *Sands of Iwo Jima* (1949), and this led him to join the Marines to fight in Vietnam.[33] More than any other branch of the armed services during the war in Vietnam, the Special Forces modelled themselves after the cinematic hero-ics of John Wayne, despite the fact that Second World War assault tactics were not appropriate in a guerrilla war. Even women were attracted by this myth: Carol McCutchean, for instance, joined the Women Marines because she was thrilled by John Wayne movies.[34]

Not surprisingly, therefore, combatants interpreted their bat-tleground experiences through the lens of an imaginary camera. Often the real thing did not live up to its representation in the cinema. The twenty-year-old Australian officer, Gary McKay, was slightly disappointed by the way his victims acted when hit by his bullets: it 'wasn't like one normally expected after watch-ing television and war movies. There was no great scream from the wounded but simply a grunt and then an uncontrolled col-lapse to the ground,' he observed morosely.[35] Or, as the fighter pilot, Hugh Dundas, admitted, his 'nasty, sickening' introduction to combat was in stark contrast to the way he had imagined it would be.[36] Others were happier with the match between fic-tional representation and their own experience: a submarine, for example, could sink just like it would in Hollywood pictures.[37] The explosion of an aeroplane during the war in Korea could make a pilot 'so damned excited' because it was 'like something you see in the movies'.[38] Nineteen-year-old Geoffry R. Jones experienced combat in Vietnam as an extension of the movies or playing cowboys and Indians as he had done only a few years before.[39] The pilot of a spotter plane in Vietnam who accurately directed artillery fire 'yipp[ed] like a cowboy' every time they hit the enemy.[40]

Killing itself was likened to film-making. In one instance during the First World War, a Royal Fusilier ordered machine-gunners entrenched in a farm house to 'cinematograph the grey devils' and to pretend that it was Coronation Day by 'tak[ing] as many pictures as possible'. He continued:

> The picture witnessed from the farm on the 'living screen' by the canal bridge was one that will not easily be forgotten. The 'grey devils' dropped down in hundreds. Again and again they came on only to get more machine murder.[41]

In 1918, one sergeant quoted in the magazine *The Stars and Stripes* said that battle was 'like a movie' with the infantry making a 'serene, unchecked advance . . . their ranks unbroken, their jaunty trot unslackened'. 'A movie man would have died of joy at the opportunity,' another sergeant exclaimed.[42] An unnamed Canadian informant during the Second World War agreed. He described training his machine-gun on thirty Germans aboard a submarine as 'like one of those movies when you see troops coming at the camera and just before they meet it, hit it, you see them going off to the left and right, left and right'.[43] In *Nam. The Vietnam War in the Words of the Men and Women Who Fought There* (1982), an eighteen-year-old radio-man confessed that he

> loved to just sit in the ditch and watch people die. As bad as that sounds, I just liked to *watch* no matter what happened, sitting back with my homemade cup of hot chocolate. It was like a big movie.[44]

Or, as Philip Caputo put it, killing Viet Cong was enjoyable because it was like watching a movie: 'One part of me was doing something while the other part watched from a distance.' Instead of focusing on mangled corpses, soldiers who could imagine themselves as movie heroes felt themselves to be effective warriors.[45] Such forms of disassociation were psychologically useful. By imagining themselves as participating in a fantasy, men could find a language which avoided facing the unspeakable horror not only of dying, but of meting out death.

As already implied in many of the quotations, films *created*, as well as represented, combat performance. So powerful were cinematic images of battle that soldiers acted as though they were on the screen. During the Second World War, William Manchester was stunned by the way soldiers in the Pacific imitated Douglas Fairbanks, Jr., Errol Flynn, Victor McLagle, John Wayne and Gary Cooper.[46] During the Vietnam War, the journalist Michael Herr commented on the performance of 'grunts' when they knew that there was a camera crew nearby: 'they were actually making war movies in their heads, doing little guts-and-glory Leatherneck tap dances under fire, getting their pimples shot off for the network . . . doing numbers for the cameras'.[47] Even in Grenada in 1983, American soldiers charged into battle playing Wagner, in imitation of Robert Duvall, the brigade commander in *Apocalypse Now* (1979).[48]

Of course, such antics often had a short life. Josh Cruze, who joined the Marines at the age of seventeen and served in Vietnam, had this to say:

> The John Wayne flicks. We were invincible. So when we were taken into . . . war, everyone went in with the attitude, 'Hey, we're going to wipe them out. Nothing's going to happen to us.' Until they saw the realities and they couldn't deal with it. 'This isn't supposed to happen. It isn't in the script. What's going on? This guy's really bleeding all over me, and he's screaming his head off.'[49]

Even worse, such fantasies could get a man killed. The combat engineer, Harold 'Light Bulb' Bryant, remembered a man with the nickname of Okie who had the 'John Wayne Syndrome'. He was itching to get into action. During his first battle, their unit was pinned down by machine-guns and Okie 'tried to do the John Wayne thing'. He attempted to charge the machine-gun, and was immediately mowed down.[50] Films, then, provided both pleasurable, and deathly, scripts.

In training camps, miles from the frontlines, men wondered 'how much resemblance there would be between the imagination and reality, between war as [we] rehearsed it day after day on the Sussex downs . . . and the real war of the trenches'.[51] Inexperienced infantrymen swapped thoughts about what it would feel like to 'run a man through with a bayonet' and swore (as did one Texan during the First World War) that they were so keen for intimate struggle that they were willing to go over the top with a penknife.[52] Like Alfred E. Bland on 30 January 1916, soldiers wrote long letters to their families describing this yearning for battle, enthusing about the 'change about to come – *real* business with *real* Germans in front of us. Oh! I *do* hope I shall visibly kill a few.'[53] When asked why they joined the corps, they would simply write: 'To Kill'.[54] Airmen interviewed by Roy R. Grinker and John P. Spiegel in *Men Under Stress* (1945) also expressed 'eager-beaver' reactions prior to going overseas. They were so wrought up that men who were prevented at the last moment from embarking burst into tears. Such keenness betrayed a blinkered view of reality. 'The men seldom have any real, concrete notions of what combat is like,' Grinker and Spiegel continued,

> [t]heir minds are full of romanticized, Hollywood versions of their future activity in combat, colored with vague ideas of being a hero and winning ribbons and decorations.

If they were told more realistic stories about what to expect 'they would not believe them'.[55] Emotions ran so high that even when not in battle, pilots acted as though they were constantly engaged in the clash of arms. Every time they 'took the air', fighter pilots were 'symbolically going into action with the enemy'. Consequently, they flew wildly, executing tight turns as they approached the runway, and did aerobatics too close to the ground. This was (according to one observer) 'seldom deliberate exhibitionism'. More to the point, they were simply demon-strating to themselves their 'capacities as fighting men'.[56] *Actual* aerial combat was often 'a cruel awakening'.[57]

Immediately prior to the first 'kill', more poignant musings might be substituted for such fantastical skylarking. Richard Hillary, the famous Second World War pilot mentioned later, recalled the 'empty feeling of suspense' in the pit of his stomach when he climbed into the cockpit of his plane for the first time in a 'real' combat situation. 'For one second,' he recounted,

> time seemed to stand still and I stared blankly in front of me. I knew that that morning I was to kill for the first time . . . I wondered idly what he was like, this man I would kill. Was he young, was he fat, would he die with the Fuehrer's name on his lips, or would he die alone, in that last moment conscious of himself as a man?[58]

Neither Hillary nor any other combatant would ever discover the answer to these questions: but combat itself would provide them with an opportunity for infusing such daydreams with a new, more frenzied passion.

Joyful slaughter

Did actual combat dent the pleasures of imaginative violence? For most combatants, the answer must be 'no'. Time and time again, in the writings of combatants from all three wars, we read of men's (and women's) enjoyment of killing. This book con-tains innumerable examples of men like the shy and sensitive First World War soldier who recounted that the first time he stuck a German with his bayonet was 'gorgeously satisfying . . . exultant satisfaction'.[59] Second Lieutenant F. R. Darrow found that bayoneting Prussians was 'beautiful work'.[60] 'Sickening yet exhilarating butchery' was reported to be 'joy unspeakable' by a New Zealand sapper.[61] Generals were praised if they managed to maintain a spirit of the 'joy of slaughter' in their troops, even if it meant arming night patrollers with spiked clubs to intimidate and bash the Huns.[62] In the words of Henry de Man:

> I had thought myself more or less immune from this
> intoxication until, as trench mortar officer, I was given
> command over what is probably the most murderous
> instrument in modern warfare . . . One day . . . I secured a
> direct hit on an enemy encampment, saw bodies or parts
> of bodies go up in the air, and heard the desperate yelling
> of the wounded or the runaways. I had to confess to
> myself that it was one of the happiest moments of my life.

He admitted that he had yelled aloud 'with delight' and 'could have wept with joy'. 'What' (he asked) were 'the satisfactions of scientific research, of a successful public activity, of authority, of love, compared with this ecstatic minute?'[63]

More than fifty years later, in the Vietnam conflict, combat soldiers confessed to similarly exhilarating emotions. Like William Broyles, Philip Caputo admitted that he never told the truth when people asked him how he felt going into combat because the truth would have labelled him a 'war-lover'. For Caputo, going into battle made him feel 'happier than [he] ever had'.[64] Similarly, although the twenty-year-old Australian officer, Gary McKay, had killed lots of Vietnamese, he clearly recalled the sensation of actually *seeing* his bullets hit a man. For him, the 'terrible power of effect of the weapon when it hit the target' overwhelmed him with 'awe'.[65] When another soldier in Vietnam went berserk and massacred many of the enemy, he remembered feeling suffused with joy: 'I felt like a god, this power flowing through me . . . I was untouchable.'[66] James Hebron, a scout-sniper in the Marines, also described the incredible sense of power he felt in combat:

> That sense of power, of looking down the barrel of a rifle
> at somebody and saying, 'Wow, I can drill this guy.' Doing
> it is something else too. You don't necessarily feel bad; you
> feel proud, especially if it's one on one, he has a chance.
> It's the throw of the hat. It's the thrill of the hunt.[67]

Killing was intrinsically 'glamorous'.[68] It was like 'getting screwed the first time' and gave men 'an ache as profound as the ache of orgasm'.[69] In the words of a black Muslim Marine, 'I enjoyed the shooting and the killing. I was literally turned on when I saw a gook get shot.'[70]

Semi-autobiographical accounts tell a similar tale. For example, in one place in James Jones's *The Thin Red Line* (1962), Doll had just butchered his first Japanese. This 'kill' pleased him, in part because he felt proud at accurately shooting 'dirty little yellow Jap bastards'. It was, he believed, like 'getting screwed the first time'. More interestingly, however, Doll's pleasure resided in his guilt. He had committed the most horrendous crime – worse, he believed, than rape – but *that* was where the heartfelt allure of killing resided. It made him feel immune from any outside power. Nobody was going to persecute him for this action. He had – literally – 'gotten by with murder', and the thought filled him with an urge to giggle: he felt so 'stupid and cruel and mean and vastly superior'.[71] People could take an immense delight in breaking the highest moral law.

As we shall see in Chapter 2, airmen were particularly liable to be enraptured by homicidal violence. In *Winged Warfare. Hunting Huns in the Air* (1918), Major William Avery Bishop thought it 'great fun' to train his machine-gun on Germans because he 'loved' to watch them running away 'like so many rats'.[72] Even a pilot's mechanic might take some of the reflected glory, bragging how 'his' pilot had 'got a Hun' while another mechanic's pilot had not.[73] During the Second World War, 'Bob' described himself as being 'elated' when he shot down a German plane, mainly because this meant that his 'score' had improved. 'Life wasn't too bad after all,' he reflected.[74] After a kill, pilots admitted that they 'all felt much better' and there would be 'a good deal of smacking on the back and screaming of delight'.[75] Although the sight of mutilated and dead Germans staining the rear cockpit of their planes might be described as 'sad and beastly', airmen admitted that they had felt 'elated then'.[76] The sense of power in the air could be exhilarating, attested a fighter pilot known as 'Durex'. With great earnestness, he enthused:

> I opened fire, the bullets roared out over the noise of the engine. They don't rattle like an ordinary Army Vickers gun. No, sir! When the 8 Brownings open fire – what a thrill! The smoke whips back into the cockpit and sends a thrill running down your spine.[77]

Story-telling

Whether or not men admitted to enjoying the slaughter, their status as men as well as combatants was enhanced by the number of enemy troops they were able to kill. Men were jealous if other units scored more highly: 'there was a race on for the most kills' admitted Lance Corporal Phil Buttigieg while based at Long Tan (Vietnam) in 1966.[78] In Vietnam, American and Australian troops competed vigorously: when American soldiers began teasing Private Peter Gates's unit for recording a low body count, Gates admitted in a letter to his family in Australia that '[i]t's getting a bit embarrassing . . . we have a hard time convincing the Yanks that we are still the best thing since canned beer'.[79] Troops were 'only happy when they've had a "kill"', Andrew Treffry reported to his fiancé from Vietnam.[80]

This emphasis on 'body counts' is frequently assumed to be a phenomenon of the Vietnam War. It is clear, however, that it was a very real goal for many combatants during the two world wars as well. Frequently this meant exaggerating the slaughter, as hinted at in the ditty composed in the frontlines by Major B. W. Bond, MC, which read:

> *We killed the enemy by the score,*
> *When coming up the Wady,*
> *But when we went to count the dead*
> *We only found one body.*[81]

It was widely acknowledged that men lied about killing. Lieutenant Roland H. Owen, in a letter to his parents on 13 October 1914, regretted that he did not have any 'interesting yarns' to tell them since 'one does not see so very much and knows still less'. He went on to say that if they wanted gory stories, they would have to consult Tommies. Indeed, one soldier had told him a gruesome story about a recent scrap, forgetting that he had been present and thus knew that it was a fabrication.[82] There was also a flourishing trade in souvenirs such as helmets, bent bayonets, and rifles for support troops to take home to bolster wild stories that they had told their loved ones.[83]

Exaggeration was most common in the air force since the status of an individual fighter pilot was directly related to the number of fighter planes he shot down (in fact, the number of destroyed enemy planes was very low). The First World War flying ace, Captain Albert Ball, promised his parents that he would have many 'good yarns' to tell them when he returned on leave although he admitted that, in the period before returning, his stories might be expected to 'grow and strengthen until they are quite ripe'.[84] In the next war, night fighter pilots at a RAF fighter station had never seen combat yet talked constantly of it. One pilot said, 'You don't see a nacelle when you're doing a split-ass turn at 400 and a Hun is sitting on your tail,' yet he had never been in battle: the language of the day fighting squadrons (who had frequently fought) was adopted by pilots irrespective of actual experience.[85]

High-blown tales of massacres belonged to a broader tradition of story-telling. The urge to kill-and-tell was fuelled by civilian curiosity. An interesting illustration of this can be found in a letter mentioned in the Introduction and written by William Willis to Miss Luttrell on 4 May 1917. Willis recognized the need to cater to the martial fantasies of his female correspondent. He admitted that he tried to make his letters 'interesting', yet, he continued, there were to be no accusations of lying: 'I always write the truth.' The same letter contained a mixture of bravado, matter-of-fact brutality, and a perverse sensitivity.

> The Huns attacked us with 15,000 men, in mass forma-
> tion, yet we numbered a few hundreds, repulsed them and
> inflicted extremely heavy losses. We killed till we grew
> sick of the sight of blood and dead men. Personally I
> accounted for that many that I eventually lost count. Our
> platoon received the order to charge and we met the
> enemy half-way. He acted up to tradition and surrendered
> or ran away. It was impossible to take prisoners so they
> were all shot. One had the opportunity to utilize his
> knowledge of bayonet work. Just imagine 32 of us opposed
> to over 300 of them and beating them the way we did.
> Don't you think we have grounds for being a little bit
> proud.

Willis's story contained fairly typical elements of many of the combat narratives related in this book: victory under great odds, the moral weakness of the enemy, mass slaughter, pride in shedding blood. Willis clearly had no scruples about admitting to killing prisoners (and he assumed that Miss Luttrell wouldn't either), yet, unlike many war stories recited in this book, he delicately distanced himself from the narrative when mentioning the bayonet – 'one' killed with the bayonet as opposed to 'I' or 'we'.[86]

Neither Willis's urge to tell combat tales, nor the evident fascination of Miss Luttrell with such stories, were unique. 'How many Germans have you killed?' was a typical question, and although some men might snap back that they had 'quite enough to do to see the Germans didn't kill [us]',[87] others revelled in the pleasures of spinning yarns which, even if referring to an actual incident, could not fail to transform the deathly encounter into a heroic narrative. Since refusal to tell such stories might throw into doubt a man's status and virility,[88] few fathers, husbands or lovers were able to resist the temptation to conform to an active warrior stereotype.

Others, like Jack Strahan in Vietnam, waxed lyrical, composing poems in praise of the 'prettiness' of destruction, the 'blasts of colour in the summer night air', the 'great rhythmic, explosions/ dancing brightly the night'. For Strahan, arclight bombs were

> *more thrilling than an earthquake, when,*
> *hit with their obscene blast wave, you laugh,*
> *not knowing war could be such fun.*[90]

William E. Merritt in *Where Rivers Ran Backwards* (1989) complained that no one had told him how picturesque war could be, particularly explosions with their waves of colour and sound which were worth paying money to watch, 'like a movie'.[91] In the navy, combatant service was also frequently discussed in aesthetic terms. In 1942, Norman Hampson was aboard the HMS Carnation. The 'long lithe body' of this ship seduced him and he struggled to express his feelings in verse:

> *Pardon if all the cleanness and the beauty*
> *Brave rhythm and the immemorial sea*
> *Ensnare us sometimes with their siren song,*
> *Forgetful of our murderous intentions.*
> *Through our uneasy peacetime carnival*

Cold sweat of death rained on us like dew;
Even this grey machinery of murder
Holds beauty and the promise of a future.[92]

Trophy-collecting

As Broyles argued at the beginning of this chapter, there was also a carnivalesque spirit to the rituals of military combat. Humour could take the form of linguistic satire: the Australian bomber, Al Pinches, recalled the delicious irony of being told to 'have a nice day' by American controllers immediately after they instructed him on whom he was to 'bomb into oblivion'.[93] Killing itself could be seen as an act of carnival: combat gear, painted faces, and the endless refrain that men had to turn into 'animals' were the martial equivalent of the carnival mask: they enabled men to invert the moral order while still remaining innocent and committed to that order. Pranks involving enemy bodies were very common in all three conflicts. Soldiers would comb the hair of corpses, shake skeleton's hands, and offer cigarettes to severed heads, attested the Revd William Edward Drury during the First World War.[94] In Bougainville (1944), John Henry Ewen of the Australian Infantry Force recalled his uncle telling him about a skeleton in the First World War which they had propped up with a dog-biscuit in his mouth. Imitating his uncle, Ewen sat a Japanese skeleton under a signpost, fixed his arm so that it was pointing up the track, added a tuft of dry grass for hair, and stuck a tin hat on his skull. He 'looked pretty good', Ewen commented, adding that he wished that he had a camera.[95] Similarly, in the Vietnam conflict, corpses were mutilated and manipulated and the unit's patch (or a card representing the unit) displayed upon the body.[96] Like 'tourists in hell', comrades clicked their Instamatic cameras while victorious warriors posed in front of corpses.[97] Rabelaisian pleasures were indulged in as men boisterously desecrated dead bodies.[98] Most famously, as represented in the film *Platoon* (1986), parties were held with the guest of honour being a recently slaughtered enemy. Ironically, such

grotesque clowning about with the dead had a double implication: *both* the dead and the jokers were being parodied.

At the root of such grotesque behaviour was the desire to
assert one's essential 'self' in the act of killing. This could also be
seen in the removal of souvenirs from the bodies of the murdered
enemy. Such practice was ubiquitous although the extent of
gruesome trophy-hunting varied according to the nationality of
the enemy ('Japs' and 'Gooks' more frequently than 'Huns'),[99]
opportunity (small patrols during the Second World War in the
Pacific theatre and in Vietnam more than in the mass entrenched
armies of the First World War), and national narrative traditions
(Americans placed more emphasis on 'scalping' their enemies
'like the Indians'). Despite these differences, it was fairly
common in all three conflicts. Between 1914 and 1918, buttons,
epaulettes, piccolos, medals, helmets and tassels from the enemies' bayonets were common trophies.[100] Even the young poet,
Wilfred Owen, sent his brother a blood-spattered handkerchief
which he had taken from the pocket of a dead German pilot.[101]
Less innocuous was the taking of enemy teeth or ears by men
such as the solicitor described by Edward Glover who loitered in
no man's land drawing the teeth of dead Germans.[102]

From 1939, collecting bodily parts became more prominent in
combat narratives or, at least, a more acceptable practice to mention. In the Pacific theatre of war, men collected breasts from the
bodies of killed (or captured) Japanese women.[103] One witness to
such souvenir-hunting was Eugene B. 'Sledgehammer' Sledge
of the 1st Marine Division. Sledge entered the Second World
War as a believing Christian and he frequently prayed in combat.
He also considered taking bodily parts. In an oral interview in
the early 1980s, he recalled the first time he saw men taking
souvenirs. It was on the Pacific island of Peleliu. Initially he was
horrified as he watched his buddies dragging a wounded
Japanese soldier around like a 'carcass', searching for bounty, but
he admitted that 'it didn't take me long to overcome that feeling'. He described how his comrades would extract gold teeth by

placing the tips of their knives under the teeth of the dead Japanese and, hitting the hilts of the knives, knocking the teeth loose. For him, it was a relatively easy action to justify. The war in the Pacific, he noted, was 'savage': mercy was neither given nor received. Death, fatigue and stress wore away even the 'veneer of civilization'.[104] The tendency to collect human trophies escalated during the conflicts in Korea and Vietnam when the bodily parts most favoured were ears, teeth and fingers but the collection of heads, penises, hands and toes were all reported.[105]

Why did servicemen take such grisly mementoes? Although there could be financial value in some types of souvenirs (especially gold teeth),[106] it would scarcely recompense men for the laborious and often hazardous work involved in their collection. More important, souvenirs were regarded as proof that a man had seen active combat and thus had proved himself on the field of battle. During the First World War, George Coppard admitted that a 'ghoulish curiosity' initially led him collect souvenirs. He recalled that the Pickelhaube, or spiked helmet, was the favourite souvenir 'as the mere display of one of these when you were on leave sort of suggested that you had personally killed the original owner'.[107] In Vietnam, Harold 'Light Bulb' Bryant testified that soldiers took ears to 'confirm that they had a kill. And to put some notches on their gun'.[108] There was, not surprisingly, resentment from combat-weary men when men who had not engaged the enemy raided the battlefields before them.[109]

Souvenirs conveyed immense power upon the serviceman. The combat paratrooper, Arthur E. 'Gene' Woodley, Jr., collected about fourteen ears and fingers which he strung around his neck. When in base camp, he found that he

> would get free drugs, free booze, free pussy because they wouldn't wanna bother with you 'cause this man's a killer. It symbolized that I'm a killer. And it was, so to speak, a symbol of combat-type manhood.[110]

Larry Heinemann's *Paco's Story* (1987) makes a similar point. In this account, Private First Class Elijah Raintree George Washington Carver Jones (known as Jonesy) had collected thirty-nine pairs of ears which he strung on a bit of black wire and wrapped like a garland around his steel helmet. Wearing this gruesome adornment, he entered base camp at Phuc Luc. The narrator described what happened next:

> you should have seen those rear-area motherfucking
> housecats bug their eyes and cringe every muscle in their
> bodies, and generally suck back against the buildings . . .
> Jonesy danced this way and that – shucking and jiving,
> juking and high-stepping, rolling his eyes and snapping
> his fingers in time – twirling that necklace to a fare-thee-
> well, shaking and jangling it (as much as a necklace of ears
> will jangle . . .) and generally fooling with it as though it
> were a cheerleader's pom-pom.

The entire combatant company (or what was left of them) were laughing hysterically.[111] It was the equivalent of what Charles Baudelaire in 'The Essence of Laughter' (1855) called the 'primordial law of laughter' or the 'perpetual explosion of . . . rage and . . . suffering' – anger against the 'housecats' who issued the orders which caused grunts such anguish. It was a laughter which 'lacerates and scorches the lips of the laugher for whose sins there can be no remission'.[112]

In one sense, the search for souvenirs enabled men to link death of the 'other', the enemy, with love of themselves. Indeed, a necklace of ears could be called 'love beads'.[113] The fact that men would also *leave* souvenirs, such as photographs of themselves, on top of the corpses or cards representing their unit in lifeless hands was also important.[114] The individual corpse came to represent a universal condition and the terror of death was diminished. Souvenirs asserted that the killed enemy was 'like us'. This is illustrated in James J. Fahey's diary entry for 27

November 1944, written after having been attacked by Japanese suicide planes:

> Parts of destroyed suicide planes were scattered all over the ship. During a little lull in the action the men would look for Jap souvenirs and what souvenirs they were. I got part of a plane. The deck near my mount was covered with blood, guts, brains, tongues, scalps, hearts, arms etc. from the Jap pilots. One of the Marines cut the ring off the finger of one of the dead pilots . . . One of the fellows had a Jap scalp, it looked just like you skinned an animal. The hair was black, cut very short, and the color of the skin was yellow, real Japanese. I do not think he was very old. I picked up a tin pie plate with a tongue on it. The pilot's tooth mark was into it very deep. It was very big and long, it looked like part of his tonsils and throat were attached to it. It also looked like the tongue you buy in the meat store.[115]

For Fahey, the mutilated Japanese corpses were both alien (the colour of skin, his tongue like an animal's) and familiar (the pilot was young and wore rings).

At a political level, these rituals of collecting and celebrating performed a crucial function in enabling men to cope with the problem of being 'handed the wrong script'. Carnivalesque rites of killing did not demand rejection of the law, but a reassertion of men's commitment to rules against extreme violence. Transgression could be enjoyable because the law was well-respected. It was (in the words of Mikhail Bakhtin) 'authorized transgression'[116]: military authorities (such as the officer, William Broyles) officially disapproved of humorous mutilation of enemy corpses but 'turned a blind eye' to such antics, accepting them as necessary for 'effective combat performance'. Despite threats by the military that stern disciplinary action would be taken if a man was caught with a part of an enemy's body,[117] souvenir-collectors were seldom even rebuked. Quite the contrary: men were encouraged to take such mementoes by their military instructors. In 1917, Cadet Ian Rashan reported that his bayonet instructor had sneered at the idea of collecting German buttons, yelling: 'Ears and other things are what yer want!'[118] In the words of a black Muslim Marine who wore a string of ears in Vietnam:

> We used to cut their ears off. We had a trophy. If a guy
> would have a necklace of ears, he was a good killer, a good
> trooper. It was encouraged to cut ears off, to cut the nose
> off, to cut the guy's penis off. A female, you cut her
> breasts off. It was encouraged to do these things. The offi-
> cers expected you to do it or something was wrong with
> you.[119]

It was generally regarded as a sign of combat effectiveness. Laughter not only enabled cruelty: it framed it.

Fantasy and experience were intertwined, but not indistinguish-able. Certainly, for combatants who were killed or mutilated, there was a radical disjuncture between the imagination and cor-poreal existence. In the act of killing, however, the two could not be easily separated and, in terms of moral survival, it was crucial that it remained that way. If combatants became disillusioned, it was because they felt that they were in the wrong film, enacting a strange script, rather than because they wanted to repudiate aggressive dialogues. When combat servicemen cried out (in the words of one Vietnam veteran) 'hey, this isn't a movie', what they meant was 'I don't recognize *this* movie'.[120] More typically, com-batants were able to construct a story around acts of exceptional violence which could render their actions pleasurable. The emphasis on the beauty of war – the colour of napalm, the shine of steel, the maternal bulk of the tank – distracted attention from the smell of burning flesh, gaping wounds and dismemberment. Carnivalesque rites and fantasies drawn from a wide range of combat literature and films enabled combatants to refashion themselves as heroic warriors. Fear, anxiety, pain; these are only too familiar in combat. But excitement, joy and satisfaction were equally fundamental emotions, inspired by imagining that they had scored a good, clean 'kill'.

Chapter 2

THE WARRIOR MYTH

How then can I live among this gentle
obsolescent breed of heroes, and not weep?
Unicorns, almost. For they are fading into two legends
in which their stupidity and chivalry are celebrated;
the fool and the hero will be immortals.

Keith Douglas, 'Sportsmen', 1943[1]

Ion Llewellyn Idriess (trooper in the Australian Imperial Force, First World War), Richard Hillary (pilot in the Royal Air Force, Second World War), and Dave Nelson (American sniper, Vietnam War) referred to themselves as 'warriors'. Despite seeming to have little in common, their retrospective constructions of combat represent three important martial myths.

Ion Llewellyn Idriess ('Jack' to his friends) was a lightly built, leathery-faced Australian patriot when he enlisted as a private in the Australian Imperial Force in 1914. He was never promoted nor did he win any medals. As Trooper 358 of the 5th Light Horse, Idriess was wounded at Gallipoli, witnessed the charge at Beersheba, and fought in the battle of Gaza. By the time he was invalided out of the army in 1918, he had been seriously wounded three times. Years after the war, he transformed his experiences as an infantryman and sniper into one of Australia's most popular warrior myths.

There was little in Idriess's upbringing which hinted at his future as a writer. He had been born in Sydney in 1889 and was raised by his father (his mother died when he was a child) in Broken Hill, Victoria. From his earliest years, he listened to his Welsh father's fantastic stories about King Arthur and Owen Tudor. He left home at the age of fourteen and, prior to the war, made a living successively as an assayer's assistant, seaman, station-hand, drover, timber-cutter, opal miner, buffalo-shooter, prospector, and pearl diver. Then the war changed his life. When he returned to Australia wounded, he decided to become a writer. His first book was published in 1927 and by the time he died in 1979, he had written forty-seven more. A number of his books explore his view of the 'warrior'. The first of these, entitled *The Desert Column* (1932), was composed from the detailed diaries he kept during the First World War. Then, after the bombing of Darwin in February 1942, Idriess devoted himself to pleading with both his male and female readers to arm themselves against the threat of a Japanese invasion. Books with titles such as *Shoot to Kill*, *Sniping*, *Trapping the Jap*, *Guerrilla Tactics*, *Lurking Death*, and *The Scout* were published in 1942 and 1943. *Sniping* is still used by army trainees at the Singleton Army Camp in New South Wales.

There were two traditions that Idriess lauded above all in his books: the bayonet charge and sniping. In some of his writings he linked the two, as in an article published in *The Daily Telegraph Pictorial* (1929), where he described lining up his sights upon a moustached Turk and firing before rushing forward, 'screaming, to bayonet him, to club him, to fall on him and tear his throat out'. More typically, the bayonet fight was portrayed as a magnificent charge of large groups of men. In lurid prose, his diary for 14 March 1917 described one such occasion: 'Then came the fiercest individual excitement', he wrote,

– man after man tore through the cactus to be met by the bayonets of the Turks, six to one. It was just berserk

slaughter. A man sprang at the closest Turk and thrust and
sprang aside and thrust again and again – some men
howled as they rushed, others cursed to the shivery feeling of steel on steel – the grunting breaths, the gritting
teeth and the staring eyes of the lunging Turk, the sobbing
scream as the bayonet ripped home.

Such forms of combat were attractive for Idriess because they
were a fair attack on equals who honoured their opponent's
prowess. In *Lurking Death* (1942), he describes two particularly
bloody bayonet fights in which, as soon as the Turks had surrendered, the Australian and New Zealander troops began
laughing. Their 'mad roar' turned into 'a roar of laughter from
heaving chests' as the Antipodeans 'slung their rifles and
panted, beckoning [to the Turks] that they were friends'. The
'dazed' Turkish survivors had their hands shaken. 'How many
others could do that?' Idriess asked, '[f]ight all day . . . right to
the bayonet point, then laugh and spare him in the height of
maddened excitement.' The warriors were bloody, brutal even,
but they fought with comparable weapons and respected their
foes.

The other man-at-arms that Idriess admired was the sniper.
During the war, he had been an effective sniper, effortlessly
transferring his extensive expertise in stalking wild animals in the
Australian outback to man-hunting. The sniper, he believed,
was a loner 'on the warpath. Out after scalps'. In a semi-
autobiographical account he appended to his book *Sniping* he
described a particularly skilful 'stalking' and sniping of a
Bedouin. After the kill, he went to examine the corpse:

Looking down on him, like a great fallen hawk in the
crushed barley, I felt no remorse; only hot pride that in
fair warfare I had taken the life of a strong man – hot
pride that this man, older and stronger physically than I,
this man reared from boyhood to regard warfare as the life

of a man and splendid sport, this desert irregular,
knowing every inch of the country, had fallen to a stranger
from a peaceful land who knew only three years of war!

Again, his emphasis was on skill, the equality of opponents, and
the honourable nature of both combatants. Like the flow of blood
following a bayonet thrust, the 'intimate contest' between two
desperate men redeemed the warrior.[2]

Idriess's skill was primarily as a writer whereas Richard Hillary's
status was derived far more from his skill as a warrior. In combat
during the Second World War, Hillary won no medals yet when
he died at the age of twenty-three he was admired by millions as
an ace pilot with five German planes (and three 'probables') to
his credit. More than 200,000 people bought his autobiography
in the eight years after it was published in 1942. According to
Lovat Dickson, Hillary epitomized 'the beauty and courage and
nobility of youth' in wartime. Arthur Koestler dedicated an article to him entitled 'The Birth of a Myth' and Eric Linklater
celebrated him as a 'national possession, an exemplar of his age,
a symbol of its *reverdissement*'. Like Rupert Brooke during the
First World War, Hillary was said to speak for 'his generation's
lack of faith' while proving 'in word and deed its innerward
unbuttressed virtue'.

Hillary's social background was integral to his later status as
an icon of British manliness. This background made the
prospect of becoming a fighter pilot possible and endeared him
to a backward-looking, romantic-minded British public.
Although, like Idriess, he was born in Sydney, at the age of three
Hillary was taken to England where he was educated at
Shrewsbury and Trinity College, Oxford. His blue eyes, 'ridiculous' nose, large pouting mouth, and athletic physique made
him almost a caricature of an English hero. He joined the
University Air Squadron and quickly became one of the 'long-
haired boys', disillusioned, disdainful, and utterly egocentric. At

the declaration of war, Hillary did what was expected of young men of his class and experience: he volunteered for service in the air force. After an uncertain start, he eventually won the admiration of his comrades for his aerial skills but the events of 10 August 1940 permanently changed his life. He was shot down, spent hours floating in the English Channel, was rescued, and then spent months in hospital undergoing painful cosmetic surgery for massive burns. He was scarred for life. It was in hospital that he conceived his autobiography, *The Last Enemy* (1942) and, in the process of writing it, recovered his fervent patriotism.

Like Idriess, Hillary articulated a particular (although subtly different) warrior myth. For the blood and gore of bayonet-fighting, he substituted the heady glory and stark virtue of aerial combat. In the air, there was no squalor, no filth. The demonstration of personal skill was once again possible. This was 'war as it ought to be', an individual contest between two men. It was exciting, individual and disinterested. Hillary arrogantly raised himself above 'mere soldiery' ('I shan't be sitting behind a long-range gun working out how to kill people sixty miles away,' he pontificated). Rather, his prey had a face, physique, emotions: was he young or fat, he asked himself before his first combat mission? Would he die with the Führer's name on his lips or would he die conscious only of himself as a man? And this first 'kill' did not disappoint him. 'I had a feeling of the essential rightness of all,' he recalled, '[h]e was dead and I was alive; it could so easily have been the other way round; and that would somehow have been right too.' The fighter pilot was 'privileged to kill well': no overly 'personalized emotions' (as in the infantry) and no childish longing for 'smashing things' (as in bombing missions). Rather, the fighter pilot was a duellist, 'cool, precise, impersonal'. In the game of kill or be killed, blood had to be shed with dignity. 'Death should be given the setting it deserves; it should never be a pettiness; and for the fighter pilot it can never be,' he concluded. In January 1942, Hillary returned to service in the RAF. Within a year he was dead, killed during a training exercise.[3]

At the other end of the social scale (and representing the seamier side of the warrior myth) was Dave Nelson, a sniper during the Vietnam War. 'I really got into the trip that I was doing,' he admitted. He aimed to become the best sniper in Vietnam; with seventy-two kills under his belt and a 'best range' of eleven hundred metres, he made a good showing.

Nelson was raised for combat. His father was a highly decorated Second World War hero who became the First Sergeant at the Jungle Training School in Panama. Nelson was taught to shoot from the age of six, killed his first deer at eight, and from the age of twelve participated in manœuvres in the jungle. He relished the drunken war stories related by his father and his buddies around the kitchen table. In 1967, when he was seventeen and his father was dying from the lingering effect of wounds received during the Second World War, Nelson decided that it was his turn to go to war. A few months after his father's death, he enlisted.

Vietnam did not initially provide him with the thrills he had imagined – that is, until he decided to become a sniper. This decision was influenced by two factors. First, he resented taking orders and having his identity submerged within a group. The highly individual and potentially lawless role of sniper appealed to him. Secondly, and more crucially, he discovered that it was impossible for an ordinary grunt to live by his concept of a warrior. The 'warrior ethos', of which he was so proud, emphasized 'respect for your enemy' and insisted that 'to kill civilians or to lose control of yourself and your concepts in life in combat is wrong'. 'That's the concept behind the warrior,' he continued, '[k]ill cleanly, kill quickly, kill efficiently, without malice or brutality.' Yet, in Vietnam, such scruples were impossible for infantrymen. Nelson quickly discovered that he would inevitably kill civilians: his first 'kill' was a young boy seen walking too close to the road while a convoy was passing and, on another occasion, he was forced to watch a young girl he had shot die in front of him, absolute hatred for him in her eyes. As a sniper, however, all his targets would be legitimate and he could ensure a 'clean hit . . . A clean hit was an accomplishment.' The warrior never abused the power of life and death. Nelson would not be forced to participate in frenzied fire-fights. Rather, his duty lay in calm, cool-headed retribution. In Nelson's words: 'I chose who lived and died because I looked through the scope and pulled the trigger.' The reverse side was his own embracing of death: he always ensured that if he were shot he would have to fight to the death (for instance, when sniping from a tree, he would tie himself to the branches so that, if wounded, he would be forced to continue fighting until he died). 'Once you've accepted your own death,' he explained, 'you can become really proficient at killing because it is no longer important if you die.' By ensuring that he was always 'scared shitless', he was able to turn himself into a more effective killer. For Nelson, sniping was the only way to remain true to his vision of himself as an honourable warrior.

However, his pleasure in being a sniper did not last beyond his second tour of duty, when he was placed in charge of a unit which suffered extremely heavy casualties, during which time he felt compelled to shoot an American pilot who had been disembowelled in a crash. This act was necessary according to the code of the warrior (an honourable fighting man puts his comrades out of their misery) but resulted in his rejection by a primarily enlisted brotherhood who held a more 'civilian' concept of the warrior ethos. 'Everybody else just got quiet and backed away from me,' Nelson tearfully recalled, 'They didn't understand . . . they looked at me like I was weird . . . because they didn't understand the concept of it.' In the final analysis, Nelson exemplified a *professional* 'warrior ethic'. In mass, conscripted armies, the old military ethic was rejected wholesale: civilian soldiers had to remain reticent. Vietnam (Nelson concluded) 'dealt a death blow to the concept of the warrior'.

Thereafter, he embraced the 'dark side', that is, killing out of a sense of hatred, revenge and frustration.[4]

These three self-designated warriors created myth out of their own memories of combat. One was a patriotic writer exulting in the brutal physical contact of bayonet fights thousands of miles from home. Another was an arrogant idealist who paid for his daring aerial duels by being grotesquely scarred and ultimately by giving his life. The last warrior hallucinated about 'clean kills' in a filthy war. These myths of honour in combat were clearly diverse and contradictory, and there was intense rivalry between men adhering to the different traditions. Yet, whether bayoneting other infantrymen, firing at fellow pilots, or sniping a long-distance foe, combatants who thought of themselves as warriors placed similar emphasis on chivalry, intimacy and skill.

Blades, planes and telescopes

'Cold steel', 'aerial cavalry', and 'man-stalking' epitomized the warrior ethos in combat. It made no difference that these three forms of fighting were all fairly rare. The bayonet charge, for instance, was not representative of most combatants' experiences of battle. Even during the First World War, long-distance shelling by artillery killed two thirds of all soldiers, while less than half a per cent of wounds were inflicted by the bayonet.[5] As the doggerel, 'My Bay'nit' (1916), complained:

> *At toastin' a biscuit me bay'nit's a dandy;*
> *I've used it to open a bully beef can;*
> *For pokin' the fire it comes in merry 'andy;*
> *For any old thing but for stickin' a man.*[6]

Neither were the fighter pilot or the sniper typical. 'Personal' aerial combat did not develop until the middle of the First World War (when machine-guns were mounted on to wings) and even then aeroplanes were more commonly used for reconnaissance, artillery observation and dropping bombs on non-combatants than shooting down fellow airmen. Sniping, too, was carried out by a small, elite group. In the entrenched warfare conditions of the Western Front during the 1914–18 war, where often no mass targets existed against which rapid machine-gun fire could be directed, on average only sixteen snipers were assigned to each battalion.[7] Immediately after the First World War, snipers were 'returned to the outposts of Empire whence they had come and were heard of no more' since army instructors believed that their importance had been due only to the unusually static nature of that war.[8] During the Second World War (when snipers were used in certain theatres of war and within specialist units such as the parachute and airborne units, commandos, and the SAS), the average number of snipers in each battalion had dropped to eight. The number fell further (to six) by the time Nelson saw combat.[9] Australian troops were not even issued with sniper weapons in Vietnam[10] and the number of kills recorded by American troops was relatively low. For instance, by mid-1969, snipers in 9th Infantry Division were recording 200 enemy killed (on average) a month.[11] This represented a minuscule proportion of total deaths.

Despite their rarity, the forms of combat idolized by Idriess, Hillary and Nelson were widely respected. The least appreciated martial technique was Nelson's. His pride in his sniping abilities was primarily a phenomenon of the Regular armed forces and specialist units: as Nelson bitterly conceded, his preferences were not shared by many civilian-soldiers. Nevertheless, snipers got immense satisfaction from their job. Idriess applauded the one-against-one nature of sniping, admitting that (like bayoneting) it gave him a thrill 'through and through'.[12] It was 'huge excitement . . . faces full of life', one combatant explained.[13] Others were proud of the notches engraved on the butts of their rifles and angrily resented the times when they were detailed merely to collect intelligence.[14] These were highly motivated killers engaged in what one sniper called a 'labour of love'.[15]

To a much greater extent, the bayonet fight represented the highest achievement of the warrior culture, particularly during the First World War and interwar years. Irrespective of the fact that bayonets were more likely to be stained with jam than blood, the martial imagination was obsessed with jabbing and stabbing. Bayonet charges were a central motif in war stories during the first few decades of the twentieth century. Fictional representations lingered on the *sound* of steel-driven terror – the way 'dripping' bayonets crashed through breast-bone and body and the 'wild, bubbling groans as the long, keen bayonets went home', as John Finnemore in *A Boy Scout with the Russians* (1915) and *Two Boys in War-Time* (1900 and 1928) vividly described it.[16] In Nat Gould's *Lost and Won. A Tale of Sport and War* (1916), the hero was exultant when he heard 'the bones crack with a sickening sound' when thrusting his sword through a Uhlan's breast.[17] This 'sickening' noise also brought a 'thrill of satisfaction', a 'heavenly joy' which was guaranteed to render Huns 'scared and amazed', novelists assured their more timid readers.[18]

Young men took such gruesome tales to heart, blurring the line between imagination and experience. For instance, warrior fictions rejoiced in outlandish feats of strength with the bayonet: in particular, in the act of 'pinning' and then 'tossing' the foe. In Captain Frederick Sadleir Brereton's *With Rifle and Bayonet. A Story of the Boer War* (1900), a kilted giant with the 'strength of a Hercules' bayoneted a 'cowardly little German' and hurled him over his shoulder 'just as a man might toss a bundle of hay with a pitchfork'.[19] A Colour Sergeant in John Finnemore's *Two Boys in War-Time* (1900 and 1928) also used his 'gigantic strength' to toss a 'riven' Boer over his shoulder.[20] In 1915 (readers were told) men of the 4th Royal Fusiliers 'had a lively time', bayoneting Huns: 'It put us in mind of tossing hay, only we had human bodies,' boasted one combatant.[21] Descriptions of real incidents echoed these fictional accounts: around 1915 an Irish marine gunner described how a 'big Turk' had been stuck by the bayonet of a fusilier (a 'little chap') and then thrown over his shoulder and down a cliff. 'It looked an impossible feat,' the gunner admitted, 'yet the little Irishman was still alive and active hours afterwards, going on as though he possessed a charmed life.'[22] Such fanciful accounts of tossing corpses over shoulders were so pervasive that an official manual published in 1915 subsequently advised instructors to discourage soldiers from 'the tendency of "hay-making"' during training. 'It is very difficult and quite unnecessary to cast your opponent over your shoulder like a sheaf of corn on a fork,' the author gravely directed.[23]

Even so, no one was prepared to detract from the glamour associated with the bayonet. In part, its status was due to men's ignorance of the realities of modern warfare. New recruits simply did not feel like real soldiers until they had fixed shiny blades to their rifles.[24] American artillerymen created 'a little hulloa' when they had their bayonets taken away prior to proceeding to the Front (for the practical reason that heavy artillery men would never be in a position to use bayonets): they had 'long cherished the hopes of getting to bayonet a Hun', these men protested.[25] For many men, bayonet fighting would always be the 'reallest' type of combat.[26] It was 'the kind of bloody fighting that rejoiced the hearts of pirates', recalled Coningsby Dawson at the time of the First World War.[27] 'How they hungered for the wild exulta-tion of the bayonet charge, the shock of man to man in deadly encounter, the pursuit of a vanquished foe!' exclaimed the Irish journalist, Michael McDonagh.[28] Miles behind the frontlines and with little chance of moving closer, men felt inspired to write fanciful letters to their families in which they swanked about how in a recent battle they 'saw red and remember nothing more' except that afterwards their bayonets were suitably stained with 'blood & hair'.[29] To some extent, the bayonet 'kill' also had important sexual overtones.[30] Men preparing for a trench raid would pass 'caressing' hands over their bayonets, all the while preparing for that gory moment of bliss when the blade would be 'driven deep' into another man's chest 'so that his heart's blood spouts in a great jet on to the muzzle of the gun'.[31] It is no coincidence that men bragged most frequently about their bayoneting exploits to mothers, sisters and girlfriends, rather than to fathers, brothers and male comrades.

In the interwar years, threats by military command to strip bayonets from corps such as the Royal Army Service Corps, Royal Army Medical Corps, Royal Army Ordinance Corps, Royal Army Pay Corps, Army Educational Corps and the Army Dental Corps (on the grounds that these corps were never in a position where the bayonet could be employed) were met with fury and dismay. After all, bayonets were the 'emblem of their combatancy and their continued use is essential to maintain the military spirit and pride'.[32] Only this 'honourable' yet old-fashioned instrument of war, it was widely believed, could defeat a beastly foe. The bayonet made British, American and Dominion troops 'irresistible'.[33] Turks and Italians abandoned their frontlines at the sight of steel.[34] The bayonet struck fear in the hearts of German soldiers.[35] Even the Japanese whom, it was conceded, were exemplary with the sword, had 'no relish' for close fighting and could become demoralized at the sight of the bayonet.[36]

Later during the Vietnam War (when there could be no doubt that massive firepower had replaced intimate combat), the myth of cold steel was still potent. Private First Class Rod Consalvo, for instance, was a rifleman, well equipped with a M14 but, on his first night on guard duty in enemy territory, he recalled:

> when it started gettin' dark I started gettin' scared . . . so I put my bayonet on my rifle. I didn't know how much good that would have done me but . . . at least it *looked* awfully tough.[37]

If the bayoneting infantryman dominates First World War accounts, the fighter pilot was his later equivalent. Indeed, aerial combat was often conceptualized as the technological equivalent of the bayonet fight. Men also spoke about aerial warfare as though it resembled an infantry advance or a cavalry charge with swords. David Lloyd George praised airmen as the 'cavalry of the clouds'.[38] Flying was 'a gratifying mixture of excitement and lawlessness; we were going to nip over and create a short hell and be away with a roar and a yell – just as the raiding N.W. Frontier tribesmen descend from the hills', in the words of Kenneth Hemingway describing his experiences in Burma in 1944.[39] In Vietnam, the New Zealander Colin P. Sisson described going to battle in one of a number of helicopters in close formation. 'My fear dissolved,' he recalled,

> and suddenly I felt inspired and greatly excited. It was as if I were in the cavalry, swords drawn, charging into the face of the enemy . . . I gripped my rifle tighter, eager to reach the battlefield and engage the enemy.[40]

Helicopter formations were easily portrayed as cavalry charges, as in the film *Apocalypse Now*.

Nowhere was the fusing of technology with a more traditional warrior ethos achieved more successfully than in the air.[41] All air crew were volunteers and highly trained. The high status and prestige of the air forces meant that promotion opportunities were superior and airmen won a higher proportion of decorations and military awards. Officers tended to have better relationships with their enlisted men than in the other services and this encouraged a higher degree of group identification. The fact that they were more than usually dependent upon their comrades, who were more immediately and clearly identifiable than they were in the infantry, meant that the *esprit de corps* of air force bases tended to be extremely high. Life in the base was often like a camp for overgrown schoolboys. Practical jokes, binges, the stripping and washing of comrades' abdomens with beer, and the tossing of furniture were all part of high spirits of 'the boys' after a successful 'kill'.[42]

Aerial warfare was very popular amongst combatants.

According to one American survey conducted during the Second World War, three quarters of combat air crew were willing to perform further combat duty compared with only two fifths of combat infantrymen. The more 'personal' the fight, the more combat air crew enjoyed their job (thus, when American aerial combat personnel were asked during the Second World War: 'If you were doing it over again, do you think you would choose to sign up for combat flying?', 93 per cent of fighter pilots, 91 per cent of pilots of light bombers, 81 per cent of pilots of medium bombers, and 70 per cent of heavy bomber pilots replied 'yes').[43] It 'amused' one light bomber to 'see the people running away from under the machine', even though he 'felt sorry when I saw the remains of an ancient city being blown up by my bombs'.[44] Aerial fighters enjoyed 'stalking' their prey. In the words of James Byford McCudden after shooting down a 'Hun':

> I think that this was one of the best stalks that I ever had. I cannot describe the satisfaction which one experiences after bringing a good stalk to a successful conclusion.[45]

Roderick Chrisholm was a night fighter in the Royal Air Force during the Second World War. On 13 March 1941, he destroyed two enemy aircraft. The experience, he wrote, could 'never be equalled':

> For the rest of that night it was impossible to sleep; there was nothing else I could talk about for days after; there was nothing else I could think about for weeks after . . . it was sweet and very intoxicating.[46]

Equally, the Spitfire pilot, Flight-Lieutenant D. M. Crook, described the 'moments just before the clash' as 'the most gloriously exciting moments of life'. He was 'absolutely fascinated' by the sight of a plane going down and could not pull his eyes away

from the sight.[47] Kenneth Hemingway dive-bombed Japanese
soldiers on the ground: 'Oh, boy . . . Oh, Boooyy!' he yelled,
describing his 'exhilaration' as being similar to the joy of drinking champagne on a sunny spring morning. He felt 'ruthlessly
happy – quite an atavistic orgy!'[48] Even anti-war poets such as
William J. Simon during the Vietnam War agreed that killing
from the air was sublime. In his poem 'My Country', he
described it thus:

> Daily riding the clouds, appointed scythe-swinger of the
>   aluminium age,
> silvery engines of slaughter: hellfire raining from above.[49]

The chivalry and skill of the warrior

All three warrior types were seen as exemplifying certain chivalrous codes (or recognized formalities, ceremonies and courtesies)
involving honourable exchange, compassion and altruism at the
same time as invoking reckless adventure and a high-minded dis-
dain of death. Combatants employed the language of chivalric
ideals to describe their actions. Thus, bayonet fighting could be
conceived as bringing back the 'barbaric nobility of war'.[50] In the
words of Harold Stainton, the battle at Wytschaete Wood (1914)
resembled eighteenth-century combat: 'Here was no steady
advance behind a creeping barrage of shell fire, but the wild rush
so dear to the Highland tradition, with effective use of cold
steel.'[51] Even the *Irish Ecclesiastical Record* raved (in 1914) against
the disappearance of 'much of the high chivalry and poetry and
romance of war', claiming that the denial of the 'personal element' in combat had a demoralizing effect: the bayonet was the
instrument to challenge such demoralization.[52]

Sniping and aerial combat linked modernity with old-
fashioned chivalry. Firing at one's foe through telescopic sights
combined the 'savagery of modern twentieth century warfare'
with the 'romanticism and chivalry of previous conflicts', according to one Australian sniper.[53] To a much greater degree, aerial

combat could be imagined as similar to medieval rituals of combat. Lloyd George rhapsodized, 'Every flight is a romance; every record an epic . . . They recall the old legends of chivalry.'[54] W. N. Cobbold, in his poem 'Captain Albert Ball, V. C., D. S. O.' (1919), waxed lyrically and at great length (in over thirty-two stanzas) about the 'knight-errant of the clouds above' who drove his chariot across the heavens, singlehandedly combating squadrons of Huns and evading scores of anti-aircraft guns.[55] In 'The Revival of Knighthood' (published in the same volume), Cobbold claimed that the spirit of the knight had been reborn in the aerial fight. The airman was equated with the cavalry of old: he was the 'glorious Horseman of the clouds', roaming amongst the clouds and speeding through fields of space. The fight was 'mount against mount' and airmen drove 'steeds of steel and wire'. Once again, there was 'single combat', a 'romantic adventure', a duel in which the true warrior either conquered or died.[56] In a less poetic vein, the fighter pilot, Cecil Lewis, echoed the words of Richard Hillary earlier in this chapter:

> To be alone, to have your life in your own hands, to use
> your own skill, single-handed, against the enemy. It was
> like the lists of the Middle Ages, the only sphere in
> modern warfare where a man saw his adversary and faced
> him in mortal combat, the only sphere where there was
> still chivalry and honour . . . You did not sit in a muddy
> trench while someone who had no personal enmity against
> you loosed off a gun, five miles away, and blew you to
> smithereens – and did not know he had done it! That was
> not fighting; it was murder. Senseless, brutal, ignoble. We
> were spared that.[57]

Airmen lived and fought 'like gentlemen'.[58] Their bases were in safe areas so that, between missions, there was ample opportunity for relaxation. Visitors to air bases marvelled at how men who 'half an hour ago had been . . . shooting men down' were now 'sitting, washed, bathed, cleanly and comfortably dressed, at a full-course dinner, with flowers on the table and a good band playing outside'.[59]

At its core, this chivalrous code contained an idealistic spirituality. Self-professed warriors insisted that the bayonet had a righteousness that other weapons lacked.[60] The bayonet was the sword that symbolized 'justice and right and the vengeance of a living God for outraged humanity', one writer declaimed in *The Incinerator* (the news-sheet for the Young Citizen Volunteers). Even men who felt squeamish about killing believed that 'a bayonet-fighter of crusader-faith' was 'nearer to God than a machine-gunner'.[61] Religious spokesmen promoted this idea more than anyone else. In a broadcast on 10 December 1939, Cardinal Hinsley spoke eloquently on the horrors of modern mechanized warfare and looked back nostalgically to the days when the weapon of war was the sword – the 'symbol of justice . . . the instrument of chivalrous defence of the weak against the strong' – now exchanged for the mine, the bomb, torpedos or gas. The 'high achievements of science', he lamented, had been 'prostituted to serve barbarism'.[62]

Most importantly, however, these ways of killing highlighted individual prowess in anonymous, technologically driven warfare. Many combatants yearned for *personal* blood-letting as against the dispiriting anonymity of shelling.[63] As Captain M. D. Kennedy put it:

In open warfare, you at least had a run for your money. In the kind of trench warfare now being experienced for the first time, you had little or none. It was a disheartening experience for troops brought up on the traditions of open warfare, in which everyone had a chance to hit back, and the killing or wounded of oneself or the enemy was in fair open fighting. What was wanted by officers and men alike was the chance for a real good 'scrap'.[64]

This was what made the bayonet fight attractive to many combatants. For instance, Charles Cecil Miller of the Royal Inniskilling Fusiliers described hand-to-hand combat in no man's land as 'immensely exciting', bluntly commenting that it was 'less exciting to be torn to pieces by a screaming piece of metal fired from several miles away.'[65] Similarly, in a letter written on 4 May 1915 to his sister, Guy Warneford Nightingale (of the Royal Munster Fusiliers) described the exhilaration of face-to-face combat using the bayonet: 'we had a great time', he wrote, 'we saw the enemy which was the chief thing and the men all shouted and enjoyed it tremendously. It *was* a relief after all that appalling sniping.'[66]

Nightingale's dislike of sniping expressed his hatred of *being* sniped at by anonymous warriors. The person *doing* the sniping, however, gained pleasure from his activities precisely because the foe was identifiable. Sniping, too, was 'personal' killing: it required that the killer looked his victim 'in the eye'.[67] It enabled men to display their individual skill in a war that provided very few alternative outlets.[68] Snipers such as F. de Margry took great pride in knowing the 'Tricks of the Trade' and being able to outwit German soldiers by surprise tactics and superior wit. He would crawl into no man's land with 'quiet confidence', his pockets bulging with his girlfriend's love-letters (which he called his 'armour').[69] Individual valour and the opportunity to display one's skills were central to all warrior stories.

There was another element in men's homage to the individual struggle. More than any other form of killing, sniping could be seen as similar to game hunting. For instance, when the Mounted Sharpshooters Corps was established for service in South Africa in 1899, recruits were deliberately sought amongst big game hunters, deerstalkers, and keepers employed on Scottish deer forests.[70] It was widely believed that the best snipers were countrymen, and particularly those of higher social status who might have had the opportunity for hunting in civilian contexts.[71]

Sniping was simply big game hunting in which the quarry shot back.[72] The prey was nothing more than 'soft-skinned dangerous game'.[73] The number of men killed became a source of immense pride and competition, identical to the way that the number of partridges 'bagged' was a matter for boasting, announced Max Plowman in 1919.[74] As the rifleman Herbert W. McBride cynically quipped while in sniper training: 'since we can no longer have an open season on Indians, about the best way to acquire the skill [of sniping] is by stalking wild game.'[75]

There was also a sense that aerial combat was seen as a 'sporting' way of killing.[76] Airmen going to their first battle were going to be 'blooded'.[77] Men watching air fights described it in terms of 'rounds' between two contestants.[78] The 'old delight of battle' was possible in fighter planes because of the speed and skill of the killing.[79] Roderick Chrisholm was a night fighter during the Second World War. In 1942, after shooting down a Heinkel, he reflected that this type of combat was

> a big-game hunt, and thought was focused on personal achievement. In the aftermath it was satisfactory to know that the enemy bomber force had been reduced by one, but immediately it was the elation of personal success.[80]

This romanticization of the individual fighter meant that the rewards were individual too, particularly for the pilot who drew the 'first blood' of the day.[81]

The premium placed on individual skill tended to generate a sense of respect between pilots, irrespective of nationality, especially in the early years of war.[82] In a sermon preached in Lenton Priory Church on All Saints' Day in 1918, Frank Bertrand Merryweather prefaced the immensely popular story about the British ace pilot, Captain Albert Ball, with the reminder that the events he was about to describe recalled the 'old days' when 'champions' approached each other in single combat as in a duel. He then told the story of how Ball discovered one day that his

German counterpart (ace pilot Max Immelmann) was in the opposing trenches. He flew over the German lines and dropped a note challenging Immelmann to a 'man-to-man fight' at two o'clock that afternoon. He swore that British anti-aircraft guns would not fire and trusted Immelmann to do likewise. At the appointed time, all guns fell silent and men on both sides watched as the two aces fought for their lives. Ball won and, to British cheers, he returned to the place where Immelmann's body was being lifted from the wreck and released a huge wreath of flowers.[83] During the Second World War, similar codes were respected. The fighter pilot, Paul Richey, was 'fascinated' watching a plane he had just hit go into a spin:

> I remember saying, 'Oh God, how ghastly!' as his tail sud-
> denly swivelled sideways and came right off, while flames
> poured over the fuselage. Then I saw, with relief, a little
> white parachute open beside it. Good.

When an enemy pilot was captured, Richey explained, 'we all felt that this German had put up a damned good show, and as a trib-ute to the spirit that all pilots admire, we determined to have him to dine with us as our guest', and they treated the pilot to the best dinner and clothes they could muster.[84] Similarly, Wing Commander Roland Beaumont discovered that he was unable to kill a German pilot that he had downed during the Battle of Britain. 'I could not bring myself to fire at this man who had just put up such a beautiful exhibition of flying,' he explained. The downed pilot was thus able to destroy his plane before saluting Beaumont and walking into captivity.[85]

This personal element may be taken a step further. Not only did mythical warriors kill identifiable individuals, they also per-sonified their machines. 'MIGs' (Soviet-built planes flown by the North Vietnamese) were described as being 'killed' as though they were people.[86] When bullets hit an aeroplane, an aerial machine-gunner might recall seeing 'his enemy stagger, leap upwards, lurch and roll'.[87] The 'long belly' of a Messerschmitt was called 'the German'.[88] A curious slippage often occurred between man and machine. A gunner simply known as 'Dizzy' reported: 'I killed the rear gunner and when I left him . . . he was pouring glycol from one engine.'[89] When pilots who had just bombed a U-boat wondered whether they had 'killed him', they meant the boat.[90] Warriors not only killed identifiable humans, they also slaughtered 'human' machines.

This was in stark contrast to aerial bombing which fighter pilots considered less legitimate because it lacked skill. No fantasies of sporting prowess could be indulged in by the dropping of bombs on unarmed civilians. In the words of one man with twenty-five years commissioned service in the RAF:

> I was a fighter pilot, never a bomber pilot, and I thank
> God for that. I do not believe I could ever have obeyed
> orders as a bomber pilot; it would have given me no sense
> of achievement to drop bombs on German cities.[91]

Reality disappoints

But these warrior myths rarely lived up to the realities of modern combat. Men were not prepared for the horror of being unable to remove their bayonets from the body of their foe, or for the stench of blood. Fighting seemed never to end: killable 'individuals' just kept on coming. Attempting to act like a true 'warrior' could cost men their lives. Men too keen on using the bayonet, for instance, would chase a retreating enemy and thus put themselves at grave risk.[92] Even aerial combat – that most romanticized form of killing – could be disappointing.[93] In the words of one pilot during the First World War, it was lucky that the 'Big Bugs' did not give the pilots and gunners time to think about the fights. 'Damnation and hell,' he continued,

> when I start to think of it I start to drink, to keep my
> blood at boiling point – ginger myself up to go out and kill

and laugh at my conquests. Great for these days of
emancipation, when man is supposed to be superior to the
beast. I don't think.[94]

In 1942, Commander B. W. Hogan observed the difficulties
experienced by pilots during an attack on Guadalcanal. Men felt
elation after shooting down Japanese planes but were profoundly
shaken when forced to fly down 'on running human beings,
opening up all the guns, and bullets spraying, killing and maim-
ing many of those unknown individuals.'[95] Or in the words of
Stanley Johnston after describing the low bombing and strafing
attacks by pilots on Guadalcanal:

Theirs was not the high-flying fights and duelling for
enemy scalps which brings fame and decorations. It was a
dangerous task . . . These flyers seldom get mention,
much less see their names splashed across newspapers.[96]

The romance of aerial combat had been tainted by the mass
destruction caused by bombing. As the historian, John H.
Morrow, put it:

the myth of strategic bombing of civilian populations,
reminded those that chose to pay attention that the air war
was truly the child of the era of total war, which conflated
civilian and military targets and deemed the bombing of
women and children an acceptable means of winning a
war.[97]

Equally, the myth of two equally matched opponents facing
one another in the air was fanciful. 'The majority of kills
achieved by the major aces were at the expense of fledgling pilots,
barely able to control their planes,' concluded the historian,
Robert L. O'Connell.[98] What genuine chivalry existed in aerial
warfare had evaporated by 1917. In the words of Stanley Johnson

in 1945: 'under no circumstances must an enemy, caught at a disadvantage, be allowed to escape . . . There is no chivalry in this war and no place for it.'[99]

Sniping, too, came to be regarded as a 'dirty' and dishonourable trade. There was something not quite proper about killing with 'eager hand'.[100] Snipers belonged to the 'Hate Squad'.[101] Arthur Emprey was a keen and aggressive killer, but he could not stand sniping. In his bloody book, *First Call* (1918), his tone changed when he discussed his six weeks as a sniper:

> To me it appeared quite all right to 'get' a man in the heat
> of battle, but to lie for hours and days at a time, waiting
> for an enemy to expose himself – then to plug him,
> appeared to me a little underhanded.

He killed two men in this fashion, felt 'sick to the stomach' and requested to be relieved. As a sniper with nerves was useless, his request was readily granted.[102] Even snipers admitted that their business was 'little better than murder' and 'a filthy sort of business. "Every bit as bad as Jerry",' according to Victor G. Ricketts.[103]

Official sources also registered a distaste for the sniper's job. Frank Percy Crozier was a harsh man, but even he could not stand the bloodthirstiness of sniping. In his memoir *The Men I Killed* (1937) he told how much he enjoyed game hunting in south-west, west and central Africa, but balked when the target was human: 'The game was dirty. I had to give it up. The cool, calculating murder of defenceless men was diabolical.' This reaction, he believed, was commonplace: only the 'most perverted' could take pleasure in such a 'cold-blooded' form of killing. Of course, he recognized the importance of snipers and so never revealed his emotions, lest the snipers under his command should 'become infected' with ethical doubts. He admitted that, later in the war, when he sniped at *tanks* with 18-pounder field guns, he

enjoyed the joke. It was different; that sense of guilt, that conscious-stricken feeling of killing a man who at the moment was not menacing you and who was brought almost within hand-shaking distance by the telescopic sights, had disappeared.[104]

Bill Howell was also a sniper, who served at Loos. He explained that there was another reason why snipers were disliked intensely by other soldiers:

The men loathed us, and the officers hated us. They could not order us out of their sector. The trouble was, we would watch through a loophole for hours, and when we were absolutely sure of a target, we would fire. No other firing went on without an order, and Jerry knew it was a sniper, and he would let everything he had loose on that sector. Of course we hightailed it out, as fast as our legs would carry us, and poor old Tommy had to take it.

So hostile were the bulk of soldiers to snipers that they were kept separate from the rest of the battalion and were exempted from the routine duties of service.[105]

In their letters and diaries, men told 'dreadful romances' of bayoneting, sniping and aerial combat.[106] These martial myths depended upon notions of intimacy, chivalry and skill for their appeal. The ability of combatants to imagine themselves as engaged in honourable combat not totally unlike that experienced by chivalrous knights of the past was crucial to their sense of pride and pleasure. Although modern slaughter was typically anonymous, dirty and banal, by conjuring up myths of chivalry, combatants were able to evoke feelings of respect and compassion for their enemy while still remaining committed to the killing enterprise. The disappointment (sometimes even horror) of being unable to enact the role of the warrior sometimes resulted in the realization that they were not engaged in 'war', but in 'bloody murder'. Despite such bouts of disillusionment, the myth of the duellist, the knight and the expert maintained an indistinguishable appeal throughout the century. Chivalry was evoked to stifle fears of senseless violence; intimacy was substituted for confusing anonymity; skilfulness was imposed to dispel numbing monotony.

Chapter 3

TRAINING MEN TO KILL

Point and parry, Port and thrust!
Make your opponent feel your might!
Yell, man! Scream, man! Wake your lust!
On guard! Withdraw! My God, man . . . fight!

Shawn O'Leary, 'The Bayonet', 1941[1]

Richard ('Rick') Edward Marks's letters to his mother and sister trace his transition from training camps to the war in Vietnam. 'Tonight is my second night at Parris Island,' he wrote in his first letter on 14 November 1964, 'and I can honestly say that the Marine Corps. has grounds for calling itself the "best". The training they give, physically and mentally, is by far the hardest imaginable.' However, as he was to discover, military training was of little help in coping with the realities of service and was useless when he faced the greatest trial of all: exactly fifteen months after his first letter he was killed in combat, aged nineteen.

Before his death, Marks's letters record the main elements of his day: 'I don't want a teenage queen/ All I want is an M14', he chanted along with his comrades; he was 'barked' at during drill; he was instructed on 'everything from how to handle yourself if captured by your communist enemy to Marine Corps. History'; and he endured endless marches, lectures on homosexuality ('they should all be shot'), and lessons in the handling of the M14 rifle and .45 automatic. But, as he reminded his mother, 'the most deadly weapon in the world' was 'a Marine and his rifle', despite the fact that these weapons were wielded by mere 'boys like myself'. This was changing. Over time, Marks observed the growth of facial hair and the loss of cranial hair, the deepening of his voice, and his swelling self-confidence. Despite these changes he reassured his family that he was still 'their' boy: yes, he observed Hanukkah and a rabbi was sometimes available; yes, male comradeship was a fine thing; but, yes, he missed his mom and sis.

There were parts of his training that he enjoyed: in particular, he revelled in bayonet exercises. In his letter on 1 December he recorded:

> Today we had our first bayonet class – it is quite a
> weapon – we all left the field thirtier [sic] than hell – all of
> us had the same idea, like a child with a new toy, to try it.
> That is the way we are all growing to feel about combat in
> general – we want a taste of it.

Combat was a long way off, however, and Marks was impatient to leave the camp. 'Marine Boot Camp is the hardest, most trying, thing one person can be subjected to,' he wrote. 'The first two weeks here you are torn down to nothing; they make you feel lower than a snake in a pit, and then the following 8–10 weeks they proceed to re-build you, the Marine Corp. way.' He lived, he said, in a 'perpetual state of shock and fear'.

By May 1965, eighteen-year-old Marks was an ammunition carrier in an M60 machine-gun team in Vietnam. 'I don't mind saying I am scared,' he stammered, and he was already apprehensive about how he would cope with his eventual return to civilian society after his tour of duty. On his nineteenth birthday he noted how his moral values had changed: 'a human being becomes so unimportant, and the idea of killing a V.C. is just commonplace now – just like a job,' he observed, admitting that this callousness 'scares me more than being shot at'. War was not only about combat, he soon discovered, and he deeply resented having to act as a 'Peace Corp worker'. He scorned officials who believed that Marines who had been trained to act in a 'blood thirsty' manner could be expected to protect Vietnamese civilians. He was happiest when armed with his rifle, counting 'kills'.

Marks was killed in combat on 14 February 1966. A year earlier, he had written his 'Last Will and Testament' in which he reminded his family that he had 'always wanted to see combat'. He requested to be buried in his Marine Corps uniform. His last words were simple: 'I love you Mom & Sue, and Nan, and I want you all to carry on and be very happy, and above all be proud. Love and much more love, Rick.'[2]

Training Rick Marks to be an effective fighter had not been an easy task for the US Marines in the 1960s. Prior to the First World War, the professional military, with its claim to a distinct field of expertise – that is, the incitement and control of state-legitimized violence – regarded the stimulation of murderous aggression in soldiers as a relatively simple matter, requiring little effort on their part. From 1916, the mass, conscripted armies, together with startlingly new technologies of war, stung them out of this complacency. Natural science – particularly physics, chemistry and engineering[3] – had created a battlefield which rendered combatants passive and stripped officers and men alike of that 'offensive spirit' believed to be essential in order to gain victory on the battlefield. It was not possible to rely upon the instinctive flaring-up of martial aggression: military training regimes needed to reflect the 'civilian' codes of the new recruits. Motivating men to kill through bayonet drill and vigorous training suddenly generated intense debate. Promoting the offensive spirit and turning men into 'thinking bayonets' (in the words of an instructor at the Royal Military College, Sandhurst)[4] became a central concern, for which the study of human psychology seemed to hold the answer.

Training for combat

The chief function of military training was to convert civilians into effective combatants. This required more than merely teaching men to name the various parts of the rifle, Bren gun or grenade: they had to imbibe essential military traits as well, including toughness, alertness, loyalty and discipline.[5] Combat could never be 'a theoretical business like pure mathematics, or pure science, or pure anything else' but was always dependent on the man and his training, observed the editors of the *Australian Army Journal* in 1956. Man-the-weapon was their catchcry.[6] Furthermore, men had to be taught to react instinctively to orders: to shoot rifles, stab with bayonets, and throw grenades as if such actions were 'second nature'.[7] The ideal soldier was epitomized by the young man of the Fourteenth Royal Welsh who was able to bayonet a German in parade-ground style, automatically exclaiming: 'In, out, on guard!'[8] It was believed to be crucial that battle habits became 'so deeply ingrained' that they would 'persist in the face of the most overwhelming provocations to rage or panic'.[9] After all, blood-lust, rage and hatred were counterproductive responses, making men's hands tremble when shooting at the enemy.[10] Military psychology decreed that men had to be taught how to become 'practical realists', conscious only of the necessity to kill or be killed, if they were to arrive at the battlefield emotionally intact.[11]

Nevertheless, in all three conflicts, the efficaciousness of military training was loudly questioned by officers in the frontlines. First, there were doubts about the *nature* of the training. It tended to be haphazard and *ad hoc*. Indeed, in the British army, there was no common approach to training until July 1918 when Sir Ivor Maxse was appointed Inspector General of Training. Prior to this time, the emphasis had been on drill and physical fitness and it was left up to each division to decide on precise methods employed. Even afterwards, when programmes were regularized, combat training was still barely adequate, as one crisis was swiftly followed by another, resulting in men being sent to the front with less and less preparation. For instance, in September 1942, of 860 reinforcements posted to the 50th Division, Eighth Army, only one quarter had done *any* field firing, seven had never fired a rifle in their lives, nine had never fired a Bren gun, 131 had never thrown a live grenade, and 138 had never fired a Thompson submachine-gun.[12]

Much more serious was the realization that no matter how thorough the training, it still failed to enable most combatants to fight. No amount of military training could deal with volunteers, conscripts, and even Regular servicemen who simply lacked that elusive 'offensive spirit'. During the First World War, it was commonly believed that only 10 per cent of soldiers could be called brave[13] and many military commentators deplored the 'live and let live' principle in which servicemen on both sides came to agreements not to shoot if the other side restrained themselves too.[14] 'Live and let live' relied upon approximate perceptions of the relative strength of each military unit and was therefore strongest when both sides were evenly matched. Refusing to 'go over the top' unless forced to do so at gunpoint, and malingering, were commonly observed.[15]

By the Second World War, concern about 'passive combat personnel'[16] reached almost hysterical levels, largely due to widespread commentary and to shocking statistical information which made it very apparent that many 'stable' men (that is, men not at risk of breaking down under stress of combat) simply did not kill. It was relatively easy to cope with the rare recruit who unexpectedly showed himself to be a 'belated conscientious objector', refusing to go through with bayonet drill or a bombing run.[17] Immeasurably worse – and infinitely more common – were soldiers who completed their training and went into battle, yet never fired their weapons. This lack of offensive spirit was frequently commented upon. The 'average Jack' was 'quite

amazingly lethargic' noted one report in 1943.[18] Lieutenant-
Colonel Robert G. Cole (the man in charge of the 502nd
Parachute Infantry which was considered to be one of the best
units in the US Army) was horrified to find that when his men
were being attacked along the Carenton Causeway on 10 June
1944 it was impossible to make them fire back. 'Not one man in
twenty-five voluntarily used his weapon,' he lamented, despite
the fact that they could not dig in or take cover so their only
protection was to ensure that the enemy kept 'his head down'.
They 'had been taught this principle in training. They all knew
it very well,' Cole continued, but

> they could not force themselves to act upon it. When I
> ordered the men who were right around me to fire, they
> did so. But the moment I passed on, they quit. I walked
> up and down the line yelling 'God damn it! Start shoot-
> ing!' But it did very little good. They fired only while I
> watched them or while some other soldier stood over
> them.[19]

Passivity was not a vice unique to land forces. A study of the
famed 51st Fighter Wing (known as the 'MIG-Killers') in Korea
revealed that half of their F-86 pilots had never fired their guns
and of those who had fired only 10 per cent had ever hit anything.[20] In the words of the fighter pilot, Hugh Dundas: 'When
it comes to the point, a sincere desire to stay alive' overpowered
any incentive to 'engage the enemy'.[21] As a Canadian military
instructor wrote in 1951, the combatant's

> sole function is to kill the enemy . . . The possible criti-
> cism that such training will result in lack of control and
> 'trigger-happiness' is not tenable. The problem is not to
> *stop* fire, but to start it, and it is far better to have some
> excess of enthusiasm than the present lassitude displayed
> by some of our so-called riflemen.[22]

In other words, servicemen stumbled into combat zones with weapons poised for action but this did not mean that they possessed 'the active attitude' essential to actual killing.[23]

If these influential commentators were capable of stirring up disquiet about the effectiveness of training regimes, even more important was the statistical work carried out by researchers employing sociological sampling techniques. During the Second World War, Colonel S. L. A. Marshall of the US Army interviewed men in 400 infantry companies in the Central Pacific and European theatres of war, only to declare that no more than 15 per cent of men had actually fired at enemy positions or personnel with rifles, carbines, grenades, bazookas, BARs (Browning Automatic Rifles) or machine-guns during the course of an entire engagement. During the actions he had examined it would have been possible for at least 80 per cent of the men to have fired and nearly all men were (at some stage) within firing distance of the enemy. Furthermore, these operations were not 'casual' ones, but actions crucial for the unit. To be counted as a 'firer', a man would only have had to fire his weapon or lob a grenade 'roughly in the direction of the enemy' once or twice. Even allowing for the dead and wounded, and assuming that in their numbers there would be the same proportion of active firers as among the living, the proportion of 'active combat personnel' did not rise above 25 per cent. The most active men were those using the heavier weapons such as the BAR, flamethrower or bazooka. The terrain, tactical situation, experience of the troops, nature of the enemy, and accuracy of the enemy fire seemed to have no effect on the proportion of firers to non-firers and there was little to differentiate well-trained and campaign-seasoned troops from the rest.

Marshall used the battle on Makin Island (part of the Gilbert Island invasion of November 1943) as an illustration. During this battle, one battalion of the 165th Infantry Regiment was attacked by Japanese troops wielding swords and bayonets; half of the American guns were knocked out and half of the men in forward foxholes were killed. Only superior firepower saved the remaining American troops. Afterwards, Marshall interviewed the survivors. Even making allowances for the dead, he could identify only thirty-six men who had fired at the enemy with all the weapons available and most of these 'active' men had been manning heavy weapons. In other words, the active firers were men who were working together in small groups and they tended to use a variety of weapons: if their machine-guns ran out of ammunition, they switched to rifles and then to grenades. Most surprisingly, Marshall found that there were some men who identified targets yet did not shoot, and there were other men who were under direct attack yet did not attempt to use their weapons to retaliate or in self-protection. Furthermore, passive combatants were not 'green' troops. The only effect of battle experience seemed to be to convince NCOs of the need to 'use direct methods to increase the fire power' (that is, for the NCO to move continually up and down the firing line) – but this was suicidal during a prolonged attack. All this led Marshall to conclude that not more than one quarter of men

> will ever strike a real blow unless they are compelled by almost overpowering circumstances or unless all junior leaders constantly 'ride herd' on troops with the specific mission of increasing their fire.

Despite the fact that Marshall did not interview as many men as he said he did, and not one of the men he interviewed remembered being asked whether or not he fired his weapon, his 'statistics' shocked the military establishment. Believing Marshall's conclusions, they were forced to re-examine their training regimes. After all, these soldiers were not, technically speaking, 'cowards' (they were there to *be* killed and 'passive combat personnel' were not more liable to quit ground sooner than their more aggressive comrades), they simply lacked the offensive spirit.[24]

Technology and personnel

What were the factors inhibiting aggressive behaviour in battle? This question has been vigorously debated throughout the century but the most important considerations can be summarized under two headings: technology and personnel. It was widely argued that new military weapons had rendered redundant earlier ways of ensuring that soldiers would fight in combat. When, prior to the First World War, military commentators emphasized that war was an art not a science,[25] they were partly correct. After all, prior to 1914 the armed forces in all three countries possessed only a small number of motor vehicles, a handful of aeroplanes, no tanks, no long-distance bombers and no gas canisters. This situation was beginning to change, however. Troops were no longer able to advance on the enemy in close order: long-distance weapons separated soldiers from each other, making the battlefield one of the loneliest places on earth and inhibiting offensive actions.[26] Very quickly, it became clear that the industrialized battlefield as devised by physicists and engineers heightened rather than diminished the importance of 'human factors'.[27]

This problem caused by the use of long-distance weapons was a central explanation put forward by Marshall for men's passivity in battle. Time and time again, men, after their first taste of combat, confessed to him that they 'saw no one' and were 'unnerved' by the feeling that they were 'fighting phantoms'. Training regimes prepared them for 'flesh-and-blood, fully mortal, and therefore vulnerable' targets – it did not prepare them for an eerie enemy 'who does not seem to be present'. Combatants even feared being rebuked for wasting ammunition on an invisible (and therefore potentially non-existent) enemy. The longer the feelings of isolation and confusion lasted, the less likely it was that anyone would act aggressively.[28]

The second important reason for passivity in battle focused on personnel. By the middle of the First World War, officers were no longer 'socialized warriors' who had experienced long periods of military training at Sandhurst, West Point or Duntroon. They could not be depended upon to 'act promptly on sound lines in unexpected situations'.[29] The British public school system was widely regarded as the most fertile recruiting ground for well-disciplined, effective officers, yet by 1945 only one third of British officer army cadets were former public school boys.[30] The problem was even more urgent when it came to training men in the other ranks, most of whom were ambivalent about acting aggressively. The Regular army was not accustomed to civilian soldiers who were older, more highly educated, and more liable to come from the middle classes than previous cohorts.[31] These 'new' recruits' images of battle (as we saw in Chapter 1) could also be a major problem for a pragmatic military. Soldiers had their own idea of how they wanted to kill the enemy – and these methods did not always agree with the expediency promoted in official training manuals. Soldiers might choose to use their revolvers as clubs to beat Germans to death instead of pulling the trigger.[32] Some were so anxious to bomb their victims that they forgot to shoot them, thus placing their own lives in serious jeopardy.[33] Queasy about making a 'mess', they preferred to strangle instead of firing their weapons at their opponents.[34] They were reluctant to actually 'hate' the enemy.[35] Crucially, these recruits were civilians first and only temporarily soldiers. As Field Marshal Earl Wavell lamented in 1945: 'the old soldier *was* tough, the modern type usually has to be toughened'.[36] The conventional uses of negative incentives (punishment and fear) were not only devalued but were also less effective amongst reluctant servicemen who regarded dismissal from the services in a positive light and were often almost eager to exchange a prison cell for near-certain death in combat. An ever-vigilant population back home frequently agreed and in wars waged in the name of democracy and freedom there were limits on the incentives that could be used to promote combat effectiveness.

Of course, the military could have ignored the sensitivities of the new generation of combatants and attempted to force passive men to act in appropriately martial ways. In many instances, they *did* attempt to do this. In Chapter 5 we will examine some of the most savage forms of 'hate training'. Basic training itself was often extremely brutal, even for conscripted recruits. The most notorious training regimes were those conducted by the US Marine Corps,[37] but even in the other branches of the armed forces, violence was a common component of military training. In all these training programmes, the fundamental process was the same: individuals had to be broken down to be rebuilt into efficient fighting men. The basic tenets included depersonalization, uniforms, lack of privacy, forced social relationships, tight schedules, lack of sleep, disorientation followed by rites of reorganization according to military codes, arbitrary rules and strict punishment.[38] These methods of brutalization were similar to those carried out in regimes where men were taught to torture prisoners: the difference resided in the *degree* of violence involved, not its *nature*.[39] Lieutenant William Calley, a participant in the My Lai massacre in Vietnam, who is the subject of Chapter 6, described training at Fort Benning (Georgia) at the Officer/Cadet School:

> One thing we were taught at OCS for twenty years we had thought was bad. To kill, and sergeant in gym shorts and a T-shirt taught it. We sat around, and he kicked another man in the kidney: a few inches lower, really, or this could be a lethal kick. It was just gruesome: a POP, and I thought, *Oh god. No one can live through that*. He really kicked or he flipped a man with karate and WHAM: he would show us the follow-up. And stomp on him right between the eyes: pretend to, and push his nose right into his brain. Or stomp on his solar plexus: his ribcage, to push splinters into his lungs. And then stomp on his heart to smash it.[40]

Such techniques were not limited to the American forces during the Vietnam War. In 1919, Private Stephen Graham described training in the British Army as a nightmare involving 'constant humiliation and the use of indecent phrases' aimed at reducing each man 'to a condition when he was amenable to any command'. He continued:

> To be struck, to be threatened, to be called indecent names, to be drilled by yourself in front of a squad in order to make a fool of you, to do a tiring exercise and continue doing it whilst the rest of the squad does something else; to have your ear spat into, to be marched across parade-ground under escort, to be falsely accused before an officer and silenced when you try to speak in defence – all these things take down your pride, make you feel small, and in some ways fit you to accept the role of cannon-fodder on the battle-ground.

He described having to hit his rifle with his hands during drill until blood flowed as a necessary hardening procedure.[41] The general attitude of the instructors can be seen in a lecture associated with bayonet training in the Australian forces. According to 'Allied Land Forces in S.W. Pacific Area Training Ideas No. 5' (1943), instructors had to insist that the

> whole basis of the new system is KILLING – with the bayonet, which implies the production of a determined and capable man, who delivers his point under complete control of himself and weapon . . . The new system should also see the finish of the 'On Guard – As you were – On Guard Etc.' type of Instructor. Smartness of movement is still catered for, and developed, but the 'motif' throughout is KILLING! 'Pansy' flicks and dabs, wild and unbalanced thrusts etc. won't do for the Japanese; what IS necessary is the aggressive, controlled and well-balanced

killer, remembering the record of the yellow barbarian in
Malaya and New Guinea and 'going straight in' with
blood in his eye and cold rage in his heart and 'Advance
Australia' as his incentive . . . It is KILLING TIME in
the S.W.P. [South West Pacific]![42]

Taunts about virility and competence could be irresistible to
young, immature men trapped in hostile environments miles
from home. Probably the most effective of these negative exhortations were the numerous petty disciplines whose purpose was
to teach men to act harshly. A typical example can be found in an
article entitled 'The Case of the Fainting Soldier' (published in
the *Canadian Army Journal*, 1958) in which men were reminded
never to go to the aid of anyone who fainted on parade. 'In
modern armies there is nothing obsolete about iron discipline,' it
began,

> But let us get the term straight. It is not synonymous with
> brutality, or callousness, or cruelty. It is, in the end,
> humane. It is a stark fact that we must train our soldiers to
> kill our enemies, and not to flinch when confronted with a
> similarly trained adversary. The fallen comrade must await
> a lull in the battle before he can be cared for by his
> friends.[43]

Even recruits who attempted to resist being turned into 'killing
machines' could find themselves forced to act savagely. For
instance, Michael Rosenfield was drafted into the US Army,
1969–70 and recalled:

> They'd give you a rifle with a bayonet and they'd say,
> 'What is the spirit of the bayonet?' And you had to yell,
> 'To kill!' And, I'll never forget, I'd pantomime. I wouldn't
> say it. So, to make us say it louder, the sergeant would yell,
> 'I can't hear you!' People would scream, 'To kill!' And

again I was pantomiming. Once, they said, 'If you don't say it louder, we're not going to give you people lemonade!' And, I'll never forget, at one point I yelled that the purpose of the bayonet was to kill – the first time I ever did that – and it was to get some lemonade because I was so exhausted and dehydrated.[44]

Such forms of training were justified by military commanders: after all, men were being training for war, not a Sunday School picnic.[45] A man who 'could not take being shouted at and kicked in the ass' (rationalized Philip Caputo during the Vietnam War) 'could never withstand the rigors of combat'. Brutalization in training camps so lowered an individual's sense of self-worth that it made him even more anxious to prove that he was 'equal to the Corps' exacting standards'.[46]

Despite the frequency of violence in training camps, the military increasingly found itself having to cope with more civilian encroachments. The vast technologies of modern warfare brought combat to everyone's doorstep. From the First World War onwards armies were numbered in the millions rather than the tens of thousands, and the military was forced to justify itself to entire populations. In exercising this daunting duty, there were two chief problems. The first was to define a clinical practice and a language to deal with men who could not stomach combat. This is expanded on in Chapter 8 but briefly, psychiatrists, clinical psychologists and social workers were highly valued for 'curing' men who experienced stress in killing. These social scientists also provided crucial changes to the military terminology: 'cowardice' (with its accompanying need for execution or punishment) became 'shell shock' and then 'anxiety states' (which called for treatment, albeit stigmatizing). By the end of the Second World War and in Vietnam, it became 'battle exhaustion' (which called for rest and elicited more sympathy). This process of redefining what the military continued to regard as cowardice would not have been achieved without the political and economic pressures of 'total war' and it could not have been done without applying concepts from pathological psychology to men's reactions in combat.

The second (and related) problem was to find a way of exchanging the language of military bellicosity for a language of civility: or, more correctly, a language of bellicose civilians. Dehumanizing and excessively brutal ways of training men for war were scrutinized by a wide range of people, including parents anxious that their 'boys' both did their duty and returned to peaceful civilian lives afterwards as better citizens, not as professional fighters. There were things that could be done to small numbers of Regular soldiers which had to be justified when applied to huge cohorts of volunteers and conscripts. The frustration-aggression formula of the psychologist John Dollard and others – which implied that by increasing frustration, aggressive behaviour could be fostered – was used to legitimate many of the more sadistic aspects of basic training.[47] The language of transference and displacement replaced cruder military words, without necessitating any change in the behaviour of instructors. Psychological jargon was also used to justify manpower decisions regarding black soldiers (the use of evolutionary psychology and then group dynamics to exclude them from combat roles, and then when combat manpower became desperately low, the use of psychoanalytic languages to integrate them).[48] Military officers consciously, and unashamedly adopted the language of psychology to justify treating civilian soldiers in a military manner when it suited them.[49]

Morale and realism training

There were a variety of responses to the problem of passivity caused by increasingly technological warfare and by a more 'civilian' pool of personnel. Technologically orientated responses included emphasis on offensive weapons as opposed to weapons of defence (on the grounds that since weapons were 'an extension of the ego', offensive weapons would encourage aggression)[50], the introduction of 'quick kill' or 'instinctive' techniques of firing weapons from 1967, and 'REALTRAIN' from 1971.[51] Personnel-orientated training responses included the promotion of *esprit de corps* and the use of reward systems such as badges and stripes.[52]

Imbuing recruits with patriotic or political conviction was also important. The traditional view that the soldier was merely 'a bundle of conditioned reflexes, a belly, genitalia, and a pair of feet' was being rejected.[53] Each soldier had to know *why* he was fighting if he was to avoid being turned into a 'thug or a psychotic with a guilt complex'.[54] The New York psychiatrist, Leonard R. Sillman argued that American soldiers must be given

> positive, fixed, passionately believed word concepts for protection. His government should give him an official, deeply emotional crystallization of the meaning of the war, integrating his welfare with hatred of the enemy and religious devotion to American ideals.[55]

However, most commentators had little faith in the usefulness of reason in enabling men to kill. Even the prominent military commentator, Charles Moskos (who believed passionately in the importance of men's beliefs), was forced to acknowledge that 'ideological sentiments' and 'patriotic rhetoric' were '*latent* beliefs', rarely actually articulated.[56] In the words of that more hard-hitting analyst, Colonel S. L. A. Marshall, the reluctance to kill was 'an emotional and not an intellectual handicap'. Therefore, 'it is not removable by intellectual reasoning, such as: "Kill or be killed".'[57]

The promotion of *esprit de corps* and ideological training only went so far. It did little to eradicate the most important factor causing men to freeze in battle: that most noxious 'virus', fear.[58] Transforming fear into anger was a chief preoccupation of military instructors. In the words of a Marine trainer: 'Fear will cause you to die' so men must 'get angry and kill!'[59] The conventional approach to fear was to teach men to act automatically, so that in battle their bodies would 'instinctively' carry out the necessary movements. In actual combat, however, the limitations of such a policy were instantly obvious: there were, in fact, very few routine procedures which would be of use in any given crisis since each fight had unique objectives, resources and environments. Furthermore, soldiers were frequently required to act without direct command from a superior officer and they needed to be taught how to exercise independent judgement in such circumstances: automatic movements and instinctive obedience to commands were inadequate – indeed, the most perfectly drilled and well-disciplined men were frequently the first to crawl into the bushes during battle.[60]

The most influential initiative to eradicate fear was realism training and the two most vocal promoters of this were John Frederick Charles Fuller (during the 1914–18 conflict) and Colonel S. L. A. Marshall (after the Second World War). Fuller was arguably the most prescient military philosopher of the twentieth century.[61] He believed that the 'spirit of the offensive' could best be inculcated by a process of slow, continuous, and almost imperceptible indoctrination. Human psychology was crucial (after all, he sensibly pointed out, 'the firer and not his weapon' was 'affected by the fear of death'). In developing his argument, Fuller drew on his own brand of instinct theory and crowd theory. According to him, the character of an individual depended on his impulses and his will which, in turn, depended for their nature on the spirit or ego. Everything that the individual came into contact with produced within him a sentiment for good or for evil. If repeated frequently enough within the individual, these sentiments became habits; if repeated frequently enough within the race, they became instincts. Although the army could not change a man's instincts, it could 'so bombard him with warlike impressions that his acquired tendencies, his reflexes, become wholly warlike'. In addition, the army itself acted as a crowd and, as such, was governed by the same laws which governed a crowd. Like the individual, crowds were ruled by voices from the past. Examining the army 'as an entity', Fuller proclaimed:

> we find that that part of it which we call its mind is swayed by that part of it which we call its soul; and that in this soul the dominating impulses are those drawn from the substratum of unconscious acts, and in particular from the inherited instincts. Under certain circumstances the conscious personality of the individual evaporates, and the sentiments of each man are focused in the same direction: a collective soul is then formed, and the crowd becomes a psychological one, and henceforth acts as an individual in place of a mass of individuals.

In other words, the 'army-crowd' would be swayed by the voice of training 'for uniformity of environment creates uniformity of character and of spirit'.

Bombarding the individual and the crowd with 'warlike impressions' was crucial, yet, as Fuller recognized, training regimes did not do this because they ignored the one vital ingredient of battle: bullets. Without this ingredient, recruits could easily be trained to act appropriately – but they would not act similarly under real battle conditions. In order to simulate the presence of bullets, he proposed using a series of coloured flags. A red flag meant that the fire of the enemy was superior to that of the attackers: blue signalled equality; and white indicated inferior firepower of the attackers. Through a series of manœuvres, recruits would have to respond in different ways to combat situations depending upon the risk. They could be taught to advance under cover, make appropriate choices of fire positions, and act calmly under pressure. In contrast to company drill, which stressed perfect order and absolute precision to the word of command, combat drill placed emphasis on tactical order and adaptation to the ground.[62]

After the Second World War, Marshall took Fuller's insights further. He also believed in the importance of realistic training if men were to act aggressively, and he agreed with Fuller that men had to be *taught* to act upon their own initiative. Automatic movements and precise drill were damaging. Officers and their men had to understand psychological processes, particularly men's need for human contact. Marshall encouraged men to touch each other ('patting another on the back . . . may turn a mouse into a lion', he observed). Human warmth – or merely being able to see another man – was essential to the offensive spirit, as was movement, even if it was merely the digging of a foxhole or administering first aid to a comrade. In Marshall's words: 'Action is the great steadying force. It helps clear the brain. The man who finds that he can still control his muscles will shortly begin to use them.'

Marshall warned against overly optimistic expectations. The passive 75 per cent of men would generally remain passive. But, he noted, even those soldiers who did not fire were crucial to the battle: their presence was essential for morale. Active combatants were too busy fighting to notice what their comrades were (or were not) doing. In fact, it was the presence of passive soldiers which enabled active soldiers to continue fighting. They contributed their weight to the mass of the attack, even if they contributed little to its velocity. Firers should be rewarded with promotion while persistent non-firers had to be moved to gun crews where the need to work as a group would force them to be active. In some instances, it might help to give passive combatants one-man weapons such as flamethrowers or BARs on the grounds that the individuality and prestige of these weapons would relieve them of the paralysing effect of anonymity.[63]

However, by the time Marshall was writing, the military had already gone much further than either Marshall or Fuller by introducing realism training, or 'battle inoculation' as it came to be called. There were many varieties (I examine an extreme form of it in Chapter 5, 'Love and Hate'), but during the Second

World War it took the form of an infiltration course which required crawling over rough ground under live gunfire. Soldiers were exposed to nearby artillery fire and took part in realistic battle manœuvres. There were lanes containing surprise targets which popped up at recruits, demanding to be quickly dispatched. 'Blitz courses', 'infiltration courses', 'village fighting courses', and 'close combat courses' all required recruits to fight under different terrains and tactical situations while exposed to loud explosions and overhead fire. Assault courses were geared to test and train men psychologically as much as physically. Pits were filled with smoke so that the bottom was not discernible; dummies were placed so that they could not be seen until the man was on top of them; men passed through two screens only to be faced with a dummy on either side of him, thus testing speed of reaction in bayoneting.[64]

Realism training was not approved of by everybody.[65] The Australian Army enthusiastically adopted realism training but recognized that if 'badly administered' it could 'produce nervous conditions identical with that of "shell-shock"'. In a report issued during the Second World War, Australian officers were reminded that the aim of training was not to 'give men frights' and 'must not become an opportunity for exhibitions of manly prowess by instructors'.[66] However, if the exercises were carried out in a 'bland matter-of-fact manner', the general view was that such forms of training provided soldiers with more realistic expectations of combat, increased their motivation to learn combat skills, taught acceptance of loud noises, gave experience in firing at moving targets, enabled men to practise psychological techniques for distracting themselves and focusing only on the task at hand, and reassured men that they could cope under intense pressure.[67] Lance Kent of the Australian forces during the Second World War described the effectiveness of realism training as carried out at the jungle warfare school at Canungra:

As for the absence of enemy opposition, this lack was

compensated for by the instructional staff. Instructors who were crack shots would position themselves a hundred yards or so on the flank of the trainees who were undergoing training, either on the bayonet assault course or in an exercise which demanded that they attack. If anyone lagged behind, they would fire at his rear foot. The bullet would land at his rear foot. The bullet would land a yard or less behind him. It was remarkable how a soldier who thought he was exhausted could muster up a sudden burst of energy when he heard the 'ping' of a shot or the 'thunk' of it hitting a couple of feet behind him and, maybe, the 'whe-e-e-e' of a ricochet droning off.

Another method of 'smartening up the troops', he continued, was to 'take half a plug of gelignite, crimp a detonator with your teeth to a short fuse, light it and throw it behind the laggards!' At ordinary ranges, the resultant explosion was physically harmless – and mightily effective.[68] Like Kent, other combatants acknowledged the effectiveness of these methods, not only in training but in battle as well: when 700 soldiers with combat experience in North Africa were asked in 1943 what kind of training would lessen the shock of battle, one third asked for more training under live ammunition and (a year later) over 80 per cent of 344 combat infantrymen in Italy believed that going through tough, realistic battle training had been a very important part of their preparation.[69]

Bayonet training

While realism training was the advanced, modern innovation, it did not replace more traditional ways of encouraging men to act aggressively in combat. The dominant method common to all three conflicts examined in this book was bayonet drill. It is curious that the bayonet remained so central to military training despite awareness that modern technologies and rapid-fire rifles had greatly limited its usefulness. After all, even before the Boer

War, it had been acknowledged that modern armaments meant that the hand-to-hand fight would no longer 'play a very prominent part in the battles of the future.'[70] Combat experience showed that bayonet attacks were seldom 'pressed home' because machine-guns, artillery and wire prevented the troops reaching their foe or because the enemy surrendered long before the bayonet stage of the battle plan could be reached. In the words of a Second World War soldier, explaining why he did not use the bayonet: 'If I was that close to a Jerry, where we could use bayonets, one of us would have already surrendered!'[71]

Percy Raymond Hall may have been a humble private during the First World War but he was perfectly correct to criticize the order by 'some fool of a brass hat' to bayonet the enemy: 'How can you use a bayonet when the enemy is across half a mile of flat land and can mow you down with machine-guns long before you can get anywhere near him?'[72]

Nevertheless, commanders and officers continued to insist that the 'decisive factor in every battle is the bayonet'.[73] The cavalry might reconnoitre areas, aeroplanes might scout, the artillery might harass, Lewis gunners could provide covering fire, and bullets could hold positions, but it was the bayonet which captured positions. According to a huge range of military professionals, even in the age of gas and tanks and aeroplanes and nuclear warheads, wars could only be won by infantrymen, skilled in the use of bayonet and rifle.[74]

Consequently, men's prowess with the bayonet was constantly tested in training camps. The British Army training manual of 1916 informed instructors that men carrying out final assault training should be brought to the assault course already exhausted to test the accuracy of their bayoneting. The sacks used to practise bayoneting were carefully constructed to simulate the difficulties men might experience in battle. Indeed, the 1916 manual insisted that

sacks for dummies should be filled with vertical layers of

straw and thin sods (grass or heather), leaves, shavings, etc., in such a way as to give the greatest resistance without injury to the bayonet. A realistic effect, necessitating a strong withdrawal as if gripped by a bone, is obtained by inserting a vertical layer of pieces of hard wood.

Men's accuracy was tested through the use of discs attached to dummies. These discs could only be 'carried' by a vigorous thrust and clean withdrawal. The assault had to begin from a trench which was six to seven feet deep, as well as from the open; soldiers were not supposed to cheer until they were close to the enemy; and dummies were to be hung from galleys or placed on tripods, as well as being placed in trenches or on parapets.[75] The vulnerable parts of men's bodies would be painted on to the sandbags.[76] Even the navy taught men to use the bayonet. Hal Lawrence described training under a Canadian naval officer who would yell at them:

> You've got to *want* to come to grips with the enemy. No captain can go far wrong by putting his ship alongside that of the enemy. Board them. Give them a taste of cold British steel. Shoot them. Club them. Subdue them!

This was training for piracy, not modern war at sea.[77]

Why did the armed forces continue to insist upon training men in the use of the bayonet? To some extent, its retention was simply due to the conservative nature of the military institution. In fact there were very few changes in combat training between the First World War and the Vietnam War.[78] Some commentators argued that bayonets had to be retained if only because Russian and German riflemen were still equipped with them.[79] In a few instances, they argued, it was still a valuable instrument of war, being a silent weapon and thus important when secrecy was paramount.[80] In forest fighting and guerrilla warfare, bayonets could be essential.[81] At night, the bayonet was a safer weapon because the intimacy of any encounter made it less likely that friends would be killed accidentally.[82] When ammunition ran out or when rifles jammed, it was the final, desperate resort.[83]

There were more fundamental reasons for the retention of bayonet training, though. In a military milieu where officers willing to carrying out offensive actions received the highest accolade, the superiority of the bayonet over the bullet was assured. Bombs, grenades and bullets encouraged men to cower from the enemy while bayonets forced men to leave their trenches and engage with the enemy. Bullets could 'drive the enemy to cover' but would not 'drive him from it', a training manual reminded platoon officers in 1919. 'Confidence in their ability to kill with the bayonet carries men to the assault', it reasoned.[84] In 1915, the War Office declared that only men confident in their 'own power to use the bayonet' were liable to 'wish to come to close quarters.'[85] Bayonets developed the 'lust for blood'.[86] Even astute commentators like Fuller, who knew only too well that bayonets had been of little use in previous conflicts, accepted the need to train men with the bayonet on the grounds that it was the weapon which enabled infantrymen to kill.[87]

This was not merely a fad of the 1914–18 years. Similar comments were made during the other conflicts.[88] The bellicose display of 'cold steel' was guaranteed to cause the hearts of the enemy to sink.[89] As Andrew G. Elliott succinctly put it in a manual for the Home Guards:

> a shot should never be fired at a single Nazi if he can be disposed of by the bayonet, or a blunt instrument, for the demoralizing effect is great when, next morning, a sentry or a patrol is found mutilated.[90]

In 1944, General Franklyn (Commander-in-Chief, Home Forces) instructed commanders of training units that the bayonet assault course was crucial in teaching men 'dash' and a

'determination to close with the enemy'. He deprecated 'feminine squeaks' who flinched from such training.[91] *Bloody
Bayonets. A Complete Guide to Bayonet Fighting* (1942) advised
that 'the boys' must not be treated as if they were 'effete'.[92] And
in the words of one Vietnam veteran reflecting on instruction in
hand-to-hand combat: 'They were selling us charms to ward off
fear.'[93]

The psychology of military training

'Charms to ward off fear' were deliberately manufactured within
the armed forces by a new group of professionals. As has already
been hinted, the empty, dehumanized battlefield created by the
physical sciences (and rejected by the arts) had provided a political and logistical space for a 'human science': social psychology.
Whereas physical scientists (in their terminology and by distancing combatants from each other) attempted to deny that
warfare was concerned with killing human beings, men at the
front were all too conscious of this fact. For them, the great
instruments of destruction were not impersonal tools, but intimate weapons. Thus, while the physical scientists pretended that
emotion was irrelevant, the social scientists took the experiences
of combatants seriously and placed excitement, fear and fantasy
centre-stage. They attempted to convince the military hierarchy
that unless officers were trained in psychology and were able to
counter the effects of mechanization and anonymity, all their
expertise with guns, ranges, and ballistics was useless.[94]

The role of behavioural scientists in enabling men to kill is
only very rarely, and then extremely obliquely (under the heading 'soldiers' morale'), mentioned by historians.[95] This is not
really surprising: it does not mark a proud moment in the history
of the profession and it is not a particularly marketable skill.[96]
The impression most modern commentators convey is that intelligence testing, manpower allocation, and man-machine
efficiency regimes consumed the energies of psychologists in
wartime.[97]

However, if modern social scientists have averted their eyes from theoretical and applied research on how to encourage 'effective combat behaviour', social scientists in the past have been far less bashful. As the psychologists, E. F. M. Durbin and J. Bowlby explained in their book *Personal Aggressiveness and War* (1939): 'just as it is the task of the physicist to study the general laws governing the behaviour of forces, such as electricity or gravitation', so too it was the task of social psychologists 'to describe and analyse the general psychological forces lying behind the timeless and ubiquitous urge to fight and kill'.[98] While it is true that squeamish psychologists, and those with pacifist inclinations, could easily find themselves a bloodless niche within which to employ their talents, those who whole-heartedly embraced the militaristic enterprise were no small group of warmongers. Indeed, during both of the world wars, it was the psychologists and their professional representatives who pleaded with an initially reluctant officer class that they be allowed to prove themselves in inciting high combat motivation.[99]

Initially, the military establishment was decidedly ambivalent and suspicious of such civilian encroachments.[100] They feared that its influence might 'dangerously affect manpower'.[101] Just as threatening, the presence of psychologists teaching new recruits how to act aggressively was liable to be misunderstood by the general public and cause despondency within the ranks.[102] Too much psychology in officer training might produce officers who were 'scientists first, soldiers second, and generals never'.[103] Furthermore, Regular army officers were sceptical about the predictive possibilities of psychology.

Until the first few decades of the twentieth century, an understanding of 'human factors' was more liable to be sought in literature. For instance, in a twelve-page-long article published in *The Australian Military Journal* just before the First World War, Lieutenant-Colonel W. L. Raws set out to define combat effectiveness – not through the sciences, but through an analysis of Rudyard Kipling, Robert Browning, George Eliot, Bernard

Shaw, John Ruskin, George Meredith, Thomas Carlyle, William de Morgan, Henry Newbolt, Matthew Arnold, 'a Japanese writer', and the Prophet Ezekiel![104] Regrettably, such commentators observed, men were not machines. Rather, combat was unpredictable: spunkless boys fought like 'lions' while bullies trembled in trenches.[105] Combat soldiers were also less than wholehearted about the implications of much of the research. Many of them preferred to interpret their actions in terms of the biological impulses: popular interpretations of evolutionary, instinct theory, with the implication of warring men being overwhelmed by some life-enhancing, animalistic force impossible to resist, held much greater appeal than the rather banal, bloodless application of all-too-human social manipulation.

Despite this hostility, by the Second World War, history and literature had been ousted and social psychology dominated similar discussions in military journals.[106] The introduction of psychology into military training regimes was a gradual process which began in the first decade of the twentieth century and was strongly influenced by the writings of Fuller who (as we have seen) read widely in crowd and instinct psychology. Fuller made some headway in the interwar years with amendments to the British Army Regulations which introduced training in psychology for officers.[107] In these years, military manuals slowly began to introduce psychological factors,[108] but psychology did not carve out a secure niche within the armed forces until the Second World War by which time military heavyweights such as Field Marshal Earl Wavell (in Britain) and William C. Westmoreland (in the United States) were arguing publicly for increased psychological training and support for officers and men.[109] During the Second World War, a majority of professional psychologists were involved in war work and social scientists replaced natural scientists on the boards of important research organizations such as the National Research Council and the Research Information Service.[110] Military psychology courses sprang up in colleges throughout Europe and the United States; and the media popularized their research. Many psychologists believed that warfare had been fundamental to the growth and status of their profession by enabling them to 'connect scientific psychology to life'.[111] As Lewis M. Terman put it in his presidential address to the American Psychological Association in 1923, war had transformed their discipline from a 'science of trivialities' into a 'science of human engineering'.[112]

The influence of psychology

Military psychology was not only used to train men to act aggressively: there were other ways of making men more 'combat effective'. From 1918, psychologists devised ways of improving men's shooting technique.[113] Through the use of motion studies, naval psychologists were able to reduce the gun crews by two men without affecting loading time or strain.[114] Films were used from the Second World War, not only as a training device but also to desensitize recruits to the noise and bloody gore of battle.[115] Psychologists were important in improving training regimes by applying principles of learning to the training of soldiers.[116] Military instructors bought textbooks on the psychology of learning and many attended psychology lectures at university.[117] Although some of the principles were of limited usefulness for the military establishment, by 1942 important techniques such as spreading the practice of a particular skill over a few days rather than excessively concentrating it into one long period of drill, renewed stress on active participation, the use of varied material, accurate record-keeping combined with positive reinforcement, and systematic lesson plans, were all mainstream training devices.[118] By 1942, a special method of instruction wing was opened at the British Army School of Education providing courses in methods of teaching.[119] Finally, personnel selection was also a crucial function of military psychologists, particularly from 1941.[120]

The range of applications was immense and the military forces adopted popular, pragmatic psychology: there was never a

coherent text and theories used in one decade could vanish and
then reappear a few decades later. In general terms, though, there
was a shift in training practices from the use of instinctive theories (which encouraged drills aimed at stimulating reversion to
primal passions and promoting automatic movements) to (by the
end of the Second World War) seeing the important variable as
character or personality (thus, drills were instigated to develop
these character traits, particularly in leaders), to environment
(most men would make suitable soldiers, but appropriate combat
behaviours needed to be instilled in every man through socialization).

Instinct theory, particularly as popularized by the founding
father of functional psychology, William McDougall, in his
Introduction to Social Psychology (1908), never lost its influence
over military instructors.[121] The Revd Stopford A. Brooke wrote
his *Discourse on War* in 1905. It was reprinted many times,
including during the First World War. In it he argued that fighting was a primary instinct in human nature:

> It comes down to us from the brutes; and linked to it, I
> can't tell why, is a sense of keen pleasure, eagerness, and
> exaltation. We cannot get rid of this hereditary passion. It
> is universal; as acute in the civilized as in the savage.[122]

In the words of the philosopher William James in his renowned
'The Moral Equivalent of War' (1910): 'Our ancestors have bred
pugnacity into our bone and marrow, and thousands of years of
peace won't breed it out of us.'[123] Since pugnacity was one of the
primary instincts, combat training was aimed at stripping the
civilized veneer from the individual psyche. The 'beast within'
was encouraged to find expression in bayonet drill and dehumanizing rituals.

There was a problem, of course: the instinct of self-
preservation was as liable to lead to 'flight' as to 'fight'. To
overcome this, military psychologists coupled instinct theory

with the crowd psychology of men like Gustave Le Bon and his English popularizer, Wilfred Trotter.[124] Humans were herd animals, with strong gregarious impulses. In a crowd – and the army was only a trained crowd – the 'group mind' would take over, endowing the individual with a sense of almost limitless power and immortality. Group solidarity led to a return to primitive forms of behaviour, including reliance on the leader as the father substitute. Crowd psychology also promoted automatic movements: group drills, with emphasis on monotony and everyone doing the same thing together, enabled men to carry out the required movements almost without conscious thought, all the time feeling 'supported by the formidableness of the group'.[125] This was the training regime that one Canadian soldier of the First World War was praising when he wrote: 'Mechanically we stabbed a dummy figure. Mechanically we would stab and stab again a breathing human frame.'

Between the two world wars, instinctive theories of aggression came to be ridiculed by academic psychologists and anthropologists such as L. L. Bernard, Ellsworth Faris, R. E. Money-Kyrle, and W. J. Perry, to be joined by Franz Alexander, Knight Dunlan, Mark A. May, and T. H. Pear during the Second World War.[126] However, they never died out completely, especially within military psychology. Instinct theory remained powerful, but primarily as a justification for the gruesome training enterprise, rather than as a practical way to encourage men to kill. This was summed up nicely in an article published in 1965 by Captain P. P. Manzie of the Royal Australian Army Medical Corps. He noted that the resolve of the army leader could be weakened by a feeling that his task was 'unnatural and immoral; that he has to pervert a peaceful being into ways of violence'. Not so! he reassured his readers, killing was part of every man's natural inheritance. The job of the military instructor was, in fact 'already half done – the polite bank–clerk strips down, not to a peaceful individualist, but to a soldier born'.[127] Similarly, crowd theories had a long life. In the period leading up to the First

World War, the ability of the crowd to instigate actions that were antithetical to individuals was regarded as a justification of the army's emphasis on leadership or the 'father figure' who could 'sway' the unit by his personality.[128] The *character* of the leader was the central feature: he had to embody aggression, courage, strength of mind and physique, and responsibility.

Increasingly, group dynamics were substituted for crowd theory: in particular, emphasis was placed on the cohesion of the wider group and the need for reassurance. What Marshall called the 'primary group concept' and 'group identification' came to dominate psychological discussions on combat motivation.[129] 'Even the timid soldier comes to feel secure by being in a powerful group and often assumes the aggressive attitude of the organization,' observed one expert from the Far Eastern Command.[130] Effective training regimes needed to reflect the motivational power of love, rather than hate.[131] In the words of Major J. O. Langley of the Royal Australian Infantry: 'man is a gregarious animal. His greatest steadying force is the touch of his fellows under battle's pressure.'[132] Intense group identification would lead to a displacement of self-love on to the group, resulting in a declining fear of self-annihilation and the consequent reduction of inhibition about the expression of aggression. It strengthened combat ideals and standards: 'even the poorly motivated soldier is literally forced to adopt the prevailing group attitude because the battle situation is hardly a place to be left alone', decided the foremost military psychologist, Albert J. Glass. And then there was the 'effect of the group on the character trait of passivity'. By identifying with the strong group, the passive soldier would be able to adopt its aggressive attitude.[133]

Behaviourism, particularly as popularized by John Broadus Watson, was greatly favoured by military psychologists. Watson's behaviourism of the 1920s argued that the human infant was a *tabula rasa* on to which the environment scrawled its cultural message.[134] His theories implied training that was in line with earlier drills involving positive and negative reinforcements. The assumptions of top–down conditioning were compatible with the authoritarian hierarchy of power within the military. From 1971, B. F. Skinner's reinforcement and learning theory was adopted, with the components of the 'job' identified and rewards rather than punishments becoming the focus of training regimes.[135]

Use of psychoanalytic concepts in training was highly controversial. Some military psychologists (particularly in the American forces) embraced them. For example, in the 1940s, Major Jules V. Coleman argued that psychoanalytic principles and frustration–aggression theories were crucial in enabling men imbued with the commandment 'Thou Shalt Not Kill' to 'move in on the enemy and destroy him'. Patriotism and idealism would never provide sufficient incentive. Effective training required the mobilization of 'free-flowing aggression' and the control of anxiety and guilt. These two principles were related, since anxiety and guilt inhibited aggression, and the harnessing of aggression helped to control inner tensions. Important ways of maintaining morale involved the provision of competent leadership, training for discipline and skill, instilling pride in the unit, promoting an aggressive attitude, and taking care of the troops. Killing the enemy would thus resemble a mythical rite in which the death of the father (represented by the company commander) could be celebrated in 'an orgy of displaced violence'. Coleman argued that this slaughter would satisfy

> deep-seated, primitive unconscious strivings derived from early childhood fantasy . . . The enemy is a sacrificial object whose death provides deep group satisfaction in which guilt is excluded by group sanction. Combat is a ritualistic event which resolves the precarious tensions of hatred created by the long drawn out frustrations of training. Without these frustrations, a group would not be a military force.

In combat, the 'hatred which has been carefully nurtured and encouraged through the irksome indignities associated with long military training' would be displaced from the men's leaders to the enemy. In other words, killing the enemy became an act of vengeance and the enemy's dehumanization helped allay feelings of guilt.[136]

Coleman was part of a large movement towards introducing psychoanalysis into the military.[137] But, despite making some inroads, psychoanalysis never gained much favour, even within the American forces. As a clinical discipline, psychoanalysis was too time-consuming, it required a highly trained personnel, and it was inappropriate given the transient nature of military relationships.[138] As John T. MacCurdy acknowledged, if individual attention was to be paid to men who had problems coping with the act of killing, time would have to be 'taken from somewhere else'.[139] Psychoanalysis was much more important in dealing with mental breakdown than with training, precisely because although violent passions lay just underneath the surface, the chief problem was that of sublimation afterwards. Despite the attractive pension implications for the military (some military psychoanalysts argued that since emotional problems could be shown to have been based in infancy or early childhood, the military should not be liable for compensation), it remained a marginal influence.

Of course, the social scientific model was not adopted in its entirety. After all, its disciplines were rejected by many soldiers: they malingered, manipulated the tests, and responded in inappropriate ways – particularly when the scientific content threatened to de-heroicize their role.[140] But the social sciences were still a powerful influence because their languages and practices were accepted by a wide range of people outside the divisions of military psychology within the British, American and Australian psychological associations. Social scientists were increasingly employed by the military as *the* professionals capable of creating effective combat soldiers, not only through the well-known processes of intelligence–testing, manpower allocation and man–machine efficiency regimes, but also by stripping the military enterprise of its primitive combat glossolalia and substituting psychological regimes of 'scientific' military training. The presence of 'passive combat personnel' continued to threaten the entire military enterprise, but with the help of psychologists 'passive' members could be put to better use within units. In this way, social scientists were integrated into the entire militaristic enterprise, engaged in developing ways to augment (rather than abate) human aggression.

Chapter 4

ANATOMY OF A HERO

Single-handed he did it – it was most amazing,
Against half a dozen, or more, it may be;
For odds count for nothing when match'd 'gainst the courage
and ready resource of O'Leary, V.C.

Lily Doyle, 'O'Leary, V.C.', 1916[1]

'I would rather be a recipient of the Congressional Medal of Honor than be president.'

President Harry S. Truman

While the highest medal given by the US government was fastened around his neck on 24 February 1981, Roy Benavidez proudly recalled Truman's declaration. This American hero, known as 'That Mean Mexican', was a veteran of many wars. Born in 1935, his parents were Mexican and Yaqui Indian, but they were dead by the time he was seven and he was brought up by poor, patriotic relations in El Campo, Texas. From his youth, Benavidez was an aggressive fighter, as keen on watching newsreels about combat during the Second World War as he was on waging his own violent war against local schoolchildren. In his own words, he was a 'tough, mean little kid' whose anger was eventually channelled into the military. From the Texas National Guard, he progressed to the US Army, following in the footsteps of his idol, Audie Murphy, the most decorated soldier of the Second World War. It was his military service in Korea which won Benavidez the admiration of his extended family and gave him the experience and maturity needed to begin training in the air force. Finally, in 1965, he was assigned to Vietnam as adviser to a Vietnamese infantry unit called the 25th Infantry 'Tigers'. Here he was to meet his greatest challenge. Unlike many of his American comrades, he had little trouble 'blending in'. In his words:

> My size, my skin color, my Hispanic-Indian, almost
> Oriental, look became an advantage. All my life I had been
> fighting against the bigotry created by these differences.
> Now they were helping to keep me alive.

But there were certain limits, he discovered, to his ability to 'blend in'. Army of the Republic of Vietnam (ARVN) atrocities involving the torture of pregnant women and villagers could not be reconciled with his military ethos. After observing one such event, he admitted that, although he had 'killed lots of men', 'I will never forget the look in those villagers' eyes.' Still, as an official observer he believed that 'there was nothing we could do . . . This was [the ARVN officer's] country, his war'. His first tour in Vietnam ended when 'walking point' (that is, walking several metres in front of everyone else, watching for signs of ambush or traps) on a patrol (his officer had 'seen too many reruns of the Lone Ranger and figured "Tonto" was the guy he wanted out there scouting on his patrols'), he stepped on a land mine and only regained consciousness in hospital.

Desperately wounded, Benavidez wouldn't give up. He fought the doctors (who predicted that he would never stand again), the military (who wanted to discharge him), and (the most difficult task of all) the pain, and taught himself to walk. He then quali-fied to wear the Green Beret. After all, he reasoned, he was 'a warrior, not a clerk'. After a gruelling training, he was sent back to Vietnam in a top-secret unit assigned to gathering intelligence from the North Vietnamese Army (NVA). The name of the mission – 'SIGMA' – has become notorious.

The event which eventually confirmed his status as a war hero occurred on 2 May 1968 when he volunteered to help rescue some members of his unit who, in the course of a highly secret mission, had been stranded in Cambodia and were surrounded by North Vietnamese troops. During the rescue, Benavidez was shot five times and had shrapnel lodged in six parts of his body. On his final attempt to rescue a comrade, an NVA soldier slammed his rifle butt into Benavidez's head. He fell and when the NVA soldier hesitated just prior to bayoneting him, Benavidez stabbed him to death with his Special Forces knife. Despite his injuries, he rescued the survivors and retrieved the bodies of his dead comrades. He even had the presence of mind to destroy secret documents and electronic gear before boarding the helicopter so severely wounded that the medics believed him dead and rolled him into a body bag. When a medical officer bent over him prior to zipping up the bag, Benavidez did the only thing he could: summoning up all his remaining energy, he spat into the doctor's face.

For this sacrifice and his courage, Benavidez was initially awarded two Purple Hearts and the Distinguished Service Cross, the second highest award given by the US military. He retired from the army on a total disability pension in 1976 and five years later, in front of forty-three members of his family, President Reagan hung the most prestigious medal – the Medal of Honor – around his neck.

His story did not end with this triumph. Three years later, he was shocked to be informed by the Social Security Administration that he had to be examined by a doctor and a psychiatrist to determine whether he was exaggerating the extent of his injuries. Although his disabilities were confirmed, his public protest at being humiliated in this way provoked a national scandal. War veterans from all over America pleaded with him to intervene on their behalf. He recalled:

> Most of the letters addressed me personally, referring to
> me as Roy. Most referred to me as a hero and indicated
> that I was the only person they knew who could help
> them. Many stated a feeling of hopelessness: If the gov-
> ernment could do this to a Medal of Honor recipient,
> what chance could they have?

Although he was uneasy about criticizing his country and his
beloved commander-in-chief, in the end he sided with his comrades and began campaigning for veterans' rights. While writing
his autobiography, he was working for LifeSupport in Houston,
Texas. 'I travel the country speaking to various groups,' he
explained, 'I tell them that freedom has a price. That price is
often human flesh and blood.' In opposition to the peace protestors, he was proud of his medals and wore them in honour of all
the 'silent heroes' who served their country but were not recognized.

> There are not enough medals of distinction to reward the
> military heroes of this country. Most must bear their hero-
> ism in their hearts. Everyone in uniform is a 'silent hero',
> and no matter how manipulated a serviceman or service-
> woman may feel in any given geopolitical military
> confrontation, his or her duty, honor, and responsibility is
> to follow the colors of individual and national freedom.
> Even to death.

To young men in inner city slums, he advised: 'If you want to
join a gang, join my gang, the United States Army Special
Forces.'[2]

Roy Benavidez's autobiographical narrative imitated mythical
tales in which the hero was customarily deserted at birth by his
parents and reared by foster parents before undertaking a dangerous journey in aid of his grateful community. By being placed

in a body bag after his heroic action, Benavidez even suffered (symbolically) the hero's honourable death.[3] It was brave men like Benavidez that the armed forces desperately needed to nurture. Although military organizations, with their old and refined traditions for 'appropriating violence', recognized that they could convert nearly all men from civilians into effective fighters, they also believed that some men were better adapted to acting aggressively than others. During the course of the twentieth century, theories about what *type* of man would constitute the best combatant were vigorously debated. Racial traits always remained powerful indicators but in other areas there was a shift from regarding combativeness as innate, instinctive, and (most importantly) individual, towards regarding it as artificially stimulated by group dynamics.

The ideal soldier

The nature of modern warfare changed beliefs about the type of recruit likely to become the most aggressive fighter. Some of the criteria advanced were absurd, such as the idea that red-haired soldiers were the most passionate killers[4] or, as a survey in 1961 revealed, that clean-shaven men and men with 'trimmed' moustaches made good officers, in contrast to men with 'clipped' moustaches.[5] Or they were banal: the younger and fitter the man, the more keenly aggressive he was likely to be.[6] Medical officers devised tests to identify men who were 'accident prone' because they were said to make the best commandos, as were borstal boys, ex-convicts, and 'full-blooded Cockneys'.[7] Marriage diminished men's combat effectiveness because 'at critical moments' the minds of married men were 'likely to flash back to a picture of a wife left husbandless, children left fatherless – an involuntary natural thought that can make a man give pause at a critical point in battle'.[8] The best killers were reputed to be sportsmen (like the famous Australian cricket player who won the Victoria Cross during the First World War for killing forty-seven Turks through efficacious use of his bowling arm).[9] The most aggressive soldier was the man who hailed from a family with a long history of sporting prowess and military service. In the words of two researchers in 1945: 'driven by the identification' with his father, the son 'seeks to recapitulate the father's experience and perhaps outshine him'.[10]

Certain occupational groups were also prioritized. Miners, for instance, were described as excelling at 'hand-to-hand killing'.[11] In contrast, artists and musicians were definitely unlikely to become heroes, despite the exemplary performance in the trenches of the poet, Siegfried Sassoon.[12] Other criteria were simply incompatible. For instance, it was argued on the one hand that rural recruits were best in combat because they were used to hardships and clever at woodcraft. They should be segregated into separate units and used for the more difficult patrol missions, according to Lieutenant-Colonel Matheson of the Australian Imperial Force in 1943.[13] But, on the other hand, urban men were ideal because they were 'less perturbed by the noise of modern battle'.[14] It was sometimes argued that the best soldiers were unimaginative, 'bulldog' types[15] and that traits such as strength, agility, courage and ferocity could be facilitated by selective breeding and 'judicious' drugging.[16] The belief that effective combat soldiers were somehow bred that way was an extremely common one. Officers, steeped in military history, embraced Carlyle's influential analysis of the hero as a man with an almost mythical awareness of what needed to be done. He was the embodiment of intuitive insight, precise action, and the unflinching pursuit of truth. In the words of Field Marshal Earl Wavell in 1945, 'toughness is partly inherited, partly produced by training, and . . . inheritance is the more important'.[17] A letter from a combat soldier to his brother, also in the trenches, is a popular version of these beliefs:

I note what you say about people imagining that you are thirsting to get to grips with the enemy etc. I have noticed exactly the same thing myself. So few people have any

guts, that fellows like you are apt to get all the dirty &
dangerous work shoved on to them. On the pretence that
you are always eager for a chance to get killed, they don't
want to disappoint you. Bah. Some people have the guts
of a louse, some the guts of a rabbit, some the guts of a
dog, and *a few* the guts of a *man*.[18]

Or, as Frank Markham's mother admonished her fearful son in
the frontlines during 1917: 'as for *you* being a coward, well, lad I
do know that is utterly impossible, *why* well *you* have not *that* sort
of blood in you, now laddie don't think I want to preach to you
but this is only what I feel'.[19]

But the armed forces needed more precise indicators. After all,
mothers, fathers, brothers and girlfriends could not be trusted to
evaluate correctly the 'guts' of their loved ones and manpower
requirements meant that it was impractical to differentiate rural
from urban recruits. Not enough borstal boys were available (with
or without red hair) and married men simply had to be recruited.
Military officers needed a 'science of combativeness' which would
be fast, cheap, and accurate. During the First World War, combat
effectiveness seemed to be inscribed on the body itself. In the
words of the medical officer, Robert William MacKenna:

There is no doubt that a definite parallelism exists
between a man's physical state and the courage he
exhibits. Bravery, unless a man be a poltroon, is more naturally expected of a man in splendid physical condition
than of a weakling; for the relation and interaction
between soul and body is an intimate one.[20]

During 1914–18, therefore, potential recruits were subjected to
a physical examination in an attempt to weed out men who would
be physically incapable of fighting, and to distinguish between
men who would perform best in the most active fields of battle
and those who would do better behind the lines. Recruits were

placed into four categories. Within each of these grades, the shape of the body was crucial: height, chest width and weight formed the basis of the decision. The grading of recruits into these four fitness categories was based as much on perceptions of the relationship between physical masculinity and combative- ness as it was on medical principles of healthiness.[21] Some commentators did not think that the armed forces had gone far enough in developing a 'science of combativeness'. For instance, two influential psychologists, Charles Bird and G. Stanley Hall (both of Clark University, Massachusetts), independently argued for the adoption by the armed forces of French morphology which divided the human race into four somatic types (abdomi- nal, respiratory, muscular, and cerebral), each with important implications for combat worthiness. Men of the 'muscular type' – easily identified by their cubical-shaped face and body and strong, well-proportioned limbs – should be sent immedi- ately to the most aggressive frontlines. Abdominal-type men – with their truncated pyramid-shaped heads, large digestive appa- ratus, long fat trunks, short limbs and small thorax – should be sent to defensive posts where they would fight, if only for stores. Cavalrymen and pilots should be chosen from respiratory-type men: they could be identified by their large thoraxes and faces which were widest at the nose. The cerebral type were small men with extraordinary powers of reserve energy; while they were, unfortunately, liable to sudden emotional collapse, they could be turned into good officers.[22]

Although such essentialist typologies did not survive the First World War, the theories were not completely rejected. In the 1920s, valour was still judged by means of a physical examina- tion. Thus, at the Australasian Medical Congress in 1927, Lieutenant-Colonel A. D. Carbery of the New Zealand Medical Corps offered the following criteria as to whether or not a par- ticular man should be admitted to the services:

Sex characteristics should be observed: the development

of the genital organs, the distribution of the body hair, especially the contour of the upper margin of the pubic hair, the scalp and the eyebrows, the size of the pelvis and the locality of fat distribution. Generally males who approach the female type in size of pelvis, hair distribution and fat disposition or are lacking in the normal development of sex characteristics, make poor soldiers.[23]

By the Second World War, such methods were supplemented with a personality interview and a brief fitness test. Researchers then devised a method for the selection of the most effective combat soldiers based on a 'brief inspection of the body build to determine characteristics of masculinity'. Under this scheme, 'normal' men (that is, men without any endocrine abnormalities) would be graded according to whether their 'component of masculinity' was strong, medium, weak, or very weak (see illustration 5). The recruit possessing a 'strong masculine component' had an angular and muscular body, narrow hips relative to shoulders, 'flatness of the mammary area' and abdomen, a space between the thighs, 'prominence of inner curvature of calves', and pubic hair running towards the navel. These men would immediately be sent to the front. Men with weak masculine components were characterized by a roundness and softness of body outline, lack of muscle, relatively greater hip breadth to shoulders, fullness in the mammary area, 'feminine abdominal protuberance', close thighs, 'greater outer curvature of calves', and lateral distribution of pubic hair. Unlike 'masculine men', they might carry their arms at an angle at the elbow (hyperextensibility). They would never do well in the fight. In 1943, this theory was tested by the Royal Officer Training Corps. Of those men with a 'strong masculine component', 41 per cent were rated as excellent officer candidates; the proportion in the medium masculine component was 11 per cent; and in the weak or very weak group, none were rated as excellent officers.[24] Combativeness was thus inscribed on the male physique.

Despite these methods, the actual experience of war discredited such reliance on the individual's body as a signifier of military prowess. By the Second World War, a new generation of military researchers descended upon training camps, number-crunching, interviewing, correlating. Two of the most influential were S. L. A. Marshall and Samuel A. Stouffer. Although Marshall's statistics were found later to be extremely dubious, his argument that combativeness was a *group* rather than individual attribute, and that training must therefore utilize psychological principles of group dynamics transformed military training regimes in Britain, America and Australia. Stouffer, and his team of researchers, were also important. Taking a sample of over 12,000 soldiers, they stressed not only the importance of the 'primary group', but also the element of pleasure. The best combatants were those who *before* combat expressed a desire to kill. For instance, 48 per cent of soldiers who later on performed particularly well in combat said that they would actually enjoy killing Japanese soldiers, compared with 44 per cent of soldiers who performed adequately and only 38 per cent of those who did not fight well.[25] Thus, by the Second World War, pride of place was given to personality and, related to this and assigned an even higher status, the nature of the '*group* personality'. The best soldier was the man who felt strongly attached to an effective group who would protect him and whose interests were identical to his own.[26] As in combat films, by 1942 heroism in battle was no longer the attribute of a single individual and/or pair of competitive men but arose out of actions performed by a unified group.[27] There was still a role for the warrior-hero, but he was now a product of the group, as opposed to a 'natural' hero who embodied the military ideal in his very blood (or guts).

Psychotics and homosexuals as potential heroes

But all was not well for the 'science of combativeness'. To the distress of military officers concerned about the status of their profession, some psychiatrists bluntly concluded that the best

combat soldiers were, quite possibly, psychotics or even homosexuals and that, logically, these men should *not* be screened out
or dismissed from the services – at least not in time of dire emergency.

It was not unusual for military men, even those as respectable
and respectful as Captain J. F. C. Fuller, to observe that the

> failures of the labour market, and even those with a crimi-
> nal taint in their characters, have often, in their blood, that
> dash of 'barbarism' which in the field carried them
> beyond their more amenable brothers.[28]

Indeed, stories of such men were a source of bemusement to
many senior officers like Dr H. W. Wills, a neurologist to the 4th
Army during the First World War, who received a note describing a serviceman who had been drinking heavily. In the words of
the note, this soldier

> suddenly made his way up the communication trench
> muttering 'Bloody Bosche'. When he reached the front
> line he suddenly scrambled over the parapet. We all
> expected machine-gun fire but a little later heard him
> accusing someone of unnatural vices. Then he appeared,
> holding a German officer by the scruff of his neck and
> kicking him. They fell into the trench and our man mur-
> murred [sic], 'Found the bugger in a shell hole', and
> passed out. Now we are all in a quandary. We can't decide
> whether to recommend him for a V.C. or a Court Martial,
> so we are sending him to you.[29]

In the words of the psychologist John T. MacCurdy at the end of
the First World War, the best soldier 'must be more or less a
natural butcher, a man who can easily submit to the domination
of intellectual inferiors'.[30] From the Second World War to the
Vietnam War, elite forces which depended upon recruiting the

most aggressive men often targeted 'cowboys', ex-borstal boys, and men who had prison records, and oral accounts acknowledged that 'the guy who gives you the most trouble in peacetime' was the best in battle.[31]

It was one thing to comment sardonically on 'bad boys', but quite another to place scientific weight behind arguments which identified heroes as psychopaths, 'obsessive-compulsives', or men recent paroled from prisons having served sentences for manslaughter or murder.[32] So important were mentally ill combatants that they should be given more public recognition for their military services, argued Major Dallas Pratt and Abraham Neustadter in 1947.[33] Psychiatrists did try to sweeten the pill by pointing out that the army could 'make a good deal of use' of the 'aggressive type of psychopath' by simply ordering him to 'go over the top; go ahead; clean up' and then rewarding him with 'a citation for bravery'.[34] Even though these men were dangerous behind the lines (where they were liable to turn their aggression against their comrades),[35] if they could be kept close to the front-lines, the military had a captive pool of keen fighters.

The positive link between serious mental illness and the enactment of heroic deeds was explained in a number of ways. The most generous school of thought argued that combat soldiers who began fearing that they would 'crack' under the strain would attempt to destroy the threat as quickly as possible and, in a fit of 'desperate aggression', often acted heroically.[36] Other commentators were less kind. The most effective combatants, they argued, were hostile, emotionally insecure, unstable men who were acting out infantile fantasies of heroism – in other words, they were lucky that killing was socially approved in war because otherwise they would find themselves in prison.[37] If they were not insecure, they were chronically hostile narcissists who strove to make the environment conform to their personality and who expressed their dissatisfaction with the outside world by aggressive and bizarre behaviour.[38]

Worse was to come: some psychiatrists dared to claim that

their best soldiers were homosexuals. Within an institution where it was agreed that all homosexuals, irrespective of their combat record, should be expelled, the idea that an individual could be *unconsciously* homosexual was a cause for much concern.[39] In 1915, Sigmund Freud's disciple and the popularizer of psycho-analysis in England, Ernest Jones, drew public attention to the importance of sexual desire in 'darkly impelling' men to enlist. These motives could range from 'the fascinating attraction of horrors' to 'the homosexual desire to be in close relation with masses of men'.[40] By the Second World War, delineating the relationship between homosexual desire and combat had many adherents. In a paper read before a meeting of the Oxford University Anthropological Society in March 1936, R. E. Money-Kyrle argued that unconscious homosexuality had two effects. Unconscious inverts who turned their aggression inwards showed self-sacrificing devotion to their comrades; unconscious inverts who turned their aggression outwards were killers. The former identified with the good elements of the parental imago; the latter with the bad elements. The usefulness of both in times of war was clear.[41] Charles Berg speculated in *The British Journal of Medical Psychology* (1942) whether war itself was 'a dramati-zation of such unconscious phantasies – a homosexual substitution . . . an emotionally all-powerful (an orgastic) dealing with *men* rather than with women'.[42] In the words of one extremely aggressive homosexual who enlisted in the Marine Corps in December 1942: 'If you can't love, you gotta hate. If you can't show affection, show aggression.'[43]

Not all homosexuals were potential heroes. The most detailed delineation of the type of homosexual liable to act heroically was published in the *British Journal of Medical Psychology* in 1945. The author, Charles Anderson (psychiatrist to the Wharncliffe Neurosis Centre), drew a distinction between active and passive inverts. Active homosexuals generally adapted well to military life because they responded to the military's demand that they express aggression externally. The active invert did not have to change the aim of his aggressive impulses: he only had to change his target from his comrades to the group's enemies. For most, this was a relatively easy alteration 'for often he hates and wants to destroy his love-objects, and so to him enemies are as often comrades as comrades are enemies'. The active invert suffered nervous collapse only when feelings of guilt broke through his self-assurance (something which was more likely to occur in periods of inactivity when he once again turned to his own group for satisfaction). In contrast, the passive invert was in a much more difficult situation because the externalization of his aggression went against his nature. The passive invert was 'expected to perform a psychic *volte-face*, to visit upon others what he wants to have visited upon himself'. In combat, the passive invert was forced to become the object of aggression, rather than its subject. These homosexuals frequently could not cope with the accumulating tension and sought relief either in anxiety neuroses or by rushing 'in an animal panic' to surrender to the enemy.[44] Passive inverts and unconscious homosexuals had most trouble adjusting[45] but, as Anderson discovered in 1945, 'curing' homosexual servicemen (even those who had voluntarily sought psychiatric help) was almost impossible because they accepted their homosexuality 'with self-possession' as 'an integral part of their make-up', and merely wanted protection from the hostility of the group.[46]

The racial factor

The assertion that psychotics and homosexuals could be particularly heroic remained a minor although significant claim. More than any other indicator, scientific racism was employed to identify the best combatants. The notion that nations, or substantial geographical areas, possessed their own individual character or personality has a long history. As David Hume wrote in his 'Of National Character' (1741–2): 'men of sense . . . allow that each nation has a peculiar set of manners and that some particular qualities are more frequently to be met with among one people than among their neighbours'.[47] The physical environment was thought to influence the temperament of a nation's 'racial blend'. The 'density of population and the nature of its occupations' affected a nation's disposition. The 'sum of [a nation's] acquired tendencies' added up to its character.[48]

When applied to war, these ideas were potentially useful to military advisers. As the Limerick-born journalist Michael MacDonagh summarized it in 1917:

> Each nationality evolves its own type of soldier, and each type has its distinctly marked attributes. As troops, taken in the mass, are the counterpart of the nations from which they spring, and, indeed, cannot be anything else, so they must . . . reveal in fighting the particular sort of martial spirit possessed by their race.[49]

The 'game' of war was played 'according to each nationality's separate national characteristics'.[50] Identifying such races could help military strategists allocate men to particular types of service and could be used by military propagandists to scare the enemy into deserting by using 'phantasy-fears about certain types of Allied soldiers' (such as 'wild Ghoums, Poles, Czechs, Gurkhas, Highlanders').[51] In the three conflicts examined in this book, some racial groups were identified as being particularly good combatants. For instance, Australian and Canadian troops were reputedly more brutal in battle.[52] Fijians were said to possess that 'jungle prowess' essential for effective combat in the Pacific theatre of war.[53] Even as late as the 1970s, it was believed that Scottish troops were consistently aggressive and victorious.[54] The martial prowess of Highlanders was supposed to come from their 'fiery Celtic blood'. In the words of one writer in 1915: 'Possibly some of their ancestors, claymore in hand, had charged down to death at Culloden. The steel had always appealed to them.'[55] No wonder Germans got the 'wind up' when they saw the kilts coming.[56] Certain types of Indian soldiers were more likely than any other group to be praised for

their racial bloodthirstiness.[57] In 1914, Saint Nihal Singh
described the Gurkhas' mode of fighting:

> The Gurkhas are famed for the night raids they make
> upon the enemy. They move with the stealth and noise-
> lessness of panthers until they are right upon their
> unwary foes, and then they make effective use of their
> *khukris* and Western arms. Their wonderful sight – almost
> cat-like in the darkness – and their amazingly keen sense
> of hearing go a long way to insure their success.[58]

Similarly, as we saw at the beginning of this chapter, American
Indians were regarded as 'natural' warriors and, as a consequence, would be asked to 'walk point' on patrols.[59]

In this chapter, however, I will focus on the relationship
between race and murderous efficiency with reference to two
specific groups: Irish and black American soldiers. Both had
strongly established reputations within the military: Irishmen
were considered to be a 'martial race' while black American soldiers were regarded as distinctly poor fighters. What was the
'science' behind such distinctions and what was its effect?

It was widely believed that the Irish were innately combative.[60] Irishmen were born either soldiers or monks, informed
one popular saying.[61] Intimately related to this was the idea that
Irish soldiers were *particularly* brave: indeed, they were acknowledged to be braver than English soldiers which was why military
commanders used them as 'missile troops'.[62] They were said to
be unsurpassed for courage, initiative and 'dash'.[63] As the poet,
Alice Cooke, put it: the Irish are 'the best in the trenches, the first
in a charge'.[64] This so-called instinctive bellicosity was also used
to explain the growth of extra-military organizations in Ireland.
The Liberal writer, Harold Spender, commented in *The
Contemporary Review* in 1916:

The young Irishmen who now held aloof from the British

> Army and the Nationalist Volunteers were not inclined to
> remain outside the great world distraction. When the
> whole world is converted into a gigantic Donnybrook Fair,
> you cannot reasonably expect Irishmen to sit by their fire-
> sides. The flame of war patriotism flickered; but the taste
> for war remained. The men who held aloof . . . began to
> drift towards one of those alternatives of violence.[65]

In other words, harboured within the breasts of Irishmen was a
reservoir of aggression which needed to be released – if not in the
legitimate armed services, then within other armed alternatives.
Even when low recruitment rates led to accusations that the Irish
lacked a certain physical vigour, the martial spirit was still
assumed to be inherent in the Irishman's 'soul' and people who
denied this fact were likely to be 'warned off the hunting fields'
and made targets of abuse.[66]

The myth of the innate combativeness of Irishmen was not
merely a quaint belief held by people in the early decades of the
twentieth century. It survived throughout the century. During
the Second World War, Irishmen were highly prized in commando units on the grounds that they were particularly
bloodthirsty (in aggressiveness they were said to be followed by
Welshmen and Scotsmen).[67] Historians frequently agreed. In his
broad-ranging book, *Modern Warfare. A Study of Men, Weapons
and Theories* (1973), Shelford Bidwell commented approvingly on
the idea that certain groups, such as Irishmen, Scotsmen and
Gurkhas, were 'martial races'.[68] Sean McCann, in his book entitled *The Fighting Irish* (1972) described the typical Irishman as
'quick-tempered . . . he has never been short of a fight, right or
wrong, through any stage of history'.[69] Even the scholarly work
of Peter Karsten adopted the idea of 'Irish bellicosity': the Irish
simply liked to fight.[70]

Like all true heroes, Irish soldiers were especially skilled with
the bayonet.[71] In the words of Father William Doyle of the 8th
Dublin: 'We should have had more prisoners, only a hot-blooded

Irishman is a dangerous customer when he gets behind a bayonet and wants to let daylight through everybody.'[72] Irish propagandists also took up the theme. S. Parnell Kerr in *What the Irish Regiments Have Done* (1916) claimed that '"close-in" fighting appeals to the Irishman. That sort of fighting is in his blood. Like the Frenchman, he is a little impatient of lying in the trench to be "potted at"; he loves to be "up and at 'em".'[73]

If Irishmen were natural fighters, lauded for their military prowess, black American soldiers were thought to exemplify all that was unwarlike and unheroic. This caricature of black American soldiers in war was not historically accurate. Indeed, black soldiers had a long and very distinguished history in the military services in America prior to the First World War. Their valour during the American Revolution, the War of 1812, the Civil War and the Spanish American War is well-established.[74] Yet, during the First World War, they were relegated to service rather than combat units. As a consequence, three quarters of black soldiers served in non-combatant units.[75] During the Second World War, the army deliberately limited its intake of blacks to the percentage they represented in the population, that is, one in ten.[76] At this time, the Quartermaster Corps, Corps of Engineers and the Transportation Corps accounted for three quarters of blacks in the army. While black soldiers constituted 10 per cent of all men in the US Army, they constituted 20 per cent of the men in service units, 15 per cent of men in the category 'overhead', and less than 5 per cent of men in combat units.[77] In the words of a First World War ballad:

> *Black man fight wid de shovel and de pick –*
> *Lordy, turn your face on me.*[78]

Although their labour in service units was highly valued, the refusal to allow them a combatant role was seen as a direct snub to their masculinity.

The racism behind this policy was explicit. In the US Air

Force, it was initially argued that black Americans were totally incapable of flying (as was said about the Japanese) and, when this was proved wrong, that they lacked the aggressive spirit in flight and were excessively fearful, even of 'light and inaccurate' anti-aircraft fire.[79] John Richards commanded a black unit during the First World War. He regarded the black soldier as a 'splendid physical specimen', ideal during the theatrics of parade-ground drill, and possessing great powers of endurance and loyalty. He will 'follow like a dog through artillery barrage and the wind of machine-gun bullets', Richards wrote. On balance, however, Richards believed that black soldiers were of very limited usefulness since they were afraid of the dark, lacked initiative, and were liable to panic during attacks. They lacked the pugnacity of Anglo-Americans and were susceptible to mass suggestion which resulted in desertion. Richards continued:

> They do what they are told, but move as if bewildered. I think they lack the free, independent spirit that stirs in the breast of the white; that rises within him when the shells are falling thick and says, 'I am a better man than any – Boche, and I am coming through'.

Such spirit was exceptionally rare amongst black soldiers and officers. Richards wrote: 'They are boys. They do not grow up, even under shell-fire.' In 1926, a professor of sociology at the University of Oklahoma agreed with Richards, arguing that the innate childishness of black troops, and their simple faith in their leaders made 'superior leadership' the most important prerequisite in exacting military effectiveness from these men. The best black units were those commanded by white officers 'of very superior intelligence and moral fiber', he insisted.[80] Similarly, General Bullard's report on the black troops during the First World War was devastating. He reported in his diary that they were a failure. It is in a quiet sector, yet can hardly take care of itself, while to take any offensive action seems wholly beyond its powers. I have been here now with it for three weeks and have been unable to have it make a single raid upon the enemy. They are really inferior soldiers.[81]

Such attitudes could even lead to confusion over who was the real 'enemy'. A survey of white southern college women between 1941 and 1945 revealed that women rated black Americans as *much* worse than Germans and, until 1944, worse even than Japanese. Black Americans were described as superstitious, lazy, ignorant, very religious, unreliable, happy-go-lucky, physically dirty and slovenly. Despite the increased integration of black men into the military services, negative stereotyping remained unchanged.[82]

Some commentators argued that the poor performance of black troops was due to their low status in the eyes of white troops and officers. Poor self-esteem, lack of self-confidence and inferior education all combined to lower the motivation to fight. In fact, the Chief Historian of the US Army noted that the lack of aggressiveness was a trait deliberately fostered within black communities in an attempt to minimize attacks from the outside. He concluded that it would be unrealistic to expect to make first-class soldiers out of second, third or fourth-class citizens.[83] More commonly, the negative stereotypes were confirmed by racist ideology. For instance, one of the criteria of combat fitness was psychological stability. Although it was generally noted that black soldiers were less liable to suffer psychoneuroses, they were (at least according to a study carried out by William A. Hunt, a clinical psychologist at a large naval installation during the Second World War) four times more likely to be diagnosed as suffering from hysteria. According to Hunt, the diagnosis of hysteria for these men was not justified. Rather, their behaviour more closely resembled that of 'an extremely suggestible, uncritical, emotionally unstable individual'. It would be more correct to diagnose black servicemen as suffering from 'emotional instability' rather than 'true hysteria or conversion neurosis'. In other words, the 'cultural primitivism' of black soldiers meant that emotional conflict was expressed through immediate, primitive expression: such men were incapable of the 'devious and complicated mechanism' of psychoneuroses.[84]

Like many commentators, Hunt's diagnosis was strongly influenced by the so-called recapitulation theory. This theory decreed that the counterpart to savagery or primitivism could be found in the white child. While white children grew out of it, the 'savage races' were trapped in this childish state. John Richards, the officer commanding a black unit during the First World War, mentioned earlier, was also influenced by such 'scientific' theories. He warned his readers to

> remember that races develop slowly! A few years ago, these men were slaves in cotton-fields. A few years before that they were children in the jungles of Africa. They are children still . . . let the white citizen remember the lovely traits of his colored brother. We have so much in power, prestige, and development which they have not. We inherit an independent spark, fostered through ages of war and upward groping. Let us hold out our hands and open our hearts to these wonderful boys who move among us, remembering that white or black lie side by side in the fields 'over there'.[85]

A similar theme was developed by John J. Niles when telling the story of a black infantryman dubbed 'Dog Star' who had a reputation for being a 'hard fightin' sonovabitch', especially when armed with his favourite weapon (a French chaut-chaut automatic). During an infiltration attack, Dog Star found himself alone and without ammunition. Niles continued with the tale:

> What could a soldier do with a funny French 'sho-shot'

and no clips! And then he suddenly found himself con-
fronted by men in greenish gray uniforms – greenish gray
uniforms and tub-shaped hats. He still held the useless
chaut-chaut in his hands. It weighed less than twenty
pounds, but backed up with his fiendish strength it was a
veritable battering ram. The ground was covered with
mausers and unexploded grenades, but the blood of forgotten races of black savages surged in his veins. He was
not a mathematician, a linguist, an intellectual dilettant.
He had reverted back to the tribesmen in the upper Nile
Valley. He no longer understood the mechanisms of
modern warfare, but his sense of aim was perfect – his
desire to live, supreme. The clipless chaut-chaut gun
swung in a wide circle, squashing tub-shaped hats down
upon greenish gray soldiers.[86]

Dog Star was a hero, certainly, but he was also a primitive who
did not possess enough reason to use the unexploded grenades
and dumped mausers. Such racist stories could be told with even
more vicious emphasis. T. Corder Catchpool, who later became a
leading conscientious objector, offered the following description
in a letter he wrote in November 1914:

The niggers are hopeless. Often they try to cut the throats
of German wounded – occasionally they have been found
with German heads in their haversacks! Their slow
thought-out movements – such as in lighting a pipe, for
example – often suggests a monkey to me almost as much
a man. And yet I am curiously fond of them. Wag your
head at them as you go by, and you win the richest smile
in the world, white teeth, thick lips, black eyes, all com-
bine in the most bewitching production.[87]

Martial reputations based upon racial criteria were problematic for *both* Irishmen and black Americans. The reputation for

aggressiveness held by Irishmen and for passivity held by black Americans had negative consequences for both groups. Irishmen paid a high price for their much-lauded prowess in war. In *A Military History of Ireland* (1996), David Fitzpatrick argued that militarism was 'one of the few Irish stereotypes which evoked almost universal approbation in a bellicose era'.[88] He was only partly correct: aggressiveness in the war period was greatly admired, but distinctions were made between different types of martial courage and, in the final analysis, the type of martial valour said to be possessed by the Irish made them profoundly unsuited to self-government. In wars which provided little outlet for individual prowess and led to carnage on a level never seen before, Irish success in the field of battle could be regarded as an indication of their lower biological and political development. In part, the contradiction between the high status associated with racial bellicosity in wartime and its low status after the war was related to distorted evolutionary ideas which emphasized that humans fought wars because of inborn instincts. Fighting was our inheritance from the 'brutes'. Although humankind could never rid itself of its animal inheritance (indeed, pugnacity was essential for evolutionary progress), the most pugnacious races were those lower in the evolutionary scale.[89] For all the admiration they won in battle, Irishmen and other so-called 'martial races', were at the same time slotted into a lower rung in the ladder of civilization.

If Irishmen were said to possess a different type of martial spirit to that of the English, these differences had important implications in terms of the theories of combat. This can be seen by looking in more detail at the work of one of the most popular commentators on the Irish soldier, Michael MacDonagh. Born in Limerick in 1860, he was educated by the Christian Brothers and became a journalist on the *Freeman's Journal* and the London *Times*. In addition to two books written during the First World War, he was the author of biographies of William O'Brien and Daniel O'Connor. By 1924, he was the chairman of the Press

Gallery in the House of Commons. More than any other single writer, his two books on the First World War have shaped the image of the Irish at war.

MacDonagh had a political agenda behind writing *The Irish at the Front* (1916) and *The Irish on the Somme* (1917). He intended his argument about Irish prowess in war to promote Irish self-government. This was stated most explicitly in *The Irish at the Front*, where he reproduced at length a letter written by a Scottish soldier who praised Irish troops (especially the Munsters and Dublins) for their bravery.

> As you know, I am not Irish, and have no Irish connec-
> tions whatever – in fact, I was rather opposed to the
> granting of Home Rule; but now, speaking honestly and
> calmly after having witnessed what I did [at Gallipoli] –
> the unparalleled heroism of these Irishmen – I say nothing
> is too good to give the country of which they are, or rather
> were, such worthy representatives. My God, it was
> grand! . . . Aye, the race that can produce such men,
> supermen, as those chaps were, to do such glorious work
> for the Empire has the most perfect right to demand, what
> is more, to get the freedom of its country and the right to
> rule it.[90]

Not everyone interpreted events in the same way. Although MacDonagh could never be accused of subtlety, he did not regard one set of martial traits as necessarily better than any other. For instance, he explicitly argued that it was wrong to say that the Irish were *more* courageous than the English, Scottish or Welsh soldier: 'They are all equally brave, but the manifestation of their bravery is undoubtedly different – that is, different not so much in degree as in kind.'[91] However, it was precisely the terms with which MacDonagh and others differentiated Irish courage from that of the English which harmed the case for Irish self-government. As Harold Spender explained in 1916:

there is in the Irish soldier a peculiar quality of electric zeal and dash. 'Missile troops' they have been called; and the phrase is eloquent of much. Those who most doubt the Irishman's capacity in civil affairs are often the readiest to admit his fury and prowess in battle.[92]

This point can be seen more clearly if we look at descriptions of other racial groups. Many of the traits said to characterize Irish soldiers were, in fact, also said to be characteristic of other groups. Cynthia H. Enloe called this the 'Gurkha syndrome'.[93] Black African and Indian soldiers were said to be 'impetuous', 'good at attack and bad in defence', and to be particularly needing of strong leadership if they were to be effective in battle.[94] As we have already seen, much writing about the First World War was imbued with popularized interpretations of recapitulation theory. Just as the growth of the embryo 'recapitulated' the evolutionary history of the species, so too nations developed in a biological fashion, passing through the stages of infancy to childhood, adolescence, and maturity.[95] The Irish, as well as Indians, black Africans and black Americans, were still in the child stage. Thus, MacDonagh wrote in *The Irish on the Somme* (1917):

It may well be that sometimes the English officers of Irish battalions are puzzled by the nature of their men – its impulsiveness, its glow, its wild imagery and over-brimming expression. It is easy to believe, too, that the changeful moods of the men, childlike and petulant, now jovial, now fierce, and occasionally unaccountable, may be a sure annoyance to officers who are formal and precise in matters of discipline.[96]

This was a surprising comment, perhaps, from a propagandist supposedly arguing *for* the Irish, but officers commanding Irish battalions often agreed. Brigadier W. Carden Roe of the Royal Irish Fusiliers confided that the Irish soldier acted like a 'naughty

child' who would 'resent' being chided by his officers.[97] Rowland
Feilding, commander of the 6th Battalion Connaught Rangers
from 1916, wrote to his wife on the 14 June 1917 saying that
'Ireland will always be Ireland. It is a land of children with the
bodies of men, and politically they do not themselves know what
they want.' Like children, Feilding believed, the Irish were
'easily made happy' and, equally, 'easily depressed'.[98] Politically,
they needed strong leadership from an older and more cautious
father-nation.

If the martial races paid a high price for their heroic reputations, the non-warrior races were in a much worse situation. As
one black soldier complained during the Second World War,
during battle 'everybody is buddies and everybody talk to you. If
you don't have a fox hole in a raid maybe a white fellow call you
to come get in his hole.' But this lasted only as long as the fight.
Immediately after the danger had past, a

> colored fellow can't get a lift in a jeep and six steps further
> on they pick up a couple of white fellows. Also colored
> men can't use the white man's latrine at the base. They
> gotta dig their own. How that make a fellow feel?[99]

In the First World War, some black regiments were 'so wrought
up' about racial inequality that they were 'next to useless as a
fighting unit'.[100] During the Second World War, there was even
an attempt to segregate the blood given to wounded soldiers
through the Red Cross into 'Negro' and 'White'.[101] No wonder
morale sank and men's performance was impaired.[102]

Racial stereotypes made it practically impossible to be promoted: blacks were simply considered to be 'incompetent
officers' who lacked knowledge of military tactics and experience in leading men.[103] With the exception of the 92nd and 93rd
Divisions, nearly all black soldiers during the First World War
were commanded by white officers. Even in the 92nd Division,
the percentage of white officers actually *increased* from 18 per

cent at the beginning of the war to 42 per cent by November 1918. In addition, black officers from the 369th, 370th, and 372nd, and graduates from the officers' training schools were largely transferred to the 92nd Division. This policy made room for white officers to be promoted in the units from which the black officers were transferred, but seriously limited the promotion possibilities for black officers in the 92nd.[104] A similar situation occurred during the Second World War. When America entered the war, around 7 per cent of white men in the army were officers compared with less than half a per cent of black men. While the proportion of white men who became officers gradually increased to around 11 per cent, the percentage for black men remained at less than 1 per cent. Educational differences between white and black soldiers could not account for this difference since the minimum requirement for becoming an officer candidate was an AGCT score of Class I or II, yet one fourth of white soldiers who met this standard became officers compared with only one tenth of black soldiers.[105] In 1943, nearly three fifths of black soldiers were serving in companies where all the lieutenants were white.[106] During the two world wars, only white officers were appointed to the high-command positions. There was only one black General (Benjamin O. David), who rose from the ranks to become a Brigadier-General immediately before the presidential election of 1940. It was widely believed that had David been a white officer, he would have been promoted long before October 1940 and his promotion was only made possible for political reasons.[107] It was hardly surprising that the notion that they were fighting a 'white man's war' spread, that some black officers were unable to conceal their contempt for the white officers, and that black privates frequently joined sick parades.[108]

In each of the conflicts examined here, as the war progressed and discrimination became increasingly well known, black soldiers proved less keen to go into combat. In a survey carried out in June 1945, Samuel A. Stouffer asked soldiers in America: 'How do you feel about being sent to a overseas theatre which is fighting against the Japs?' Sixty-four per cent of black American servicemen replied 'The army should not send me at all', compared with 41 per cent of white soldiers.[109] An earlier survey carried out by Stouffer showed similar results. In March 1943 when a cross-section of black troops were compared with a matched sample of white troops and asked 'If it were up to you, what kind of outfit would you rather be in?', only 16 per cent of the black troops chose a combat outfit overseas, compared with over a third of the matched white sample. When asked what kind of job they would choose if their unit was sent overseas, only 28 per cent of the black soldiers chose an actual fighting job compared with 45 per cent of the matched white sample. Black soldiers who were young, had volunteered, were more highly educated, and held rank were more liable to choose a fighting job. According to Stouffer, neither white nor black soldiers were particularly keen to see combat, but the white soldier was more liable to value traits such as bravery and success and to be afraid of accusations of cowardice if he failed to live up to expectations. The black soldier, on the other hand, had little to fear from other black soldiers. He experienced less stigma if he failed in battle and accusations of cowardice were ascribed to racism rather than to any personal flaw. The fact that many of the whites regarded him as inferior anyway did not provide any incentive to fight for the wider group of comrades.[110] Indeed, the black civilian population was more liable to castigate a black soldier for being a 'crazy nigger' by donning the uniform and 'fighting for white trash'.[111]

Ironically, the most militant black soldiers were those most keen to see actual combat. Although these men were a minority, they believed that only by proving themselves and their race in combat would the position of blacks in civilian society improve. One black soldier during the Second World War argued:

> The reason why I prefer combat is because we all are supposed to be American citizens and there aren't any of us

Negro people fighting in this war. Since we are citizens we should be granted the privilege that the rest are getting because we are just as good as the next man . . . it will better our status after the war.

Another soldier was concerned about the post-war implications of the fact that black soldiers were given primarily non-combatant jobs, fearing that some people might taunt him with the sneer: 'Didn't his white brother (?) [sic] die on the front line, while he was comparatively safe in the rear echelon?'[112] The promise of being rewarded in the post-war world for their actions in combat was deliberately invoked by certain military officers. One black soldier, Captain John Long, recalled being addressed by General George S. Patton (Commander of the 3rd Army):

Men, you are the first Negro tankers ever to fight in the American army. I would never have asked for you if you were not good. I have nothing but the best in my army. I don't care what color you are as long as you go there and kill those Kraut sons-of-bitches. Most of all, your race is looking forward to your success. Don't let them down, and, damn you, don't let me down![113]

Captain A. Gates also recalled this speech, adding that the black soldiers 'whooped' because they had been told to go out and shoot white men.[114]

For many blacks, however, service in the armed forces proved particularly disillusioning. Captain Charles A. Hill, Jr., belonged to the 332nd Fighter Group. When he entered this Group, he sincerely believed that American democracy was the protector of all citizens. This was what motivated him to kill Germans. But his experience in the army eventually led him to ask himself: 'Could a joke have been on me for being naive enough to believe in my government?'[115] A southern black soldier who saw combat in New Guinea and other islands was reported as being unable to

see much sense in dying for his country as at the end of
the war members of his race would be treated as badly as
before. The fact that he was now fighting for his country
indicated that he was just as much an American as anyone
else and yet here he was in a segregated unit expecting to
return to his former life, characterized by discrimination
and prejudice.[116]

The 'enemy' of black servicemen was frequently white soldiers
and officers, rather than Germans or Japanese troops.[117] Others
protested, for example, during one notorious parade at
Thanksgiving in 1918 when a 3,000-strong black regiment which
had won the Croix de Guerre for bravery in combat only a
month earlier (they had suffered 1,100 casualties) refused to sing
'My Country, 'Tis of Thee'.[118]

Anti-heroes

Although numerous men were encouraged to join the armed
forces in the hope of becoming a hero and despite the favourable
treatment given to such men upon return to civilian society,[119]
the mantle of heroism was not so attractive to men in the frontlines.[120] The typical view (at least for men *in* combat, rather than
those merely fantasizing about it) was: 'I did not volunteer for
France to win a V.C.'[121] Indeed, only one third of men entitled to
medals at the end of the Second World War actually bothered to
claim them: only servicewomen and sailors, that is, people
unlikely to have seen active combat, showed a little more interest
in collecting these mementoes. Most of the men who did not
collect their medals said that they could not be bothered (41 per
cent) or that the medals were a waste of money or of no value (20
and 17 per cent respectively). Those who did collect their medals
generally gave their reason as simply being 'entitled to them' or
'as a souvenir'. Only 16 per cent said that they 'would like to have
them'.[122]

In part, this lack of interest in medals was a reflection of the application of civilian values to a combat situation: many servicemen recognized that the hero was the most effective killer – and this was not something they thought should be lauded. Indeed, by the First World War, medals were being awarded less and less to men who rescued the wounded and increasingly for bloody deeds. In 1918, the Canadian War Records Office's report on the Victoria Cross even stated explicitly that 'deeds of valour must show material rather than sentimental results; the duty that inspires the deed must show a military rather than a humane intention'.[123]

It was widely rumoured that 'it was not always the best men who volunteer for the more spectacular forms of soldiery',[124] and concern was expressed about the legitimacy of encouraging 'fanatical fighting men . . . imbued with some ruthless and impelling political or religious idea', in the words of the commander of the 14th Army in 1945.[125] The actions for which men won medals were often reprehensible. The link between savagery and heroism was commented upon by men in the field, who regarded 'heroes' as inhuman and unreliable.[126] As one self-styled 'hero' from the Vietnam War acknowledged:

> I became a fucking animal. I started fucking putting fucking heads on poles. Leaving fucking notes for the motherfuckers. Digging up fucking graves. I didn't give a fuck anymore. Y'know, I wanted –. They wanted a fucking hero, so I gave it to them. They wanted fucking body count, so I gave them body count.[127]

In a more delicate tone, Brian Sullivan expressed his own ambivalence about having been nominated for a Silver Star in a letter to his wife on 2 March 1969:

> you know I'm lying if I say I'm not pleased. I am, I'm proud, but only the worst part of me. My better part is just so sad and unhappy this whole business started.[128]

The Vietnam veteran poet, Chet Pedersen, put such feelings into his poem, 'Wastelands' in which he vowed that

> *. . . THEY*
> *would nevermore*
> *consign me to*
> *a status board*
> *or pin a medal*
> *on my chest*
> *for services*
> *at best*
> *abhorred.*[129]

Awarding medals for murderous efficiency was also said to be unfair and misplaced. They rewarded officers who needlessly risked their men's lives. A serviceman during the Second World War recalled a senseless attack in which a number of Marines were killed:

> Yeah, some goddam glory-happy officer wants another medal, I guess, and the guys get shot up for it. The officer gets the medal and goes back to the States, and he's a big hero. Hero, my ass; gettin' troops slaughtered ain't being no hero.[130]

Medals were merely 'ego inflaters and career enhancers for military personnel who required such adulation', sneered Dennis Kitchin, the less-than-proud holder of the American Commendation Medal.[131] The word 'hero' was used to 'describe negatively any soldier who recklessly jeopardizes the unit's welfare. Men try to avoid going out on patrols with individuals who are overly anxious to make contact with the enemy,' observed a military sociologist in 1970.[132] In the words of a Korean War veteran:

> The hero exposed others to unnecessary risk, thought of

himself (if anyone) first, and his buddies second, and
forgot that buddies must fight together or not at all.[133]

Despite the high status of the Commandos (which was based
largely on very effective self-promotion) most servicemen
regarded them warily. In 1941, a report on the selection of volunteers for Special Service Units (Commandos) revealed that a large
proportion of men arriving at training centres had not been aware
that they were volunteering for the Special Service and promptly
asked to be returned to their unit. In one intake of 250 men in early
1941, 82 immediately demanded to be returned to their unit.[134]

There was also the issue of equity. Medals were distributed
unevenly, often for actions that were the result of confusion rather
than conscious valour.[135] Men were given medals for simply
'being there', rather than for any special action.[136] Black soldiers
in particular were less liable to receive medals. As an army specialist wrote in a letter to *Sepia*: 'I know a brother who burned up
two M-60 barrels on Viet Cong and saved many fellows from their
deaths and all he got was a slap on the back. But a white guy was
given the Silver Star and a promotion to Sgt.'[137] A lesser indignity
was reserved for Australian troops who were dependent upon
Britain for awards and honours. This was not changed until 1966
when a Vietnam service medal was introduced. In the words of the
editor of *The Australian*, 17 January 1966:

> The decision to award Australian troops fighting solely at
> Australia's direction with the British General Service
> Medal was incredible. It represents a hankering after the
> old associations with the British Empire, a distaste for the
> Commonwealth of independent nations and a triumph for
> Colonel Blimp, who still thinks of the Australian Army as
> a colonial appendage of the British Army.[138]

Heroic acts were often not performed by men possessing traits of
courage, fortitude, and independence, but were the actions of

men scared to death. In the words of nineteen-year-old Lieutenant Graham who won the Victoria Cross during the First World War:

> What else could I do? 'Skedaddle', I suppose? I never thought of it. It would have wanted some pluck. 'Stuck to my gun?' I stuck to [my machine-gun] to save my skin. It was all I had to hang on to. I can tell you, I should have had cold feet without it.[139]

The 'temporary lapse' of the coward was widely believed to be similar to the 'momentary heroic impulse' of the highly deco- rated hero.[140] Letters of recommendation for medals themselves devalued 'actual' combat experience since they had to be (in the words of Colonel Rowland Feilding) 'couched in the flamboyant language of the Penny Dreadful' or in 'glowing "paint-the-lily" style' if they were to persuade authorities far behind the lines.[141] The Vietnam soldier Robert E. Holcomb could not even remem- ber how he won his medals.[142] Like Roy Benavidez at the beginning of this chapter, Sherman Pratt who served in the Korean War noted:

> For every man who gets a medal, there are probably five or six who also deserve one but never get written up. A com- mander has to hear about the brave act, and then he has to write the man up for an award. But many acts of heroism are simply not witnessed. Or the commander, in the press of his combat duties, just doesn't have time to stop and write up the award. Or by the time he does get around to it the witnesses are gone.[143]

By the time of the Vietnam War, the value of medals had fallen even further as over one quarter of servicemen (and 6 per cent of servicewomen) received a combat medal.[144]

As suggested earlier, even the military had problems with

heroes. Instead of obeying orders to 'get down', heroes were men who were 'so excited' that they ran forward, as George Wilson described the deed for which he won the Victoria Cross during the First World War.[145] The highly praised 'martial races', such as the Irish, were criticized for this failing: they were moody, impetuous and useless at defence.[146] In modern warfare, where anonymity was the norm, fantasies of heroism could actually be psychologically crippling. In the words of Dr. Gustav Bychowski, lecturing to the New York Psychoanalytic Institute in 1943, men who entered the services determined to act outstandingly but were faced with a war in which individual valour was inappropriate would often come to feel 'miserably insignificant. With this may come the belief that he is a coward, and the intensity of the feeling may well be in proportion to the extent of his heroic phantasies.'[147] Heroism simply could not survive the horror of twentieth-century warfare.

Heroism was ambivalent and the distance between the hero and the anti-hero narrow. There was no 'seeing eye', no 'moving heart', no mythical sense of a greater life existing beyond what was actually taking place (as Thomas Carlyle in *On Heroes, Hero Worship and the Heroic in History* would have had us believe).[148] Instead, there was damage and fear – and for those like Roy Benavidez, there was only frantic immersion in the chaos. Even the lauded 'martial races' were disparaged for the same failings as the so-called unwarlike, unheroic races. In the first half of the twentieth century, those commentators who continued to emphasize men's physique and 'masculinity rating' were attempting to retain the possibility of individual valour long after it had been lost. Even from *within* the armed forces, the new and powerful discourse of psychiatry was pathologizing the most aggressive combat soldiers. Citations for medals, written in *Boys' Own* style or the language of the penny-dreadful, further parodied the individual hero. In addition, heroic actions were often ugly. As an unnamed Canadian soldier grumbled:

why [do they give] medals for the bloody types? Why were the heroes ones who killed and killed again? Yes, I admit killing is what war is about, the end result, but there is something else.[149]

For the individual attempting to cope with the brutality and anonymity of modern conflicts, what that 'something else' actually was could not be determined.

Chapter 5

LOVE AND HATE

This is no case of petty right or wrong
That politicians or philosophers
Can judge. I hate not Germans, nor grow hot
With love of Englishmen, to please newspapers.
Beside my hate for one fat patriot
My hatred of the Kaiser is love true.

Edward Thomas, 'This is No Case of
Petty Right or Wrong', 1915[1]

On Ascension Day, 13 May 1915, Captain Julian Henry Francis Grenfell was hit by shell fragments in the head. 'I think I shall die!' he gaily exclaimed: within thirteen days, his prophecy was vindicated. He was twenty-seven years old.

Grenfell was born into an aristocratic family in 1888 and educated at Eton and Balliol College, Oxford. He could shoot from the age of seven and his holidays were spent riding, fishing and hunting. He had always wanted a military career and in his early twenties obtained a commission in the 1st Royal Dragoons. Prior to the First World War, he served in India and South Africa where he discovered the delights of boxing, buck-stalking, and pig-sticking ('pig-sticking is beyond my dreams, I can't tell you what it means to me', he wrote home to his mother). Rumours of war thrilled him.

Don't you think it has been a wonderful and almost

incredible rally to the Empire; with Redmond and the
Hindus and Will Crooks and the Boers and the South Fiji
Islanders all aching to come and throw stones at the
Germans. It reinforces one's failing belief in the Old Flag
and the Mother Country and the Heavy Brigade and the
Thin Red Line, and all the Imperial Idea, which gets
rather shadowy in peace time, don't you think?

A fortnight after returning to England on 20 September, he
sailed for France as part of the Third Cavalry Division of the
Fourth Army Corps.

War pleased Grenfell and he proved himself to be a good cav-
alryman. During the Battle of Ypres, he was responsible for
successfully sniping and killing a number of Germans. He
described to his family how he had to plead with his commanding
officer to be allowed to engage in some private sniping of the
enemy. He crawled close to the 'Hun trench' and, peering over the
parapet, he saw a German soldier. 'He was laughing and talking,'
Grenfell wrote, 'I saw his teeth glistening against my foresight,
and I pulled the trigger very slowly. He just grunted, and crum-
pled up.' He repeated the sniping over the next few days and for
these exploits (and for warning his unit of an impending attack)
he earned the Distinguished Service Order. He recorded his kills
in his game book alongside partridges 'bagged'.

On 15 October 1914 he wrote: 'It is all the most *wonderful*
fun; better fun than one could ever imagine. I hope it goes on a
nice long time; but pigsticking will be the only tolerable pursuit
after this one or one will die of sheer ennui.' Four days later, the
first Battle of Ypres began and, as he wrote breathlessly to his
mother on 24 October: 'I've never been so fit or nearly so happy
in my life before; I adore the fighting.' His attitude towards his
foe was ambivalent, though. As he 'thought of the dead', he felt
intense hatred for the German prisoners and could not prevent
himself from scowling at captured German officers. Yet, when a
German prisoner 'looked me in the face and saluted me as he

passed . . . so proud and resolute and smart and confident, in his hour of bitterness', Grenfell admitted that he felt 'terribly ashamed'. So many of his enemy were also simply 'poor dead Huns'. On 3 November 1914, he observed that war was

> the *best* fun. I have never, never felt so well, or so happy, or enjoyed anything so much. It just suits my stolid health, and stolid nerves, and barbaric disposition. The fighting-excitement vitalizes everything, every sight and word and action.

More to the point, he continued, 'one loves one's fellow-man so much more when one is bent on killing him'. Within six months, this aristocratic, idealistic young man had been killed by his fellow men.[2]

Love for one's comrades was widely regarded as the strongest incentive for murderous aggression against a foe identified as threatening that relationship. In contrast, hatred was less significant in creating 'combat effective' servicemen. Was the best soldier the man who killed with passion or in cold blood? Neither extreme seemed ideal: attempts to stimulate hatred in men like Grenfell proved remarkably ineffectual, and yet 'cold-blooded murder' was guilt-ridden. Both love and hatred had a transformative power which could not be easily harnessed. Exhortations to love one's comrades tended to have too diffuse an effect, occasionally spreading even into enemy territory. Hatred, too, was transitory, and in actual combat as opposed to behind the lines, it was frequently overwhelmed by feelings of empathy.

Love and comradeship

Whether called 'mateship', 'the buddy system', or 'homo-erotic relationships', the power of love and friendship in enticing men to kill has been widely commented upon. Although frequently exaggerated,[3] throughout the conflicts discussed in this book, combatants reported that they were able to kill because of the love they felt for their comrades. The importance of comradeship in enabling men to 'carry on' is at the heart of most histories of 'life at the front': so much so that it has become a cliché of military, cultural history. During the conflicts discussed in this book, the importance of affective relationships in stimulating murderous aggression was also a central belief of sociologists and psychologists. In a survey of 568 American infantrymen who had seen combat in Sicily and North Africa in 1944, men were asked what was the most important factor enabling them to continue fighting. Leadership and discipline, lack of alternatives, vindictiveness, idealism, and self-preservation ('kill or be killed') were scarcely mentioned. Rather (after simply desiring to 'end the task'), combatants cited solidarity with the group and thoughts of home and loved ones as their main incentives.[4] Captain Herbert X. Spiegel, a psychiatrist and medical officer during four major engagements in North Africa, repeatedly made the point that aggressive action was motivated by 'a positive force – love more than hate'.[5] 'The men are fighting for each other,' concluded Roy R. Grinker and John P. Spiegel in their study of combat pilots, and ('since so large a portion of his self interest has been transferred to the interests of the group') morale could remain high despite soaring casualty rates.[6]

Combatants themselves frequently attested to the immense love they felt for their comrades-in-arms.[7] It was 'such friendships' which made the 'horrors of war endurable', wrote the First World War gunner, Richard C. Foot or, as Jack W. Mudd told his wife in 1917, 'out here, dear, we're all pals, what one hasn't got the other has . . . You wouldn't believe the Humanity between men out here.' As the American soldier, Allen Hunter, recalled after killing his first Vietnamese: 'It was the first enemy I had killed, and everyone patted me on the back. I wasn't a "cherry" anymore, but accepted because I had proved myself. I was so hyped up and felt good.'[8] The Vietnam veteran, William F. Crandell, also described his complex love for his comrades – a love which condoned any number of atrocities. He knew 'decent and courageous men' who had raped and murdered large groups of civilians yet these were 'men for whom I would take serious risks'. He confessed that it 'has taken me twenty years to work out my feelings of guilt for loving them'.[9]

The love for one's comrades was often discussed in loosely veiled, homo-erotic language. 'Jack' served in Palestine during the First World War and wrote in a letter to Miss D. Williams on 23 November 1917:

> Hell I think I must close now as my *wife* is in bed (if you can call it so. 1 blanket on the ground not bad eho) & wants me to keep her warm but its only a Palestine wife. another Sussex boy. & we are both Jacks so there is nothing doing . . . I really must ring off now as my mate keep saying some nice words which I do not understand. getting cold I suppose. it is very chilly at night right now. & it is nice to have someone to keep each other warm.[10]

Less commonly, this love was explicitly physical, as in a poem written by Jack Strahan about a machine-gunner who served alongside him in Vietnam:

> *Such a friend, that we could touch so often.*
> *The other men would tease, out of envy only,*
> *that we could love or touch, in teasing back,*
> *in happy celebration; could joke or laugh about,*
> *in Monsoon rains, or difference and our play.*
> *His handsome body, full of pride and light;*
> *his easy, even soft and gentle movements; his grace,*
> *deserved my awe when I had seen his clean and witty*
> *precision as he fired that sleek and swift*
> *and deadly accurate Machine Gun.*[11]

The military establishment deliberately set out to promote

certain forms of affection. Soldiers were instructed to treat their rifles with the same delicacy as they would caress their only child.[12] The 'Song From the Trenches' (1914) exhorted soldiers to

> *love your gun – as you haven't a wench –*
> *And she'll save your life in the blooming trench –*
> *Yes, save your life in the trench.*[13]

'Tips for Gunners' (1933) included the sage reflection that guns were like women and should be treated as such.[14] Men 'caressed' and 'took motherly care' of their weapons.[15] Weapons could be both the paternal and maternal phallus. The chant 'this is my rifle,/ this is my gun;/ this is for fighting,/ this is for fun' was recounted in countless Second World War and Vietnam War memoirs and mythologized in films such as *Full Metal Jacket* (1987) where a Marine sergeant informed recruits that their rifle was the 'only pussy you people are going to get . . . You're married to this piece, this weapon of iron and wood, and you will be faithful!'[16]

Much more important than this was the encouragement of brotherly love between enlisted men and their officers. In the words of Colonel Harry Summers of the US Army War College, 'a man who cannot love cannot command'.[17] This love was portrayed, primarily, as parental. The military was very aware of the importance of the biological father in enticing men to kill – after all, combatants frequently admitted that they had joined in imitation of their father[18] or that, in combat, their father would appear in spirit, offering comfort and reassurance.[19] In imitation of such relationships, the military establishment rhetorically evoked the symbolic parent. Furthermore, they evoked not only the father, but the mother too. During the First World War, the English officer was advised to be 'the guide, philosopher, father, mother, and nurse of his men' and a good officer was praised for resembling 'a harassed but efficient mother of a reckless family'.[20] During the Second World War, gender metaphors were also complex, with officers being described as 'giving birth' to a battery while also being exhorted to 'father' their men.[21] Norman Copeland in his book *Psychology and the Soldier* (1942) counselled officers to take on a paternal role in similar terms. Just prior to attack, the officer should play father to his child-soldiers, mentioning each man's name in turn and ensuring that everyone felt that they were receiving special attention.[22] As one Vietnam officer described it:

> I became the mother hen. You know 'C'mon, c'mon, c'mon, c'mon, c'mon, get over here, get over here. Stay down. All right . . .' I was watching the other five guys like they was my children.[23]

A detailed study of this phenomenon was undertaken by Roy R. Grinker and John P. Spiegel and published in *Men Under Stress* (1945). According to them, if men were to prove effective in combat, they needed to be lavished with fatherly affection. A combatant's desire to please his father was largely unconscious, they explained, and was based upon strong love and affection developed in early life. The most important factor in the process of identification was the assurance that he was adored by a man wielding considerable authority. When this occurred, the desire to conform to the demands of the group become almost irresistible. The 'father' had to be strong, decisive and technically competent so that his men would feel protected. He had to demonstrate good judgement and had to play the role of a 'just and impartial father', rewarding and punishing men appropriately. So long as the soldier retained strong ties of affection for the group and identified with its ideals, he would be protected from overpowering anxiety and would be able to kill without qualm.[24] It was the successful establishment of family ties which led one soldier during the Italian campaign in the Second World War to describe transferring to another unit as similar to becoming 'an orphan . . . you feel lost and lone-some'.[25]

'Love thine enemies'

This emphasis on familial ties could be taken further. Oscar Wilde's *The Ballad of Reading Gaol* (1896), Abel Chapman's *On Safari* (1908), W. R. D. Fairbairn's *From Instinct to Self* (1937) and Eli Sagan's *Cannibalism* (1974) all proposed that people 'love what they kill'. Chapman and Fairbairn also used hunting metaphors to illuminate this fusion between love and hatred. According to Chapman, the sportsman 'loves game as though he were the father of it', and, almost thirty years later, Fairbairn observed that the 'fox-hunter's affection for the fox which he kills' provided a clue to the 'unfortunate zest imparted to war'.[26] Religious spokesmen approved of such metaphors, often exhorting men to love their foe while they were intent on slaughter.[27]

For most combatants, this was not possible. As one American officer wrote in his diary during the First World War: 'Love your enemies! Great Scott! I can't do that and try to kill them at the same time. I don't hate them, I admit, but love them! No, I can't do that.'[28] While 'love' might be too strong a word, many combatants were surprised to find themselves feeling intensely affectionate towards their foe. Lieutenant-Colonel James Young admitted that war was 'strange', writing in his diary on 13 July 1915:[29]

> On the one side you have all the signs of excessive hate
> and unbridled passion that show the innate madness that
> still lurks in the human soul. On the other you have all the
> signs of unselfish devotion and kindliness of spirit, even
> towards the man whom you have just struck in your hate,
> that show that there is, somewhere, a reserve of saving
> grace that rescues man from utter degradation.

Or, in the words of a First World War rifleman: in 'the actual heat of battle', men killed 'remorselessly and indiscriminately', but afterwards 'animosity' was forgotten and the prisoners were 'treated very much as one of our own'.[30] Men could attack each other with passion, but would take these same enemy cigarettes when they were wounded and honoured them in funeral rites.[31] Combatants were 'as careful to collect souvenirs from his dead enemy as from the girl he left behind him', the British Psychological Society were informed in 1938.[32] Even the sniffing of corpses had a 'peculiar fascination' for some soldiers.[33] The enemy exerted a 'magnetic attraction' to fighters. For instance, during the First World War, the fighter pilot Cecil Lewis noted that combatants did not 'love' the enemy 'in the conventional sense', but they certainly honoured and respected them. He continued:

> Besides, there is, as everybody who has fought knows, a strong magnetic attraction between two men who are matched against one another. I have felt this magnetism, engaging an enemy scout three miles above the earth. I have wheeled and circled, watching how he flies, taking in the power and speed of his machine, seen him, fifty yards away, eyeing me, calculating, watching for an opening, each of us wary, keyed up to the last pitch of skill and endeavour. And if at last he went down, a falling rocket of smoke and flame, what a glorious and heroic death. What a brave man! It might just as well have been me.[34]

Affection between opposing combatants was also portrayed at a more symbolic level. Soldiers have described killing a foe with the bayonet as 'getting home' and frequently claimed to have seen two corpses embracing, each bayonet piercing the body of the other man.[35]

In contrast to such surreal brutality, men at the front were remarkably ambivalent about their foe. As we saw earlier in Stouffer's data, men without any combat experience hated the enemy more than actual fighters did, and servicemen who had not left the country hated more than those overseas.[36] Combatants at the frontlines frequently expressed respect for their opponents – even to the extent of fraternizing with them.[37] Fraternization between German, British and French soldiers during that first Christmas 1914 became such an important war myth that practically every soldier claimed to have defied High Command to participate. When, in later years, opportunities for fraternization were limited and more rigorously policed, soldiers lamented the fact. Gerald Dennis, for instance, recalled that on Christmas 1916, he would

> not have minded fraternizing as had been done the previous two years for in a way, the opponents on each side of No Man's Land were kindred spirit. We did not hate one another. We were both P. B. I. [Poor Bloody Infantry] we should have liked to have stood up between our respective barbed wire, without danger and shaken hands with our counterparts. Why not?[38]

Fraternization tended to occur on religious holidays, such as Easter and Christmas Day,[39] and when men were wounded ('Each had his arm round the other's like lovers,' Clifford Nixon wrote to his father).[40] Sometimes fraternization occurred when a soldier was forced to look at the man he was about to kill. Cecil H. Cox was unable to kill cold-bloodedly. During a campaign in Northern Italy, he was ordered to take an island:

> I saw a young German coming towards me and at that moment I just could not murder him and lowered my gun, he saw me do so and he followed suit, shouting 'What the h— do you want to kill me for, I dont [sic] want to kill you.' He walked back with me and asked if I had anything to eat? At once the relief inside me was unspeakable, and I gave him my iron rations & my army biscuit.[41]

Homo-erotic love/hatred

The extent to which homo-erotic love was predicated upon violence has been debated by many writers on warfare. W. Stekel's
Sadism and Masochism: The Psychology of Hatred and Cruelty
(1953) linked hatred with sexual pleasure: 'Man is cruel for the
sake of the pleasure which the barbarous act produces,' he suggested.[42] In 1964, the anthropologist Derek Freeman, noted that
expressions of pleasure (such as laughter) were commonly associated with witnessing destructive acts or inflicting pain.[43] In a
lecture delivered in 1968, the psychoanalyst Ralph R. Greenson
also drew his listeners' attention to the complex ways in which
people reacted to violence: dread and guilt coincided with titilla-
tion and fascination.[44] More recently, Adrian Caesar in *Taking it
Like a Man. Suffering, Sexuality and the War Poets* (1993) has
drawn a relationship between the wartime experiences of men
like Rupert Brooke, Siegfried Sassoon, Wilfred Owen and Robert
Graves and their latent sado-masochism, contending that their
responses were typical amongst servicemen. He asked why men
had to attempt to kill each other 'in order to express gentleness,
tenderness, loving kindness, and love for each other'.[45]

The recurring use of sexual metaphors in combat stories also
drew men's attention to the relationship between 'blood lust'
and 'body lust'. By the 1960s, psychoanalysis had so permeated
literary language that it was almost *de rigueur* to use such
metaphors in fictionalized accounts of combat. Two unexceptional examples (from America and Australia) can be found in
Daniel Ford's *Incident at Muc Wa* (1967) and John Carroll's
Token Soldiers (1983). In Ford's novel, killing is portrayed as rape
in which the two combatants 'strained together, like lovers':

Then they fell . . . [Stephen] took the bayonet from his
scabbard and drove it into Charlie's side. But it hit a bone
or some damn thing, and the poor bastard twisted and
kicked, tearing with his fingernails at Stephen's hand . . .

> [Stephen] pulled the bayonet free and used it to cut
> Charlie's throat. Blood pulsed across his hand.[46]

Equally violent, but with the emphasis on the orgasm rather than
the act of penetration, Carroll described his character's 'serene'
expression immediately after slitting the throat of an old man:

> You could see that Savage was thrilled, almost overcome
> with what he had done – like he had just had the fuck of
> his life, and didn't want to come out of it just yet. Yes,
> that's what that shining look was – pure pleasure, in the
> sexual sense.[47]

Sexual intercourse and murder were linked in every literary
genre, from prosaic military publications to vivid autobiographies. In a training manual for the Home Guard, entitled *Rough
Stuff* (1942), readers were informed that just as successful lovers
were those men who were capable of giving 'free rein' to their
desires, so too successful bayonet fighters were men who held
nothing back, allowing hatred full expression.[48] Linking sex and
death was one way of conveying the intensity of combat to an
unblooded, voyeuristic public. Philip Caputo's description of
leading an attack in Vietnam which resulted in an 'ache as profound as the ache of orgasm' has frequently been cited.[49] Other
combatants, in other conflicts, made similar observations.
Anthony S. Irwin, a subaltern in one of the first battalions of the
British Expeditionary Force, for instance, described in 1943 the
terrorizing experience of being dive-bombed by six German
planes, all of whom (he believed) were attempting to pierce his
body. After they flew away, he described lying on the ground for
half an hour:

> I felt like a man who has just had a perfect and shattering
> union with a woman. I sweated and wanted more. The
> earth seemed to shake and shudder under me, the trees

> seemed to be alive, the gently waving branches stroked
> me, and I wanted to sleep.

In this state of post-coital bliss, he was unable to doze off until he placed his tin helmet over his belly, in a gesture which both protected his manhood and simulated pregnancy.[50] Combatants clung to affective relationships which were both 'fatherly' (that is, hierarchical and empowering) and 'mothering' (inspiring and comforting).

Hatred

Killing generated many strong emotions, including love and desire. However, hatred, defined as 'an enduring organization of aggressive impulses toward a person or class of persons . . . composed of habitual bitter feeling and accusatory thought', was also frequently assumed to be a normal part of combat behaviour.[51] Many historians, psychologists, and military commentators shared the assumption that hatred was crucial in inciting the desire to kill and enabling individuals to act upon this urge.[52] The pleasurable sensations accompanying hatred were thought, by some, to be a powerful incentive to declare war, while others contended that an 'irreducible minimum of hatred' lurked within the human psyche and had to be expelled – and what better outlet than armed conflict?[53] Even pacifists registered their preference for hate-induced killing on the grounds that the thought that men might be capable of killing without passion was too deplorable to countenance.[54] The military establishment tended to accept that hatred was a desirable emotion for men in combat and (as we shall see) training regimes were developed to encourage it.[55] Many psychologists writing about combat confirmed this belief: instinct theorists – in particular, followers of the psychologist, William McDougall – argued that hatred was an important outlet for the feelings of fear, disgust and self-abasement that arose during combat.[56] Throughout the century, virulent hatred was believed to stimulate pugnacity, which was the most effective antidote to fear and anxiety (the chief obstacles to combat effectiveness).[57] For instance, in the middle of the Second World War, Gregory Zilboorg pointed out to the American Psychiatric Association that the fear of death could be mastered through hatred of the enemy. Indeed, the very essence of 'morale' consisted of the degree to which fear could be converted into 'murderous hatred': combatants could only overcome death 'by means of murder'.[58] In the same year, Leonard Sillman, psychiatrist at Columbia University, repeated his view that hatred was part of the 'psychological mobilization' necessary for effective combat performance. It was a 'steel helmet for the mind'. Paradoxically, if Americans were to fight fascists without becoming fascist themselves, they had to be taught to hate so that their aggressive impulses could be externalized against the enemy. The side which hated the most would be victorious. Sillman regretted American resistance to hatred (Americans were too individualistic to hate without encouragement, he believed) but insisted that it was possible to stimulate hatred through government-sponsored propaganda.[59] During the war in Korea, Colonel John J. Marren, psychiatrist at the army hospital in San Francisco, agreed that hatred dulled the effects of fear and enabled soldiers to wreck vengeance on the enemy. Like Sillman, he believed that it was difficult to indoctrinate American soldiers with feelings of hatred (because of their cultural background), but he insisted that it was a 'highly desirable' trait to inculcate.[60]

Hate training

Since hatred was considered to be important in stimulating 'appropriate' combat behaviour, it was not surprising that the military establishment considered ways to promote it. 'Reasoned' propaganda explaining to combatants why they were fighting and emphasizing the diabolical nature of the foe were incapable of inducing virulent enmity: true hatred resisted such processes of reasoning. Furthermore, combatants resented being herded together for these 'pep talks': they regarded congregating in

groups as dangerous, they were too weary to be stimulated by lectures on the evils of fascism or communism and, preoccupied
with the intimacy of combat, they sneered at political plati-
tudes.[61]

More robust forms of military training had to be used to
encourage men to hate. During the First World War, Second
World War, and (even) the Vietnam War, the bayonet drill was, as
we have seen, widely used to 'awaken savage instincts'.[62] There
was another method of inciting hatred, however, which needs to
be examined here. By the Second World War, 'hate training' had
taken on a much more radical dimension. Immediately before the
declaration of war, in a series of weekly conversations with
Oxford undergraduates, Humphry Beevor argued that 'decent
people' could not be expected to kill 'other decent people against
whom they have feelings neither of personal malice nor of moral
indignation' unless they had been subjected to some form of hate
training. Especially if a war was prolonged, it would become an
'essential part of the policy of those who direct the fighting
machine' to

> brutalize those who are to do the actual fighting. It will
> not do for your soldiers to regard themselves as the chival-
> rous champions of law and order; they must be properly
> inoculated with the blood-lust, they must desire to kill for
> killing's sake.[63]

The most comprehensive experiment in 'hate training' was
carried out in the battle schools for the British Army in 1941 and
1942. As part of their training, recruits had to go through a gruelling, mile-long assault and obstacle course. Loudspeakers
chanting 'Kill that Hun. Kill that Hun' and 'Remember Hong
Kong. It might have been you' taunted and disorientated recruits
following the course. There were explosions and, as the soldiers
waded through water and mud pits, they were shot at with live
ammunition. They were instructed to fire their own weapons at

imitation German and Japanese soldiers. When they arrived at the section which involved a bayonet charge, recruits were showered with sheep's blood. At another stage in their training, recruits were taken to slaughterhouses and they were exposed to a 'hall of hate' consisting of pictures of German atrocities in Poland.[64]

This gory experiment in 'hate training' was widely rejected. General Sir Bernard Paget, Commander-in-Chief (Home Forces), reminded army commanders that there was a difference between artificial hate and the 'true offensive spirit combined with will-power'. He directed that the use of strong language and attempts to produce blood-lust or hate were to be stopped.[65] By May 1942, the training had ceased, but not before a public outcry. 'Hate training' was revealed to the public during a peak-time BBC broadcast on 27 April 1942. Wasn't there enough hatred in this world without it being deliberately stimulated by the War Office, queried R. J. Davies in the House of Commons?[66] Others questioned whether such training was Christian or even 'English': it was undignified, critics argued.[67] Writers to *The Times* registered their disapproval. 'What is the use of hating one's enemy?' thundered Thomas Howard from Winchester. 'Nothing wears one out so much as hate; and it is lasting power that is needed in war.' He claimed that when he was in battle it was only his 'calm and determination' which had enabled him to 'pick out leader after leader and so do my bit to win those particular fights'.[68] Psychologists were even more dismissive, arguing that anything that made recruits faint and vomit was dangerous and they suggested that comparing the battlefield to slaughterhouses was neither relevant nor inspiring.[69] It was noted that previously keen students became depressed.[70] Such training was 'so crude and artificial that it could only be a product of an abnormal and infantile mind' and would be more liable to stimulate unconscious guilt and depression than to heighten morale.[71] Attempts to 'work up hot blood in cold blood' were psychologically damaging: the use of real blood was even worse.[72]

Blood training undermined the 'foundation stones of morale – human self-respect'.[73] The schools were based on an incorrect notion of human nature as innately sadistic and bloodthirsty.[74] As the British Directorate of Army Psychiatry in 1942 commented wrily:

> In the course of most wars, individuals or small groups in training or back areas not infrequently become convinced that we must learn to hate the enemy and that blood lust is an important component of combatant morale.
> Fascinating as this idea is to officers and men who are chafing over inactivity or struggling with boredom, experience shows that attempts to arouse primitive passion – even if they are successful in overcoming the British soldier's sense of honour – have not been found useful as a method of increasing combatant efficiency.[75]

Hatred of the enemy

'Combat effective' servicemen were divided over their attitude towards their foe. Some denied feeling animosity while others revelled in loathing. Combat in theatres where the enemy was of a different race was portrayed as particularly hateful and liable to involve atrocities.[76] Hatred of the enemy was extreme, for instance, during the war in the Pacific. Eugene B. 'Sledgehammer' Sledge described the 'brutish, primitive hatred' which overpowered all other feelings.[77] One major survey revealed the much greater hatred of the Japanese when compared with the Germans.[78] The Australians hated the Japanese even more than the Americans did, in part due to historical tensions. As one official put it, becoming a 'cool and calculating killer' should not be difficult for Australian snipers when dealing with Japanese enemies who were potential invaders – 'our racial prejudices will be hotly aroused', he explained.[79] Lieutenant Uzal Ent of the 25th Division in Korea admitted that he did not feel 'any personal animosity' for the enemy (after all, he reasoned, 'whoever he might be, North Korean, Chinese, Russian, whatever, was serving his country and following orders, just as I was doing for my country'), but believed that his attitude was not typical. 'Many men fought with a visceral hatred of the enemy. Maybe the fact that they were Orientals had something to do with it,' he pondered.[80] In Vietnam, Philip Caputo claimed that he

> burned with a hatred for the Viet Cong and with . . . a
> desire for retribution. I did not hate the enemy for their
> politics, but for murdering Simpson [a comrade] . . .
> Revenge was one of the reasons I volunteered for a line
> company. I wanted a chance to kill somebody.[81]

In contrast, other combatants during these conflicts claimed that they did *not* hate their enemy. Typically, Lieutenant Frank Warren during the First World War observed that his comrades 'have no special hate of the Hun as an individual, are inclined to respect him as an equal in some ways at least where Prussian arrogance is absent, but would only give him a sporting chance at the other end of rifle or bayonet'.[82] Military trainers frequently failed to teach young soldiers to hate the Germans.[83] This was the case even in Vietnam: indeed, the South Vietnamese allies were often violently despised while the designated 'enemy' was accorded respect. The young Australian officer, Gary McKay, admitted that he did not hate the Viet Cong, although he had a 'vested interest' in killing as many as possible. Indeed, he confessed, he actually respected them as fighters because they displayed traits that he rated highly, particularly tenacity, toughness, and resolution.[84] Echoing Edward Thomas, the poet quoted at the beginning of this chapter, a soldier in Vietnam wrote the following words in his diary on 19 February 1968:

> *I hate not these people,*
> *I hate not the land,*
> *I hate but the person*

with his peace waving hand
starts the war and wants
everyone else to fight it.[85]

Another officer in Vietnam complained that the 'problem with this war is there's nobody to hate . . . Nobody like the Japs and the Germans.' An air force captain agreed:

In World War II, you felt that if you failed, your homes
and country were at stake. Here you don't feel that way. It
hasn't come across to most men why they're here . . . In
order to hate, you have to be emotionally involved.[86]

As one anti-socialist writer quipped in 1891, more hatred was expressed by trade unionists during an average strike than by men in the midst of battle.[87]

Such qualitative evidence provides little clue as to what proportion of combatants felt hatred. There is some survey evidence which attempts to quantify this emotion. Most investigations were small in scope: for instance, in a series of interviews of successful combatants in the US Eighth Air Force during the Second World War, it was found that less than 30 per cent expressed any sense of personal hatred for the enemy which made them want to kill.[88] During the Vietnam War, John Helmer questioned ninety Vietnam veterans about how they felt about the enemy after their first contact. While 27 per cent hated the enemy more after battle, 38 per cent said that they either hated the enemy less (10 per cent) or even respected the enemy more (28 per cent). Twenty-nine per cent experienced no change or felt neutral.[89]

In 1944–5, a unique study in scope and range was carried out. Samuel A. Stouffer's survey of American infantrymen was a detailed investigation of men's experiences in wartime. At one point in the survey, he asked infantrymen: 'When the going was tough, how much were you helped by thoughts of hatred of the enemy?' Overall, approximately one third admitted to being strongly impelled by hatred, although this proportion rose as high as 38 per cent amongst enlisted infantrymen in the Pacific and fell as low as 27 per cent of men in the Mediterranean and in Italian theatres of war. 'When the going was tough', hatred of the enemy was considerably less important than prayer and a desire not to let one's mates down. For instance, while 38 per cent of the enlisted men in the Pacific were helped by hatred of the Japanese, almost twice as many (61 per cent) kept going because they did not want to let their comrades down.

In another part of his survey, Stouffer approached the question of hatred from a different angle. He asked American infantrymen what they thought should happen to Japanese and German people after the war: should they be 'wiped out altogether'; should they be made to 'suffer plenty'; or should only the leaders of these countries be punished? Sixty-seven per cent of enlisted infantrymen who had not left the United States wanted the Japanese race 'wiped out' altogether, compared with 29 per cent who wished a similar fate on the Germans. An interesting difference can be seen in comparing the attitudes of enlisted men in the Pacific with those in Europe. Servicemen in both theatres of war shared similar attitudes towards the Germans (in both 22 to 25 per cent wanted the Germans 'wiped out' after the war, 6 to 8 per cent wanted the Germans to 'suffer plenty', and 65 to 68 per cent wanted the leaders to be punished, but not other Germans). Attitudes to the Japanese varied much more widely. A much lower proportion of enlisted men in the Pacific wanted the Japanese 'wiped out' compared with enlisted men in Europe. Forty-two per cent of soldiers in the Pacific wanted the Japanese 'wiped out' compared with 61 per cent of soldiers in Europe.[90] In other words, servicemen who had not left America hated the enemy most, and men serving in Europe hated the Japanese more than did the men actually killing Japanese troops in the Pacific.

The quantitative and qualitative statements suggest that the expression of hatred was related to two things: the physical and psychological proximity of the foe and combat experience. The physical proximity of the enemy was clearly crucial, with the most anonymous forms of killing (such as the artillery) being the least conducive to hatred because their way of meting out death was too impersonal.[91] Even though most infantrymen recognized that it was not so much hatred as the 'instinct of self-preservation' which impelled the bayonet thrust,[92] they generally agreed that they were more hateful than 'machine fighters'.[93] The First World War field-gunner, Huntly Gordon, admitted that he felt little sympathy for the Germans, but noted that he could not hate them either. He believed that this was because he was not required to creep about in no man's land with a spiked truncheon.[94] During the Second World War, Norman Copeland noted that when a man 'strove with man in thrust and parry', it was relatively easy to hate:

> One's opponent was standing less than three feet away; he is hissing imprecations in a strange tongue; he looked what he was – a foreigner; his eyes were dilated with fear or anger; and he was displaying every evidence of an insensate desire to kill.

But automatic rifles, machine-guns, and high explosives gave 'death a much wider range', rendering it impersonal, disinterested even. 'It was not easy to feel pugnacious or angry about an enemy one rarely saw,' Copeland concluded.[95] Similarly, the artilleryman, Lieutenant-Colonel Kenneth H. Cousland, never came 'face to face' with a living German and consequently 'never thought in terms of killing others'. He described combat as

> a strange impersonal feeling; they were merely targets. I do not recall having hard feelings against them. In fact we respected them as brave soldiers, but our job was to defeat them and win the war.[96]

Social and psychological connections with the enemy were equally important. Individual combatants came to war with unique histories which could make hatred difficult. Particularly during the war in Europe, many servicemen had some knowledge of their foe in peacetime and this affected their attitude. The aerial gunner, A. G. J. Whitehouse, could not forget that the men he shot down were once his friends. He mused at great length on the first man he killed. 'They were Germans, of course, but somehow I felt sorry for them,' he reflected before describing the entire event as 'a dream – unreal. I didn't hate them, and I remembered Vic Seaman, Charlie Rothnagel, Mr Snyder, a school-teacher, and oh, a lot more, who were Germans back in New Jersey.'[97]

In addition to civilian pasts, recent combat experiences also made a great difference in men's emotional responses to the enemy. Civilians were more prone to articulate virulent hatred toward the enemy, leading many commentators to conclude that reading or writing about killing was more likely to stimulate hateful feelings than actual participation in the slaughter.[98] While pilots and their crew admired and respected their victims, often burying them with full military honours, civilians felt only contempt for the enemy, declaring courteous rites as analogous to according respect to Jack the Ripper.[99] Captain Albert Ball, the ace pilot personally responsible for downing forty-seven German planes, was cheered on by his father who encouraged him to 'let the devils have it'. Ball replied that he did not see the enemy as devils: 'I do not think anything bad about the Hun. He is just a good chap with very little *guts*, trying to do his best,' he informed his father.[100] On 4 May 1915, J. H. Early sent a letter to his family in response to a press cutting. He rebuked them with the words, 'You get no "brutal Huns" here.' Then admitted:

Probably we ought to feel more apoplectic, and it may be due to really outrageous things being over-shadowed by our immediate surroundings and also to a lurking feeling

for the poor dogs who must be living the same silly sort of
life that we are, behind the sandbags over there. War is a
disgusting waste of time, and your neighbours here would
only think you an ass if you got fervent about it on any
other lines![101]

Furthermore, civilians who had most experience of war (such as
people subjected to aerial bombing) were *less* liable to demand
reprisals. Indeed, one major survey found that the demand for
reprisals for the bombing of British cities came most strongly
from rural areas such as Cumberland, Westmorland, and the
North Riding of Yorkshire where bombs had not been dropped,
leading the British Institute of Public Opinion to conclude in
April 1941 that favourable attitudes to reprisal bombing were in
reverse ratio to the individual's experience of bombing.[102]

   Two non-combative groups were regarded as particularly susceptible to hatred of 'the enemy': women and Home Guardsmen.
As one Vietnam War novel succinctly put it in 1967: there was 'no
more bloodthirsty creature on the face of the globe than a well-
educated young woman with liberal convictions.'[103] In 1939,
Maurice Wright, physician to the Tavistock Clinic in London,
expressed his concerns about the flood of hatred unleased at the
outset of war:

   I have never seen such an appalling release of primitive
   hate, cruelty, and sadism . . . The release of sadism was
   evident in its most primitive form in the press and in the
   conversation of ordinary people – most evident, I think, in
   women, in whom the related aggressive instincts are more
   deeply repressed in normal times and situations.[104]

Denied weapons themselves, women satisfied their aggressive
urges by pestering their menfolk to act on their behalf and decimate the enemy.[105]

   The training manuals of the British Home Guard contained

some of the most 'hate-filled' material of the Second World War.[106] Tom Wintringham (who had fought in the Spanish Civil War) published his training lectures for the Home Guard in 1941, in which he taught them to kill noiselessly with knives, exhorted them never to take prisoners, and insisted that there were no rules in combat.[107] The same year, Norman Demuth in *Harrying the Hun* set out to teach men how to kill quietly. His book was classified as an 'A' publication by the Commander-in-Chief, Home Forces, that is, a highly recommended publication which could be purchased using the training grant.[108] Demuth warned against feeling any 'qualms' about killing. 'Remember the concentration camps,' he exclaimed, 'remember the utter bestiality of the Nazi mind':

> Remember that if you do not do this when they come
> here, the most damnable barbarians since Attila will lay
> our country waste, indulge in mass executions, and turn
> our women into whores for themselves . . . Have at him,
> because if you do not he will have at you. Ask yourself
> how the prisoners in Germany are faring – and stab hard,
> and stab to kill. IT IS THE ONLY THING TO DO.

He emphasized the need for Home Guardsmen to 'accustom themselves' to blood and advised instructors to take their men 'to your local slaughter-house, and let them gaze to their hearts' content. At first they will hate it, but the exercise must be repeated until there are no signs of repugnance.' He suggested that Home Guardsmen tested the 'resistance of a body' by using their 'killing knives' on the carcases ('They will be surprised at the strength of stab required').[109]

The next four textbooks, which were all written in 1942, share Demuth's tone. In *Total War Training for Home Guard Officers and N.C.O.s*, Major M. D. S. Armour agreed with Demuth's training methods and encouraged Home Guardsmen to visit slaughterhouses in order to practise using their knives on the carcases and rubbing the blood on their hands and faces.[110] The same year, Lieutenant E. Hartley Leather of the Royal Canadian Artillery wrote a book entitled *Combat Without Weapons* which declared its intention to teach members of the Home Guard how to 'exterminate Germans' efficiently. In Leather's words:

> No matter how unsavoury the job may be, it has got to be done, and done thoroughly; remember, the German is always thorough: we will only beat him if we play his own game, because he will never play ours. Sportsmanship and decency are entirely foreign to his nature. Kicking a man when he is down does not appeal to the average Briton, but we must forget our long-learned niceties when dealing with the Boche; the Nazi is congenitally incapable of being decent. It is not a matter solely of beating the Hun; it is even more than that – a matter of saving your own life and those of your wife and your children.[111]

Such explicit viciousness was not exceptional within this genre. In *Rough Stuff*, a book written by Sydney Duffield and Andrew G. Elliott, Home Guardsmen were exhorted to ignore instructors who argued that bayonets were superfluous in modern warfare. On the contrary, there was never a greater need for this weapon of intimate murder, especially for combat at night, in fog, in streets, or whenever ammunition was running low. As further protection, these authors recommended that all men purchase daggers. They agreed with Leather that the Home Guard had to be brutal: their job was simply to 'destroy the enemy' and, to do this effectively, they had to jettison 'British' niceties of warfare (examples given of outdated 'British' practices included taking prisoners and all notions about what was 'sporting').[112] Elliott contributed to another training manual for the Home Guard in 1942. Entitled *The Home Guard Encyclopedia*, it was written by Elliott in conjunction with 'J. B.' and 'Scientist' and was even more aggressive, encouraging men to ignore any attempts by commanding officers to impose a moral code of combat: the enthusiastic bloodiness of Home Guardsmen was not to be restrained by the Regular army. For example, the authors advised Home Guardsmen to file the back of their steel helmets to a very sharp point in order to kill invading Germans and to conceal this sharp point by painting over it 'so that your C. O. will know nothing about it'. Scepticism about the military establishment reached greater heights when weapons of war were discussed. *The Home Guard Encyclopedia* sneered at 'older officers' for their resistance to unorthodox ideas. Elliott then described how such officers 'pooh poohed' his idea of how to kill a sentry:

> I remarked that as this must usually be accomplished silently, I felt that there was no better weapon than a cross-bow and steel-headed arrow. Instead of having to creep right up to the sentry, he could be despatched from ten or even more yards by this means . . . owing to the risk of discovery before the attacker is within reach, I would prefer to order a man to attack a sentry with a bow and arrow than to throttle him.[113]

In 1943, the social psychologist Mark May argued that civilians were more liable to hate because they were exposed to campaigns of buying bonds, collecting 'scrap to kill a Jap', restrictions, enduring material discomfort, and the dread of the death of their loved ones.[114] But most of these points applied just as much to men at the front. In the same year, the American Psychiatric Association was informed by Gregory Zilboorg that the expression of hatred for the enemy amongst civilians was their way of repressing their own fear of death. But, in Zilboorg's words:

> It is not the degree of its repression, however, but the degree of the conversion into murderous hatred that is the

main ingredient of what we call 'morale' among those who
are vicarious participants in the war, at a safe distance
from the scene of battle.[115]

The guilt of not fighting themselves was displaced on to the
enemy. Female civilians had their menfolk taken away from them.
Warmly ensconced at home, there was no need to acknowledge
the humanity of the enemy and their victims' cries went unheard.

Soldiers at the frontline were profoundly aware of the discrep-
ancy between the demonized pictures they were shown of their
foe while in training and those they encountered in battle.[116]
Combat experience taught servicemen to be distrustful about
what they were told.[117] Face to face with the enemy, it was often
impossible to believe the atrocity stories. Hiram Sturdy described
his first view of the dreaded Hun. Although he had fought in the
Battle of the Somme, his first sighting of a German soldier
occurred after the battle when he went to look at the prisoners:

> one can't very well form an opinion of a people when he is
> seen, after lying for months in sodden ground, dead, and
> then uncovered. Such was our only view, up to the pres-
> ent, of our enemy. But now we were to see the men who
> cut the breasts of Belgian woman, and put bayonets into
> others in a certain condition. The Hun, the brute. The
> batch arrives, and I get one of the greatest disappoint-
> ments of the war . . . Our prisoners were young men,
> bandaged and battered, who . . . furtled and jumped, a
> solid bunch of nerves, after having, I suspect, been
> through that hellish bombardment. The most savage com-
> ment I heard while watching the prisoners, came from an
> infantryman. That was 'Poor buggars'.[118]

This contrast between the mental image of the enemy in battle
and the real images encountered afterwards were frequently

commented upon. Sidney Rogerson described his reaction after seeing some German prisoners:

> The enemy early became a legend. The well-wired trenches that faced ours frequently at a distance of only a few yards, gave shelter, we understood, to a race of savages, Huns, blond beasts who gave no quarter, who crucified Canadians and bayoneted babies, raped Belgium women, and had actually built kadaver works where they rendered down the bodies of their dead into fats! . . . But were these pallid, serious youths really capable of such enormities?[119]

With occasional exceptions most servicemen killed the enemy with a sense that they were performing a slightly distasteful but necessary job. They regarded themselves as craftsmen, professionals, or merely as ordinary men fulfilling their duty.[120] In 1942, John J. Floherty described the reactions of Sergeant Russell Brown, a flight engineer forced to take over the guns after the gunner and two other members of the crew were killed:

> The hot anger that had burned within Brown [after seeing his comrades killed] gave way to a deadly cold determination. The excitement of combat had left him, he was now a skilled workman doing a routine job with the tools he knew so well.

Brown then described how he gazed into the eyes of the men in the German plane before shooting them down.[121] When the highly decorated Sergeant Ehlers was asked whether he killed Germans out of hatred, he replied:

> Sir, I don't hate anybody, and I don't like to kill anybody. But if somebody gets in my way when I have a job to do so

I have to kill him so I can get on with it, why then I kill him, and that's all there is to it.[122]

The Maori soldier, Paul Thomas, described a similar reaction prior to going to fight in Vietnam. 'The light at the end of the tunnel was to go to Vietnam to kill the enemy,' he recalled. There was 'no emotion attached to the act of killing; it was just a pro-grammed response'.[123]

There was also an acknowledgement that the enemy was suf-fering in similar ways to oneself. In the words of trooper William Clarke, during the First World War:

face to face with them you couldn't feel a personal hatred, they were soldiers like ourselves, manipulated by states-men and generals and war-mongers. We were – they were – cannon fodder.[124]

Particularly during bombardments, soldiers often felt intense pity for German soldiers while still declaring their intention to kill as many as possible.[125] At the front, it was easier to draw a distinction between enemy combatants and their leaders, as the first two stanzas of a First World War poem published in *Aussie* declared:

> *I don't desire that Hans and Fritz*
> *Should all be blown to little bits;*
> *It is a Peace I want to see –*
> *Their pieces are no use to me.*
>
> *But I would hang the Kaiser high*
> *With all his private company –*
> *As high we'll say, as his name smells;*
> *And shoot him down with Archie shells.*[126]

This kind of feeling was particularly representative in the

European theatres of combat, rather than in the more 'alien' theatres (such as during the war against the Japanese).[127]

Love or hate?

It was much more desirable to kill for positive than for negative emotions. Hatred might actually reduce 'combat effectiveness' while love might enhance it. Technological innovations and the increased use of bombs and grenades progressively undermined the value of passion: a cool head and steady eye were more essential in combat than virulent enmity.[128] Blood-lust, rage and hatred were liable to make men's hands tremble when shooting at the enemy.[129] In noisy, fatiguing and dangerous surroundings, it was imperative that servicemen maintained 'cognitive orientation and a commitment to task-completion', wrote one psychologist, belabouring the point.[130] Not only was it difficult (if not impossible) to maintain any sense of hatred or anger towards the enemy for prolonged periods,[131] but troops who were too aggressive were liable to overrun their objectives and sustain heavy casualties.[132] The development of hate was always at the expense of self-control, making soldiers little better than an armed mob.[133] Psychological conflicts over hatred might be detrimental. Chronically hostile men who yearned to kill their hated enemy found it difficult to adapt their aggressive personalities to the needs of teamwork and co-ordination.[134] Indeed, hatred of the enemy and sadistic gratification from killing frequently led to an escalation in those emotions that actually hampered successful combat.[135] Servicemen who overindulged in aggression were more liable to suffer nervous breakdowns as a result of combat situations.[136]

More to the point, hatred reduced a sense of the rightness of the cause. People hated only those they feared: this was why any 'hate the Hun' movement could be emotionally dangerous, explained one padre.[137] Psychologists argued that the stimulation of hatred would only lead to disruption and frustration. For instance, although Edward Glover admitted that some men needed to hate in order to kill, he believed that the absence of hatred revealed a healthy confidence in the rightness of one's cause and an assurance of winning. In other words, hatred was indicative of uncertainty.[138] Confidence in the political aims of the war allowed fighting to become part of the individual's ego-ideal, thus diminishing psychological conflict over killing: in contrast, the bitter emotionalism of hatred was an indication of tension and could result in emotional breakdown.[139] Furthermore, hatred tended to diminish the effect of important rituals of war, particularly that of group or 'tribal' identification. According to this view, although war led to a burgeoning sense of 'tribal identification' amongst both civilians and combatants, only men of a particular status were given the honour of killing members of the opposing tribe. For this privileged group, the absence of hatred was a highly valued trait. Their pugnacity had to be separate from 'common anger'. In the words of the philosopher, William Ernest Hocking, the combatant

> has a sterner and weightier as well as a larger task . . .
> Even our instincts are aware that war is a relation not
> between persons but between States . . . The fighting-
> instinct understands that it is in the service of the social
> instinct. And not uncommonly the feeling of crowd-
> loyalty, together with the equally instinctive love of
> adventure, quite submerge the sense of hostility or
> resentment.[140]

During the Second World War, a similar discussion could be heard, with an emphasis being placed upon different kinds of crowds or 'herds' (therefore, according to Lord Moran, English instincts had no place for 'hymns of hate' because the English crowd resembled socialized animals in contrast to German society which exemplified the wolf-pack).[141]

Finally, hatred reduced the civilizing sense of chivalry in combat. Killing chivalrously meant avoiding ugly feelings of hatred and acknowledging the humanity of the enemy.[142] The absence of hatred enabled soldiers to 'kill men without being murderers' or, in the words of Robert William MacKenna, what distinguished murder from military killing was the absence of 'personal hatred'.[143] Chivalry, or the 'spirit of the warrior', demanded killing without hatred.[144] Since soldiers were motivated to kill not through hate but through love of their comrades and leaders, the most effective training regimes had to reflect this motivation.[145]

It was precisely the individual's ability to transcend hatred and to be transported by emotions of love and empathy which facilitated extremes of violence. Killing as a response to negative emotions like hatred contained none of the power and pleasure of killing as a response to dreams of love and friendship. Hateful aggression was liable to lead to a disintegration of the personality and disorder; in contrast, individuals were ennobled by loving the men they murdered. Emotions such as love and desire could not be manipulated by the military. Unlike hatred, love possessed a transformative power, flourishing where and when it was prohibited. Refused one outlet, it found another.

Chapter 6

WAR CRIMES

tell them shove it, they're
not here, tell them kiss
my rear when they piss about
women and children in shacks
we fire on . . .
those are the Enemy.
waste them all.

Walter MacDonald, 'Interview With a Guy
Named Fawkes, U.S. Army', 1989[1]

'Rusty' Calley felt no remorse for slaughtering hundreds of old men, women and children in one day in March 1968: after all, 'what the hell *else* is war than killing people?' From the start, he just could not understand why such a fuss was being made. When he was first accused of mass murder, he was incredulous:

I couldn't understand it. I kept thinking, though. I thought, *Could it be I did something wrong?* I knew that war's wrong. Killing's wrong: I realized that. I had gone to a war, though. I had killed, but I knew *So did a million others*. I sat there, and I couldn't find the key. I pictured the people of My Lai: the bodies, and they didn't bother me. I had found, I had closed with, I had destroyed the

VC: the mission that day. I thought, *It couldn't be wrong or I'd have remorse about it.*

But Lieutenant William L. Calley was tried by a jury of six combat veterans, charged with violating Article 118, Murder, of the Uniform Code of Military Justice. At the end of March 1971, after hearing the testimony of over a hundred witnesses, he was convicted and sentenced to loss of all pay, dismissal from the army, and confinement with hard labour for life for premeditated murder.

The massacre had begun just after eight o'clock on the morning of 16 March 1968, when 105 American soldiers of Charlie Company, 11th Brigade of the Americal Division, entered the small village of Son My (known to the Americans as My Lai and thought to be the base of the 48th Viet Cong Local Forces Battalion) in the San Tinh District, Quang Ngai Province, on the north-eastern coast of South Vietnam near the South China Sea. By the time Calley and his men sat down to lunch, they had rounded up and slaughtered around 500 unarmed civilians. Within those few hours, members of Charlie Company had 'fooled around' and laughed as they sodomized and raped women, ripped vaginas open with knives, bayoneted civilians, scalped corpses and carved 'C Company' or the ace of spades on to their chests, slaughtered animals, and torched hooches. Other soldiers had wept openly as they opened fire on crowds of unresisting old men, women, children and babies. At no stage did these soldiers receive any enemy fire or encounter any form of resistance save fervent pleadings. Yet, they were 'only' obeying orders, doing their duty, and – they reasoned – even little babies could be Viet Cong ('I thought,' Paul Meadlo testified, 'they had some sort of chain or a little string they had to give a little pull and they blow us up'). After the massacre, the men of C Company burned their way through a few other villages, eventually reaching the seashore where they stripped and jumped into the surf. A year later, Private First Class Michael Bernhardt remembered that there had been

no sense of hangover in the company, no brooding over
rights and wrongs. If you had told them a year ago that
they were going to be on trial, maybe for their lives, they
wouldn't have believed you. It would have been so fantastic.

Of course, some men had been shocked by what they had done or
seen, but 'war was war' and there were other battles to fight.
However, Lieutenant Calley was very definite about his duty to
obey orders. A useful insight into Calley's attitude can be taken
from his autobiographical account of the massacre, *Body Count*
(1971). He recalled that at one stage during that bloody morning,
he came across Dennis Conti forcing a young mother to give
him oral sex. Calley ordered Conti to 'Get on your goddam
pants!', but admitted that he did not know 'why I was so damn
saintly about it. Rape: in Vietnam it's a very common thing.' He
continued:

> I guess lots of girls would rather be raped than killed any-
> time. So why was I being saintly about it? Because: if a GI
> is getting a blow job, he isn't doing his job. He isn't
> destroying communism . . . Our mission in My Lai wasn't
> perverted, though. It was simply 'Go and destroy it'.
> Remember the Bible: the Amalekites? God said to Saul,
> 'Now go . . . and utterly destroy all that they have, and
> spare them not; but slay both man and woman, infant and
> suckling, ox and sheep, camel and ass. But the people took
> the spoil –' and God punished them. No difference now:
> if a GI is getting gain, he isn't doing what we are paying
> him for. He isn't combat-effective.[2]

What Calley omitted to say in his memoir was that he immedi-
ately murdered the mother and her child: he was obeying orders.
  Calley was not alone in this belief: all the participants in the
My Lai massacre claimed that they were 'only' doing what they

had been told. In the briefing prior to entering My Lai, Colonel Henderson had taunted the officers for their poor performance in earlier attacks and their lack of aggression which enabled 'men, women, or children, or other VC soldiers in the area' to escape. Men left the briefing feeling resentful and furious. William Calvin Lloyd recalled 'we knew we were supposed to kill everyone in the village' and Robert Wayne Pendleton remembered that as they cleaned their weapons the night before the attack people were 'talking about killing everything that moved. Everyone knew what we were going to do.'[3]

The person who eventually reported the atrocity had not even been present on that day, but he had heard enough to be convinced that a terrible crime had been committed. A year after the atrocity, Ron Ridenhour (a helicopter door gunner from Phoenix, Arizona) took it upon himself to ensure that the reports were taken seriously. The army instructed Lieutenant-General William R. Peers to investigate the allegations and, two years after the massacre, sixteen officers and nine enlisted men were charged with criminal offences ranging from dereliction of duty to premeditated murder. Charges against all men save 'Rusty' Calley were dropped 'for lack of evidence' or 'in the interests of justice', although several were reduced in rank and stripped of their medals. Only Calley was found guilty.

Calley was an archetypal 'ordinary executioner'. He was a short, podgy man, aged twenty-five at the time of the massacre. He had grown up on the south coast of Florida in a relatively prosperous middle-class household (his father sold heavy construction equipment). Admittedly, he was less intelligent than average, but he had found a relatively successful niche within the US Army. He told reporters that he was an 'expert on the horrors of war' and planned to make a combat movie 'so realistic and grotesque that the audience would lurch from the theatre and vomit'. As we shall see, his ideas about what was legitimate conduct in war were widely shared and his beliefs about the enemy were not aberrant. Like Meadlo, he had no doubt that even babies could be 'the enemy': 'The old men, the women, the children – the *babies* – were all VC or would be VC in about three years', he asserted, continuing, 'And inside of VC women, I guess there were a thousand little VC now.' In reality he was barely punished for slaughtering this supposedly omnipresent enemy. Although sentenced to life imprisonment, this sentence was gradually reduced through appeal to the courts and to President Richard Nixon. Two days after the sentence was passed, the president freed him to be confined in his apartment in Fort Benning. In total, he served less than three years under house arrest before being granted parole on 10 September 1975.[4]

Calley did not particularly 'enjoy' killing, there was something about the 'cold blooded' mass murder of unresisting people which disturbed even the most banal of men. Yet atrocious behaviour was a feature of combat in the two world wars, as well as in Vietnam. The horror of acts such as those carried out by the majority of men in Charlie Company that day in March 1968 resists rational analysis. The worst accounts[5] degenerate into pornographic prose flush with sadistic and voyeuristic sensation-alism, which attempt to deny that the orgasmic pleasure experienced in My Lai that day by 'double veterans' (that is, men who raped and then killed their victims) was anything other than momentary, muted and infused with self-loathing. The withdrawn pose of much legal and historical analysis is equally brutalizing: empathy is accorded to the executioners (who are 'just like us') rather than the victims. As we shall see, however, the perpetrators of atrocities are familiar figures within the martial imagination, and many of those who resisted them turn out to be implausible moral agents whose detachment could also be atrocious. In combat, moral law existed, but only as passionate commitment.

'War is atrocious'

In the twentieth century, international law, judicial decisions, military regulation and moral convention prohibited the calculating and purposeless killing of civilians and prisoners of war and (from 1944) insisted that combatants who witnessed infringements should inform their superiors. The corpus of the law of war is now immense and has been analysed in numerous texts.[6] For the purposes of this chapter, the Hague Conventions (1899 and 1907), the Nuremberg Principles (1946), and the Geneva Convention (1949) are the most important. According to the Hague Conventions, the life of an enemy combatant who had 'laid down his arms, or having no longer means of defence' was to be spared. Every prisoner of war 'must be humanly treated', they decreed. The sixth Nuremberg Principle defined a war crime as including murder and the ill-treatment of civilian populations or prisoners of war. The Geneva Convention of 1949 also insisted that 'persons taking no part in the hostilities, including members of the armed forces who have laid down their arms and those placed *hors de combat* by sickness, wounds, detention, or any other cause, shall in all circumstances be treated humanly' and it specifically forbade 'violence to life and person, in particular murder of all kinds, mutilation, cruel treatment, and torture'.[7]

In addition to international law, the armed forces had their own regulations. All armed forces during the two world wars prohibited the gratuitous slaughter of civilians and unarmed or wounded personnel. The situation in Vietnam was more complex. Prior to 3 March 1966, the US Military Assistance Command in Vietnam (MACV) which was responsible for the command, control and support of US personnel in Vietnam, had only published war crime directives designated to apply to violations of Geneva Conventions inflicted by forces *against* the Americans. In 1966, however, MACV Directive 20–4 was published to include war crimes committed *by* American personnel. It unequivocally stated that the

wilful killing, torture, or inhuman treatment of, or wilful causing great suffering or serious injury to the body or health of persons taking no active part in the hostilities,

including members of the armed forces who had laid
down their arms or who were not combatants because of
sickness, wounds, or any other cause, was a war crime.

In addition, the maltreatment of dead bodies, firing on localities
which were undefended and without military significance, and
plunder, were defined as war crimes and it became incumbent
upon all military personnel who had knowledge that a crime had
been committed to report it to his commanding officer 'as soon
as practicable'.[8]

The most difficult issue to be resolved, however, was not what
constituted a major war crime but who was responsible.[9] The
plea of *respondeat superior* ('just obeying orders') was commonplace: but was it valid? In Britain, the military code of 1749 had
decided that troops were only compelled to obey lawful orders.
However, the first edition of Lassa Oppenheim's *International
Law* (1906) stated that 'in case members of forces commit violations ordered by their commanders, the members cannot be
punished, for the commanders alone are responsible' and paragraph 433 of the 1914 edition of the *Manual of Military Law*
required combatants to give absolute obedience to *all* commands
issued by superior officers. This remained unchanged until 1944
when the idea of *lawful* orders once again became mandatory,
making individual combatants liable for actions which violated
'unchallenged rules of warfare' and outraged 'the general sentiment of humanity'.[10] In America, the military code did not refer
to the issue of superior orders until the 1914 edition of the *Rules
of Land Warfare* which granted immunity to individuals within
the armed forces who broke the laws of war under orders from
their government or commanders. Again, this decision was
reversed in 1944 when a new Section 345.1 declared that individual combatants were liable, although the fact that a particular
action had been carried out under orders could be 'taken into
consideration in determining capability.'[11] The US Army Field
Manual of 1956 agreed that the defence of superior orders could

never be valid unless the accused individual 'did not know and could not reasonably have been expected to know that the act was unlawful.'[12] The US *Manual for Courts Martial* (1969) similarly commented that homicide 'committed in the proper performance of a legal duty' was justifiable, but not when the acts were 'manifestly beyond the scope of his authority, or the order is such that a man of ordinary sense and understanding would know it to be illegal'.[13] On an international level, the Nuremberg Principles (1946) decreed that 'any person who commits an act which constitutes a crime under international law is responsible thereafter and liable to punishment' and 'the fact that a person acted pursuant to order of his Government or of a superior does not relieve him from responsibility under international law, provided a moral choice was in fact possible to him.'[14]

Despite the weight of law, regulation and convention, atrocious behaviour in war flourished. While the conflict in Vietnam cannot be characterized as a ceaseless orgy of butchery, and while many men who had been in Vietnam (particularly those serving in the standing armies and in the Australian forces) denied that combat there was unusually vicious,[15] what happened at My Lai was neither a unique nor an isolated incident. Numerous servicemen (both American and Australian) readily admitted to excessive savagery. Rifleman Barry Kavanagh was sent to Vietnam from Avondale Heights (Melbourne, Australia) and, in incredulous tones, described 'the bad days' in Vietnam when he served alongside a 'big tough newly arrived Aussie all keen to empty two magazines into a teenage boy carrying water. Sure he was probably VC, but two magazines?' Kavanagh went on to describe a night when his platoon heard 'some scuffling in the bushes', and so they opened fire. In the morning, 'we discover it's a party of schoolgirls who had been missing from a nearby village. The big Aussies shot the ones who were still alive so no one would start a nasty scandal,' he recalled.[16] Others ruefully observed a more gradual erosion of their morals. One Australian warrant officer recalled that when he first joined an ARVN unit, he 'used to raise a bit of a stink' every time an officer killed a prisoner. Eventually, however, he changed:

> Now, the officer just looks at the prisoner, and looks at me with one of those long knowing looks. I go for a walk for a few minutes and when I come back they tell me the man was shot trying to escape.[17]

Americans in Vietnam were even more relentless. 'A lot of us wiped out whole villages,' a black Special Forces paratrooper recounted, 'we were [all] afraid that we were gonna be the next ones that was gonna be court-martialled or called upon to testify against someone or against themselves.'[18] One veteran expressed his feelings in a poem called 'My Lie':

> *The possibility, of course,*
> *existed*
> *she was elemental.*
> *But I had orders.*
> *And no allowance*
> *was made*
> *for things real or beautiful.*
> *So,*
> *before she wandered off*
> *and got lost,*
> *I did*
> *what*
> *I had to do.*[19]

Rape, torture and murder were frequently reported.[20] By the late 1960s, when bitter disillusionment with the conflict was increasing, numerous Vietnam veterans 'bore witness' to atrocities in mass public rallies. In one such event – known as the 'Winter Soldier Investigation' of January and February 1971 – over one hundred veterans publicly confessed to witnessing or taking part in atrocities. Their message was clear: war crimes in Vietnam did not start (or finish) with Charlie Company in March 1968 but were common practice within other army and marine divisions as well.[21] Sociological studies attempted to quantify these statements, confirming that all men who had been engaged in 'heavy combat', around a third of those who had been involved in 'moderate combat', and 8 per cent of those who had seen 'light combat' had witnessed atrocities or had helped murder non-combatants.[22] Even senior military officials were forced to admit to a failure to enforce the regulations: a survey of 108 US Army general officers who had served in Vietnam revealed that less than one fifth believed that the rules of engagement had been 'carefully adhered to throughout the chain of command'. Almost 15 per cent claimed that they were 'not particularly considered in the day-to-day conduct of the war'.[23] Certain groups were more liable to report seeing or participating in atrocities than others. Striking differences emerge from a comparison between members of Veterans of Foreign Wars (who tended to be conservative) and men belonging to Vietnam Veterans Against the War (who were more anti-war and left-wing). Veterans of Foreign Wars were undecided about how frequently atrocities occurred, although 43 per cent claimed that the My Lai atrocity was 'an isolated incident' compared with only 7 per cent of members of the Vietnam Veterans Against the War. Eighty-six per cent of Vietnam Veterans Against the War believed that what happened at My Lai was common or 'one of many similar incidents', while only 27 per cent of the other group agreed. When asked whether they had ever witnessed the shooting of civilians, twice as many (40 per cent) of Vietnam Veterans Against the War replied 'yes'.[24]

Undeniably, the Vietnam War was especially lawless, but nevertheless one should not imagine that there was too wide a chasm between 'atrocious Vietnam' and other, 'more civilised' conflicts. There are three reasons why the Vietnam War has been singled out as unusually bloodthirsty. First, combatants who served in Vietnam were much more willing to admit to atrocities, whether boastfully or humbly. The fashionably self-conscious, psycho-analytical style of war memoirs emerging in the 1960s encouraged a more detailed, more individual, and more confessional rendering of battle stories. In reminiscences, the wanton slaughter of non-combatants during the two world wars could be treated as evidence of the beastliness of war, best not dwelt upon: during the war in Vietnam, it became an 'atrocity' committed by recognizable consciences seeking relief.[25] Secondly, the chief difference between the avid attention given to the killing of non-combatants in Vietnam compared with the disinterested mention of atrocities committed before 1945 was that the world wars could be portrayed as 'holy' or 'just' wars: in contrast, the widespread disenchantment with the conduct of the war in Vietnam rendered the suffering caused by that conflict painfully audible. Finally, many groups and institutions within society had a political and moral agenda which could be furthered by portraying the Vietnam War as particularly gruesome: for the armed forces it deflected attention from the generally atrocious nature of warfare; for veterans (particularly those active in political organizations such as the Vietnam Vets Against the War movement) the horrors they witnessed and participated in during that war shifted the blame for excessive violence from their bloodied hands to the government; for the public 'back home', who were becoming increasingly uneasy about the conduct of the war, the emphasis on a vaguely understood place called 'Vietnam' enabled them to argue that the problem was the bestial nature of a particular conflict rather than questioning their own hawk-like (or dovish) assumptions. For all commentators, 'the problem' became 'Vietnam' rather than themselves.

It is clear, however, that although Vietnam has come to represent 'the atrocious' in warfare, massacres have a long history in each of the three cultures examined in this book – particularly in the context of conflict with indigenous peoples and in the colonial empires. Even if we focus only on face-to-face atrocities (this chapter does not discuss other forms of illegitimate killing, such

as the Allied bombing campaigns of 1942–5 and the bombing of
Vietnam, although a strong case can be made that they constituted atrocities),[26] it is not surprising that during the two world
wars British, American, and Australian troops also illegally and
'in cold blood' slaughtered unarmed people. As the Vice Chief of
Staff of the US Army, 1968–73 ominously argued: 'Americans
did indeed commit war crimes in the course of the protracted
Vietnam War, but no more in proportion to the numbers of
people involved than have occurred in past wars.'[27]

The killing of prisoners had always been an important part of
military expediency. During the First World War, it was agreed
that the good soldier was the one who did not take prisoners.[28]
The magazine of the Young Citizen Volunteers, the *Incinerator*,
offered this advice:

> If a fat, juicy Hun cries 'Mercy' and speaks of his wife and
> nine children, give him the point – two inches is enough –
> and finish him. He is the kind of man to have another nine
> 'Hate' children if you let him off. So run no risks.[29]

In the words of Captain Guy Warneford Nightingale (of the
Royal Munster Fusiliers), writing to his sister on 4 May 1915
from Gallipoli: 'we took 300 prisoners and could have taken
3,000 but we preferred shooting them'.[30] There were many reasons why the killing of prisoners might be countenanced. Robert
Graves (who believed that the killing of prisoners was the main
form of atrocity perpetrated by the Allies) emphasized revenge
for the death of comrades, jealousy that the prisoner would be
given a comfortable prison camp in England, martial enthusiasm,
or (probably the most common reason) simply because combatants were too lazy or impatient to escort prisoners to safety.[31]
Fear and greed were also issues (prisoners could overpower the
guards and they consumed scarce food and water) as was sympathy (a severely wounded prisoner of war might be put out of his
misery).[32] The most powerful motive, however, was revenge. A.

Ashurt Moris described an attack in his diary entry for 16 June 1915. He noted that he was 'so excited that [he] was all trembling' and was the first out of the trench. He shot at fleeing soldiers, then wrote:

> At this point, I saw a Hun, fairly young, running down the trench, hands in air, looking terrified, yelling for mercy. I promptly shot him. It was a heavenly sight to see him fall forward. A Lincoln officer was furious with me, but the scores we owe wash out anything else.[33]

During the Second World War, there were a few highly publicized atrocities committed by Allied troops: for instance, on 14 July 1943, American troops of the 45th Infantry Division massacred around seventy Italian and German prisoners of war at Biscari (Sicily). At the trial of one of the men accused of ordering the slaughter of over forty of these prisoners, the defence lay with Lieutenant-General George S. Patton's address to the officers of the 45th Infantry Division. Prior to battle, Patton had rallied his men with the following words:

> When we land against the enemy, don't forget to hit him and hit him hard. We will bring the fight home to him. When we meet the enemy, we will kill him. We will show him no mercy. He has killed thousands of your comrades, and he must die. If you company officers in leading your men against the enemy find him shooting at you and, when you get within two hundred yards of him, and he wishes to surrender, oh no! That bastard will die! You must kill him. Stick him between the third and fourth ribs. You will tell your men that. They must have the killer instinct. Tell them to stick him. He can do no good then. Stick them in the liver. We will get the name of killers and killers are immortal.[34]

Certain officers interpreted these words as orders, and prisoners were slaughtered *en masse*.

What happened at Biscari was particularly vicious but again was not exceptional.[35] Training manuals illegally advised soldiers to employ prisoners in clearing booby-trapped houses.[36] After American troops entered Germany, they engaged in orgies of rape and murder.[37] German prisoners were shot because 'I couldn't spare the men to take them to a prisoner of war cage', confessed one British lieutenant.[38] The Australian sergeant John Henry Ewen recalled that at Bougainville, men killed prisoners 'in cold blood'. He had initially taken a 'dim view of it' but 'I do it myself now.'[39] Retired Major-General Raymond Hufft admitted that he ordered his troops to 'take no prisoners' when he led a battalion across the Rhine: 'if the Germans had won, I would have been on trial at Nuremberg instead of them', he observed drily.[40] In protest against Calley's conviction, veterans of the Second World War and the Korean War tried to surrender to police in several cities on the grounds that, 'if this man is guilty, he is guilty for the same things we did. We shot up villains under orders and killed civilians too.'[41] The wholesale slaughter of prisoners was a particularly serious problem in the Pacific theatre of war: in August 1944 when German prisoners were arriving in the United States at a rate of 50,000 a month, it was found that between December 1941 and July 1944, only 1,990 Japanese prisoners had been captured *in total*.[42] In part, this was due to the reluctance of Japanese troops to be captured (it was considered shameful), but it was also due to the propensity of Allied troops to kill anyone who attempted to surrender. As one confidential intelligence memo on 22 July 1943 noted, it was necessary to bribe troops with promises of ice-cream and three days leave before they could be persuaded to keep prisoners alive.[43]

Military complicity

Servicemen of all ranks were unperturbed by most of these acts of lawless killing. George MacDonald Fraser's response to atrocities carried out during the Second World War was unexceptional. Fraser eventually became famous for his hilarious stories about the exploits of Private McAuslan, his bestselling Flashman novels, and his scripts for *The Three Musketeers*, *The Four Musketeers*, and the James Bond film, *Octopussy*, but in 1992 he decided that he had another story to tell – a 'true' one which would 'set the record straight'. He decided to publish his memories of life in the jungles of Burma, serving in the largely Cumbrian unit of the 14th Army. He explained that when war was declared, he was just nineteen years old and an 'obedient cog' in the great war machine. He came from a family steeped in military tradition: his mother's uncle had served in the Crimea; his great-uncle (whose ring he wore throughout the campaign) had accompanied Roberts to Kandahar and was buried somewhere in Afghanistan; two of his aunts had lost lovers in the First World War, two uncles had fought in the trenches, and his father had been wounded in East Africa. Fraser prided himself on his 'killer instinct . . . the murderous impulse of the hunter' and his memoirs contained numerous descriptions of killing (the 'jolt of delight' he felt each time he hit a 'bastard'). Neither he nor his comrades felt any remorse for the killing – after all (he rationalized), the Japanese were a 'no-surrender' enemy who would not hesitate to kill them and 'Japs' were even further down the evolutionary scale than Europeans. Even as late as 1992, he unashamedly admitted that he preferred not to sit next to Japanese tourists.

At one stage in his memoirs, however, he described an Indian unit who were serving alongside him in Burma. One night, these men callously slaughtered all the Japanese POW patients. Fraser reacted to the massacre at the time in a way he later characterized as typical. He admitted that it was a war crime, but shrugged off any moral censure. After all (like Lieutenant Calley), soldiers were in the business of killing. War crimes, he claimed, 'do come in different sizes': it was right to react with revulsion and rage to Belsen but not to the slaughter of wounded prisoners in a

Burmese hospital. Of course, he acknowledged, he could have reacted differently. He could have investigated the atrocity ('like a good n.c.o.'), informed his superiors, or even written to his Member of Parliament. At the time, none of these options crossed his mind:

> I probably grimaced, remarked 'Hard buggers, those jawans', shrugged, and forgot about it. If I had made an issue of it with higher authority, I'd have been regarded as eccentric. I'd have regarded *myself* as eccentric.

He observed that the massacre simply did not matter to him. Although he admitted that the killings were 'well beyond the civilized borderline', he excused his actions thus:

> being of my generation, in the year 1945, towards the end of a war of a particularly vicious, close-quarter kind, against an enemy who wouldn't have known the Geneva Convention if it fell on him, I never gave it a second thought. And if I had, the notion of crying for redress against the perpetrators (my own comrades-in-arms, Indian soldiers who had gone the mile for us, and we for them), on behalf of a pack of Japs, would have been obnoxious, dishonourable even.

It was 'inconceivable' (his phrase) that anyone would be charged for committing war crimes – the evidence was scattered or non-existent, the perpetrators were men 'of no eminence', and they were the victors. To anyone who criticized the practice of shooting prisoners, Fraser advised: 'Get yourself to the sharp end, against an enemy like the Japanese, encounter a similar incident . . . and let me know how you get on.' Fraser's chief claim was that only combat soldiers had a right to judge men's actions in warfare. There was no 'law' higher than the knowledge and instincts possessed by the fighting man.[44]

This insistence that only men who had 'been through it' had a right to judge other combatants – and that these men would exonerate the perpetrators – was repeated time and time again. In the words of one Vietnam veteran in conversation with a comrade who had been accused of shooting prisoners, 'neither of us was really sure whether he had done it, but we thought we would probably have remembered it. The moral certainty common to both hawks and doves back home seemed to be a luxury to the footsoldiers in Vietnam.'[45] Even an Australian training memorandum, published in 1946, which denounced the killing of enemy prisoners, argued: 'how can anyone judge who has never seen his buddies mangled or been shot at himself?'[46]

Such confidence in the complicity of fellow servicemen was not misplaced. From senior officers to 'diggers', 'tommies' and 'grunts', there was widespread acceptance of killing civilians and prisoners. Indeed, it could bring a sense of pride, as in the case of Drill Sergeant Kenneth Hodges, who was one of the men who trained Charlie Company for combat. He was 'pleased' with their performance that day in March 1968 and even boasted:

> They turned out to be very good soldiers. The fact that
> they were able to go into My Lai and carry out the orders
> they had been given, I think this is a direct result of the
> good training they had.[47]

More typically, the killing of civilians and enemy prisoners was regarded as simply an additional, unavoidable, though abominable, act of war. Combat commanders regarded the laws of warfare as 'unnecessary' and 'unrealistic' restraining devices which could forestall victory.[48] Officers tended to simply 'accept' that atrocities would take place. As one colonel admitted during the First World War: 'I've seen my own men commit atrocities, and should expect to see it again. You can't stimulate and let loose the animal in man and then expect to be able to cage it up again at a moment's notice.'[49] All atrocities could be justified on the grounds that the more people that could be killed, the less threat there was of *being* killed, as Lieutenant-General George S. Patton explained to his wife on 4 April 1944: 'Some fair-haired boys are trying to say that I killed too many prisoners. Yet the same people cheer at the far greater killing of Japs. Well, the more I killed, the fewer men I lost, but they don't think of that.'[50] When queasy combatants attempted to bring atrocities to the attention of the authorities, they were ignored. For instance, Billy Conway (of the 1st Cavalry Division) found that his platoon sergeant simply 'didn't want to hear' about how three NVA nurses had been gang-raped by his unit and then had flares shoved into their vaginas and lit.[51] The Marine, Ed Treratola, also described extensive military complicity regarding atrocities. When his unit was on a long patrol, they would simply slip into a village, kidnap a woman, and gang-rape her. Depending on their mood, they either freed her afterwards or killed her. Sometimes they would do this every night and, Treratola admitted, 'the villagers complained'. 'Sometimes,' he continued,

> the brass would say, 'Well, look, cool it for a little while', you know, 'at least let it happen with little more time in between.' But we were never discouraged.[52]

Torture and 'strategic executions' were particularly common in guerrilla and counter-revolutionary work; the culture of the Special Forces explicitly presented themselves as containing the toughest, the most lawless, and the most aggressively masculine combatants; they frequently worked alone or in small cohesive groups where they were relatively independent from main command structures; and intergroup pride and intense dependency meant that any atrocities that did occur were less liable to be reported. Killing civilians would send a warning out to guerrilla troops (and people suspected of helping them), that the troops would fight to win. It was 'really a very good tactic if you stop to think about it', retorted one officer in 1969 when asked his opinion about the My Lai atrocity, 'if you scare people enough they will keep away from you . . . Aw, I'm not saying that I approve of the tactic . . . I think it's an effective tactic.'[53] It was also an 'effective tactic' in inciting men to kill. Senior officers recognized that aggressive behaviour of troops towards prisoners behind the lines encouraged the offensive spirit and that punishing offenders reduced the likelihood of subsequent aggressive behaviour when it *was* legitimate (in battle). Brutal behaviour behind the lines was essential practice for future combat. In the words of James Jones in *The Thin Red Line* (1962), it was important not to 'jeopardize the new toughness of spirit'. That spirit was 'more important than whether or not a few Japanese prisoners got kicked around, or killed'.[54] All's fair in war.

Men who actually did the fighting also recognized that the killing of prisoners and civilians might be necessary. For instance, John Eugene Crombie had left Winchester College as a schoolboy to join the Gordon Highlanders. In April 1917 (just one month before he was killed), he wrote a letter to a friend in which he described 'mopping up' a captured trench: they threw smoke bombs into the trench and since the officers refused to spare any soldiers to escort prisoners back to camp, when the Germans surrendered, they were immediately bayoneted. It was a horrifying thing to have to do, Crombie felt, although he knew that it was 'expedient from a military point of view . . . if we decide to beat the German at his own game, we can only do it by being more Prussian than the Prussian'.[55] It was widely agreed that they were just 'doing the job'.[56] The breaking of rules were widely condoned by combatants since, in battle,

who is going to debate niceties of design, degrees of ferocity then? Flame-throwers, napalm, phosphorous, crossbows, poisoned stakes, shumies – don't expect men caught in the desperate straits of war, crushed with a thousand hellish decisions, to resort to Marquess of Queensberry tactics then.[57]

Indeed, in certain contexts, combatants were actually instructed in the best way of killing prisoners. The Special Forces soldier, Donald Duncan, attended a class on 'Countermeasures to Hostile Interrogation' in which the instructor tackled the question of prisoners. On the positive side, a prisoner might provide essential information and there was a certain propaganda value in setting the prisoner free once everything had been disclosed. There was even a chance (albeit small) that the prisoner might be converted to the anti-Communist cause. However, the negative effect of taking a prisoner far outweighed these considerations. The prisoner's 'value as an informant' had to be weighed against the 'liabilities of having him on hand' (these liabilities included reduced troop mobility, the sharing of vital supplies such as food and water, and the need to assign guards). Despite the fact that once the enemy knew that he would be killed if he surrendered, he would fight more determinedly to evade capture, immediate execution was essential. Men incapable of killing prisoners were 'psychologically unsuited' to wearing the green beret and had to be reposted.[58] James Adams described similar training in the Marines. At the age of twenty-one, he was already a hardened veteran able to recite his instructor's advice:

> In reference to enemy wounded: 'If you should pass an
> enemy soldier wounded, lying on the deck, you should
> never leave any wounded that you know about laying
> around alive.' You would just, like, if you had fixed bayo-
> nets, reach down, and in the instructor's own words again,
> 'Cut his head off' or 'Pump a few rounds into him for
> good measure'.[59]

Numerous servicemen admitted that they had been told by their instructors that 'we could rape the women' and they were taught how to strip women prisoners, 'spread them open', and 'drive pointed sticks or bayonets into their vaginas' afterwards.[60]

Some combatants were guilt-ridden for participating in

atrocities: others found them enjoyable. Jimmy Roberson was one man who was ashamed of massacring civilians. Roberson was nineteen years old when he decided to leave his comfortable life in Washington, DC, by enlisting in the army. He ended up in Vietnam and, one day, his superior officers decided to raze a village, killing men, women and children. A year or so later, in halting tones, he painfully recalled the shock of participating in mass murder, the nightmare of his sudden surge of empathy, and his uncontrollable, numbed obedience to the norms of the group:

> Like, you know, as far as myself, you know, I happened to
> look into somebody's eyes, a woman's eyes, and she – I
> don't know, I looked, I mean, just before we started firing,
> I mean. You know, I didn't want to. I wanted to turn
> around and walk away. It was something telling me not to
> do it. Something told me not to, you know, just turn
> around and not be part of it, but when everybody else
> started firing, I started firing.[61]

Roberson's response was not shared by everyone. For others, there was a great deal of pleasure to be got from committing an atrocity, particularly if it involved 'higher status' victims.[62] A man who participated in the gang-rape and murder of a Vietnamese 'whore' described his action as 'making love', gleefully boasting that it was the first time he had ever 'made love to a woman with his boots on'.[63] Twenty-year-old Chuck Onan recalled that troops training alongside him in the Marine Corps Special Forces 'liked the idea' of torturing, raping and killing prisoners: 'Many volunteered to go to Vietnam. The sergeants made it seem attractive – in a sick way, you know – you'll get a chance to kill and all this.' The fact that Marines were allowed to rape women was 'an inducement to encourage Marines to volunteer for Vietnam'.[64]

Because there was widespread endorsement of certain atrocities within the armed forces, the conviction of Calley outraged the military community. Senior officers bombarded him with supportive letters and (the night after the conviction) one hundred GIs paraded outside the stockade at Fort Benning, chanting: 'War is Hell! Free Calley!'[65] The national commander of the Veterans of Foreign Wars, Herbert Rainwater, was appalled, informing journalists that 'there have been My Lais in every war. Now for the first time in our history we have tried a soldier for performing his duty.'[66] Many believed that the higher command should take some responsibility for what happened: after all, the case of General Tomoyuki Yamashita (a Japanese commander during the Second World War who was executed because he knew an atrocity had occurred and failed to do anything to prevent further atrocities or to punish the perpetrators) had made the point that a commander – General William C. Westmoreland (the US Army Chief of Staff) in this case – could be held liable for war crimes committed by his troops which he did not order but which he failed to take measures to prevent.[67] Even left-wing, anti-war veterans argued that the prosecution of Calley should not have gone ahead. The Citizen's Commission of Inquiry into US War Crimes in Vietnam in 1970 represented this concern most strongly with veterans such as Michael Uhl testifying to the 'deep resentment' of ex-servicemen to this prosecution, and protesting against the idea that any soldier should be 'held responsible and prosecuted' when they had been 'compelled to conduct and carry out military policies such as search and destroy, free fire zones, no prisoners, pacification and reallocation and the like'. Blame had to be justly allocated: the 'highest levels of civilian and military leadership' should be indicted, not lowly combatants.[68]

Civilian endorsement

American and Australian civilians – safely quartered thousands of miles from the battle zones – tended to respond to news of the slaughter of 'enemy' non-combatants in two ways: denial or resignation. When the My Lai atrocity hit the headlines, many

people even refused to countenance that there had been a massacre. 'Our boys wouldn't do this. Something else is behind it,'
declared one man, or, in the words of the governor of Alabama, 'I
can't believe an American serviceman would purposely shoot any
civilian . . . Any atrocities in this war were caused by the
Communists.'[69] Others accepted the news with a show of moral
neutrality. A *Time* poll in 1970 found that two thirds of people
questioned denied being upset when they heard the grisly story
of the My Lai massacre: 'incidents such as this are bound to
happen in a war', most of them reasoned.[70] In more recent times,
historians have tended to adopt a similar stance. The Australian
historian, Kenneth Maddock, for instance, argued that

> the rules of war are like the rules of the road: any honest
> and realistic person will expect them to be broken, but
> some drivers will commit more frequent and more serious
> violations than others, and there may be other drivers who
> very rarely offend.

He added that 'more than one code can exist even within the
same country's armed forces'. For instance, Australians who
served in covert programmes like Phoenix were not

> 'driving' by the same rules as the men of the infantry bat-
> talions, and must often have acted criminally by the
> standards the latter were required to observe. Judged by
> their own standards, however, the covert operators were
> not necessarily exceeding any 'speed limit'.[71]

There was nothing in such pronouncements which indicated
awareness of the rules of warfare – whether defined in terms of
international law, military code or moral convention.

The conviction of Calley for premeditated murder, however,
provoked an entirely different response. Denial and resignation
gave way to fury. Draft boards in Arkansas, Florida, Kansas,

Michigan, Montana and Wyoming resigned in protest; flags were flown at half-mast in state capitals throughout the nation, and veterans organizations such as the American Legion and Veterans of Foreign Wars collected money to appeal against the conviction. At a revival meeting at the Columbus, Georgia, Memorial Stadium, the Revd Michael Lord proclaimed that 'there was a crucifixion 2,000 years ago of a man named Jesus Christ. I don't think we need another crucifixion of a man named Rusty Calley.' President Nixon received over 100,000 letters and telegrams within twenty-four hours of the announcement, practically all demanding that Calley be released. Hundreds of thousands of 'Free Calley' stickers were stuck on to car bumpers and a recording company in Nashville released a 45 rpm single entitled 'The Battle Hymn of Lieutenant Calley' (sung by an Alabama vocal group called 'C Company' with a disc jockey speaking the words to the background music of 'The Battle Hymn of the Republic'). The words went like this:

> *My name is William Calley*
> *I'm a soldier of this land;*
> *I've vowed to do my duty*
> *and to gain the upper hand;*
> *but they've made me out a villain,*
> *they have stamped me with a brand,*
> *As we go marching on.*[72]

The single sold over 200,000 copies on the day of its release, and reached the million mark within a week. When Calley was eventually released by presidential order into house arrest, the House of Representatives applauded. No wonder one book on Calley was entitled *The Making of a Hero*.[73]

These reactions were not out of the ordinary. Immediately after the conviction, a Gallup telephone poll reported that only 9 per cent of Americans approved of the trial while nearly 80 per cent disapproved. Of this latter group, one fifth felt that the massacre was not a crime.[74] Once passions had 'cooled off', and the press began reporting more positively on the trial, such high levels of disapproval declined although they remained high. Two months after the conviction, Herbert C. Kelman and Lee H. Lawrence interviewed just under 1,000 people about their reactions to Calley's trial and conviction. Only around one third of those interviewed approved of the trial while 58 per cent disapproved. By far the most important reason for disapproving of the trial (given by 45 per cent of those who disapproved) was: 'It is unfair to send a man to fight in Vietnam and then put him on trial for doing his duty.' Others felt that Calley was merely being used as a scapegoat for the failures of his superiors and that since many other men had acted as Calley had done, he should not be singled out.[75] Such a high degree of support for Calley was not restricted only to American society. Leon Mann surveyed a sample of 1,435 people in Sydney, Australia, and found that 66 per cent thought the soldiers who participated in the My Lai atrocity should go unpunished. The actual proportion in the entire country holding this view was likely to be higher than indicated by this survey since it was biased towards younger and more highly educated people.[76]

More importantly, the rationale that a combatant accused of war crimes was 'only obeying orders' was regarded by the vast majority of civilians as perfectly sensible. In the interviews Kelman and Lawrence put forward a hypothetical situation where soldiers in Vietnam were asked to shoot all the inhabitants of a village, including old men, women and children. Sixty-seven per cent of the respondents said that most people would follow orders and shoot while only 19 per cent said that most people would refuse to shoot. When they were asked 'What would *you* do in this situation,' slightly more than half said that they would shoot and one third said that they would refuse to shoot. Kelman and Lawrence were particularly struck by this latter response:

Since it was a hypothetical question, it would have been

easy enough for respondents to give themselves the benefit of the doubt and to say that they would refuse to shoot.
But the important point . . . is that for many people it is
not at all clear that this is the socially desirable response.

In other words, a majority of respondents felt that the desirable
response was to follow orders. These respondents were 'not necessarily admitting to moral weakness; for many of them, in fact,
this response represent[ed] what they would view as their moral
obligation'.[77]

Numerous other surveys produced the same responses. In
Australia, Leon Mann found that almost one third of respondents admitted that they would shoot civilians if ordered to do
so.[78] In America, interviews conducted by Edward Opton in
December 1969 asked people what they would have done if
ordered to line people up and kill them: only 27 per cent of men
said that they would refuse (in contrast to 74 per cent of women).
There was little difference by age, although older respondents
were slightly more willing to suggest that such orders ought to be
disobeyed.[79] In another survey, two researchers from the City
University of New York questioned forty-two randomly selected,
male undergraduates aged between seventeen and twenty-four
years of age. The questionnaire listed thirty-two situations, in
which a soldier had to decide whether or not to shoot. The students were asked to indicate on a seven-point scale the likelihood
that they would shoot if they were that soldier. Would they be
more liable to shoot if fire was coming from the house, if there
were soldiers or civilians inside the house, if they were ordered to
fire, if their comrades were firing, or if the enemy was a long way
away? They found that the decision to fire was most influenced
by whether or not they had received orders to do so.[80] These
findings were consistent with experimental investigations into
aggression. Most famously, the studies of Stanley Milgram were
able to show how easy it was to persuade people to cause severe
pain in others, particularly if *ordered* to do so by a figure of

authority (such as a Yale-based social scientist).[81] The legitimacy of committing atrocious acts if ordered to do so made sense to a large proportion of people.

Institutional weaknesses

The military establishment blamed poor leadership for what happened at My Lai. Few of the officers involved in the massacre had any significant combat experience: only two had been in combat prior to serving in Vietnam and none were Vietnam returnees. The platoon leaders were particularly ineffective. They were regarded as 'nice guys' and 'buddies' rather than leaders (indeed, the platoon officers were afraid of their men). Calley was criticized most ferociously for his patent stupidity, his fumbling insecurity, and his inability to inspire confidence in his men. The Peers Commission concluded that

> if on the day before the Son My operation only one of the leaders at platoon company, task force, or brigade level had foreseen and voiced an objection to the prospect of killing non-combatants, or had mentioned the problem of non-combatants in their preoperational orders and instructions . . . the Son My tragedy might have been averted altogether, or have been substantially limited and the operation brought under control.[82]

Incompetent leadership was not a problem unique to My Lai but applied to the Vietnam War in general. The US Army had instituted one-year tours of duty (thirteen months for the Marines) which prevented officers from gaining experience and establishing coherent relationships within units. The new officers were simply not of the same standard as in the past, the Army Chief of Staff claimed.[83] The breakdown in relationships of respect between officers and the men was most viciously enacted by the number of 'fraggings', or attempts to murder officers with grenades. In the three years after 1969, the US Defense

Department admitted to 788 fraggings and, if attempts to murder officers with other weapons (such as rifles) were added to this total, over 1,000 officers and NCOs were killed by their own men in Vietnam.[84]

In a situation where many officers were ineffectual and lacked the support of their men, atrocities were committed because men were frightened of the consequences of disobeying. Men serving in Vietnam tended to be poorly educated and immature (the average age of those who saw combat in Vietnam was nineteen compared with twenty-seven in the Second World War): they overreacted to threat.[85] Their fears of what would happen if they disobeyed a command was probably most powerful in the context of their immediate comrades: men who did not participate in massacres were regarded as judgemental and as lacking loyalty to the group. Ostracism was rightly dreaded: not only did it deny men what little comfort could be grasped in a terrifying environment, it could also be deadly. Even during the Second World War, Marines labelled 'eight balls' (that is, lacking the offensive spirit) could suffer the most severe penalty.[86] And there were lesser – yet potent – punishments. One man who participated in the rape and killing of a Vietnamese woman said that he was 'afraid of being ridiculed'. He was particularly terrified of being derided as 'queer' and 'chicken' – so he followed after his comrades in raping the woman.[87] Soldiers who had been at My Lai feared being fragged if they reported the atrocity.[88] Private First Class Michael Bernhardt had refused to take part in the massacre at My Lai but did not report it, deciding that 'it was dangerous enough just fighting the acknowledged enemy'.[89] Greg Olsen had also been there but was unprepared to talk about what had occurred to anyone in authority. He claimed to be unaware of any 'avenue to do it' yet his real concern was that 'you certainly had second thoughts about taking that kind of stand . . . You got to remember that everybody there has a gun . . . It's nice to face your accuser, but not when he's got a gun in his hands.'[90]

There was also the military to worry about: few people

imagined that they would punish a man for reporting an atrocity, but there was little doubt that military superiors would hit out harshly if an order was disobeyed. What would happen to a combatant who refused to obey an order to kill a prisoner of war? The Vietnam veteran, George Ryan, had no doubts. No matter how atrocious the order, he would never have disobeyed it. If he *had* refused to murder a wounded North Vietnamese soldier, 'I wouldn't have been court-martialled but I would have been blacklisted, put on KP back at base camp, and labeled a coward. That I couldn't take,' he explained.[91] For others, a court martial was a distinct possibility. This was what Calley feared: he knew that he could be court-martialled or shot for disobeying an order.[92] Another army sergeant recalled informing his commanding officer that a mass slaughter of Vietnamese civilians had taken place, and being threatened with prison if he did not keep quiet. As a skilled technician in artillery and radar, and extremely dependent upon the good will of his superior officers and the men under him, he decided to say no more:

> I knew there was no chance of ever having anything done and I was just jeopardizing everything and didn't want to get hassled or thrown into the brig . . . so I just forgot it and I didn't do anything more about it . . . I just tucked it away in some dead space . . . and went on functioning.[93]

Similarly, during the Second World War, a Pathfinder navigator confessed to having felt guilty about the legality of indiscriminate bombing. 'I always . . . thought of women, children, hospitals and suchlike,' he admitted:

> But to whom could you express such doubts? Raids on our cities helped to still the small voice of conscience but it worries me still to this day. Had the Germans won the war, should we or ought we to have been tried as war criminals? If we believed it morally wrong, should we have

spoken out to our squadron commanders and refused to participate? What would have been the result? Court martial! It would have needed much more courage to have spoken out on this matter than the mere fact of continuing to fly on operations.[94]

There were lesser punishments, such as demotion. When Calley's unit were failing to register a body count (because they could not find anyone to shoot), Colonel Barker warned him: 'You'd better start doing your job, Lieutenant, or I'll find someone who can.'[95] Bomber personnel who refused to bomb residential areas would be accused of 'Lacking Moral Fibre' and demoted.[96] The questioning of commands would not be countenanced: an officer-candidate who filed a report disagreeing with his commander, 'may as well resign. Or spend twenty years as a captain.'[97] Speaking out against unnecessary brutality also carried with it serious penalties, as Jeff Needle listed:

a lot of people would think I was a traitor to my country because I didn't believe in the war anymore, it meant some of the people in the company and outside the Army would hate me because they wouldn't understand why I had changed my mind, it meant I would get a dishonourable discharge, it meant I would find it hard to get a job, it meant losing the privileges of the G.I. Bill for schools and hospital care, it meant hardship on my parents.[98]

Australian rifleman Barry Kavanagh attempted to explain why he did not report the numerous atrocities committed by his comrades, by saying:

Look I'm no hero. I wasn't prepared to take on the whole army. I didn't want to be court-martialled and be stuck in some gaol up there. I used to say 'what a shit war' and wait for home-time and the housing grant.[99]

As friends of one soldier who eventually refused to carry a rifle in the combat zone told him: 'You don't push the Army around like that.'[100] In a situation where there was (in the understated tones of the Peers Commission) a 'permissive attitude towards the treatment and safeguarding of noncombatants', it would take enormous courage, especially in battle, for a soldier to refuse to obey an order on the grounds that it was unlawful.

Fears about the consequences of not following the crowd were exacerbated by widespread ignorance about the precise nature and status of the laws of war. L. C. Green fought in the Second World War and claimed that the 'ordinary soldier' was given scarcely any information about the laws of engagement. He knew that if he was taken prisoner, the Geneva Convention of 1929 required him to give his name, number and rank, but he had no idea about the rights of enemy personnel nor how they were to be treated.[101] The fact that guidelines and procedures had been published tended to be regarded as sufficient to exonerate the military command from further responsibility. As General William C. Westmoreland (the US Army Chief of Staff) argued: 'orders were clear. Every soldier had a card on how to treat an enemy in his hands.'[102] The issuing of the appropriate papers was perceived to be the end of the matter.[103]

Although the Rules of Engagement for US soldiers were republished every six months, the distribution of these rules to the lower ranks was uneven and often inadequate.[104] Many combat personnel (both officers and men in the other ranks) confessed to being uncertain about the legitimacy of particular acts of war. In a 1974 survey of over one hundred American army general officers who had served in Vietnam, 17 per cent claimed that (prior to the My Lai atrocity) the rules of engagement had been 'frequently misunderstood throughout the chain of command' while only 29 per cent said that they were 'well understood'. Most officers admitted that they merely relied on 'common sense'.[105] Members of Charlie Company admitted that, since arriving in Vietnam, they had never been admonished to spare civilians' lives or to respect Vietnamese customs and on the only occasion when soldiers had been chastised for abusing civilians (three soldiers had brutally raped a Vietnamese woman), they were only lightly punished.[106] Calley recalled that during his officer training, there was never any mention of civilians in combat zones. Only on the day they left for Vietnam were they were told to be polite and not to assault Vietnamese women.[107] He was never told that there was a distinction between legal and illegal orders but was taught that 'all orders were to be assumed legal'.[108] Even men like Ronald Haeberle, who was at My Lai on 16 March 1968 taking photographs, claimed that he had never heard of the MACV order of 1967 ordering personnel to report any war crime to the commanding officer.[109] The men involved in the My Lai massacre had undergone 'accelerated training', receiving 'only marginal' instruction in the provisions of the Geneva Conventions, rules of engagement, and the correct treatment of non-combatants.[110] Indeed, two months prior to embarkation, they had been assigned around fifty replacement personnel who neither underwent full unit training nor were securely integrated within the unit. Even fully trained soldiers would only receive one hour of training in war crime regulations, compared with hundreds of hours of instruction on how to kill efficiently and to obey orders.[111] Furthermore, this training was carried out by skilled combat veterans who followed their lectures on the rules of engagement with lurid tales which broke every one of those rules: 'cherries' lapped up such stories avidly.[112] In fact, some senior officers believed that teaching men the laws of warfare could put the whole killing enterprise in jeopardy and was the job of lawyers, not combatants.[113]

The situation was exacerbated by the fact that the rules which were supposed to be applied were inherently obscure. In Vietnam,

it was morally right to shoot an unarmed Vietnamese who was running, but wrong to shoot one who was standing or

walking; it was wrong to shoot an enemy prisoner at close
range, but right for a sniper at long range to kill an enemy
soldier who was no more able than a prisoner to defend
himself; it was wrong for infantrymen to destroy a village
with white-phosphorus grenades, but right for a fighter
pilot to drop napalm on it,

explained one confused combatant.[114]

Guerrillas

Other factors could make combat 'atrocity-prone'. The mismatch
between the fantasy and the reality of contact was never more
pronounced than in the guerrilla-type warfare conducted in the
Pacific theatre of war during the 1939–45 conflict and during the
Vietnam War. The enemy seemed to be everywhere – and
nowhere – and men hit out blindly with frustration and passion.
Combatants felt impotent: they were viciously assaulted but were
rarely able to attack in return (in Vietnam, American forces had
the initiative in only 14 per cent of engagements).[115] In Vietnam
the only 'measure' was the number of corpses at the end of the
day. After all, 'if it's dead, it's VC' chortled numerous hardened
combatants. This was the attitude the Peers Report was criticizing when it concluded that 'given the competitive nature of
command assignments and the general tendency to evaluate
command performance on the basis of tangible results', officers
in Charlie Company 'probably viewed the Son My operation as a
real opportunity to overcome their past failures (or lack of opportunity) to close effectively with and defeat a major identifiable
enemy force'.[116] When the 'minimal psychological satisfaction' of
wreaking bloody revenge upon an enemy that had slaughtered
one's comrades was denied to men, all 'meaning' was lost, and
rage mounted.[117] In the words of Philip Caputo:

our humanity rubbed off of us . . . We were fighting in the
cruelest kind of conflict, a people's war. It was no orderly

campaign, as in Europe, but a war for survival waged in a
wilderness without rules or laws; a war in which each soldier fought for his own life and the lives of the men beside
him, not caring who he killed in that personal cause or
how many or in what manner and feeling only contempt
for those who sought to impose on his savage struggle the
mincing distinctions of civilized warfare.[118]

In such conflicts, the wholesale killing of prisoners and civilians often followed what was seen as the senseless, unheroic
deaths of comrades: even the dedicated non-combatant, David E.
Wilson, confessed that he considered lashing out at prisoners
after seeing the way booby traps dismembered his comrades. He
recalled that prisoners were routinely slaughtered 'after we were
hit with booby traps' because 'the guys were frustrated and
angry . . . It seemed like there was nobody else to take it out on,
so they just killed people.'[119]

Charlie Company certainly had difficulties along these lines
immediately prior to the My Lai massacre. In the previous three
months, they had lost one quarter of their men to sniper fire,
mines and booby traps. There was a widespread feeling that their
lack of aggressiveness accounted for this death toll. As Sergeant
Gregory T. Olson put it, 'the attitude of all the men, the majority, I would say was a revengeful attitude, they all felt bad because
we lost a number of buddies prior to My Lai'. Everyone was
'psyched up'.[120]

Gooks and commies

Racism in all its forms (cultural ethnocentrism, scientific racism
and broadly-based ideas about 'national character') was a key
factor in the prevalence of atrocities in certain theatres of war.
The most vicious and widespread atrocities carried out by
British, American and Australian troops occurred in circumstances, in all three conflicts, where the enemy was considered to
be racially very different (as in the war in the Pacific between

1939–45 and in Vietnam). Prejudice lay at the very heart of the military establishment (for instance, during the Second World War, drill instructors told recruits: 'You're not going to Europe, you're going to the Pacific. Don't hesitate to fight the Japs dirty'[121] and, in the Vietnam context, Calley was originally charged with the premeditated murder of 'Oriental human beings' rather than 'human beings'), and undeniably, men who carried out atrocities had highly prejudicial views about their victims. Calley recalled that on arriving in Vietnam his main thought was, 'I'm the big American from across the sea. I'll sock it to these people here.'[122] Even Michael Bernhardt (who refused to take part in the massacre) said of his comrades at My Lai: 'A lot of those people wouldn't think of killing a man. I mean, a white man – a human so to speak.'[123] After an horrific recital of rape and murder, Sergeant Scott Camil of 1st Marine Division explained that 'it wasn't like they were humans. We were conditioned to believe that this was for the good of the nation, the good of our country, and anything we did was okay. And when you shot someone you didn't think you were shooting at a human. They were a gook or a Commie and it was okay.'[124] By classifying the Japanese or Vietnamese as inhuman, they all became fair game. Furthermore, such racism contained an element of fear, as the historian John W. Dower has pointed out in his exhaustive survey of racial attitudes in the Pacific theatre of war: Japan was the first non-white country to industrialize and become an imperial power, it was the first to claim a place among the Great Powers (at the Paris Peace Conference), the first to beat a western power at war (Russia in 1905), and the first to raise the idea of Asia for the Asians.[125] These people needed to be put in their place.

Explanations

Widespread military and civilian complacency about the killing of non-combatants, failures in military leadership, fear of punishment, ignorance of the rules of engagement, a belief in the value of unquestioned obedience, guerrilla tactics, and racism can all be blamed for the incidence atrocities and the grotesque behaviour indulged in by combatants. Taken together, it was surprising that more atrocities did not occur, yet presented singly, each factor failed to 'explain' why a particular atrocity took place. For instance, the atrocity at My Lai was probably one of the worst that occurred in Vietnam but an attempt to explain it in terms of the personal characteristics of the men in C Company has only limited usefulness. There was little to distinguish Charlie Company from other American units in Vietnam.[126] Whatever leadership problems did exist were commonplace in all wars and were often simply a way of displacing blame for uncontrolled violence from one group to another. Even Lieutenant Calley was not remarkable, despite the sneering comments made during the trial and in most subsequent analyses. He may have been an incompetent leader, but there was little to differentiate him from thousands of other officers hurriedly trained and sent prematurely to a very 'foreign' and frightening environment.

Furthermore, it is too easy to claim that soldiers were ignorant of the laws of warfare. At the time of the My Lai atrocity, the American command had issued fourteen directives concerning the prevention and reporting of war crimes, thousands of pamphlets entitled *Soldier's Handbook on the Rules of Land Warfare* had been distributed, and pocket-sized cards outlining the rules were given to every soldier entering Vietnam.[127] The US Army required that all personnel arriving in Vietnam received information cards entitled: 'The Enemy in Your Hands', 'Nine Rules', 'Code of Conduct', and 'Geneva Convention'. These cards 'stressed humanitarian treatment and respect for the Vietnamese people' and stipulated that each individual had to comply with the Geneva Conventions of 1949.[128] Although Calley was hazy about what he had learned about the Geneva Conventions, when he was asked whether he had received instructions about the taking of prisoners, his answer was immediate: 'Yes, sir. Treat them with respect, humility. Don't humiliate them. Keep them silent. Keep them separated and keep them closely guarded, sir.'[129] More to the point, men did not have to be 'taught' that it was wrong to shoot babies or people who were obviously unarmed and unresisting.

Fears of punishment were grossly exaggerated. Indeed, the extent to which violators were *not* reprimanded is disturbing. For instance, Marines who gang-raped and murdered 'a Viet Cong whore' were simply given 'a light slap on the wrist'.[130] In the case mentioned earlier when two helicopter pilots fired 2,000 .30-calibre machine-gun bullets and seventy-four rockets into two 'friendly' hamlets, the perpetrators were not court-martialled but were simply told off. One senior army lawyer defended General Ramsey's decision not to take the matter more seriously: after all, 'in practice no commander in his right mind is going to impair the efficiency of his combat unit by trying people who think they're doing the right thing . . . It's a little like the Ten Commandments – they're there, but no one pays attention to them.'[131] More commonly, men were *praised* (not punished) for excessively aggressive action. The after-action report filed by the commander of Task Force Barker immediately following the My Lai atrocity implied that it had been a great battle with 128 enemy soldiers killed in action, inspiring General William C. Westmoreland (the Army Chief of Staff) to send a message to Charlie Company hailing the fact that they had 'dealt enemy heavy blow. Congratulations to officers and men . . . for outstanding action.'[132] Private First Class Michael Bernhardt had chosen not to obey orders to shoot during the My Lai massacre, but he was not disciplined for his refusal.[133] Nor were the other resisters. The military was unwilling to punish men who committed atrocities: aware of the My Lai massacre, they covered it up and even when forced to act through public protest were reluctant to punish anyone.[134] Of the thirty individuals who were accused of criminal behaviour either during the massacre or in the cover-up, only Calley was convicted. Excluding the My Lai trials, there were only thirty-six court martials for war crimes committed by American troops between January 1965 and August 1973.[135] Thirty-eight per cent of the court martials were for assault, 28 per cent for murder or attempted murder, 17 per cent for manslaughter, 16 per cent for rape, and 1 per cent for mutilation of corpses.[136] Military priorities are strikingly exposed when Calley's light sentence is compared with the sentences handed out to servicemen taking part in peace marches (for instance, David Miller served twenty-two months in a federal prison for burning his draft card and Dr Howard Levy served two years for refusing to teach medical techniques to Green Berets because he believed that they would use such techniques to harm rather than help people).[137] Once forced to act, the military simply focused their attention on My Lai and performed token acts of repentance and punishment: as the My Lai survivors complained in 1974, their 'international celebrity status . . . diverted attention away from other atrocities' committed in Vietnam.[138]

The 'punishment' argument did not work at the level of individual interactions either. Rather than being boycotted for *not* participating in an atrocity, some combatants ostracized men who *had* 'needlessly' shot civilians.[139] The racism explanation, although crucial, also should not be exaggerated. During the Vietnam War (and in contrast to the two world wars), government and military institutions devoted relatively little propaganda against the NVA and the Viet Cong and many combatants viewed them with considerable respect (in contrast to the attitude towards members of the ARVN).[140] Racism was present in most military conflicts, even in the 'civilized' European wars. Indeed, Herbert Kelman and Lee Lawrence found that nearly half (47 per cent) of their sample justified Calley's action on the grounds that it was preferable to kill some Vietnamese civilians than to risk the lives of any American soldiers, but even more people (53 per cent) were prepared to justify killing German civilians rather than risk one American life.[141]

Non-participants

Another factor that needs to be taken into account is that a number of soldiers who were equally immersed in the military institution and the environment refused to take part in mass slaughter or the committing of atrocities. There were two distinctive responses possible for men choosing not to act atrociously: passive avoidance or active intervention. Those who simply withdrew from the scene of crime were far more typical. At My Lai, for instance, some soldiers (such as Richard Pendleton) concentrated their attention on shooting animals. James Joseph Dursi had killed a woman and child when he first arrived in My Lai but when he realized that the villagers were unarmed he refused to fire upon anyone else ('I can't! I won't!' he cried).[142] Ron Grzesik, Robert Maples and Harry Stanley also disobeyed direct orders to shoot unresisting women and children. 'We had orders,' Stanley recalled,

> but the orders we had was that we were going into an enemy village and that they were well armed. I didn't find that when I got there. And ordering me to shoot down innocent people, that's not an order – that's craziness to me, you know. And so I don't feel like I have to obey that.[143]

During other scenes of atrocity in Vietnam, combatants also found ways to avoid participating. Interrogators such as John Tuma and Harlan H. John would not work alongside interpreters using torture to extract information. Harlan John simply 'refused his next prisoner and the next. Until the Major understood he really meant it' and reassigned him.[144] The rifleman Reginald 'Malik' Edwards from Phoenix, Louisiana, confessed that he always found it 'hard . . . to just shoot people'. When his unit was ordered to destroy a village which had harboured a sniper, he 'just kind of dealt with the animals. You know, shoot the chickens' while villagers were slaughtered.[145] Soldiers ran ahead to warn villagers of impending atrocities.[146] Others would only fire over people's heads.[147]

These ways of avoiding killing unarmed people were used during the two world wars with equal effectiveness. It was often 'very easy to go in the other direction. Say you get lost, get sick, get hurt. By the time you get back to your outfit, a couple of days have gone by,' advised Robert Rasmus in the context of the Second World War.[148] In the early 1940s, the Long Range Desert Group took pride in their murderous effectiveness, but rather than obey an order from the Commanding Officer to kill seventeen Italian prisoners, Lieutenant-Colonel David Lloyd Owen drove the prisoners eighty miles into the desert and then abandoned them with sufficient food and water to enable them to reach their own lines (by which time the information they had gathered would have been of no use).[149] Many soldiers were squeamish about killing prisoners – not, perhaps, on moral grounds but because they could not bear hearing the pleas of the captured men.[150] That such avoidance behaviour was common has been attested to in the context of the First World War by the historian Tony Ashworth in *Trench Warfare, 1914–1918* (1980): 'live and let live' was sometimes as common as 'kill or be killed'.[151]

Avoidance-type behaviour was usually not inspired by deeply-held moral considerations, however. Rather, it was often a form of moral detachment or vacuousness. Richard Pendleton was one soldier at My Lai who only shot animals, yet he felt the need to defend this non-action as though it represented a form of disloyalty to the group. In his words: 'I didn't do it. But nobody saw me not doing it. So nobody got on me.' He just 'didn't feel like it' at the time.[152] Charles Connie had 'avoided' participating in the wholesale slaughter of wounded troops in a trench in Vietnam, but his response was not moral:

> I didn't shoot anybody myself. I couldn't do it. I just never
> went to the front of the line. It was ridiculous, like a game

played when you were a kid. We proceeded down the
trench single file and the lead man kept switching. When
you killed a guy, you went to the back of the line.[153]

Richard Boyle, a journalist working for a major daily paper in
Vietnam, exemplified this martial ethos of detachment:

The worst they could call a reporter was 'sob sister'; the
worst sin was to become involved, get committed, protest.
'Play it cool', I was told; 'be objective and you'll go far.' So
I was well trained as a war correspondent. I could watch a
burned infant trying to nurse from its dead mother's
breast, see young men with their faces blown away, witness
a boy deliberately gutted . . . and never protest.[154]

The moral responses of combatants who withdrew from
scenes of massacre may be examined in closer detail by looking at
one man: Private First Class Michael Bernhardt. He had come
from a relatively well-off, middle-class Catholic family and was a
lifelong Republican. He accepted war as inevitable: each generation of men were forced to prove themselves in battle, it was
simply a 'natural stage in the life cycle'. In 1967, he dropped out
of the University of Miami in the middle of his junior year, keen
to 'test his courage' under fire in Vietnam. Military life suited
him. He performed exceptionally well in training, but an administrative mistake meant that he failed to be admitted to helicopter
training, which had been his ambition. Instead, he was sent to
Charlie Company and at the age of twenty-one found himself at
My Lai. While hundreds of old men, women and children were
being massacred, he had kept his rifle slung on his shoulder,
pointing at the ground: 'I just didn't have any use for it at the
time,' he said. He did not report the massacre. When news of the
massacre eventually broke, he was working as a drill sergeant at
Fort Dix and decided to tell his side of the story. He was struck
by other people's inability to understand why he had not fired

that day. Even his family were divided in their response. Bernhardt's own moral stance was confused. He was forced to recognize that 'just about anybody' could slaughter women and children. 'Maybe this was the way wars really were,' he reflected. He admitted that it was not compassion which led to his passive protest, but a sense of how ridiculous and illogical their actions were: 'I wasn't really violently emotionally affected. I just looked around and said, "This is all screwed up,"' he recalled as he closed his eyes to the atrocity.[155]

Combatants who allowed atrocities to take place were not freed of responsibility for what occurred. Sergeant Michael McCuster recalled one time when his Marine platoon went into a village and gang-raped a woman (the last man to rape her, shot her). He recalled that their sergeant 'took no part in the raid. It was against his morals. So instead of telling his squad not to do it, because they wouldn't listen to him anyway, the sergeant went into another side of the village and just sat and stared bleakly at the ground.'[156] Similarly, a former Mormon minister who was platoon leader might 'condone the rapes. Not that he would do them, but he would just turn his head to them because who was he in a mass military policy?'[157] Equally, combatants of whatever rank who failed to protect the innocent could be seen to be acting atrociously. Terry Whitmore was a young, black Tennessee man. In one 'search and destroy' mission, he noticed a young child with her brother watching the Americans decimate her village. 'We left her under the tree,' he stammered, 'I don't know – I'm scared to say – but I think they killed that kid. I left. I don't want to see it. I knew some guy was going to come along and shoot her while I was standing there – and I couldn't stand to see that. So I left.'[158] In fact, there was a great deal that he could have done to save the young child and her brother: instead, he walked away.

Other combatants took a more interventionist stand against the killing of civilians, although such men were exceptionally rare. At My Lai, however, there was one 'active resister': Warrant

Officer Hugh C. Thompson, pilot of an observation helicopter flying over the village on 16 March 1968. He could scarcely believe his eyes when he witnessed unresisting civilians being rounded up and shot, and he quickly recognized the futility of dropping green smoke grenades on the wounded to indicate that they needed medical attention: no one was going to help them. Shortly after ten o'clock that morning, he spotted a group of women and children running towards a bunker, followed closely by heavily armed soldiers. Realizing that another massacre was about to take place, he landed his helicopter between the bunker (which the villagers had reached) and the soldiers, and went to question the lieutenant in charge. The lieutenant told him that the only way to get the civilians out of the bunker was with a hand grenade. Thompson acted passionately, and irrationally. Yelling at his two helicopter gunners to train their arms on the *American* soldiers, he walked (unarmed) into the bunker and persuaded the people inside to come out. He then transported the civilians to safety. By this stage, Thompson was furious. He persuaded his crew to return to the ditch where they had witnessed unarmed people being killed, to see if anyone was alive. After wading through the blood and gore, they found a two-year-old child trapped under the dead bodies. They flew the child to the civilian hospital in Quang Ngai City.

Thompson had done more than merely 'avoid' the atrocity: he had attempted to prevent civilians being killed. He also fulfilled the moral obligation to report what had taken place. That afternoon, Thompson filed an operational report and informed his section leader and then the commanding officer of Company B, 123rd Aviation Battalion of the atrocity (both officers felt that he was 'overdramatizing' the situation). He spoke to the division artillery chaplain who promised to publicize the atrocity through 'chaplains' channels'. Thompson was not opposed to killing *per se*: after all, on the morning of the massacre he had caught sight of a man who he thought was a VC and had pursued him in his helicopter, firing at him.[159] But atrocities could not be reconfigured as combat in his mind, and he responded with moral indignation and passion.

Defining what constituted an atrocity was difficult for everyone from international lawyers to the men behind the guns, but the intimate and gratuitous slaughter of unarmed people was commonly accepted as wrong despite high levels of complicity amongst combatants and civilians alike. Undeniably, atrocious behaviour flourished at all levels, but especially in relation to the treatment of enemy prisoners. Nevertheless, combatants distinguished the most vicious acts of slaughter from 'ordinary' killing in war and were generally unable to fashion the former acts into a pleasurable combat narrative. Furthermore, even those combatants who did not participate in atrocities themselves rarely protested against them. Moral outrage like that expressed by Hugh Thompson at My Lai seemed to carry with it an old-fashioned, sweetly quaint odour inappropriate within a pragmatic military. It was within this 'hard-nosed' military context that the morally detached combatant who merely withdrew physically or psychologically from grotesque scenes of carnage could come to epitomize the 'honourable' resister in war. In contrast, combatants who actively intervened to stop the carnage were profoundly disquieting, exposing as they did the possibility of humanity in hellish circumstances, and reminding people that although we have a language for murder and for detachment, we lack words to encapsulate either the passions of the active resister or the pain of the victims.

Chapter 7

THE BURDEN OF GUILT

Merry it was to laugh there –
Where death becomes absurd and life absurder.
For power was on us as we slashed bones bare
Not to feel sickness or remorse of murder.

Wilfred Owen, 'Apologia Pro Poemate Meo',
1917[1]

Sergeant Bruce F. Anello – 'Buddy' to his friends – was a 'combat effective' soldier in the US Army serving in Vietnam. He was awarded the Purple Heart, the Silver Star, the Military Merit Medal, a medal from the Army of South Vietnam, and a marksmanship citation. He was a bit of a rebel, too: sloppily dressed with his girlfriend's red suspenders hanging out of his pockets and a huge colourful pawn proclaiming 'peace' painted on the back of his flak jacket, in honour of his favourite record, Bob Dylan's 'A Pawn in the Game'. Anello openly admitted that he was pleased to be given the task of searching underground tunnels for hidden supplies of weapons and food since this allowed him to avoid participating in the atrocities being committed above ground. He would fervently beg his guardian angel to 'let up a little bit and cut me a break'. All escape routes were contemplated, including deliberately getting wounded.

Experience has taught me to keep my head down. It also

has taught me to hold a leg up. A shrapnel wound can get
you out of the field for two weeks. Two purple hearts
get you out of the field permanently.

He even considered running away but understood that this
would constitute deserting his comrades, rather than escaping
the army. He despised his superiors and sneered at them for
attempting to 'violate [our] minds' and for being 'kill crazy' while
risking only *other* people's lives. He dismissed his own medals as
part of some 'silly military war game': 'Are you a hero or are you
not a hero? Medals don't buy more bread or clear a conscience,'
he confessed in his diary.

It is not obvious from reading his diary that Anello was a
superb killer, highly valued by his superiors who recognized
that, despite his 'unmilitary' bearing, 'when it came to the battle
he was the best one to lead the men. He knew what he was
doing.' Anello's problem was that he was a sensitive young man.
His mother had died when he was four years old and his father
had sent him and his brothers to an orphanage. Starved of
affection, he would begin sobbing whenever he received a letter
from his girlfriend. 'It was so beautiful,' he wrote about one
letter,

I can't even express how it made me feel. A lot of words
wouldn't mean half enough of how good I feel. I was
gonna read the envelope for three days, then open it, read
the heading for the next three days, and one sentence per
3 days. It should last me until the next letter.

Combat was changing him, however; he was becoming 'kind of
hard . . . Screaming on little kids.' He pitied and defended a
Vietnamese woman who was having her rice confiscated, but
admitted that he was often so tired that he did not care whom he
shot, particularly if it could be done anonymously. As his buddy
remembered, they both thought that the

grenade launcher was good because . . . even though people are getting killed by what you're firing it's not a direct thing. It's not a bullet per bullet. You didn't have to put your eye on a particular person and shoot him and kill him. You pulled the trigger and the grenade went out there and blew up. If it hit someone they got hurt or got killed. He liked that idea – the fact that the grenade launcher didn't kill somebody directly.

The other problem Anello faced was that he could see no reason for the war, and openly admired the Viet Cong: 'They have a cause to die for, it's their country. We have nothing to gain. We don't even want the country,' he observed. In the immediate aftermath of one bloody encounter, he described himself, in a poem, as a 'fool/ . . . caught in/ the eyes of destruction'. He tried to 'tell of the feelings I felt/ in one day of hell' but could not:

> *The rest remains in terrified eyes*
> *and in the hollows*
> *of each man's mind.*

'Too much blood,' he began another poem, ending it with the exclamation: 'I kill for no reason!'

As these poems suggest, Anello was consumed with guilt for all the bloodshed. He describes the grotesque pranks played upon corpses, the rapes, and the way platoons were 'willing to kill any body' simply in order to beat another platoon's 'kill record'. It was all 'grossness, ridiculous and senseless killing. And no conscience whatsoever.' Even worse, he found himself acting in similar ways (see illustration 13). In his diary entry for 2 May 1968, he described what happened one day when he had been 'walking on point':

Saw a man about 20 years old, so I yelled, 'La day' (meaning come here). He turned and saw me. His eyes went

big – and he tore off running – so I shot him. He ran a
hundred yards down some trails with his guts in his
hands. The thought of what I did made me sick . . . I'm
not proud of what I did.

The next day he struggled (and failed) to legitimize this murder
in his diary:

I am what I am or what's left is what I am, and what I am
is strung up – wound up, beat up, put up, – Ahh! Shut up.
What I really am is nervous due to my leave coming up in
a few days.

Fears of retribution were consequent upon this guilt. Time and
again, Anello dreamed of walking through elephant grass up a
hill by himself, and he kept walking and walking and walking, but
just could not walk out of Vietnam. Another time, he dreamed
that he was standing on top of a hill with a Viet Cong soldier.
Both were 'looking out over the land, looking at the other hills
and the blue sky and stuff', whereupon Anello said to the Viet
Cong, 'You have a really beautiful country here' and the
Viet Cong said, 'Well, thank you. I think it's pretty nice, too.'
Then the two soldiers turned and looked at each other, suddenly
realizing that they were enemies. In the dream, Anello said, 'We
shot each other because we were enemies.' A few days later,
climbing a hill in search of a grenade factory, Anello was killed.
He had lasted only seven months in Vietnam and was not yet
twenty-one years old.[2]

For combat servicemen like Bruce Anello, the impossibility of
reconciling bloody warfare with sincere moral beliefs was
ruinous. Yet it was precisely the struggle to construct a moral
self through the application of multiple and contradictory laws
that enabled men to continue fighting. The commandment
'Thou Shalt Not Kill', with its semblance of resounding

finality, was in fact the great law of slaughter, the ultimate prohibition which inflamed desire. Religious maxims and legal precepts lent their voice to a chorus of customary regulations orchestrating laws of morality and were central in moulding effective combat performance. During wartime, certain acts of killing were forbidden; yet it was only by designating a forbidden zone that killing could be countenanced and ultimately condoned.

Individual guilt

Yet what part does individual guilt play in military history? The subjective sense that a particular act of extreme violence was reprehensible has been silenced within modern military storytelling. At the end of the slaughter, public pleas for national atonement were commonplace while the search for personal forgiveness had to be carried out in dank chapels and darkened bedrooms. Historians have been reluctant to mention issues of individual responsibility in war. Some even argued that it would be 'very dubious' to question the 'morality of the individuals' who carried out actions of mass destruction in war.[3] Such scruples would be entirely correct, were it not for the fact (as we shall see in this chapter) that combatants themselves constantly raised issues of personal responsibility. Indeed, they insisted upon it.

In contrast, military spokesmen have been outspoken about the issue of responsibility and guilt. Except in the case of atrocities, individual grunts, they declared, shared no responsibility for killing if they were obeying legitimate orders. The military further maintained that guilt should never be attached to appropriate killing. Many even went so far as to refuse to believe that combatants ever *did* feel remorse.[4] Robert William MacKenna, a medical officer during the First World War, asked his fellow doctors if they had met soldiers who had been 'perturbed in the slightest degree' by the knowledge that they had killed. All replied in the negative, as did padres when asked whether any soldiers had confessed to possessing a guilty conscience due to the bloodshed.[5] The American psychologist, Charles Bird, asserted that during an attack the soldier ceased being governed by the ethical code accepted by civilians: 'He does not question the morality of his acts, he never thinks of it.'[6] When guilt was acknowledged, little weight was given to killing as the precipitating factor. 'Survivor's guilt', or guilt for having lived when one's comrades had been killed, was frequently discussed; 'killer's guilt', or remorse for having killed, was subdued. For instance, in Irving N. Berlin's article, 'Guilt as an Etiologic Factor in War Neuroses' (published in *The Journal of Nervous and Mental Disease*, 1950), it was admitted that guilt was an important factor in war neuroses, yet Berlin only mentioned remorse as a reaction to the death of a buddy (survivor's guilt) rather than as a response to being the avenging angel.[7] The emphasis on survivor's guilt was stated most strongly by Major Jules V. Coleman, writing in 1946. He asserted that guilt could most frequently be observed in men who narrowly escaped sharing a death which took the life of a comrade. In contrast, there was 'little guilt associated with killing the enemy', he concluded.[8]

Admittedly, guilt was generally not felt in the heat of battle. Many soldiers experienced a kind of separation from the self – including the moral self – during battle. 'I am not feeling at all like myself,' mused Private John Doran in a letter to his mother on 3 July 1916, immediately after his first bloody encounter.[9] Ivone Kirkpatrick went into greater detail, recalling how his

body and soul seemed to be entirely divorced, even to the extent that I felt that I no longer inhabited my body. My shell at the bidding of purely automatic forces, over which I had no control, ran hither and thither collecting men, hacking its way through the scrub with a rifle, and directed the fire of my platoon and in short struggled with all the duties which I had been taught to perform. But my mind was a distinct and separate entity. I seemed to hover at some height above my own body and to observe its doings and the doings of others with a sort of detached interest.[10]

The mind, according to others, became blank without memory or self-consciousness during battle.[11] Furthermore, certain forms of killing were less liable to incur guilt. This was particularly the case in the most anonymous forms of modern warfare. In aerial warfare, for example, there was a strong correlation between altitude and guilt, with B-52 pilots and crews less liable to experience remorse than men on fighter-bomber missions who, in turn, were less guilt-ridden than men flying helicopter gunships where the victims were clearly visible.[12] For this reason, bomber pilots and artillerymen could kill 'untold numbers of terrified non-combatants' without regret.[13] Such anonymity in the slaughtering process even enabled the navigator of the *Enola Gay*, that dropped the atomic bomb on Hiroshima, to deny any adverse emotional reactions: he claimed to have 'come off the mission, had a bite and a few beers, and hit the sack, and had not lost a night's sleep over the bomb in 40 years'.[14]

But feelings of guilt inspired by breaking the sixth Commandment cannot be dismissed so easily. Even long-distance killing could inspire guilt – if only in the form of feeling guilty for *not* feeling guilty. This was the sense in which the bomber pilot, Frank Elkins, experienced his sinfulness. He knew that his bombs were killing hundreds of Vietnamese civilians each raid. His anguish is apparent in a diary entry for 1 July 1966:

The deep shame that I feel is my own lack of emotional reaction. I keep reacting as though I were simply watching a movie of the whole thing. I still don't feel that I have personally killed anyone . . . Have I become so insensitive that I have to see torn limbs, the bloody ground, the stinking holes and guts in the mud, before I feel ashamed that I have destroyed numbers of my own kind?[15]

Elkins was not alone: for most men, military training followed by combat dulled, but did not eradicate, remorse over shedding blood.[16] In the words of a poem from the Boer War:

> *I killed a man at Graspan,*
> *I killed him fair in fight;*
> *And the Empire's poets and the Empire's priests*
> *Swear blind I acted right . . .*
> *But they can't stop the eyes of the man I killed*
> *From starin' into mine.*[17]

The moral reaction might be delayed, as in the case of R. H. Stewart after he bayoneted a German during the Battle of the Somme. 'It was,' he confessed,

> the first time I had to kill a man at close range and I did it with a fixed bayonet. It was not very light and he was a shadow but as I twisted the bayonet clear he squealed like a stuck pig. It was not till I was on my way back that I started to shake and I shook like a leaf on a tree for the rest of the night.[18]

An American who joined the Foreign Legion recalled bayoneting a 'young Fellow . . . as delicate as a pencil'. For months afterwards he could not sleep 'for remembering what that fellow looked like, and how my bayonet slipped into him and how he screamed when he fell'.[19] Sometimes, consciences circled guiltily around and around fears of retribution: 'no fox hole was deep enough to protect him from an avenging fate'.[20] This was the case with the Spitfire pilot 'Bogles' when he shot down a German plane. Immediately afterwards, he saw two 109s 'streaking down hell-for leather' to attack him: 'Christ!!!!' he exclaimed, 'they saw me do it! they saw me shoot him down! saw him crash, now they are going to kill me for it.'[21] Jerry Samuels had enlisted in the US Army in October 1968, determined to 'make my wife and my mother proud of me'. Instead, he found himself killing unresisting civilians and participating in the rape of women. On one occasion when the women were being killed after being gang-raped, Samuels was anxious to insist that although he had participated in the rapes, he had not helped murder the women:

> I felt like a big bolt of lightning was supposed to come out of the sky with Uncle Sam's name attached to it and strike me dead. But it didn't . . . I was hoping for some kind of reprimand, somebody to say, 'You just murdered innocent people'. But nobody did.[22]

When the perceived legality of any particular military action was breached men were even more likely to feel blameworthy. As one authority commented in 1949:

> Guilt reactions in combat are not usually severe, as long as the soldier follows the rules of warfare. Rules of war are accepted by the soldier, and group approval helps to prevent the feeling of guilt. It was in those cases where the soldier did not play the game that a sense of guilt became most disturbing. Although in many of these cases the surface picture was one of severe depression, amnesia, or a mixed neurosis, the underlying difficulty was a tormented sense of guilt.[23]

It was not surprising, then, that Vietnam veterans who admitted participating in atrocities were more liable to be consumed with remorse.[24] Men like 'Fred' were particularly susceptible. He was engaged in illegal operations in Cambodia as part of a SEAL (Sea, Air and Land Capacity) team. During these operations, he was frequently ordered to massacre women and children, but 'after each mission I would vomit for hours and beg God to forgive us for what we were doing'.[25] Guilt was often present even when the breaching of a code was unintended, as in the case of the Hawaiian soldier, John Garcia, during the Second World War, who inadvertently killed a woman and her infant. Forty years later, he confessed that the killing 'still bothers me, that hounds me. I still feel I committed murder . . . Oh, I still lose nights of sleep because of that woman I shot. I still lose lots of sleep.'[26]

The military consequences of such fits of conscience could be serious. While, as we have just seen, many military and psychological commentators denied the importance of guilt, an equal number were forced to admit that such feelings sometimes precipitated war neuroses.[27] Anxiety and guilt inhibited aggression.[28] The exercise of personal consciences could seriously threaten an entire military enterprise by weakening automatic obedience to orders, promoting pity for the 'would-be-enemy prisoner with whom the army dare not encumber itself', and taunting men during long, sleepless nights with the chant 'Thou Shalt Not Kill'.[29] At the very least, remorse was damaging to morale.[30] In some wars, guilt even resulted in waves of self-mutilation, as in Italy immediately after the end of the Second World War.[31] Understandably, the military establishment attempted to alleviate these problems. The American publication, widely distributed in both Britain and Australia, *Psychology for the Fighting Man* (1944), emphasized the need for combatants to admit to feelings of guilt. As they put it, unless combatants 'face it squarely, they may head into trouble, because killing is the main job of a combat soldier'. That committee concluded:

The cure for the anxiety that results from this kind of conflict between conscience and reason is to understand it. Once a man realizes that the feeling is natural in men brought up, as the average American is, to respect human life, this particular worry won't haunt him so much. He may have a few bad dreams, but that won't interfere with doing the job ahead, disagreeable though it may be.[32]

But acceptance was not always straightforward. Although men strove to be causal agents, decisions made by moral actors in the heat of battle were inevitably confused. Simple adherence to the legal laws of warfare was insufficient. Indeed, these rules were so contradictory, nebulous and subtle that they were often of little help to servicemen in the heat of combat. Combatants responded by developing their own 'rules of thumb' to differentiate legitimate killing in wartime from guilt-ridden murder. These rules were not identical to legal militarist formulas; in contrast, they were flexible, consolatory and contradictory. They were, however, widely applied. For combat servicemen, guilt could be allayed through recourse to five types of justifications. The first four were weak rationalizations: obedience, reciprocity and revenge, depersonalization and sportiveness. It was no coincidence that these four rationales were approved of and promoted by the military hierarchy. The last way of coping with killing was looked at askance by the military establishment but was the most important for combatants: personal responsibility. As we shall see, these categories were not mutually exclusive, and these five justifications enabled men to impose an ordered, 'sensible' narrative (rationalization) on what was inherently chaotic violence, while retaining the remorse-laden integrity of their moral selves (responsibility).

Rationalizations

Personal feelings of guilt were often alleviated by reminders that combatants were merely obeying orders, and that these orders had been handed down by a legitimate authority. The authority could be the local commander (whose 'orders is orders' might be used to justify flinging a bomb into a dugout when there was 'no time' to collect prisoners) or, less convincingly, the nation (as one Vietnam veteran admitted, he had done a few things he 'shouldn't have . . . But as I sit here I say I never did anythin' wrong other than obey my country's orders. I never did anythin' bad personally').[33] Since the authority of military law was weak in alleviating the sting of guilty consciences when compared with orders given by officers in the field who were in closer proximity than legislators in The Hague or Geneva, men formally accused of war crimes typically highlighted their devoted observance of the commands of their superiors. Men like Lieutenant William Calley during the Vietnam War, who justified orchestrating the massacre at My Lai by arguing that 'personally, I didn't kill any Vietnamese that day: I mean personally. I represented the United States of America. My country.'[34] 'Obeying orders' was an efficient way of minimizing emotional conflict and generated the 'appropriate' response: that is, murderous aggression. This was widely recognized by military instructors who insisted on instantaneous obedience to orders so that each man might be able to 'sleep like a child and awaken refreshed – to kill and fear not'.[35] The unfortunate consequence – that officers would experience more 'collective guilt' about the war than privates – was rarely commented upon.[36] By obeying orders, killing could be re-conceptualized as something other than murder.

Even in the absence of direct orders, combatants were able to legitimize their aggressive behaviour by appealing to notions of reciprocity: 'kill or be killed'. This rationale was applied, in varying degrees of intensity, from the nation to identified strangers ('women and the weak') to friends and, finally, to oneself. As the level of abstraction decreased, the legitimacy of killing increased. Thus, for soldiers on active service (as opposed to propagandists well behind the lines) the legitimacy of killing was least convincing when the threat was to the nation.[37] At one level closer to the soldier's experiences, there were identifiable strangers whose deaths required avenging. The Second World War Spitfire pilot nicknamed Bogles could not decide whether or not to fire at a German gunner who was attempting to bale out of his pilotless plane. Realizing that the gunner was stuck, he was suddenly overwhelmed with guilt – until, that is, he conjured up a vision of 'the people down below, wives, young mothers, kiddies, huddled in their shelters, waiting for the "All Clear"', so he killed the man.[38]

Retribution was even more important in alleviating feelings of remorse. Most men only became willing to take another human life after seeing their wartime companions slaughtered. Such sights gave soldiers a 'big score to settle' and so they 'snapped into action'.[39] The sniper, Thomas Ervine, explained his ability to engage in face-to-face killing in the following way:

> Someone fired a gun and hit me in the leg. I fired back at him and hit him in the face, and I could see the blood shooting out of his face, in gushes like, coming out of his cheek. He was a nice looking young man but he was a sniper, and if I hadn't got him God knows how many people he would have killed.[40]

Similarly, in the Vietnam conflict, the justice of vengeance was frequently invoked. In the words of one soldier after seeing his friends killed:

> I felt a drastic change after that . . . I really loved fucking killing, couldn't get enough. For every one that I killed I felt better. Made some of the hurt went away. Every time you lost a friend it seemed like a part of you was gone. Get one of them to compensate what they had done to me.[41]

Or, in the words of another Vietnam soldier, every time a comrade was killed, he would personally take revenge, all the time talking to the ghosts of his comrades: 'Here's one for you, baby. I'll take this motherfucker out and I'm going to cut his fucking heart out for you.'[42] What seemed like senseless and wholesale slaughter of comrades had to be rendered significant by 'getting back at the enemy, and, if one is unable to engage him in the manner desired, to create an enemy out of defenceless peasants'.[43] In this way, grief was converted into rage.

Revenge was employed most effectively when the soldier was forced to face his own mortality. As Neil J. Smelser pointed out in his study of the determinants of destructive behaviour: 'one of the most profound aspects of evil is that he who does the evil is typically convinced that evil is about to be done to him'.[44] Repeatedly, men reiterated that their choice was to slay or be slain.[45] As the dedicated killer Sydney Lockwood drily commented after knocking a German's head off: it 'was not nice but one of us had to die that night'.[46] The sniper, Victor Ricketts, agreed:

> It's not too pleasant to have a fellow human in one's sights, with such clarity as to be almost able to see the colour of his eyes, and to have the knowledge that in a matter of seconds, another life has met an untimely end. However, one had to be callous, after all it was, an eye for an eye, a tooth for a tooth.

He then added, 'anyway, it could very well have been in reverse'.[47] A similar justification was used by men fighting in Vietnam, as in the case of the twenty-year-old Australian private who described to a newspaper how he machine-gunned a Vietnamese woman: 'It was her or me . . . If I had been chivalrous I would be dead now.'[48] In fact, it was a rationale which facilitated *any* atrocity – even of the magnitude of My Lai. For instance, in the *San Francisco Chronicle* on 31 December 1969, four sergeants defended committing atrocities in Vietnam on the grounds that: 'I want to come home live, if I must kill old men, women, or children to make myself little "safer", I'll do it without hesitation.'[49]

If the notion 'him or me' could justify the most brutal acts of violence, atrocities committed by the enemy could also be used to justify particularly violent combat. Combatants were reassured that the enemy was too evil to warrant survival. During the First World War, the publication of atrocity stories became a lucrative industry. Indeed, the largest single category of propagandist books published in Britain during this conflict concerned enemy atrocities.[50] In William Le Queux's *German Atrocities. A Record of Shameless Deeds* (1914), the author summarized his case by arguing that the Hun has a 'terrible and overwhelming record which makes the very heart sick with horror, and the blood run chill'. In the 1939–45 conflict, the most notorious atrocity stories were recited over the radio by Sir Robert Vansittart. In seven broadcasts, to an audience of attentive listeners, Vansittart characterized German civilization as the paramount example of reverse evolution and portrayed the German psyche as composed of three traits: envy, self-pity and cruelty. He described the Germans as 'butcher birds' who felt no compunction about committing the most vile atrocities, on which he elaborated lavishly. Indeed, Vansittart claimed, Germans would machine-gun cows if they could not find children![51] By the Vietnam War, atrocities were more effectively presented in gruesome photographs. A typical catalogue may be found in *Vietcong Atrocities and Sabotage in South Vietnam* (1966), distributed widely by the South Vietnamese Embassy. The Viet Cong were seen kidnapping, torturing, mutilating, and murdering large numbers of South Vietnamese. They were 'ruthless men' prepared to use 'all means, no matter how brutal' to 'achieve their ends. And the people suffer'. Gory scenes were displayed on every page, including photographs of beheaded women, men hacked to death with machetes, a baby whose body was 'riddled' with submachine-gunfire; the bodies of priests and a 'dedicated doctor and great humanitarian'; breasts sliced off a nurse; the corpse of a tortured teacher; and a dead mother complete with nursing baby. Readers were told to walk through this 'gallery of horrors' and 'look carefully at the pictures, and realize what the South Vietnamese must suffer . . . The Communist Viet Cong guerrillas are at the root of the evil that besets Vietnam.'[52]

Combat soldiers agreed that atrocity stories heightened their eagerness for the fray and reduced their sense of remorse for killing.[53] The first use of gas on the Western Front – widely (although not for long) regarded as contrary to the laws of warfare – led many combatants to swear vengeance. In the words of

Harold Peat who belonged to the Canadian Contingent which was exposed to the first gas attack on the Western Front:

> we who inhaled less of the filthy noxiousness grew black with a deadlier hate. Then, with what strength we could gather, did we kill and kill and kill. More, we butchered savagely. Sharp, twisting bayonet points, clubbed guns, a knife dropped by some colored man, snatched quickly . . . buried to the haft . . . swift death . . . swift gorging of hate . . . lust of battle . . . revenge . . . madness![54]

Knowledge of the concentration camps of the Second World War had a similar impact. The existence of these camps was downplayed by the western governments, although they had been well documented by November 1942, and they did not become a powerful incentive for revenge until late in the war. By then, the thirst for revenge was unquenchable. A black soldier, Captain John Long of an American tank division, recalled liberating one of these camps:

> From this incidence on Jerry was no longer an impersonal foe. The Germans were monsters! I have never found any way to find an excuse for them or any man who would do to people what I saw when we opened the gate to that camp and two others. We had just mopped them up before but we stomped the shit out of them after the camps.[55]

The Australian war correspondent, John Bennetts, commented that the Vietnam War was 'unsporting' yet, he continued, 'sportsmanship plays little part in a war against guerillas who think nothing of killing women and children with knives or flame-throwers'.[56] Rumours that the enemy mutilated corpses and tortured captured soldiers led to feelings of 'terror' – but at the same time 'if the Vietnamese did not act like human beings, then they did not have to be treated as such. All laws of civilization were suspended. And when you shot someone you didn't think you were shooting at a human.'[57] Thus a vicious circle was established: atrocity fed atrocity and could therefore be justified.[58]

Accusations of atrocities were linked with the process of dehumanization. The dehumanizing rationale had two aspects: on the one hand, the combatant was no longer a 'civilized' person and, on the other, the enemy was not human either. It was essential that the combatant should be described as having lost control. Over and over again, battle narratives insisted that men were not 'really' killing: they had been 'taken over' and returned to their 'real selves' afterwards. Once a soldier's bayonet had been anointed with the blood of his first enemy, in the words of the student-soldier, Donald Hankey, he

> 'sees red'. The primitive 'bloody-lust', kept under all his life by the laws and principles of peaceful society, surges through his being, transforming him, maddening him with the desire to kill, kill, kill![59]

A soldier's murderous attack might be forgiven on the grounds that he 'lost his head completely'.[60] Or because his 'blood was up'.[61] A soldier who bayoneted a prisoner confessed: 'I could not help it, sir, the feeling came over me; I tried not to do it, but I had to; I killed him.'[62] Charles Alexander could excuse using his rifle and bayonet with 'fiendish joy' on the grounds that 'all primitive instincts were on top but I thought it was a glorious morning'.[63] Killing was regarded as part of the human inheritance but only exceptional circumstances resulted in this instinct overpowering the 'civilized' man.

Equally, it was important to encourage the fiction that those killed were not really human.[64] They were animals – baboons, rats, vermin, wild beasts.[65] They were simply targets on a tactical range.[66] Or just dim shadows.[67] They were a vaguely designated 'enemy' or an 'ideology'.[68] As the Australian trooper, Simon Cole, said,

When I was killing the enemy I was killing a commie . . . Oh,
maybe the first time I saw a dead North Vietnamese I
flinched a bit but after that they just became dead animals. It
was either he'd shoot me or I'd shoot him and I wasn't shooting at a person. I was shooting at a bunch of ideologies.[69]

Dehumanization worked best with an obviously 'foreign' enemy,
like the Japanese or Vietnamese, who could be characterized as
gooks, dinks or zipperheads. In the words of the Vietnam veteran
Harry O'Connor, the 'gook syndrome' was 'very real in
Vietnam . . . I've seen men bat around people, hit them on the
head with rifles, act like gods, do anything they want with human
beings'.[70]

The 'body count' of the Vietnam War formalized psychological processes of dehumanization: as one Australian gunner
enthused in a letter to his sister in December 1965, he was
hoping that he had killed some Viet Cong on the previous
evening 'because that might make our total up to about 200'.[71] It
became the main index of military prowess, promising rewards as
valuable as cold beer and furloughs. Such indexes led to the
exaggeration of body counts by as much as 100 per cent (or
higher).[72] But reliance on the body count was defended by experienced military officers and advisers such as Lieutenant-General
Julian J. Ewell and Major-General Ira A. Hunt who, in 1974,
expressed concern over criticisms that it was leading to a 'more
cold blooded approach than would otherwise have been the case'.
Determined to tackle their critics 'head on', Ewell and Hunt
argued that what they called the 'Constant Pressure Concept'
was based on the premise that the best way to defeat the enemy
and protect the South Vietnamese was to use maximum force
against 'the entire Communist system'. They denied that their
approach had led to a 'brutalization' of the conflict, claiming
instead that it had produced a higher proportion of prisoners of
war, led to fewer civilian casualties, and enabled pacification to
proceed more quickly.[73]

Notions of sportsmanship were frequently used to rationalize particular battles. This was done in two ways. First, killing *was* sport – a justification in itself – and, second, because it was a sport, it allowed for the possibility of 'fair play'. The notion of combat as a game was extremely common and was epitomized by the black Aboriginal soldier, Reg Saunders, who went to look at the body of the first man that he had deliberately shot and recalled feeling 'terribly sorry about it . . . I wish I could say, "Come on old fellow, get up and let's get on with the bloody game."'[74]

More than any other sport, as we have seen, battle was like game hunting.[75] The link between blood sports and war was neatly encapsulated in a picture published by the *Morning Post* on 13 November 1933 which showed fox-hunters reverentially observing the Silence for the Dead.[76] Sergeant J. A. Caw, a former employee of the Canadian Bank of Commerce, wrote in August 1916 about a recent fight of the 'fiendish delight' of killing: 'for excitement, man-hunting has all other kinds of hunting beat a mile'.[77] In both wars, the ideal infantryman was portrayed as a poacher.[78] Dead Japanese soldiers were laid in rows on the ground as if they were 'part of the "bag" after a mixed shoot at Pertwood'.[79] Other men compared favourably their experiences of fox-hunting in Britain with chasing Japanese soldiers.[80] In the play about the Vietnam War, *Tracers*, a black platoon leader stood on the stage and spoke directly to the audience, using both these images:

They call 'em patrols. I call them hunting parties. That's what we do, you know . . . hunt 'em, kill 'em, and count 'em. If we lose any, we count them, too. Then we call in the count and we get points. Where does it all go? I think it goes to a big computerized scoreboard, and every day the big brass go in and they look at it. They nod their heads and they say, 'Ah, very good hunting, boys.' How do I feel? It's my team against his. And a kill is just a touch-down.[81]

Such metaphors were not the sole property of foot-soldiers. In the Second World War, even tank warfare was likened to hunting. The tank was the 'quarry' to be hunted and ingenuity in stalking was crucial: 'the spectacle resembled that of an ugly old boar standing at bay with a pack of snarling hounds standing back at a safe distance, unwilling to risk a jab from the treacherous tusks', according to one Major.[82] Even at sea, a ship might be described as 'the relentless hunter speeding after its prey' or being 'dead on the tail of our quarry'.[83]

For many engaged in warfare, the game-hunting metaphor held an attraction. It ennobled fighters by linking them to traditionally upper-class activities and it allowed a certain degree of emotional distancing. Furthermore, it tied into popular ideas about human nature and warfare: it was man's instinct to kill. There was no point in feeling guilty for what was inherent in human nature.

In addition, by linking killing with sport, it allowed for the concept of 'fair play', so that, for example, the actions of British soldiers could retain an honourable element. Thus, killing on major religious holidays, such as Easter and Christmas, and even (for some) on Sunday was frowned upon.[84] The enemy had to have a 'fair chance' – as a soldier from the Boer War put it, although he may have killed 'a score', he was haunted only by one because 'this one wasn't a chance-shot home,/ From a thousand yards or more./ I fired at him when he'd got no show; /We were only a pace apart.'[85] Sleeping men could not be killed: they had to be woken up first.[86] And it was only fair to be killed by a 'like' killer – for instance, airmen had to be killed by airmen. J. B. S. Haldane told a story of a Turkish airman who had developed a flair for shooting down observation balloons. A British officer retaliated by sending up a balloon filled with gun-cotton which then blew up the airman:

For this deed he was severely reprimanded by the local officer commanding R. A. F. for unsportsmanlike conduct.

This gentleman, doubtless, felt little objection to
bombing, for example, Turkish transport columns, consisting mainly of non-combatants and animals . . . But he
objected to airmen being killed except by other airmen.[87]

Equally, airmen who fired at men on the ground were revolted by
their actions. In the words of a fighter pilot forced one day to
open fire on ground troops; it 'almost made him sick. Killing was
his business, but it was killing an opponent in the air that he
liked.'[88]

The guilt of the survivor

None of these rationalizations could eradicate guilt altogether:
they could neither withstand the weight of violence in military
conflict nor the resilience of the modern conscience. Combat
simply *was not* sporting, no matter how hard men tried to make it
fit into civilian or chivalrous codes. If warfare was like game-
hunting, as thousands of men alleged, then it was the most
unskilful, unsatisfying form of this sport. As one commentator
argued in 1919: 'What [soldiers] objected to about this war was
the particular method of killing and being killed, which made the
cause of death mechanical, and the effect usually accidental.
They are agreed that it was an unsportsmanlike war.'[89]

The justification that 'it was him or me' was equally uncon-
vincing: long-distance artillery, sniping, orders not to take
prisoners, and unequal opponents were the norm, not exceptions. Even when a combatant sincerely believed that it was 'his
life or mine', they might still be consumed with guilt – men like
Private Daniel John Sweeney who stumblingly explained to his
fiancé at the beginning of November 1916, that:

The German that I shot who died afterwards was a fine
looking man I was there when he died poor chap. I did feel
sorry but it was my life or his, he was speaking but none of
us could understand a word he said, to tell you the truth I

had a tear myself, I thought to myself perhaps he has a
Mother or Dad also a sweetheart and a lot of things like
that, I was really sorry I did it but God knows I could not
help myself.[90]

Guilt could not be eliminated by recourse to a vocabulary of 'kill
or be killed': it could merely be blunted.

Obedience to higher authorities was also fraught with difficulties: what *was* the 'appropriate authority'? In his foreword to
Lieutenant William Calley's memoir, Louis Heren attempted to
allocate blame for the My Lai massacre. He admitted that Calley
should bear some of the burden but 'the responsibility does not
end with him':

His company, battalion, brigade, and divisional commanders, and the men who sent him to officer candidate school,
are also guilty. But in spite of the moral law of individual
responsibility, the major guilt surely rests with the then
Secretary of Defence who decided that the war would be
fought with indiscriminate firepower of megaton propor-
tions and inhuman practices such as free fire zones and
body counts.[91]

Although the financial and moral investment in the atrocities-
publicizing industry was substantial, the extent and effectiveness
of such propaganda should not be exaggerated, even at the frontlines. In one survey carried out in 1943 and 1944 only 13 per cent
of American infantrymen in the Pacific and Europe had seen
Japanese or German soldiers using methods of fighting or treating prisoners which they regarded as 'dirty or inhuman', and
less than one half had even *heard* of such stories.[92]
Dehumanization worked quite well in basic training; not so well
in battle. In combat situations, where human slaughter was ubiquitous, atrocities were difficult to define and were often simply
ignored.[93] It was impossible to maintain the fiction that the

enemy was any different from oneself for very long: even in Vietnam, combatants were frequently forced to recognize that 'we were out there fighting because we were told to, and they were out there fighting because they were told to'.[94] Furthermore, when atrocities were witnessed by combatants, they did not necessarily result in more feverish aggression. The War Office might circulate a story that the bodies of British soldiers were being collected by Germans and sent to a factory for conversion into fats but, in the words of the machine-gunner, George Coppard, 'if the object of the story was to work the British troops into a state of fighting frenzy, then it was a complete and utter washout. Tommy was giving all he could, and no more was left, except his life.'[95]

Worse still, atrocity-reporting could be counter-productive. As the philosopher, William Hocking noted in 1918:

> it is never wise to make him out less than human. For anger . . . runs in the opposite direction; it personifies and attributes conscience to even inanimate things. If we dehumanize the foe we remove him from the reach of instinctive indignation.[96]

In other words, portraying the enemy as a lower species diminished the sense that the enemy should be held accountable for his actions, yet it was precisely this accountability that sustained condemnation. During the Second World War, and particularly in the war against the Japanese, the use of atrocity stories ended up being questioned by certain sections of military command on the grounds that they were making combatants frightened of combat or of having to bale out of hit aeroplanes.[97] Dehumanizing the enemy could increase levels of fear by transforming him into 'mysterious wraiths': servicemen yearned for the reassurance that their foes were 'flesh and blood' men, even if this induced feelings of remorse.[98]

As a way of reducing incapacitating feelings of guilt,

dehumanization also had its limits. As we have seen throughout this book, during combat the enemy was often humanized, only to be killed. Thus, a young soldier at My Lai came across a small child with one arm already shot off, and he immediately recognized that this child was the same age as his own sister. He wondered: 'What if a foreign army was in my country and a soldier was looking at my sister just as I'm looking at this little boy. Would that foreign soldier have the guts to kill my sister?' The answer was clear: 'If he'd have the guts, then I'd have the guts,' and he pulled the trigger.[99] Furthermore, depersonalization could strip the killing enterprise of its moral value. As J. Glenn Gray noted in *The Warriors* (1970), viewing the enemy as a beast 'lessen[ed] even the satisfaction in destruction, for there [was] not proper regard for the worth of the objects destroyed'.[100] It was the ultimate failure of the dehumanizing process in alleviating guilt and providing pleasure in killing that led to the opposite approach: the acceptance of personal responsibility. Indeed, a paradoxical effect of the common rationalizations discussed above was that by blunting the devastating *impact* of guilt, they allowed feelings of remorse to remain.

Responsibility

Attempts by senior military officers to minimize, if not eradicate, feelings of regret for killing were not appreciated by most combat soldiers who tended to regard guilt as an endorsement of their essential goodness. If killing was rendered too 'easy', men 'just couldn't take it'.[101] Victors felt more guilt than the defeated not only because they had killed but because they had been rewarded for it.[102] Paradoxically, soldiers maintained their ability to kill by stressing that they *retained* a moral faculty. The insistence that men were causal and moral agents was crucial. Combatants strongly believed that they *should* feel guilty for killing: it was precisely this emotion that made them 'human', and enabled them to return to civilian society afterwards. Men who did not feel guilt were somehow less than human, or were insane: guiltless killers were immoral.[103] In the final analysis, men had to take responsibility for their actions. Corporal Tom Michlovic, in response to Lieutenant William Calley's attempt to disclaim responsibility for the massacre at My Lai, objected to three assumptions:

> That I lost my moral responsibility by my symbolic 'step forward';
> That a soldier has no conscience, and acts only under orders;
> And that if subjected to the same circumstances, I too would have coldly slaughtered those villagers.

He then asserted:

> I, like Calley, am an individual; and if during my sojourn as an infantryman in Vietnam I lost all sense of moral perspectives, it was my fault, not the military's and certainly not my country's.[104]

Or, as another Vietnam veteran put it: 'if you accept that you did it and you accept that it is wrong, then you've got to accept some guilt too'.[105]

As we saw at the beginning of this chapter, military psychiatrists encouraged combatants to repudiate feelings of guilt and insisted that veterans' difficulties were merely 'problems in adjustment'.[106] After the Vietnam War, American jargon labelled this process 'deresponsibilizing' – that is, persuading veterans that their actions had been the result of external causes and that any bad feelings they might have about them were caused by 'survivors' guilt' or remorse about 'the expression of aggressive impulses', that is, remorse about generally acting aggressively rather than about killing.[107] In this way, there was nothing to distinguish guilt arising from awareness of forbidden aggressive impulses and that resulting from slaughtering human beings.[108]

In the words of a former military chaplain: 'if one has blown away innocent women and children but experiences no unpleasant feelings about it, then there is no issue left to resolve'.[109]

As we shall see in the next chapter, the strength of this view in military psychiatric circles was hardly surprising. Government agencies funded most psychiatric and psychological research and thus had a strong interest in drawing attention away from the irrational violence of combat and ensuring that former combatants 'adjusted' back to civilian society as quickly (and as cheaply) as possible. Even independent researchers found that they could attract funding more easily if veterans were portrayed as victims rather than as moral agents. In addition, psychological categories proved inadequate in dealing with human consciences. Freudian concepts, which dominated military psychology by the time of the Vietnam War, regarded morality as a type of social intervention, shaped by the prohibitions of childhood, and 'fundamentally alien to the individual ego'. In this way, guilt became something 'to escape rather than something to learn from, a disease rather than . . . an appropriate if painful response to the past'.[110]

From the 1960s, however, a group of psychiatrists, psychologists and social workers began responding to the emotional problems affecting Vietnam veterans by insisting on the healing potentialities of guilt. This insistence on taking responsibility for one's actions became part of the standard therapy for disturbed veterans after the Vietnam War. Therapists like Arthur Egendorf, Robert Jay Lifton, Peter Marin and Chaim F. Shatan began arguing that the 'talking cure' (whether with a therapist or in a group) was not sufficient: guilt demanded that a public demonstration was made. In the words of Shatan in 1973:

> By throwing onto the steps of Congress the medals with which they were rewarded for murder in a war they had come to abhor, the veterans symbolically shed some of their guilt. In addition to their dramatic political impact,

these demonstrations have profound therapeutic meaning. Instead of acting under orders, the vets originated actions on their own behalf to regain the control over events – over their lives – that was wrestled from them in Vietnam.[111]

Guilt, and rites of repentance, brought ritual back to slaughter: it enabled killing by embracing guilt minus its most maddening sting.

Fighting men were not merely the avenging arm of the state nor were they simply pawns in an omnipresent moral universe against which it was impossible to struggle. They created their own moral universe and, to the extent that they did so, they bear that measure of responsibility for the killing they participated in. This view was in contrast to that of the military establishment which regarded guilt as an irritating (and dangerous) inconvenience which had to be minimized, if not eradicated altogether. Admittedly, although combatants applied moral criteria, they did not do so consistently and, in combat, a terrified soldier might fail to act upon his belief of what constituted legitimate killing. However, it was the differentiation between legitimate and illegitimate killing that maintained men's sanity throughout war and helped insulate them against agonizing guilt and numbing brutality. The rules applied by the soldiers were not necessarily shared by civilians, politicians, and non-combatants, but they were crucial if actions which, in other contexts, would have been regarded with horror and repugnance, were to be perpetuated and, eventually, accepted.

Chapter 8

MEDICS AND THE MILITARY

And I –
I mow and gibber like an ape.
But what can I say, what do? –
There is no saying and no doing.

Shawn O'Leary, 'Shell Shock', 1941[1]

On 7 July 1916, Arthur Hubbard painfully set pen to paper in an attempt to explain to his mother why he was no longer in France. He had been taken from the battlefields and deposited in the East Suffolk and Ipswich Hospital suffering from shell shock. In his own words, his breakdown was related to witnessing 'a terrible sight that I shall never forget as long as I live'. He told his mother that

we had strict orders not to take prisoners, no matter if wounded my first job was when I had finished cutting some of their wire away, to empty my magazine on 3 Germans that came out of one of their deep dugouts. bleeding badly, and put them out of misery. They cried for mercy, but I had my orders, they had no feeling whatever for us poor chaps . . . it makes my head jump to think about it.

There is little to differentiate Arthur Hubbard's letters to his family from those written by hundreds of other privates around the time of the Battle of the Somme. His active military career in the 1st London Scottish regiment had lasted just three months, from May to July 1916. In his early letters, he was cheery and reassuring: 'I am with the best of fellows,' and 'we shall all return back safely together and before this year is through'. But as he moved closer to the front and to battle, the tone of his letters started to change. Rain, mud, lice, rats and 'very tedious work' frustrated him. Isaacs, a friend he had been with since the beginning of the war, started to look like 'an old man . . . it is a pity he gets so nervous'. Arthur began speaking of life at the front as 'a proper hell . . . one cannot imagine unless one was here to witness things' and a new and bitter edge crept into his letters as he imagined his family 'sitting around the table about 8.30 enjoying a good breakfast and me miles away in this miserable place which is being and has been blown to hell by the Huns'.

He was not the only man trembling under the strain: a few days before the battle, he described going to the aid of a man who had shot himself in the foot in order to avoid the anticipated slaughter. He admitted to feeling 'miserable' but confessed to his sisters that

> I don't feel inclined to tell you a pack of lies, if the truth
> was told a bit more often, I don't suppose the war would
> be on now, when you land over here, they have got you
> tight and treat you as they think.

Two days later, Hubbard went over the top. He managed to fight as far as the fourth line of trenches, but by 3.30 that afternoon practically his whole battalion had been wiped out by German artillery. He was buried alive, dug himself out, and during the subsequent retreat was almost killed by machine-gun fire. In the midst of this landscape of horror, he collapsed.

Being buried alive and witnessing the mass slaughter of his

friends (including Isaacs) clearly contributed to his breakdown, but his own guilt-ridden aggressiveness in slaughter was also given its due weight. For his anxious family, the sudden shift in the way he described the enemy must have been poignant. Whereas in his previous correspondence he spoke only of 'Huns', when describing face-to-face killing, the three prisoners pleading for their lives became 'Germans'.[2]

Private Arthur H. Hubbard's traumatic emotional response to killing was not rare. Although, as we have seen, most combat servicemen coped well with the murderous demands placed upon them, a minority of men (like Hubbard) found that killing made their 'head jump'. These sufferers, and the medical and military personnel who ministered to them, struggled to impose some logic and meaning on the disorder of psychological collapse. Applying civilian values was pointless and in the end, only the militarist ethos itself, which was intent on destroying men's sensibilities, proved resilient enough to provide an inverted sense of order and integration for fractured personalities.

Military attitudes to breakdown

The emotional strain of warfare and combat cannot be quantified. Even if it could, the data provided by medical personnel is unhelpful, and frequently avoids any kind of psychiatric analysis, preferring 'safer' diagnoses based upon organic categories.[3] Army commanders encouraged this approach out of fear that psychiatric casualties reflected badly on their leadership abilities.[4] Poor and often non-existent psychiatric training and the necessity for hurried diagnoses in wartime conditions, dampened enthusiasm for accurate record-keeping.[5] In some hospitals, such as the 71st Evacuation Hospital, Pleiku, during the Vietnam War, not even a pretence was made that 'psych. casualties' would be treated by staff with psychiatric or psychological training.[6] The statistics were unreliable (a headache might be labelled concussion, combat exhaustion, or malingering) and reflected differences in evacuation procedures, variations in climatic conditions and terrain (for instance, a sudden increase in cases of trench foot might cause wholesale eviction of psychiatric patients from the hospital), and the presence of a wound in addition to emotional collapse (in which case, the wound would take precedence).[7]

Medical personnel frequently disagreed about the nature of the illness. The extent of such confusion was revealed most dramatically in 1947 when one researcher asked fourteen military psychiatrists to re-examine 200 randomly selected psychiatric cases from a military hospital. Isidore S. Edelman found that four out of every ten patients were placed in entirely different generic classes to their original diagnosis. For 14 per cent of the patients, the psychiatrists agreed on the generic class but not the specific disorder. In only 44 per cent of cases was there any agreement about the specific disorder. The chief disagreement, common to one third of the diagnoses, was over whether the patient was suffering from a neurosis or a psychosis and, in two instances, one psychiatrist diagnosed a psychosis while another found no mental illness at all.[8] The situation was exacerbated by the fact that decisions were made with pension requirements in mind. The critical psychoanalyst, William Needles, who served in the Marine Corps during the Second World War, recalled being bewildered by the pressures placed upon him to diagnose neurotic men as 'constitutional psychopaths'. The reasons for this were many, he discovered, but 'fear about the national debt' was 'uppermost'. After all, his colleagues argued,

> why qualify a man for pension rights by attaching the label of neurosis to him when the facts indicate that his symptoms are of life-long duration? Why not designate him constitutional psychopathic state, which rules out compensation?[9]

In other words, military psychiatrists made their diagnoses with economic and administrative repercussions firmly in mind.

The nature and causes of breakdown

Despite these limitations, it is clear that in wartime large numbers of men were rendered militarily useless because of mental breakdown. Levels of collapse varied dramatically by conflict, theatre of war, and unit. During the 1942–3 campaign against the Japanese in Arakan, for instance, everyone in the 14th Indian Division was said to be a psychiatric casualty. In contrast, psychiatric breakdown amongst the Australian troops during this war was thought to be as low as 5 per cent.[10] Overall, however, 25 per cent of all discharges during the 1914–18 war and between 20 and 50 per cent during the 1939–45 war were labelled 'psychiatric casualties'. Men fighting in Korea were twice as likely to become psychiatric casualties as to be killed by enemy fire, since over 25 per cent of combatants were diagnosed as being 'severe' psychiatric casualties while only 12 per cent of combatants were killed.[11] Levels of breakdown during the Vietnam War were initially surprisingly low. Frequent rest periods, limited tours of duty, the nature of the combat (brief skirmishes followed by a lull), effective evacuation of the wounded, the absence of prolonged artillery bombardment, and the success of the psychiatric policy of 'immediacy, expectancy, simplicity, and centrality' meant that less than 2 per cent of men *in service* suffered psychiatric breakdown. In contrast to the other conflicts, though, Vietnam servicemen were much more likely to suffer after their return to civilian society. Estimates varied widely, but the National Vietnam Veterans Readjustment Study, which surveyed 829,000 veterans, found that one quarter were suffering some degree of Post-Traumatic Stress Disorder (PTSD). Fifteen per cent of male veterans and just over 8 per cent of female veterans were suffering from full-blown PTSD and another 8 to 11 per cent of male and female veterans were suffering symptoms which adversely affected their lives but did not quite fit the diagnosis of PTSD.[12] It was also found that Chicano and black American veterans suffered higher levels of stress than did their white counterparts.[13]

For each person, the suffering was unique: every headache was harboured within a separate, hidden sector of a particular unconscious; each flashback was peopled with different, recognizable landscapes and demons; each sleepless man tossed to a very personal rhythm. The type of killing that had been engaged in could dictate the nature of the turmoil. Soldiers who had bayoneted men in the face developed hysterical tics of their own facial muscles; stomach cramps seized men who had knifed their foes in the abdomen.[14] Snipers lost their sight.[15] Combatants typically suffered uncontrollable diarrhoea.[16] Terrifying nightmares of being unable to withdraw bayonets from the enemies' bodies persisted long after the slaughter.[17] The dreams might occur 'right in the middle of an ordinary conversation' when 'the face of a Boche that I have bayoneted, with its horrible gurgle and grimace, comes sharply into view', an infantry captain complained.[18] An inability to eat, or sleep, after the slaughter was common.[19] Nightmares did not always occur during the war. First World War soldiers like Rowland Luther did not suffer until after the armistice when, he admitted, he 'cracked-up' and found himself unable to eat, deliriously reliving his experiences of combat.[20] During the Second World War, a twenty-three-year-old infantryman took emotional refuge in hysteria after stabbing to death an enemy soldier with his bayonet: 'that bothered me,' he stammered, 'my father taught me never to kill'.[21] During the same conflict, William Manchester described his experience of killing a Japanese soldier. 'I sobbed', he recalled, muttering 'I'm sorry':

> Then I threw up all over myself. I recognized the half-digested C-ration beans dribbling down my front, smelled the vomit above the cordite. At the same time I noticed another odor; I had urinated in my skivvies.[22]

Even so, the examples given above were not 'typical' psychiatric casualties. Most soldiers who collapsed had never killed anyone. During the First World War, a sample of British soldiers suffering from neurosis showed that only 20 per cent had been under fire.[23] Of the British servicemen discharged on psychiatric grounds between September 1939 and June 1944, it was estimated that 'war service' was a 'cause' in only 35 per cent of cases. There were 'constitutional' and 'disease' considerations in 40 and 15 per cent of patients, respectively.[24] Similar conclusions were drawn by Australian psychiatrists. In a 1945 study published in the *Medical Journal of Australia*, 60 per cent of psychiatric casualties had never been in contact with the enemy and during the campaign in New Guinea, two thirds had not encountered combat or aerial attack.[25] This study emphasized problems which predated war, such as childhood stresses and domestic or sexual maladjustment. Furthermore, these doctors argued that amongst those psychiatric casualties who had seen battle, it was fear of dying rather than guilt over (or fear of) killing which precipitated the crisis.[26] Indeed, as we have seen throughout this book, what struck most commentators was the *ease* with which men were able to kill. As one psychiatrist observed in 1918:

> At the present time there are millions of men, previously
> sober, humdrum citizens, with no observable traits of
> recklessness or blood thirstiness in their nature, and with
> a normal interest in their own comfort and security, not
> only exposing themselves to extraordinary hazards,
> but cheerfully putting up with extreme discomforts,
> and engaged in inflicting injuries on human beings,
> without the repugnance they would have shown in per-
> forming similar operations on the bodies of dogs and
> cats.[27]

Men unable to cope with killing were an aberrant group.

Frighteningly, psychiatrists recognized that more men broke down in war because they were *not* allowed to kill than under the strain of killing. Anonymous warfare offered men 'little personal satisfaction', little scope for personalized killing. As the psychologist, John T. MacCurdy put it at the conclusion of the First World War: in past conflicts, men exposed themselves to the risk of death, but they were

> compensated [for it] by the excitement of more active
> operations, the more frequent possibility of giving some
> satisfaction in active hand to hand fighting, where they
> might feel the joy of personal prowess.

Soldiers became demoralized and 'quite incapable of developing any excitement' once they recognized that 'all individuality in the struggle was lost'. The modern soldier was pitted against anonymous agents and his aggression was also incognito. Consequently, psychologists warned, there was a risk that they might develop a sense of pity for the enemy ('which is naturally an emotion most incapacitating for a soldier').[28] In combat, human emotions could not cope with the loss of the active warrior role.

The absence of any outlet for aggressive tendencies put soldiers at risk of psychological disorders, argued numerous psychiatrists in later conflicts.[29] Major Marvin F. Greiber found that non-combatants within the military suffered the highest level of psychological breakdown because the pacific nature of their jobs brought them little satisfaction.[30] Wars which involved movement and enabled combatants to externalize aggression, like the Second World War, were less liable to cause stress than 'static' contests.[31] Indeed, combat fatigue, it was argued, was frequently due to the blocking of elemental 'fight or flight' responses. If soldiers could not fight and were not permitted to flee, they responded by becoming profoundly anxious – a crippling emotion.[32] During the Arakan campaign, a psychiatric officer reported that 'people accommodate themselves better to the more natural strain of hunting and being hunted than they do to the strain of heavy shelling and bombing'.[33] Well-adjusted soldiers expressed their anxiety and hostility by attempting to

destroy its source (the designated enemy): combatants who were
unable to express aggression towards the enemy exhibited neurotic symptoms, concluded two researchers in the final year of
the Second World War.[34] In the words of one modern observer of
military operations: 'passivity in the midst of threat induces distress'.[35]

However, in the early years of the First World War, when shell
shock was believed to be the result of a physical injury to the
nerves, physical traumas such as being buried alive or exposure to
heavy bombardment were naturally considered as plausible
explanations for 'nervous' collapse, while fear and guilt had little
role to play in the development of the neurosis. Increasingly,
though, medical officers began emphasizing psychological factors
as a sufficient cause for breakdown. Once these factors were
admitted, fear and the act of killing itself suddenly became
important. As the pre-eminent popularizer of Freudian psychology and president of the British Psycho-Analytic Association,
Ernest Jones, explained: war constituted 'an official abrogation of
civilized standards' in which men were not only allowed, but
encouraged

> to indulge in behaviour of a kind that is throughout
> abhorrent to the civilized mind . . . All sorts of previously
> forbidden and hidden impulses, cruel, sadistic, murderous
> and so on, are stirred to greater activity, and the old
> intrapsychical conflicts which, according to Freud, are the
> essential cause of all neurotic disorders, and which had
> been dealt with before by means of 'repression' of one side
> of the conflict are now reinforced, and the person is com-
> pelled to deal with them afresh under totally different
> circumstances.[36]

Consequently, the 'return to the mental attitude of civilian life'
could spark off severe psychological trauma.[37] This was particularly the case amongst infantrymen since they were more liable to

witness the outcome of their destructive impulses, unlike, for example, naval personnel.[38]

Although the act of killing was still not considered as a likely cause of emotional collapse, psychoanalytically orientated psychiatrists in particular did sometimes link the two. John T. MacCurdy, for instance, was an influential adviser in both America (where he was a lecturer in Medical Psychology in Cornell during the First World War) and in Britain (he taught psychotherapy in Cambridge, England, during the Second World War). In his *The Psychology of War* (1917), he stressed the importance of sublimation in enabling men to kill without qualms:

> The sensitive individual who cannot develop a pleasure in
> killing – to put the matter brutally – is bound to be a
> victim of a double strain, and quickly develops an uncon-
> querable hatred of the task that will soon lead to fear.
> Once fear appears, surrender or illness is the only
> escape.[39]

Other psychiatrists developed the argument along similar grounds. In 1919, Dr Edward W. Lazell explained to the Washington Society for Nervous and Mental Disease that even if a soldier did manage to convince himself that he was right to kill, his 'infantile ego-ideal' would resist such a sudden and radical about-turn and would declare war on his conscious mind. In battle, the individual was forced to acknowledge his murderous urges. The sudden removal of the taboo on aggressive and sadistic impulses, together with the inability to sublimate or repress the awareness of death, led to unbearable stress for some men.[40] Joseph D. Teicher (1953) and Gregory Zilboorg (1943) emphasized the importance of Oedipal fantasies, wherein the male child's murderous, guilt-ridden hatred for his father, as the competitor for his mother's love, was disrupted by the wartime requirement to kill. Soldiers entered combat with consciences over-burdened with guilt for a fantasized murder which antedated their military service. If their Oedipal sense of guilt was strong, they would identify with the dead and, fearful of being punished by castration or death, retreat into passivity and fantasy.[41]

Men's inability to sublimate or repress the terror of killing was thought to be substantially more likely to occur under certain conditions and amongst certain groups. Clearly, since most men were only motivated to continue fighting through loyalty to their comrades-in-arms, group morale and the 'offensive spirit' had to remain high if psychiatric casualties were to be low.[42] Sleep deprivation, general exhaustion, and ideological disillusionment were particularly prone to reduce men's tolerance to the psychological strain of killing. In certain theatres of war, sleep was a serious concern – as in Italy during 1945 when one third of riflemen averaged four hours (or less) sleep a night.[43] It was discovered that these riflemen could be active for twice the amount of time without serious psychiatric breakdown if they were given regular periods of rest away from the frontlines.[44] Patriotic disillusionment also predisposed men to emotional collapse. During the Second World War, this was an especially serious problem in the air force where the reality of combat, in contrast to romantic myths of aerial combat, proved too disillusioning for many young pilots.[45] Finally, committing, rather than merely witnessing, atrocities dramatically increased the risk of severe psychological breakdown.[46]

Two types of men, it was generally believed, were more liable to collapse in combat: cowards and 'womanish' men. Many medical officers believed that psychological breakdown was a form of cowardice.[47] Indeed, some men *did* fake madness in order to get out of combat.[48] Edward Casey was a veteran malingerer who managed to delay further service during the First World War by pretending to be shell shocked. He described how the medical officer hypnotized him:

I had to tell him every thing I remembered before the

barrage. Talking and telling him lies while he wrote every word I spoke, in a book, telling me my complaint of shattered nerves was becoming very prevalent.[49]

During the war in Vietnam, the black rifleman, Haywood T. 'The Kid' Kirkland, was equally successful in being given an honourable discharge after pretending to have been driven crazy through combat.[50]

Even though such dissemblers were exceptional, it was widely thought that diagnoses like shell shock gave 'fear a respectable name' and encouraged the 'weaklings' to malinger 'in cold blood'.[51] Regular army officers held the harshest view. As one 'old stager' told the neurologist to the 4th Army: 'if a man lets his comrades down, he ought to be shot. If he's a looney, so much the better!'[52] Medical officers were deeply conscious that, in the military, illness could be desirable. The exaggeration of neurotic traits, or the invention of them, could exempt a man from dangerous duties if not from the military service altogether.[53] In the words of one psychological consultant to the Royal Navy in 1947:

> The neuropsychiatrist is constantly confronted with the dilemma whether to keep a man at duty who is liable to prove unsatisfactory, or by invaliding such individuals to make escape too easy, thus creating a bad precedent.[54]

Attempts to distinguish the malingerer from the genuine neurotic were inconclusive and generally rested upon pragmatic considerations of manpower requirements and unit morale.[55]

Finally, psychiatrists never tired of implying that the man who collapsed under strain of combat was 'feminine'. In his classic textbook, *War Neuroses* (1918), John T. MacCurdy described the suffering of one twenty-year-old private. MacCurdy noted that although this soldier had not exhibited 'neurotic symptoms' before the war, he still 'showed a tendency to abnormality in his make-up'. The proof of this lay in the fact that he was

rather tender-hearted and never liked to see animals
killed. Socially, he was rather self-conscious, inclined to
keep to himself, and he had not been a perfectly normal,
mischievous boy, but was rather more virtuous than his
companions. He had always been shy with girls and had
never thought of getting married.[56]

In other words, 'normal' men were psychologically capable of
killing because they were tough, did not mind seeing animals
killed, were gregarious and mischievous as youths, and were
actively heterosexual.

By the Second World War, such stereotypes were even more
prevalent. 'Socially and emotionally immature soldiers' who
'shrunk from combat with almost feminine despair and indignation' were disparaged by the highly respected Madison-based
psychiatrist, Philip S. Wagner, in 1946. He contended that their
passive 'insulationism' was as selfish as Nazi egocentricity. The
words Wagner used to describe these men were harsh: they were
narcissistic 'poseurs', excessively dependent on mother figures,
and concerned only with 'self-pleasure'. Worried that such
'socially and emotionally stunted' individuals were rewarded by
being excused from combat, he recommended that they should
be immediately forced back to the battlefields and threatened
with disciplinary action if their symptoms reappeared.[57]

The role of the psychiatrists

Within a context where men were failing to perform their duty
because of weakness (whether cowardice, effeminacy or 'genuine' madness), the psychiatric profession was important from
two points of view: they were there to provide ethical legitimation for war but they were also essential in 'curing' dissenters.
The first function should not be underestimated. Psychologists
and psychiatrists did not eschew moral pronouncements: they
embodied them. This function had traditionally been fulfilled by
the clergy but many of their responsibilities were increasingly

being transferred to social scientists who consciously adopted padre-like roles and languages.[58] Military training itself became 'treatment for an unadjusted conscience' according to a textbook entitled *Psychology for the Armed Forces*, published by the National Research Council in 1944.[59] During the war in Korea, it was medical officers who provided 'spiritual guidance'.[60] By the Vietnam War, psychology and psychiatry had largely taken over the functions previously assigned to religion.[61] In the words of one co-ordinator for the Veterans' Outreach Program: 'We aren't just counselors; we're almost priests. They come to us for absolution as well as help.'[62]

From the military point of view, shifting some of the ethical responsibilities from the shoulders of padres and into the hands of data collectors and 'mind doctors' had obvious advantages. Psychologists had a far less secure base within civilian society during this period and they willingly submitted to military hierarchies. As one sociologist writing in *The Army Combat Forces Journal* (1955) complacently agreed, the information provided by social scientists to the military must always remain 'under the supervision and control of the professional soldier who will use it wisely and constructively'.[63] These social scientists were happy to adopt combat ethics. The military desperately needed these men if (in the words of the English psychologist, Dr Charles Stanley Myers) the 'air pilot's objection to bombing women and children' was to be countered and if pensions were to be refused to men suffering psychological disturbances as a result of combat (on the grounds that pensions rewarded them for their emotional inadequacies).[64] A lecture entitled 'Reactions to Killing', which was circulated by psychologists during the Second World War, provides a good example of their ethical function. In this lecture, the killing of prisoners was taken for granted; military psychologists and other officers were simply told that if men expressed reservations about killing prisoners, they were to be advised to alleviate their guilty consciences by transferring moral responsibility to a higher authority: 'obeying orders', in other words. Guilt-ridden men were to be reminded that the act of slaughtering prisoners was 'shared by the group' and was necessary to safeguard not only the individual and his comrades, but also 'civilized ideals'. Above all, any hint that killing prisoners was an 'expression of blood–lust' had to be removed.[65] For military psychologists, then, the killing of non-combatants was merely a fact of modern warfare rather than a moral problem. Whatever the practice of *religious* advisers in the armed services regarding the legitimacy of shedding human blood, they were tutored in the theological concepts of 'just war'. The social sciences possessed no similarly authoritative moral law. Indeed, in the major branches of these sciences – instinct theories and psychoanalysis, for instance – killing was regarded as an essential, inescapable part of the human psyche. Other branches – behaviourism, for instance – preached a pragmatism which was greatly favoured by the armed forces. Such assurances were not part of the religious discourse, although it is clear from reading the diaries of padres that many of them were embracing psychology and jettisoning traditional languages of repentance and forgiveness.[66]

Nevertheless the legitimation of the warring enterprise was less important than the psychiatrists' job of curing men unable to cope with battle. The sheer numbers of sufferers horrified pensioning authorities who turned to psychiatrists to effect (in the words of the British Secretary of State during the First World War) 'a very large saving to the nation in the ultimate pensions bill for ex–Service personnel'.[67]

Methods of treatment

From the start, the purpose of treatment was to restore the maximum number of men to duty as quickly as possible. During the First World War, four fifths of men who had entered hospital suffering shell shock were never able to return to military duty: it was imperative that such high levels of 'permanent ineffectives' were reduced.[68] A massive mobilization of medical personnel to advise military officers and to cure men of their afflictions thus began, so that by the Second World War almost 40 per cent of practising physicians in America had been called into active duty to care for about 8 per cent of the nation's manpower, that is, the twelve million men in uniform.[69]

During the First World War, the shift from regarding breakdown as organic to viewing it as psychological had inevitable consequences in terms of treatment. If breakdown was a 'paralysis of the nerves', then massage, rest, dietary regimes, and electrical treatment were prescribed. If a psychological source was indicated, the 'talking cure', hypnosis, and rest would speed recovery. In all instances, occupational therapy (which constituted nearly 30 per cent of treatment during the First World War)[70] and the 'inculcation of masculinity' were highly recommended. As the medical superintendent at Bootham Park, York, put it in 1920, although the medical officer must show sympathy, the patient 'must be induced to face his illness in a manly way'.[71] Men needed to be 'hardened' through manual labour.[72]

As we have seen, by the Second World War exhaustion was increasingly regarded as a primary cause of collapse, inspiring the establishment of exhaustion centres from 1942 and the formulation of policies which limited the amount of time a man spent in combat without rest periods.[73] In conjunction, techniques involving interviews and psychotherapy were widely employed and highly popular. From that time, drugs such as insulin and barbiturates were increasingly seen as helpful.[74] Narcoanalysis involved interviewing or psychoanalysing the patient while under the influence of hypnotic drugs such as sodium amytal or sodium pentothal. The patient would be told that he had been given the drug to make him sleepy and help him to remember what he had forgotten. Once the patient was tranquillized, strong suggestion to make the subject believe in his own recovery would be used. According to one account, it was successful in 95 per cent of patients and was extremely useful in distinguishing 'genuine' hysterics from malingerers.[75] 'Suggestion' could be used in more insidious ways to enable men to return to the killing fields. Thus, in one exhaustion centre, a manic soldier was retained in the hospital because his exuberant calls to be allowed to go back to the front in order to kill some more 'bastards' infected less eager, but highly suggestible, patients.[76]

Either separately, or in conjunction with drugs, electroshock therapy was also employed, especially for the treatment of psychoses, schizophrenia, and affective states. The therapy was conducted through small portable generators. In the words of Private Frederick Bratten, a conscript in the RAMC, this was 'a very unpleasant experience to undergo as no anaesthetic was given . . . Occasionally a jaw bone, or bones in the legs and arms were fractured, owing to the powerful voltage.'[77] The treatment was said to benefit not only the patient. In the words of Lieutenant-Colonel Louis L. Tureen and Major Martin Stein in 1949, it

> made possible the treatment of psychotic patients with
> reasonable safety and without elaborate physical facilities.
> It cut destruction of property and injury of patients and
> personnel to the minimum. It made evacuation of patients
> far easier. It markedly reduced the use of drugs and
> restraints.[78]

Australian psychiatrists continued to rely heavily upon prescription drugs, especially anti-depressants and tranquillizers, and psychotherapy,[79] but generally, as from the end of the Second World War, and especially during the Vietnam War, treatment could be categorized in one of three ways: immediate treatment, continued proximity to the battle (based on the belief that the further men were removed from the fighting, the less chance that they could be transformed back into combatants), and constant reassurance of rapid healing.

Attitudes of civilians and the medical professionals

It seems that sufferers of breakdown had no choice but to recognize the stigma of cowardice and acknowledge that their reputations as soldiers and men had been dealt a severe blow: sympathy was rarely forthcoming.[80] After a major bombardment or particularly bloody attack, if the combatant had acquitted himself adequately, signs of emotional 'weakness' could be overlooked,[81] but in the midst of the fray, the attitude was much less sympathetic. 'Go 'ide yerself, you bloody little coward!' cursed one Tommy at a frightened soldier during the First World War.[82] In Vietnam, a sobbing man would be kicked, shaken and beaten by fellow servicemen, terrified that cowardice would spread like a virus.[83] Psychiatric casualties could not expect much sympathy when they returned home either. Men arriving at the Netley Hospital for British servicemen suffering shell shock during the First World War were greeted with silence and people hung their heads in 'inexplicable shame'.[84] It was observed that the families and friends of men who became psychiatric casualties in the Second World War had

> little sympathy with their difficulties, since they customarily hold to the conventional view that it is the duty of all young men in wartime to be 'heroes' and not hesitate to kill or risk being killed when called upon.[85]

Like the padres observed in the next chapter, psychologists sneered at men who were overwhelmed with guilt about killing. Captain James Henry Dible wrote that neurasthenic men 'exasperated' him 'beyond endurance' and he wanted to 'kick them out' of his presence 'with a few well rubbed in words of worldly wisdom'.[86] The US Army Medical Department admitted during the Second World War that doctors tended to be unsympathetic towards neurotic soldiers, in part because the doctor was resentful of being in the army: 'If I can't get away with it, by the gods, neither will you,' they swore.[87] In the words of one Vietnam veteran struggling to explain why he did not turn for help to a psychiatrist:

> You see, I was scared to death I might be punished . . . I was really afraid that: 'Boy, if I tell . . . what happened, I'd end up in Fort Leavenworth busting rocks . . . or maybe even hung, or shot at firing squad.' I was afraid of that because . . . I had murdered somebody.[88]

Furthermore, psychiatrists and other medical officers became increasingly hostile to men who broke down under the strain of killing. During the First World War, there tended to be a slightly more understanding attitude amongst doctors. Indeed, the authors of one of the standard books on shell shock went so far as to point out that a soldier who suffered a neurosis had not lost his reason but was labouring under the weight of too much reason: his senses were 'functioning with painful efficiency'.[89] The combatant who broke down generally possessed a higher degree of 'civilization', according to instinct theorists. By the Second World War, however, the inability to act aggressively was itself regarded as a psychiatric disorder. It was believed that men who were unable to kill were 'dull and backward'. They were men who lacked the ability to understand 'complex ideas', such as 'patriotism, appreciation of the alternative to winning the war, tradition', and, having been brought up with the 'christian attitude', did not possess 'the capacity to adjust to what [was] . . . the antithesis of this attitude'.[90] Men who experienced emotional conflicts in killing were 'psychologically inadequate individuals' or were 'ineffectives' who required 'salvaging'.[91] If they broke under the strain, they were 'childish', 'narcissistic' and 'feminine'.[92] Pacifism was a 'morbid phenomenon . . . the rationalization of self-destructive wishes,' according to Franz Alexander. In an article in *The American Journal of Sociology* in 1941, he argued that anyone who is blind to the ubiquitous manifestation of human aggressiveness in the past and present can be rightly considered a man who does not face reality. If he is not of subnormal intelligence – unable to grasp events around him – his inability to face facts must be of emotional origin, and he must be considered a neurotic.[93]

Such individuals needed to be 'cured' of this affliction and forced to react to killing in 'a human, rather than an animal, way'.[94] Men had to be prepared to give not only their limbs or life for their country, but also their 'guts' and 'nerves'.[95] Psychiatrists had to consider the needs not only of the patient, but also of the 'youngster still up there on the hill, carrying on the fight'. Captain Eugene R. Hering of the US Marine Corps summed up the attitude of his medical colleagues during the Korean War:

The motto of the Navy Medical Department is 'To keep as many men at as many guns as many days as possible.' The Army states its aim more concisely: 'To conserve fighting strength' and the Air Force: 'Keep 'em Flying'. These are not hollow phrases; let us reaffirm them in our daily handling of patients under all circumstances.[96]

Justifying this attitude was not difficult. It was obvious that, in wartime, the nation had a right to demand that servicemen gave their 'nerves' for their country, as much as their limbs, eyes or lives. In contrast to civilian practice, 'subjective complaints' had to be ignored: if a diagnosis based upon 'sound pathologic and physiologic concepts' could not be made, there was no excuse for delaying returning the patient to duty immediately, an American physician advised.[97] In addition, forcing a combat-exhausted man back into the frontlines was in his own interests since, if he was evacuated, 'he would be tempted to maintain his sickness as part of a masochistic penance for having failed to return to his unit and his duty'.[98]

The group rather than the individual was paramount.[99] 'The first duty of a battalion medical officer in War is to discourage the evasion of duty,' Captain J. C. Dunn lectured, and this duty had to be done even if it resulted in the 'temporary hurt of the individual'.[100] A decade later, a similar comment was made by Lieutenant-Colonel Philip S. Wagner, professor of psychiatry at the University of Madison Medical School and consultant at the Veterans Hospital at Perry Point. He reminded his readers that military psychiatrists had only one aim and that was to determine whether an individual still possessed 'additional combat usefulness'. The psychiatrist was not to concern himself with '"cure", nor with solicitude for the psychic pain [the patient] would have to endure to serve a few more combat days, nor even with speculations on the eventual consequence to his personality'.[101] Eli Ginzberg summed it all up when he advised social scientists in war to 'temper their humanitarian approach to the individual patient with a concern for the impact of their professional decisions on the morale and efficiency of the group.[102]

The militarization of the medical corps

Even if they had wanted to, there was a limit to the extent to which medical officers could act kindly. Within the armed forces, psychologists found themselves with very little independence. Financial backing for research in psychology was firmly tied to military needs and even the influential Human Resources Research Organization was threatened with the withdrawal of funding if they deviated from narrowly 'military' aims.[103] Similarly, the Information and Education Division of the War Department (the division which employed psychologists and sociologists), restricted research to 'concrete problems' which necessitated 'specific, concrete findings which seemed to have a definite, *immediate* bearing upon administrative policy and practice'.[104] Psychologists were forbidden to comment on appropriate punishments for military offenders unless the offender was 'quite definitely' unfit to stand trial on mental grounds.[105] There was no contest between what the military sociologist Morris Janowitz called the 'enlightenment model' of research and the 'engineering model'.[106]

Furthermore, the military simply silenced anything they did not like. Like other servicemen, medical officers were subjected to disciplinary action if they were thought to be 'lax'. For instance, a Lieutenant G. N. Kirkwood, medical officer to the 11th Border Regiment 97th Infantry Battalion during the Great War, who supported the men in his battalion when they claimed that they were too emotionally and physically exhausted to carry out a raid, was rebuked, disgraced and dismissed from the service. In the words of the commander of the 32nd Division:

> Sympathy for sick and wounded men under his treatment
> is a good attribute for a doctor but it is not for an M.O.
> [medical officer] to inform a C.O. [commanding officer]
> that his men are not in a fit state to carry out a military
> operation. The men being in the front line should be
> proof that they are fit for any duty called for.[107]

In early 1943, when more soldiers were being discharged for psychiatric disorders than the military were recruiting into the service, an order was issued under which psychiatrists were no longer permitted to recommend discharge from the service for any psychiatric reason, except psychosis.[108] By the Vietnam War, this authoritative attitude of the military led Marines like Donald B. Peterson to urge psychiatrists to act tough and uncompromisingly because if they did not 'conserve the fighting strength', non-medical military officers would.[109] Activist physicians during the Vietnam War found that when they attempted to resist their military role and help enlisted men avoid active duty, military commanders simply began sending them only those men they wanted removed from their units.[110]

In practice, however, these restrictions caused few problems for psychologists. Military psychology did not possess a set of techniques (as did experimental psychology), nor was it concerned with a common set of problems (as was developmental psychology). Rather, it was defined solely in relationship to an institution which it served as a subordinate discipline. As a young profession blissfully aware of the possibilities presented by access to this vast laboratory for applied social science with its huge, obedient source of experimental subjects, psychologists conformed to military requirements.[111] While some modern commentators explain the contradictions between the profession's ethics and military needs in terms of compartmentalization,[112] it is more plausible to recognize that there was, in fact, no 'role conflict' to begin with. Social engineering *was* their game.

More to the point, medical officers tended to be as bellicose as other military officers. Although some admitted that they got 'profound satisfaction of working to save life when everyone else had the duty of destroying it',[113] many yearned for a more aggressive role in combat. Certainly, they all acknowledged that they were healing men only to enable them to kill.[114] In training, they were reminded that 'you're in the Army to kill Germans and Japs, and win the war. Don't forget it!'[115] and when given the opportunity to carry weapons (as in the parachute regiments of the Second World War) a majority of medical officers opted to be armed and to do 'their best to enter into the spirit of the battle'.[116] According to a report during the Second World War, this conflict between their 'personal urge to be combatant' and their professional duties placed a great strain on them and was one reason why many medical officers responded by either becoming excessively harsh or too pampering of the soldiers.[117] During the Vietnam War, armed medical officers were the norm, not the exception. As Jack Strahan recalled, the medic who 'revelled in being treated as both a killer and a healer at the same time' did not regard the two worlds as morally exclusive.[118]

Although the vast majority of medical officers in wartime

experienced no conflict between their profession and their ethics, some attempted to mitigate the harsher side of their job, and thus acted in ways the military regarded as subversive. A few medical officers were persuaded to commit dissenters (such as Siegfried Sassoon in Britain or Ernst Toller in Germany) to insane asylums or mental hospitals in order to protect them from being court-martialled. Others lied to military authorities. M. Ralph Kaufman, for example, was the Consultant Psychiatrist in the South Pacific during the Second World War. His commanding officer was notoriously brutal towards emotionally disturbed officers. Kaufman responded by diagnosing all distressed officers as suffering from organic disorders.[119] During the conflict in Vietnam, some doctors refused to train paramedics for the army and they generously handed out draft deferments on medical grounds.[120] Gary Gianninoto, a medic in the navy during the Vietnam War, advised reluctant recruits of the best place to shoot themselves and even provided some with morphine to ease the pain of self-mutilation.[121] Some physicians turned a blind eye to military offenders.

An even larger number of psychologists felt uneasy about 'curing' men of the 'delusion' that their enemy was a sentient human being.[122] Sarah Haley worked for the Boston Veterans Administration Hospital during the Vietnam War. Shortly after she took up the job, a patient was admitted who had participated in the atrocity at My Lai. She was surprised to hear that this patient had been labelled 'paranoid schizophrenic':

I voiced concerns. The staff told me that the patient was obviously delusional, obviously in full-blown psychosis. I argued that there were no other signs of this if one took him seriously. I was laughed out of the room.

'These professionals denied the reality of combat!' she exclaimed. 'They were calling reality insanity!' Haley fought the diagnosis, confident in her own judgement since her father had told her about similar atrocities carried out while he was serving in North Africa during the Second World War.[123]

Other radical medical officers fought back. In 1946, Major William Needles of the US Army accused a 'goodly proportion' of military psychiatrists of being authoritarian personalities (or worse, sadists) who regarded malingering as ubiquitous, took pride in the number of patients they returned to duty (despite having no idea about how these men coped in service), misused drug therapy, and physically, emotionally and psychologically punished men for exhibiting neuro-psychiatric symptoms. With finely toned sarcasm, he described the 'memorable experience' of watching a 'dynamic, chest-thumping psychiatrist, who had never been exposed to anything more devastating than a toy pistol' harangue combat soldiers about the need to 'stand up like a man'. Needles acknowledged that military psychiatrists had two allegiances, but believed that most did not give the patient even the slightest consideration.[124]

Finally, there were small groups of peace-psychologists who refused to co-operate with the armed forces and devoted their energies to ending war, rather than adapting men to it.[125] In the context of the Vietnam War, the most important alternative psychological movements were the 'rap' groups established by veterans in which radical psychiatrists, psychologists, and social workers participated. These professionals allowed the veterans themselves to work out their own cure through talking to each other, rather than to psychiatrists.[126] Old masculine role identities were to be replaced by a new 'softer' masculinity. Veterans were encouraged not only to examine their own souls, but also to heal themselves by protesting against militarism.

Emotionally 'stitching up' men so that they could return to the frontlines as soon as possible had its civilian counterpart in some areas of civilian psychiatry (particularly in industry where the psychiatrist was in the service of the employer rather than the patient). In wartime, the position of social scientists was even more ambivalent. They spoke of peace while providing statistics for the war-mongers; they healed men in order to return them to combat to be killed; and they were in conflict with the military while at the same time exploiting its research possibilities. The medical corps, however, was a fully integrated part of the military establishment and one which accepted that the 'customer' was the commanding officer and not the patient; military psychiatrists and psychologists were therefore 'captive professionals'.[127] In time of war, clinical psychology and psychiatry took on a distinctive kind of practice: inciting the urge to kill rather than controlling people's violent urges. The role of the psychiatrist with the military was thus scarcely veiled: to stimulate men to commit violent acts without remorse.

Chapter 9

PRIESTS AND PADRES

He knows that, loving human life,
God strongly disapproves of strife
and doesn't care a damn for guns
except if they are British ones . . .
The new commandment's 'Thou shalt kill
in order to effect God's will'.

Ian Serraillier, 'The New Learning', 1966[1]

At the age of thirty, a man described as 'a cross between a leprechaun and a bantam prophet' won the dubious honour of becoming known as the Catholic priest who blessed the bomb that killed 140,000 people in Hiroshima and 73,900 people in Nagasaki. Two years earlier, Father George Zabelka had enlisted as an idealistic twenty-eight-year-old chaplain, proud to be following in the footsteps of his Austrian father and anxious to be seen to be doing his share to defend America. As the chaplain of bombers on Tinian Island, the men he called his 'kids' spent their time busily dropping napalm, conventional bombs and, finally, what they called the 'gimmick bomb' on Japanese civilians and servicemen.

'Okay, they've engaged in killing and fighting,' Zabelka recalled, 'but it didn't register. I believed it was perfectly okay.' For decades afterwards, he insisted that he had blessed 'our boys' not the bomb, but no one seemed to pay any attention. He admitted that his knowledge of St Augustine's 'just war' theory should have made him more sensitive to the theological problems involved in killing civilians, but he excused himself by passing the buck: after all, he reasoned, his religious superiors had remained silent on this issue. Indeed, he had been present at the huge mass intoned by Cardinal Spellman on Tinian Island towards the end of the war, in which Spellman had spoken in exalted terms of the need to continue the fight for freedom and justice. In fact, not one of the religious representatives on the island had anything to say about the morality of the bombing: it was merely a terrible necessity which had prevented a million American soldiers from perishing. When the bomb was dropped, Zabelka exclaimed: 'Gosh, it's horrible, but gosh, it's going to end the war. Finally the boys will get home.' Only later, when he discovered that Nagasaki was a predominantly Catholic city, did he muse guiltily about his Catholic 'boys' who had been 'piloting the plane, dropping the bomb, killing our fellow Catholics'.

Father Zabelka eventually came to feel ashamed of his wartime bellicosity, when, in the aftermath of the attack, he spoke to the survivors. He visited hospitals where innocent children lay dying. 'Many of them were very quiet, very silent. They were just quiet, just dying,' he observed with shock. Instead of returning to the United States, Zabelka chose to work as a chaplain in northern Japan. When he finally returned to America, he found that nobody wanted to talk about what had happened. 'The war's over, forget it,' he was told. But he could not. Reports from Korea and then Vietnam reminded him of the horrors he had witnessed and convinced him that he could no longer remain silent. He became an active peace campaigner.[2]

There was nothing exceptional about Father Zabelka's struggle to come to terms with mass violence in war, although his decision to dedicate himself to the peace movement was not common. In all three countries, Britain, America and Australia, religious men admitted that reconciling love for God and man with deliberate slaughter was problematic. Their position was exacerbated by their social status; both secular and religious authorities regarded clergymen foremost as moral spokesmen with the unique power to help incite the 'spirit of victory' and to bestow forgiveness afterwards. Although some pious men, like Father Zabelka, finally took a stand against war, the vast majority not only embraced the military enterprise as an opportunity for ministry but also yearned for a more active and a more bloody role, earning bitter scorn in some quarters. On the grounds that they actively encouraged other men to employ the bayonet, while refusing to stain their own hands crimson, they were taunted with chants of 'cowards, cowards, cowards all',[3] and yet their blood-lust was described as grotesque; prayer followed by slaughter was obscene.[4] The right of the churches to represent Christian principles was questioned after prominent religious officials were heard to defend killing publicly.[5] Until the Vietnam War, however, such protests were registered by a relatively small group of dissenters.

Christianity and war

'It is always unlawful to kill a man whom God willed to be a sacred animal,' pronounced Lactantius (d.320), tutor to the son of the Emperor Constantine.[6] The early Fathers of the Church, including Justin, Tatian, Irenaeus of Lyons, Tertullian, Origen, Athanasius and Cyprian, agreed. During the fourth century, this anti-war stance ebbed away as Christianity triumphed over the Roman world and a strong bond was forged between state and Church. After the Synod of Arles in 313, Christians who refused to fight could be excommunicated and by 416 only Christians were allowed the privilege of serving in the army.

Since the thirteenth century, when chaplains joined the British Army, religious representatives have always accompanied troops to war, albeit in small numbers (during the First World War there was only one clergyman to every 1,000 British and American servicemen).[7] Although smaller sects (such as Quakers, Christadelphians, Jehovah's Witnesses, Mennonites, and

Plymouth Brethren) have consistently spoken out against war
(and suffered punishment for their views) and many individual
clergymen opposed involvement in the Boer War and in Vietnam,
during the two world wars clergy from all the major denominations united against a common foe. The First World War was
portrayed as a crusade; the Second as a just war. The justifications for armed conflict were recited in almost identical terms in
both conflicts: war paved the road to peace; it promoted civilization and nurtured high idealism; virtues such as courage,
strength, patience and self-sacrifice could flourish only during
crisis; and materialism was to be swept away as entire nations
underwent spiritual reawakening. In addition, the Church had a
duty to see the moral issues which led to conflict resolved.[8]

When theologians turned explicitly to issues of killing (rather
than merely 'declaring war' and 'fighting'), there was little to
distinguish their remarks from those circulating within secular
contexts.[9] In 1917, the fundamentalist preacher, Billy Sunday,
summed it all up when he announced that Christianity and patriotism were synonymous, as were hell and traitors.[10] Jesus was no
pacifist, thundered pious war-mongers.[11] Killing in war was justified if it prevented a greater evil, such as allowing 'a band of
deluded assassins to run amok in the ranks of civilization', practising their marksmanship on 'the gentlest of women and the
noblest of men'.[12] It may not have been a Christian act to kill, but
'neither is it Christian to allow tyrants to use their power to kill',
Rolland W. Schloerb informed his American congregation in
1943.[13] Soldiers were distinguished from murderers because they
acted under legitimate authority and in the interests of entire
communities.[14] Since the shedding of human blood was portrayed as incidental to the real object of warfare (which was to
'conquer'), Paul B. Bull informed his parishioners at the
Community of the Resurrection at Mirfield in 1917 that 'as soon
as the Germans yield to what we believe is our righteous will, the
killing will stop, because the end or object of just warfare is to
make a right use of force by restraining a man or a State from

doing evil'.[15] In a sermon preached to the King's Naval and Military Forces at the start of the 1914–18 conflict, the Dean of Canterbury reminded his audience that killing was justified for the same reason that it was lawful to put men to death for great crimes like murder and treason: to attain a just world. Indeed, it was God himself who placed the sword of justice into the hands of human authorities, to punish sinfulness.[16] In 1915, the priest-in-charge at St Saviour's in Ealing, the Revd A. C. Buckell, compared good soldiers to good Christians, arguing that both needed to develop the habit of obedience in order that they would do 'the right thing instinctively by mere force of habit in time of crisis'.[17] In a 'just war', it was legitimate to kill. In the words of one Protestant minister:

> I would have gone over the top with other Americans, I would have driven my bayonet into the throat or the eye or the stomach of the Huns without the slightest hesitation and my conscience would not have bothered me in the least.[18]

An interesting example of the secular arguments used by padres to legitimate killing can be seen in the sermons of the Revd Geoffrey Anketell Studdert Kennedy, enormously popular during the First World War and practically an icon for padres during the war of 1939–45.[19] Kennedy lectured at training camps alongside men whose only title to fame were the number of Germans they had bayoneted.[20] Throughout the First World War, Kennedy provided forthright and practical instruction on killing to recruits at the Headquarters of Physical and Bayonet Training. In his sermons, he spoke in glowing terms about a lecture he had attended exhorting 'The Spirit of the Bayonet'. Kennedy agreed that this 'Spirit' was essential to victory. Warfare was a matter of 'Kill! Kill! Kill!', he reminded recruits, therefore the lecture was 'honest . . . [and] practical. It meant what it said, and said what it meant.' Reconciling the 'Spirit of the Bayonet' with the Spirit of the Cross was problematic but, Kennedy explained, he had resolved the tension after seeing some battle-weary soldiers carrying a wounded German soldier to safety.

> The two spirits united in these men because they were in the best sense sportsmen. The sporting spirit, at its best, is the highest form of the Christian spirit attainable by men at our present stage of development.

The 'Spirit of the Bayonet' had to be sustained by the 'Spirit of the Cross' and good British sportsmanship represented the best Christian ethics.[21]

Not all religious representatives would have agreed with Kennedy's secular analysis. There were some specifically spiritual justifications for killing which were occasionally heard from pulpits. Understandably, Old Testament religion, with its emphasis on 'an eye for an eye' was more often quoted than the Sermon on the Mount or the Ten Commandments.[22] Biblical prohibitions on killing (as in the Commandment 'Thou Shalt Not Kill') were easily dismissed on the grounds that 'killing' in this context meant murder or slaughter without legitimate authority, unlike killing in wartime.[23] Similarly, St Paul's command to forgive one's enemies 'seventy times seven' was said to be binding only once the enemy showed repentance.[24] Although war might be evil, it was used by God for His purposes, in a similar way that He made use of plague, famine and pestilence (that is, to prompt a resurgence of Christian virtues). The Revd J. E. Roscoe thus argued in *The Ethics of War, Spying, and Compulsory Training* (1914–18) that the use of ambushes to kill the enemy was legitimate because it displayed the allegedly Christian traits of forethought and prudence.[25] Harry Emerson Fosdick (of the Young Men's Christian Association during the First World War) and William Temple (the Archbishop of York during the Second World War) reconciled violence and Christianity by drawing a distinction between 'personality' (God's 'sacred trust' to each individual) and 'physical existence'. The latter could be sacrificed for the sake of a higher cause. In the words of Fosdick in *The Challenge of the Present Crisis* (1917):

> Any day the exigency may arise where, with no deprecia-
> tion whatsoever of my estimate of personality's absolute,
> unrivalled worth, I may, for a woman's safety or a child's
> life, have to strip some man's physical existence from him,
> if I can, and trust God that in the world unseen his abid-
> ing personality may be recovered from his sin . . .
> Bayonets do not reach as far as personality; they reach
> only physical existence, and the problem of personality
> passes far beyond an earthly battlefield.[26]

In a similar vein, nearly a quarter of a century later, the Archbishop of York made much of the fact that since the sacrifice of one's life was not the greatest injury to suffer (since it repre- sented the fulfilment of an individual's personality), it could not be the greatest injury to inflict.[27]

Of course, the bayonet-wielding soldier had to safeguard the purity of his own soul or personality, ensuring that it was not cor- rupted by bitter feelings towards the enemy. In this way, the manner in which a soldier took another man's life was important. Sin could only be avoided by keeping the 'spirit of mercy' alive, even if this merely meant adhering to the Hague Conventions or to vague notions of 'gentlemanly behaviour'.[28] Crucially, Christians practised righteousness in battle by killing without hatred. Thus Father Grayson during the Vietnam War advised soldiers that it was legitimate to kill 'but not with hatred in your heart'.[29] The same instruction had been offered during the First World War when Edward Increase Bosworth in *The Christian Witness in War* (1918) reminded his readers that the 'Christian soldier in friendship wounds the enemy. In friendship he kills the enemy . . . His heart never consigns the enemy to hell. He never hates.'[30] Or, as E. Griffith Jones advised a conference of

Wesleyans in 1915, soldiers should be like shepherds – intent on destroying wolves, while avoiding the 'wolf spirit' at all costs.[31] As they plunged their bayonets into human flesh, clergymen encouraged soldiers to murmur, 'this is my body broken for you', or to whisper prayers of love.[32]

In this way, it appeared to be possible to kill without sin. Clergymen went further, however, arguing that Christ himself endorsed killing – although (rather inappropriately given twentieth-century armaments) the chosen instrument was the bayonet. When pacifists asked their audiences whether they could imagine Christ sticking his bayonet into another man,[33] some parsons answered with a resounding 'yes'. In a book entitled *The Practice of Friendship* (1918), Lieutenant George Stewart and Henry B. Knight (YMCA director and former professor at the Yale Divinity School) admitted with characteristic aplomb that initially they had preferred to think of Christ wielding a sword instead of a crude bayonet. Upon reflection, though, they recalled that in the time of Christ, swords had been the pre-eminent instrument of mutilation and death rather than the ornamental weapon of the modern imagination. In modern times, the bayonet had replaced the sword. Christ was stripped of his white gown and razor-sharp sword to be clothed in 'a garb of olive drab which was stained with blood and mire, and in his hands a bayonet sword attached to a rifle'. It was a 'vision' which propelled them into war work with renewed enthusiasm.[34] Killing was not merely sanctioned, it was sanctified.

This is not to deny that many religious spokesmen were actively peace-loving. Particularly in the twenty years after 1918, most clergymen denounced war (pacificism), although far fewer were willing to take an absolute pacifist line. Numerous anti-war religious organizations (generally denominational) were set up, including the Christian Pacifist Crusade, the Methodist Peace Fellowship, Unitarian Peace Fellowship, Church of Scotland Peace Society, Presbyterian Pacifist Group, and Church of England Peace Fellowship. The experience of the First World

War had prepared the way for increased pacificism as many clergy became aware of the way in which they had been duped into endorsing grossly exaggerated (if not entirely fictive) atrocity stories and as the gains of that war began to seem less and less remarkable. Their calls for peace never achieved political momentum though, and total British membership in the denominational peace organizations peaked at only about 15,000.[35] Furthermore, as the Second World War approached, Christian pacifists increasingly found themselves attacked by theological philosophers such as Reinhold Niebuhr who accused them of neglecting social responsibility in the name of sacrificial love. The Second World War was easily portrayed as a war of the righteous against evil incarnate. By the time of the war in Vietnam, even the leading pacifist movements had become active supporters of armed struggle, particularly those forms of militancy represented by the National Liberation Front.[36] Although in all three conflicts influential churchmen opposed specific acts of war (most famously, obliteration bombing and the atomic bomb during the Second World War),[37] and although specific religious bodies (such as rabbis within the American military prior to 1968)[38] consistently saw themselves in opposition to the military authorities, the attitude of religious men and women to war generally consisted of regretting the need for armed aggression, while at the same time applauding engagement in 'just' war.

'Fighting padres'

The non-combative status of ordained clergymen had been firmly established by the twentieth century. Indeed, in the countries under consideration here – Britain, America and Australia – they were formally forbidden from bearing arms. This had not always been the case. Ordained men had been denied the use of the sword since the Middle Ages, although they were permitted to use the mace. It was not until 1350 that 'fighting padres' were prohibited and, in 1899, this prohibition was inscribed in the Laws of War at the Hague Peace Conference.

13th Royal Fusiliers (with war souvenirs) resting after the attack on
La Boisselle on 7 July 1916.

American snipers in action (28th Infantry regiment), at Bonvillers,
22 May 1918.

Black troops practising bayonet drill at Gondrecourt on
13 August 1918.

Unarmed combat: Lieutenant C. Trepel attacks the
Commando Depot PT Instructor from the rear with a knife
during a demonstration in 1943.

C. W. Heath's 'masculinity component'. The man at the top has a 'high masculinity component' in contrast to the man at the bottom who has a 'low masculinity component'.

(C. W. Heath et al., 'Personnel Selection', *Annals of Internal Medicine*, vol. 19, 1943)

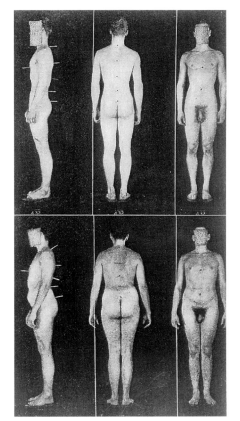

American troops throwing hand grenades in a front-line trench during World War One.

'Australians at work at Anzac two days before the evacuation took place.'

Two American riflemen plunge through the jungles of Bougainville.

'Bayonet Instructor Corporal Charles Batchelor lunges at the dummy. He says, "This is the way the Home Guards should get their teeth into Jerry's pants."'

'Home Guard Corporal Charles Batchelor considers the bayonet a capital weapon. Lovingly he fingers his bayonet, inducing the correct "mood" for this kind of fighting.'

A prisoner emerging from a foxhole, during the Second World War.

A dead man and an American soldier during the Vietnam War.

Bruce F. Anello in front of a Viet Cong he killed.
(Indochina Curriculum Group, *Front Lines*, Cambridge MA, 1975)

A US Navy chaplain leads Mass on Saipan Island.

Wren armourers testing a Lewis Gun during training at a naval air station, Lee-on-Solent, October 1942.

'An ATS Spotter chalks her slogan on the Gun Flash-Board at a Mixed Heavy AA site near London, 1943.'

Of course, although members of the various chaplains' departments of the armed forces were prohibited from bearing arms, clergymen could still enlist in the ranks. Many smaller religious groups chose to leave the decision of whether to enlist as a combatant up to individual consciences. Thus, despite their strong stand against war, over one third of British Quakers of military age enlisted.[39] Innumerable lay preachers from the Methodist, Baptist and Presbyterian churches also chose to take up arms. Members of the Salvation Army merely changed the design of their uniforms: 'Blood and Fire . . . the Blood that cleanses, the Fire that energizes' was their clarion cry sung as loudly in Hyde Park as in the bloodstained battleground where rifles and guns 'spit fire'.[40]

The Roman Catholic Church and the Church of England were less willing to allow individual priests and preachers to make up their own minds. Throughout the period, religious leaders remained divided over whether it was proper for clergymen to volunteer for combatant service. In 1916, the Revd Reginald John Campbell (a highly respected nonconformist minister who joined the Church of England in the same year as he began his tour of the Western Front) admitted that he could not 'find it in [his] heart' to 'turn any able-bodied man back who wants to exchange a cassock for a khaki coat in order to minister to them as comrade to comrade on the bloody field'.[41] From the theological department of Yale University, Henry Hallam Tweedy also defended clergymen who felt impelled to join the armed forces. In 1918, he wrote admiringly about a 'brother of all men' who refused to 'be limited by his cloth, who cannot preach to the soldier without drinking the soldier's cup and being baptized with his baptism of mud and of blood'. It was, he believed, the 'spirit of a true Christian preacher who cannot urge Christian laymen to "go over the top" unless at least some Christian ministers go with them'. Such 'intimate fellowship' in suffering 'knows no secular and no sacred', he asserted.[42] For such spokesmen, it was manifestly untrue that a combative priesthood might cast a shadow over the reputation of the Church. After all, it was noted that the conscription of French priests into combative service had won only praise and honour for religion.[43]

Nevertheless, the Church of England and the Roman Catholic Church attempted to maintain the prohibition on the carrying of arms by their clergy, even if it meant dampening the martial fervour of many of their own religious leaders. For J. G. Simpson (Canon and Precentor of St Paul's), the issue was straightforward: quite simply, it was part of the 'character' of the priesthood that they did not bear arms. This special 'character' was not founded on ethical considerations: there was only one Christian ethic and it applied to all individuals, whether cleric or lay. Rather, the prohibition lay in the nature of the padre's job which, in all circumstances and against all temptation, must consistently show mercy and exemplify the conscience of mankind. War belonged to secular society and had no place in the Kingdom of God. While sacrificing one's life could be a Christian act, killing was not, Simpson declared. Padres had a duty to represent ideal humanity while sharing the hardships of fallen mankind. They were representatives of the spiritual world and, as such, must survive the slaughter in order to bring about lasting peace. Simpson argued that since clergymen were entrusted with the Word of Life, they had no right to send any man into the presence of his Maker 'unhousel'd, disappointed, unaneled'.[44] In 1915, Simpson was supported by the Revd Arthur Foley Winnington Ingram, the Bishop of London, who refused to ordain men of military age and fit for service.[45] Within a year, Winnington Ingram was joined by the Bishops of Carlisle, Chester, and Manchester.

The Bishop of Oxford, Charles Gore, agreed with Simpson, with one exception. In 1914, faced with the disquiet amongst many of his clergy, Gore urged them to remember their vows and affirmed that the position of a combatant in the military was incompatible with the position of a man who had taken Holy Orders. Admittedly, he conceded, the status of men who were merely candidates for Holy Orders was more complex. Since these men had publicly professed a vocation for the ministry, was it legitimate for them to postpone ordination and don the uniform of the combatant? Traditionally, Gore noted, such men had not been allowed to enlist: indeed, ancient canons had prohibited former soldiers (as well as magistrates who had condemned men to death) from ordination and even candidates who had been compelled to serve had to seek dispensation from the Roman Catholic Church. In the end, Gore compromised by agreeing to ordain candidates when they returned from the battlefield if they had entertained no doubts about where their duty had lain, but he strongly encouraged candidates to 'abide by their realized call and to pursue their preparation' even when they were taunted with accusations of cowardice.[46]

By 1916, military crises and conscription infused further heat into the debate. From pulpits, and in the periodical press, numerous clerics had their say. These commentators included men like the Revd John Sinker who, in a series of sermons delivered in the Lytham Parish Church, testified to the enthusiasm for war among the younger clergy anxious to share the same dangers and excitements of other men of their age. He warned against translating this enthusiasm into enlistment and reminded waverers that it was their duty to use spiritual rather than physical weapons.[47] More to the point, in *The Contemporary Review* (1916), Edw. Winton protested that what was at stake was the relative worth of prayer over fighting. If prayer was valued, the small contribution that the addition of a few thousand more men would make to the armed forces was extremely short-sighted. The comfort of the bereaved, preparations for peace, and spiritual succour for servicemen were much more important. If the ordained man was merely another 'citizen', he should be required to fight, but he was not 'ordinary'. Every clergyman had a 'special commission and a social vocation' as a representative of God, and a living testimony to the ideal Christian state.[48]

Unfortunately for such religious spokesmen, secular opinion

was tipping in the other direction. Prior to 1916, when wars were fought primarily by regular soldiers and volunteers, the right of clergymen to be automatically excused from armed service was rarely questioned. During conscription debates, however, large sections of the public began to resent the fact that clergy were exempted from combatant service by Act of Parliament, as borne out by the support Ben Tillett gained at the Trades Union Congress at Birmingham in September 1916 when he had a resolution passed condemning the exemption of clergymen from conscription.[49]

By 1918, the traditionalists had lost their battle. In Britain, the Military Service Act of 9 April 1918 allowed the conscription of clergymen. Although the Act was eventually withdrawn for fear of another Irish rebellion if Roman Catholic priests were conscripted, the fact that the Archbishop of Canterbury had already given the Act his blessing placed clergymen in an extremely difficult position, as they were no longer able to rely on the authoritative blessing of their religious superiors for their exemption. In response to the crisis, the Archbishop of Canterbury met with seventeen diocesan bishops on 22 April 1918 and agreed to encourage their clergymen to enlist voluntarily, either as non-combatants or as armed men. By this stage, however, most clergymen who were young and fit had already voluntarily enlisted as non-combatants in the chaplains' departments.

Clergymen in the armed forces

Clergymen who served in the chaplains' departments were formally non-combatants. This did not mean that once they had enlisted into the armed forces they were spared contact with militarist symbols and rites of destruction: padres wore the insignia of the officer class;[50] their jeeps had Vickers guns attached; and batmen were appointed to carry arms in order to protect them.[51] Many carried revolvers which, as with members of the medical corps, they were allowed to use in self-defence.[52]

Though outwardly bearing the symbols of militarism, clerics still had to make individual decisions about their role within the military: each man had to examine his own conscience and ask whether he should enlist as a cleric or a combatant. It was a difficult decision. During the First World War, popular padres such as Father J. Fahey, the Revd John J. Callan, and the Revd Dennis Jones shamefacedly confessed that the idea of joining in the fray was exciting, but they exhorted clergymen like themselves to resist the temptation on the grounds that their spiritual job was more important.[53] Twenty-five years later, in the war of 1939–45, J. Fraser McLuskey (known as the 'Parachute Padre') debated for a long time whether to enlist as a chaplain or be trained for combat 'as my twenty-nine years and two hundred pounds qualified [me]'. When friends eventually persuaded him that he would be of greater value to the cause as a padre, he experienced a keen sense of anti-climax.[54] The Revd R. L. Barnes of the Community of the Resurrection revealed that he envied men who had given up the idea of being ordained in order to march into battle: he lamented being 'a lonely, wandering, troubled Bo-Peep' while combatant clerics were 'veritable fathers of their flock'.[55] The chaplain, Joseph McCulloch, changed his mind halfway through the Second World War when he could no longer reconcile his military and religious roles. His belief that it was hypocritical for a Christian to be employed by an institution devoted to killing eventually led him to resign his commission.[56]

Other clergymen made a different choice, deriving their identity more from augmenting than entombing the dead.[57] These violent men of God rose to great prominence in popular myth as 'fighting padres'. One of the most popular fighting songs of the Second World War featured a combative clergyman as its subject. The ballad described the exploits of a priest or 'sky pilot' (probably the Revd William A. Maguire) who fought alongside his men in 1941 defending the *California* from Japanese attack:

> *Down went the gunner, and then the gunner's mate,*
> *Up jumped the sky pilot, gave the boys a look,*
> *And manned the gun himself as he laid aside the Book,*
>   *shouting,*
> *Praise the Lord, and pass the ammunition!*
> *Praise the Lord, and pass the ammunition!*
> *Praise the Lord, and pass the ammunition!*
> *And we'll all stay free!*[58]

Not to be outdone, but in a far less racy style, fighting clergymen also prompted poetic outpourings during later conflicts. Take Larry Rottmann's terse lines on a chaplain in Vietnam:

> *The chaplain of the 25th Aviation Battalion*
> *at Cu Chi*
> *Prays for the souls of the enemy*
> *on Sunday mornings*
> *And earns flight pay as a helicopter door gunner*
> *during the rest of the week.*[59]

Myths aside, it was widely acknowledged that young priests (particularly those who fitted into other combat traditions, such as nonconformist Welsh chaplains)[60] yearned for a combative role.[61] In the field, chaplains had to be constantly reminded that they were prohibited from bearing arms (the US Chief of Chaplains fearing that if armed chaplains were captured, they would be punished severely for violating the Geneva Convention and could be used as scapegoats for ignoring other chivalrous codes of war).[62] Lieutenant Robert Furley Calloway was one example of a mild padre whose experiences of war led him to ask for a transfer to a combative role. While in combat training in 1916, this former missionary priest described a lecture he had recently attended. 'It was extraordinarily good,' he recounted to his wife,

but to me the interest of the lecture lay not so much in the lecture itself as in what the lecture stood for – the entire conversion of our whole attitude of mind as a nation. For it was instruction as to how best to *kill* (with the bayonet), and every possible device that had been found by experience useful to enable a man to kill as many Germans as possible, was taught. As one writes it down, it sounds the most hideous brutality, and yet yesterday I don't suppose there was an officer or man present who did not agree that if the war was to be won we must fight to kill.

While admitting that he personally did not relish the idea of bayoneting anyone, he no longer believed that it was morally wrong to do so. His sense of revulsion was not some instinctive spiritual response but was merely the 'natural instinct' shared by all educated people, he claimed. Eleven days after writing these words he was killed, aged forty-four years.[63]

Calloway was one, unfortunate example of a clergyman who was keen to get the war over with, even if it meant staining his own hands with blood. In this, he was not unusual. In all three conflicts, men of God very quickly became frustrated by the slowness with which the various chaplains' departments were sending them into battle and their zeal posed severe problems for religious leaders.[64] Many clergymen simply refused to wait for religious authorization. On the declaration of war in 1914, nearly 400 of the 1,274 students enrolled in Anglican theological colleges in Britain withdrew immediately and ordinations fell steadily throughout the war from just over 600 in 1914 to 114 by 1918.[65] Similarly, in the Australian army during this period, fifty-one Anglican priests enlisted in the ranks. Given that in the entire country there were only 1,400 clergy (of varying ages and degrees of fitness), this was a significant drain.[66] During the Second World War too, theological colleges emptied as 'young men of real worth' unequivocally opted for bazookas over bibles.[67]

There were many reasons why clergymen wished to fight

actively but two powerful incitements seemed to be boredom and the desire to prove themselves. In his book *Disenchantment* (1922), Charles Edward Montague described the typical military chaplain as a man who had been a pleasant young curate and the 'prop and stay of village cricket-clubs'. For such men, the war promised escape and excitement: 'a heaven after the flatness, the tedium, the cloying security and the confounded moral problems attending the uninspired practice of professional brightness and breeziness in an uncritical parish', Montague observed.[68] Certainly, this was the motivation behind Canon J. E. Gethyn-Jones's decision to remove his collar and insignia and holster his revolver – that is, in addition to his desire to 'prove myself to myself'.[69]

There were also intense external pressures placed upon clergymen to become combatants. As one Scottish minister (who was eventually killed leading his men at Loos) admitted when explaining why he enlisted – if he had remained in safety, he simply would not have been able to look the 'lads' of his congregation in the face.[70] Accusations of hypocrisy – that they encouraged other men to kill, while highmindedly refusing to bloody their own hands – were hard to bear and the Church's advice that they must lay responsibility for the prohibition on to the shoulders of their bishops convinced no one.[71] In the context of late nineteenth- and early twentieth-century theology, the emphasis placed on a 'participative ministry' was also a significant consideration. Some chaplains regarded enlisting in the military as an extension of their interwar ministry in the settlement movement in the East End of London; that is, the best way to serve God was to immerse oneself in the lives of 'ordinary' people.[72] Combat did seem to promise men 'golden opportunities of exercising a true Christian ministry'.[73] This was why (during the Second World War) the Catholic priest, the Revd R. M. Hickey, and the senior chaplain with the Second Army, the Revd J. R. Youens, insisted that chaplains should endure physical training and rifle training alongside combatant troops because they believed that this would draw them closer to the men and they would therefore be in a better position to give spiritual advice.[74] In Vietnam, the chaplain David Knight insisted that their 'job' in the armed forces was 'to kill or capture the enemy. Because of this, I went in with the men while on these missions . . . I was . . . able to witness to the love of the Lord.'[75]

If, *prior* to battle, the desire to share in the everyday lives of one's congregation prompted some clergymen to take up arms, the actual experience of war was even more important in leading them to renounce a non-combatant role. In the wild excitement of battle, some padres simply forgot the rules as they retrieved discarded weapons and took over machine-gun sites,[76] and shortages of personnel sometimes forced military officers to order padres to carry ammunition or even lead attacks.[77] Ministers who – to their own surprise – found themselves in the heat of battle eagerly firing their rifles at fleeing Germans might not take much persuading to renounce their vocation in order to re-enlist in the ranks.[78] The desire for retribution also inspired much bellicosity. After hearing of the death in action of his younger brother, one United Church minister during the Second World War acknowledged that he had lost the 'loving spirit' essential to the chaplaincy. Now, he only wanted to kill Germans 'with his bare hands' and his request to be transferred to a combat unit was accepted.[79] Particularly when fighting non-European enemies (Japanese or Vietnamese troops, for example) the brutality and lawlessness of the enemy was often used by combat padres to justify breaking the Geneva code by taking up arms.[80] 'Even the Padre Has Strapped on a Revolver' blazed one newspaper headline during the Vietnam War.[81] As the New Zealand fighter pilot Bryan Cox put it in the context of the Second World War, padres in the European theatre would have found it unacceptable to be found manning machine-guns, but in Bougainville (Solomon Islands) it was 'a case of kill or be killed, and none of us felt much emotion on seeing a Japanese corpse'.[82]

Nor were some clergymen immune to the general sense of

exhilaration on the eve of battle and while engaging the enemy. Chaplain Ernest N. Merrington was struck by his own feelings of glee at the thought of 'going under fire'. In 1922, he recalled how immediately prior to the Gallipoli landing both he and his comrades mentally 'rubbed [their] hands with a strange and fearsome delight at being at last "really" up against it'.[83]

Priests like Father William Doyle also clearly relished the bloodiness of battle, and his letters to his father described in lurid detail the murderous activities carried out by the soldiers to whom he ministered.[84] Indeed, padres who lacked this feeling or were squeamish about killing had no place in any of the armed services. As the Revd J. Smith put it, if a clergyman felt that he had 'conscientious objections to fighting' then he was of 'little use in any military work of any kind, the religious and medical work not excepted'. Abhorrence of 'shedding of some other man's blood' rendered a man 'doomed as a Padre'.[85]

Many padres experienced no problems reconciling their religious ethic with the moral universe of the military. Throughout Christian history, the language of combat has permeated religious writing.[86] Religion was the 'Expeditionary Force' of a man's soul and Christ was the Great Captain calling Tommy to be His private.[87] In 'So Fight I'. Thoughts Upon the Warfare in Which Every Soul is Engaged and in Which There Can Be No Neutrals (1917) the Revd G. C. Breach related a story of a company much harassed by an effective sniper. A corporal eventually located where the sniper was hiding and the area was decimated by the artillery. The incident becomes a spiritual mission in Breach's account:

The enemy had been discovered and destroyed. Have you TRAINED YOUR SPIRITUAL VISION TO DETECT A FOE? You may discover danger in some innocent thing, when God has educated your spiritual sight; but never mind what it is, good or bad, if it's hurting your soul DESTROY IT. TRAIN YOUR BATTERIES UPON IT.

Kill that habit, that liking, that influence, before it
becomes a spiritual danger. Christ will find the range for
you![88]

The shift in language from destroying a spiritual enemy to
butchering a human foe was effortless.

There was another 'ethic' shared by the military and the
clergy and that concerned ideas about manliness. It was all too
easy to conform to an idea of masculinity that valued celibacy,
meekness, patience and the cultivation of the mind over the body
in peacetime. With the declaration of war, a wider brotherhood
lauded bellicosity, patriotism and physical prowess – and clergymen became particularly sensitive to accusations of effeminacy,
with good reason. During all three conflicts the clergy's non-
combatant status was a source of derision. In a poem entitled
'The Holy War' (1915), an elderly priest with a son at the front
launched a scathing attack on clergymen for cowering beneath
non-combatant skirts. He accused 'smugly quartered'
Archbishops of encouraging vicars to fiddle idly within deserted
parishes. For clergymen (with their 'apron and the gaiters, and
the twisted hat'), could anything be more humiliating, he asked?
The poem exhorted a change of attitude:

> So to all my brother-Priests first I say be Men
> Do you call yourselves Englishmen, if so, then
> Do your duty to your country, as Frenchmen do.[89]

Although this verse was probably read by very few clergymen,
the sentiments expressed touched a raw nerve within 'stalwart
young curates' who resented being told by their elderly superiors
that they were forbidden to fight.[90] They protested against accusations of 'softness' and cowardice, declared (in the mode of
muscular Christianity) that they could 'give a remarkably good
account of themselves either with or without gloves', disputed
accusations that they were 'lisping youth fit only to recite at a

penny reading, and to play at croquet with the girls', and loudly dismissed the implications of the song 'I was a pale young curate then'.[91] They agreed that the military chaplain had to possess 'the look of a man, the courage of a man, [and] the soul of a man'.[92] Even within the hallowed walls of seminaries, military prowess and manliness were linked. For instance, in the closing lecture to the senior class of divinity students at Glasgow University in 1916, Professor H. M. B. Reid praised the fact that men from the Divinity School had enlisted in large numbers because it 'proved their manhood'. They were living proof that Divinity Hall was 'no refuge for slackers or detrimentals, but proportionally the most martial section of the University' and Reid 'humbly anticipate[d] that the return of our warrior-students will bring together elements of unusual potency to these benches'.[93] John Smith, an Australian Presbyterian minister who enlisted and fought in the ranks during the First World War, poured scorn on padres who adopted a 'mothers' meeting spirit'. In virile tones, he defended ministers like himself who went to battle:

> when the blast of war has called forth the nation's best,
> yea, when the Church itself has realized that the cause is
> God's, and has sent forth her men to slay, are there not for
> the Padre also certain muscles of the mind which should
> stiffen, should there not be some sting put into his senti-
> ment, into his conversation and preaching with some
> slight bearing on the work to be done by the men he is
> there to inspire.[94]

In an environment which emphasized 'muscular Christianity', the 'manliness' of individual clerics was highly valued.[95]

Morality and morale

There were some dissenting voices who did not believe there was a role for padres in the armed forces. Fears were expressed that converting troops to Christianity might reduce their aggressiveness in combat,[96] or that the clerics would make fighting men 'morbid or think too much'.[97] Some officers argued that there was no need for military padres at all because in wartime Christian countries ceased to be Christian. Brigadier-General Frank Percy Crozier brazenly announced that padres were 'quite out of place during a war', although they might usefully be employed distributing cigarettes.[98]

Overall, though, chaplains were considered central to the killing enterprise: they were moral arbiters, morale-boosters and confessors. Field Marshal Douglas Haig even regarded a popular chaplain as equal in value to an effective general in ensuring victory.[99] Pulpits throughout the country served as enlistment agencies. During the First World War, for instance, the Bishop of London boasted that he was responsible for adding 10,000 men to the fighting forces.[100]

In accordance with this view the spiritual world was evoked at every level – from training camps (the 'Christian Churches are the finest blood-lust creators which we have and of them we made free use' boasted one senior officer in charge of military training) to preparation for battle (padres prayed that 'these lads of mine would be successful in "splashing" the enemy').[101] Bolstering morale was an essential part of the clergy's job. During the First World War, General Sir R. C. B. Haking announced to a large gathering of chaplains that their position within the military provided 'golden opportunities of raising the morale and moral tone of the men, and, in consequence, make them more fitted for the tasks they are called to accomplish'.[102] Similarly, the President of the United States' Committee on Religion, which reported in 1950, commended the chaplains' service for inspiring men to 'greater performance of duty'.[103] They had a role in helping combatants defeat despair and develop the 'will to conquer', announced the Revd Geoffrey Anketell Studdert Kennedy in his *Rough Talks* (1918).[104] Alongside discipline, the will to win, self-respect, loyalty, and a high sense of honour, an official textbook for military chaplains, published in 1943, defined chaplains' duties as inculcating the 'fighting spirit'.[105] General B. L. Montgomery of the Eighth Army commended an article on the role of the chaplain in battle, published in *The Chaplains' Magazine. Middle East* (1943), which reminded chaplains that their duty was to

> spotlight . . . the Conflict of Good and Evil, the opposing standards of Right and Wrong, the just foundations of our Cause, the Presence of Christ, the value of prayer, the glory of sacrifice, the gift of eternal life, and such things as deepen their sincerity and confidence as soldiers gathered to battle by God.

On the eve of battle, chaplains were told to 'pitch' their message in 'a major, not a minor key', consecrating resolve, appealing to strength, and reminding troops that Christ had 'overcome the world'. The correct spirit to inculcate was 'Let God arise and let his Enemies be scattered' and the theme should be 'Praise for a just Cause, Prayer to be worthy of it, and Commendation to God of all the issues for us and ours.'[106] During both world wars and during the Vietnam War, chaplains performed a similar function, instructing men on the correct interpretation of the commandment, 'Thou Shall Not Kill'.[107]

In terms of morale, the most important duty for military chaplains was in counselling soldiers who confessed to disquiet about killing.[108] Padres had to 'put stiffness into the muscles of the military mind' and 'summon up his own, and the blood of all whom he comes into contact . . . for a blast of righteous hate'.[109] Even a 'fine fellow' might need reassurance about the legitimacy of thrusting a bayonet through another man: 'Am I to feel that if I have a German at my mercy I am to run him through with the bayonet with that merciless hate?' one soldier asked the Revd A. Irving Davidson during the First World War. Davidson replied confidently:

> You've come over here with the high ideal of doing your
> duty as a soldier for your country. You've learned the art
> of soldiering and of fighting. Go into it keeping to that
> ideal along with all the other ideals of the old home and
> when the moment for decision comes you will know
> exactly what your duty is and you'll do it.

The young man was satisfied with this response, did his duty,
and became 'one of the finest soldiers'. Davidson praised his
own success in 'hold[ing] the balances and keep[ing] men to the
noblest ideals in the midst of much that would have spoiled
them'.[110] Fifty years later, a chaplain to the RAF was faced with
a similar need to justify mass slaughter to the bombers in his
unit. He recalled being shocked when he overheard young RAF
cadets expressing reservations about the morality of accepting an
assignment to a V–Bomber (a nuclear-armed plane). The chaplain
immediately considered it his duty to inform these young pilots
of the 'morality of deterrence' and he went on to organize a programme of special training for other chaplains to counter similar
reservations.[111]

The Church also took on a responsibility to provide moral
guidance for the actual conduct of combat. As already mentioned, pulpits resounded with exhortations to 'play the game'
and fight for God and country. The sportsman 'puts right before
might. He hits, and hits hard, but never hits below the belt', soldiers in training at the Headquarters of Physical and Bayonet
Training were reminded by their chaplain.[112] Interestingly
though, exhorting combatants to be 'sporting' was as deep as the
advice went. Although the theological concepts of 'just war' were
delineated (Had the war been properly declared? Was it being
fought for a just cause? Was it a last resort?), many chaplains
expressed extremely cynical views about whether the adherence
to 'rules' was ever an expedient policy in battle. As one chaplain
in the Australian Imperial Force put it: 'hit anywhere you can and
as hard as you can . . . Really there can be no rules for the

conduct of war properly so called which implies and implies only the doctrine of the survival of the fittest. Manifestly an Outlaw scheme of things. So! Well then there is no need for apologies.'[113]

It was such banal theology which led the Australian historian, Michael McKernan, to remark drily that clergymen 'used their scholasticism to establish the conditions for a just war but refrained from applying their theology to particular moments of war'.[114] As the Revd Joseph McCulloch ominously observed in his book *We Have Our Orders* (1944): 'What I fear has happened is that the Army has converted the Chaplains.'[115]

One area where chaplains failed to provide adequate moral guidance was in their advice concerning the killing of prisoners and civilians. Clergymen were well aware that the gratuitous killing of non-combatants was proscribed by both legal precept and religious law, yet clergymen during all three conflicts were remarkably lax in their condemnation of such practices. Padres regaled recent arrivals in Vietnam with lurid accounts about the 'treachery' of Vietnamese women and children who were prone to ambush unwary soldiers, while omitting to draw attention to the presence of innocent civilians.[116] Throughout the Vietnam conflict, the words of the Revd E. W. Brereton (rector of Hollinwood, Essex) fifty years earlier were echoed:

> we are fighting for dear life against enemies who are not Christians, not human beings, but reptiles. We claim the right to fight these fiends not with kid gloves. I scorn the humanitarians who object to reprisals.[117]

Religious leaders advised combatants that because enemy civilians approved of the atrocities carried out by their soldiers, they were therefore legitimate objects of attack: even children could be killed because it was impossible to separate them from their guilty parents.[118] In 1917, the modernist theologian and editor of the *Modern Churchman*, the Revd H. D. A. Major, had been even more forthright, asserting that if the only way to protect adequately an English babe is to kill a German babe, then it is the duty of the authorities, however repugnant, to do it. More particularly is this so when we reflect that the innocent German babe will in all probability grow up to be the killer of babes himself.[119]

The extent to which padres were willing to condone the killing of non-combatants was examined in the 1960s when Gordon C. Zahn interviewed seventy-three chaplains about how they felt about the slaughtering of enemy prisoners.[120] Four refused to answer or said that they were not 'qualified' to respond. Although the other sixty-nine chaplains believed that it was wrong to kill prisoners, four were prepared to accept the commanding officer's decision if he could 'justify' the order to their satisfaction. One chaplain with fifteen years' service reasoned that he would accept the killing 'if it was a case of military necessity, if it meant saving the lives of our men' and seven others claimed that they would register a protest against the order, but would take their complaint no further than to the commanding officer himself. In other words, even though these chaplains disapproved strongly of any order to kill prisoners, one in six would keep the affair within the local confines and submit to the authority of the commanding officer. Of the chaplains who were prepared to carry their protest higher (forty-six chaplains), half would only take the matter as far as their own superiors in the Chaplains' Branch.[121]

Surprised by such levels of complicity with an ethic which was questionable in both religious and military terms, Zahn asked the chaplains what they thought combatants should do if ordered to refuse to accept surrenders and found that 42 per cent would accept the commanding officer's decision. If the proportion of respondents who would be willing to accept the practice in the guise of 'military necessity' was added together with those willing to 'keep their protest within channels', 90 per cent expressed only minor ethical qualms about violating this law of war. In justification, one chaplain observed that once a surrender had been accepted, the military was faced with the responsibility for looking after the prisoner and this was often not a desirable option. He added that his service during 'the Jap campaign' enabled him to 'appreciate the position' of the commanding officer who did not want to be burdened with this responsibility. This chaplain's advice was concise and to the point: 'Leave it to the good sense of those in command, knowing that such a decision was inevitable and hurt like hell.' Given such attitudes, it came as no shock to discover that nearly half of the clergymen interviewed by Zahn were incapable of even *conceiving* of a situation in which it would be their duty as chaplains to advise soldiers against obeying an order on the grounds of Christian ethics, and another 15 per cent were willing to consider such a situation as 'conceivable' but unlikely. All the other respondents were hesitant and qualified their responses.[122]

Such attitudes had tragic consequences – most infamously during the Vietnam War. At the Commission investigating the massacre at My Lai, it was revealed that one participant (Warrant Officer Thompson) had been taking instruction from Carl Creswell, his Divisional Artillery Chaplain at the time of the massacre. Creswell admitted that immediately after the massacre Thompson had been 'terribly upset' and had asked him what to do. Creswell advised him to protest through official channels and promised to take up the issue himself through 'chaplains' channels' (this was an inappropriate medium: chaplains' channels were for technical matters only). Creswell did take the issue to his religious superior, Division Chaplain Lewis, but left the matter there. After extensive investigation, the Commissioners concluded that Lewis might have mentioned the massacre as part of his informal visits with the military command but had not 'made any timely effort to transmit the information' to the command group of the Americal Division. Indeed, by his own account, it was ten days before Lewis even made his first 'informal' call on the Chief of Staff and, at that time, he was assured that the civilian casualties had been 'inadvertent' and were a 'natural consequence of the type of combat units faced in inhabited areas'. The Commissioners were damning. They accused both chaplains of taking inadequate steps to bring the charges to the attention of the relevant authority and concluded: 'It should have been evident to both these chaplains that the idea of conducting an investigation of a war crime through chaplain channels was preposterous.'[123]

Finally, the third task or responsibility entrusted to clergymen was the granting of absolution to servicemen when they returned from the bloody field of combat. Forgiveness was freely given. The padres' authority in alleviating feelings of guilt for killing was highly valued by some combatants. For instance, Keith Irwin recalled Christmas Day 1942, when he was serving with the 36th Australian Infantry Battalion in New Guinea. The troops were demoralized and huddled in trenches with nothing to eat or drink. Suddenly, the padre slipped into their hole and gave them a bottle of communion wine to drink, ruefully saying 'I don't think the Good Lord will mind.' Irwin immediately felt reassured and forgiven for the 'dreadful things' that he 'had been doing to the Japs': his 'faith in human nature was somewhat restored', he admitted.[124] In 1945, artilleryman John Guest was attached to an AA Battery in Italy and felt profoundly guilty for firing the guns, and was unable to stop thinking about his victims. There was a sense of relief, however, as he recorded in his journal how he was 'glad to hear, in the service we had this morning, a prayer for forgiveness, penitence'.[125]

Even psychiatrists acknowledged the power of spiritual relief. Robert L. Garrard, for instance, told the National Association of Private Psychiatric Hospitals (American Psychiatric Association) in May 1948 that 'the aid of chaplains' should be enlisted to help men who felt guilty for killing. He observed that

some patients obtain through confession a relief of guilty feelings. Confession, even though unaccompanied by punishment, is regarded as a righteous act and tends to lessen the feeling of guilt. Confession also affords relief by giving a feeling that responsibility has been shifted or shared.

Personally, Garrard believed that psychotherapy was more effective, but he recognized that for many men 'religion may also have therapeutic value'.[126]

Ethics vs military authority

Why did clergymen suborn their role of providing genuine moral leadership to the needs of the military? It is often said that the emphasis on the 'royal' and 'army' (as opposed to 'chaplains') in the Royal Army Chaplains' Department was due in part to the status of the Church within a class-based society. There is perhaps some truth in this comment. In Britain, the established Church (and individual clergymen) maintained their status through the government. Archbishops and bishops were appointed by the prime minister and their appointment confirmed by the Crown. They also sat in the House of Lords, and the leaders of the Anglican Church in particular were chosen from the ruling classes (in 1920, over half of the bishops in the House of Lords were linked with the peerage and the landed gentry, and in the century after 1860, of 242 bishops studied by one historian, all except twenty had attended public schools).[127] Although nonconformists and Roman Catholics hailed from lower social classes, they were anxious to move away from their marginalized position (at Irish or Italian missions in the case of the Roman Catholics) and consolidate their newfound power within society and within the military (it was only after 1915 that they had even been allowed to appoint chaplains to the forces). All the major denominations believed that they could do more good from within the system than outside it and they feared persecution by the state if they stood against government policy. This was true at the individual as well as the institutional level. Protest could (and, in the case of Bishop G. K. A. Bell's opposition to obliteration bombing during the Second World War, did) hamper or halt any chance of promotion within the religious establishment.[128] As the historian Albert Marrin, who studied the Church of England during the First World War, argued, the Church was

> imbedded in, and derives sustenance from, the social matrix and its associations with the state . . . The power and prestige these institutions enjoy are conditional, being conferred with the tacit understanding that they will never stray too far from society's prevailing values and interests.[129]

Just as crucially, clerics were concerned that an anti-militarist stance on their part would place too heavy a burden on the consciences of combatants: soldiers would be forced to choose between the state and the Church and, in the end, would be compelled to side with the state.

But as the historian Andrew Chandler warned in his powerful analysis of the responses of the Church of England to aerial bombing during the Second World War, it is important not to place *too* much emphasis on the Church's position as part of the state apparatus. Although he admitted that the leaders of the Anglican Church endorsed the status quo, they also believed that they had a moral part to play in government and an ethical duty to provide spiritual leadership. At important periods in history, clergymen *did* protest (the most prominent examples being over the People's Budget of 1909 and the Parliament Act of 1911). Chandler was concerned to emphasize the clergy's commitment to justifying themselves ethically, despite their class background and links with the state. In a Church which always emphasized the rights of individuals – even individual clerics – to speak according to their conscience, clergymen did have considerable freedom of speech.[130] Other factors need to be examined to explain why clergymen did not make more use of this freedom.

Most obviously, despite numerous official statements according to which the military chaplains' first duty was to the
Church,[131] the weight of military authority was omnipresent.
The will to protest against violations of moral codes was undermined by the fact that chaplains were sent to battle zones as
'military virgins'. Prior to their posting, much of their time
would have been taken up with baptizing babies and comforting
the elderly dying. Suddenly, they were transported to a horrifying environment where the provision of religious comfort was an
overwhelming need: to turn around and rebuke frightened, weary
and guilt-ridden men for 'seeing red' in battle would be unimag-
inably petty. In addition, clergymen's ignorance of military
matters was a severe handicap. The training of chaplains was
minimal and general training in military etiquette and procedures was almost non-existent, despite calls since the First World
War for the establishment of 'Chaplains' Bombing Schools'.[132]
They were given no information on how military commands
were to be interpreted. Most were sent out directly from
parochial work 'knowing nothing of military procedure or drill or
custom and very often still less of how to deal with men' and
although chaplains were increasingly integrated into the military, this was done at the expense of their status as spiritual
advisers.[133] Most chaplains simply regarded the conduct of war
as outside their purview: the 'experts' were the strictly military
officers and politicians.

Furthermore, clergymen were in a curiously ambivalent position *vis à vis* other military officers. In 1939, the Revd R. L.
Barnes of the Community of the Resurrection, in his *A War-
Time Chaplaincy*, reminded readers that while the chaplain in the
forces was responsible to two authorities (God and the armed
forces), he had to ensure that his actions did not hinder the cause
of war. The chaplain was a member of the army and he had to
follow its rules; he was attached to a unit, so he must give
fellowship to his comrades; he was an officer, so he owed a 'due
observance of its customary acts of fellowship' to the cadre of

officers.[134] The difficulties in maintaining a balance were noted in Everard Digby's *Tips for Padres* (1917): he conceded that the relationship of a chaplain to his superior authorities was 'obscure' and was based largely on the assumption that he was a 'man of honour'. On the one hand, Digby advised chaplains to retain a firm hold on their spiritual authority and to resist attempts by commanding officers to usurp their role. On the other hand, they had to acknowledge their officers' superior authority: political matters were not to be broached in services; chaplains should not interfere in matters of discipline; and they should always remember that they were not experts in the 'highly technical subject' of modern military conduct.[135]

If their military training was inadequate preparation for the demands placed upon them during wartime, their spiritual education fared little better. For one thing, by the Second World War, theological colleges bombarded chaplains with psychological jargon: they were more liable to interpret difficulties men expressed about killing in psychological than in theological terms.[136] As the historian Albert Marrin argued in *The Last Crusade* (1974), most clergymen simply did not possess the necessary philosophical and theological training for their role in wartime. Moral education was practically absent from the syllabus at theological college. Even highly educated, erudite clergymen like Bishop Winnington Ingram (known as 'the Bishop of the Battlefields') relied for his sermons on quotations lifted from the *Daily Mail*. To make matters worse, religious training encouraged them to 'theologize' issues, casting them in terms of 'absolute, divinely sanctioned moral imperatives'. Thus, religious training, coupled with 'violent enthusiasm for an ideal', predisposed clergymen to see the world in terms of traditional radical antitheses: the eternal struggle between good and evil. Once the enemy was turned into the antiChrist, and one's own side was assumed to be leading the forces of good, there was no turning back.[137]

Finally, military chaplains inevitably experienced what some sociologists call 'role tension'. Gordon Zahn argued that in the struggle between religious mores and the military traditions, the military had the whiphand because the chaplain identified first with the men to whom he owed spiritual services. As a result, the chaplain became more likely to refer his actions to them than to any remote ecclesiastical establishment (like Father George Zabelka, who introduced this chapter, who viewed the men he served as 'lads of mine').[138] According to this line of thought, there were three ways in which an individual could reconcile his conflicting roles: he could abandon one of the roles, rationalize one of them, or compartmentalize the contradictory behaviours. The inability to do one of these three things would result in aberrant or 'unhinged' behaviour. Waldo W. Burchard interviewed sixty-one American military chaplains who had served during the Second World War and were living in the San Francisco Bay area. Over half of the respondents denied that any conflict existed between military requirements and their religion. Three respondents even denied that military regulations involved any moral values at all. In other words, they compartmentalized their roles so that while they were performing religious duties, they judged their own actions (and those of others) according to religious precepts and while they were performing military offices, they judged their behaviour (and that of others) according to military requirements. Just under half of the chaplains believed that the killing of enemy soldiers was a righteous act and the rest regarded it as justifiable. None felt that the individual soldier had any moral responsibility in the matter, except to serve his country. Only 7 per cent ascribed *any* moral content to the act of killing and, even in these cases, the burden of guilt was said to be shared by the entire nation rather than resting on the shoulders of individual soldiers. When scripture was cited, the chaplains were less self-assured. None of the chaplains felt any conflict over the commandment 'Thou Shalt Not Kill', since it was interpreted to prohibit 'murder' rather than slaughter in battle. However, one fifth of the chaplains agreed that the commandment to 'turn the other cheek' was incompatible with warfare. The remainder adopted a relativist position, drawing a distinction between the individual and the nation, stating that the need for self-defence overrode all other concerns and arguing that pacifism could not be upheld in the modern world. Most interesting, though, was the fact that none of the chaplains mentioned (of their own volition) that wartime killing might pose religious problems and, when questioned about it, they responded on a pragmatic rather than a theological basis. In Burchard's words, their response

> violated one of the assumptions of the study, namely, that chaplains, being educated men and philosophers who are concerned with the consistency of self, would have sought solutions to such a dilemma. Instead, it appears that the greater the dilemma, the greater the tendency to withdraw from it – to compartmentalize role behaviors and to refuse to recognize conflicting elements.

Rationalization would have demanded facing the dilemma while compartmentalization simply entailed a refusal to face it.[139]

Significant though such studies were, there was one crucial point they tended to overlook. Writing from perspectives which were hostile to the armed forces, both Burchard and Zahn imagined that chaplains and the religious establishment possessed a different interpretation of moral actions to military officers and their institution. As this chapter has suggested, it was possible that there was no perceived difference of ethical belief between religious and military hierarchies. This argument is supported by a survey of chaplains carried out by Clarence L. Abercrombie in the 1970s. At one point, he asked both civilian and military chaplains to indicate how each of thirty value-laden attributes would be evaluated by, first, an ideal military officer and, second, by an ideal Christian clergyman. Although the chaplains did disagree on specific points, *every* clergyman saw clerical and military values as almost identical. In other words, they genuinely did not perceive any conflict between their roles.[140] In this context, it is not surprising to find that during the 1914–18 conflict, the lay sons of the clergy supplied 30 per cent of the officers of the army,[141] nor that many priests felt comfortable in the military. Indeed, there were many similarities between the religious and the military life, particularly for Roman Catholic priests and ministers straight from the seminary; long-serving Regular army officers and military chaplains shared a dedication to a higher authority and service, a concern for welfare, and were accustomed to an all-male environment.[142] Religious values, such as self-sacrifice, courage and discipline, were held in common. In other words, there was no need for religious men to change their attitudes or compartmentalize their roles. Well over one third of chaplains had served in the armed forces before becoming padres, so they were familiar with what would be required and were in agreement with those values.[143] The question of whether a chaplain should protest to religious authorities as opposed to military ones did not arise because, until the war in Vietnam, it was understood that religious ethics were in line with the military.

It is important to note, however, that many *combatants* perceived a conflict between the military and spiritual functions of padres. It seemed absurd drama or just plain hypocritical when, immediately after a rousing speech by officers instructing them to massacre hordes of Huns, chaplains were heard to counsel mercy.[144] The 'skill' with which the 'pulpit professional' ignored the Christian commandments amazed the poet Wilfred Owen (as he wrote to his mother while incarcerated in the 41st Stationary Hospital suffering from shell shock).[145] Padres were even accused of pressuring conscripted conscientious objectors to fight.[146] When chaplains told worried young soldiers that it was right to kill, the soldier might 'muse upon the perspective of a religion that frowned upon the sins of the flesh and then condoned the taking of human life'.[147] Sincere Christians might refuse to take communion because they were learning how to kill.[148]

The tortuous internal conflicts which killing inspired are revealed in the letters and diaries of the journalist, Charles Edward Montague. He admitted to his wife on 26 November 1917 that he was unsure about how to interpret Christ's opposition to warfare: 'it must be either the enemy's victory or ours' and if the enemy won 'the "Kingdom of Christ on the earth" would be more remote than it will be if we win'. At the end of 1917, he confessed in his diary that he no longer believed that taking part in war could be reconciled with Christianity:

> But I am more sure of my duty of trying to win this war than I am that Christ was right in every part of all that he said, though no one has ever said so much that was right as he did. Therefore I will try, as far as my part goes, to win the war, not pretending meanwhile that I am obeying Christ, and after the war I will try harder than I did before to obey him in all the things in which I am sure that he was right. Meanwhile may God give me credit for not seeking to be deceived, and pardon me if I mistake.[149]

Others were even more disillusioned by the failure of the Church to speak loudly enough against senseless slaughter. As a radio operator who was engaged in a mass raid on Hamburg wrote to Canon Collins:

> It was a nightmare experience looking down on the flaming city beneath. I felt sick as I thought of the women and children down there being mutilated, burned, killed, terror-stricken in that dreadful inferno – and I was partly responsible. Why, Padre John, do the Churches not tell us that we are doing an evil job? Why do chaplains persist in telling us that we are performing a noble task in defence of Christian civilization? I believe that Hitler must be defeated; I am prepared to do my bit to that end. But don't let anyone tell us that what we are doing is noble.

What we are doing is evil, a necessary evil perhaps, but
evil all the same.[150]

During the Vietnam War, the attitude of combat soldiers
towards military chaplains hardened as the war became increasingly difficult to justify on political and spiritual grounds. Mike
Pearson was profoundly disturbed by killing an eight-year-old
boy armed with only a penknife, but the Catholic chaplain told
him not to worry: he was only doing his duty.[151] The black rifleman, Haywood T. 'The Kid' Kirkland was dismissive of the
chaplain's role in giving men a right to kill: 'As long as we didn't
murder, it was like the chaplain would give you his blessing. But
you knew all of that was murder anyway.'[152] Or, in the words of
Andrew Treffry of the Royal Australian Regiment Light
Armourer, in a letter to his fiancé in 1968, it 'really confuses one
when you find a man of God amongst a group of trained killers.
It's real hypocritical when you think of it.'[153] In the ranks, hostility toward military chaplains was especially pronounced, as
the psychiatrist Robert Jay Lifton discovered:

the men felt that it was one thing to be ordered by command to commit atrocities on an everyday basis in
Vietnam but another to have the spiritual authorities of
one's society rationalize and attempt to justify and legitimize that process. They felt it to be a kind of ultimate
corruption of the spirit.[154]

In what was clearly a dirty, unChristian war, even clergymen
began to doubt their role (it was 'a little like the way it was for
chaplains in Nazi Germany', mused one chaplain).[155]

From their pre-eminent position as moral arbiters in civilian
society, during times of war the chaplaincy reached an all-time
low. Although some religious representatives consistently stood
out against shedding human blood and there were pained con-

sciences over the cutting of ties with religious communities in hostile countries, reconciling combat with Christianity was remarkably non-controversial. Indeed, in wartime, many clergy-men longed to take part in the blood-letting and, frustrated in their ambitions, sanctified the slaughter and surrendered them-selves to the cult of arms. It was possible to kill without sin, and without guilt, they assured combatants. Spiritual guidance was in perfect harmony with military needs. Some men of God went further, leaving the ministry in order to enlist in the ranks. For these men, a combatant role was seen as more than merely an escape from dull routines: it was an assertion of their manliness and (by extension) the virility of the Church.

Chapter 10

WOMEN GO TO WAR

O, damn the shibboleth
Of sex! God knows we've equal personality,
Why should men face the dark while women stay
To live and laugh and meet the sun each day.

Nora Bomford, 'Drafts', 1914–18[1]

'Ne shanni!'

An English soldier of the 2nd Infantry Regiment, Serbian Army, took aim.

'Balli!'

She fired. As her bullet found its mark in the body of an enemy soldier, Flora Sandes permanently relinquished her role as a Red Cross nurse dedicated to saving lives in order to join a fraternity devoted to killing. This jolly, buxom daughter of a retired vicar living in the peaceful village of Thornton Heath in the Surrey countryside was an unlikely candidate for the warrior role but, on the outbreak of war, her elementary medical and military training in the Women's First Aid Yeomanry Corps and St John's Ambulance provided her with a chance to escape from dull routines. On 12 August 1914, in the company of six other eager nurses, the thirty-eight-year-old Sandes steamed out of London bound for Serbia. Her transition from nurse to soldier took eighteen months to achieve but during the great retreat to Albania, Sandes finally exchanged bandages for the gun. It turned her moral world upside down: instead of empathetic identification with suffering, she admitted to feelings of joy when the savage explosion of her bombs was followed by a 'few groans and then silence' since a 'tremendous hullabaloo' signalled that she had inflicted 'only a few scratches, or the top of someone's finger . . . taken off'. She enjoyed 'stalking' the enemy and admitted that throughout her seven years' service 'romance, adventure and comradeship' compensated abundantly for 'incessant fighting, weariness indescribable'. Most importantly, she felt accepted by her comrades. In one breath, Sandes was 'Nashi Engleskinja' (or 'Our Englishwoman'); in the next, she was 'Brother'. She was welcomed into the regiment as a representative of England. Indeed, nationality was more important than gender: on the same day that she enlisted as a private, a young Greek man also offered his services but was turned away on the grounds that the regiment would 'have no foreigners'. The photographs she took after each bloody encounter were to become souvenirs of this sense of belonging.

Where did this sense of fulfilment come from? On her own admission, becoming a soldier was the culmination of her desire to be a man – a feeling that dated back to her earliest childhood when she would kneel by her bed, praying that the morning would see her transformed into a boy. The fact that Sandes was a woman attracted the attention of the British authorities and general public, both of whom eventually supported her decision to fight. However, the occasional suggestion that she should revert to being a nurse infuriated her and she curtly informed one correspondent on 10 November 1916 that if people thought she ought to be a nurse instead of a soldier, they should be told that 'we have Red Cross men for first aid'. She insisted on acting as a soldier, and being treated as such; therefore, like male combatants, she cared for the wounded, but only 'between shots'.

In contrast to the attention she attracted in civilian society, when on active service her gender only rarely drew comment. Admittedly, commandants swung from treating her as a soldier to patronizing her as a woman, but this was probably as much to do with the fact that she was an *English*woman as with her femaleness. After all, they did not treat the Serbian women who served as privates in the army as anything except soldiers. The unfortunate effect of these rare instances of discrimination was to sting her into proving her military prowess by remaining longer at her post than her male comrades, although such exceptional dedication did not help her when she was wounded and risked being sent to a hospital for sick nurses rather than being admitted into the military hospital. But she had the support of her comrades-in-arms: it was the soldier who took her to the hospital who insisted that she be treated as a 'wounded Sergeant' rather than a 'sick sister'. The only other type of discrimination she experienced was of her own choosing: she fired a light carbine instead of the longer, heavier rifle of French military pattern carried by her comrades. Despite occasional problems, she believed that the lighter carbine served her well throughout the campaign. Her martial valour was were recognized in June 1919 when a special Act of Parliament made her the first woman to be commissioned in the Serbian Army.

The military was a wonderful life for Sandes, offering her previously unimagined freedom. Consequently, reverting to civilian life was frustrating.

> I cannot attempt to describe what it now felt like, trying to get accustomed to a woman's life and a woman's clothes again; and also to ordinary society after having lived entirely with men for so many years. Turning from a woman to a private soldier proved nothing compared with turning back from soldier to ordinary woman.

It was, she continued, 'like losing everything at one fell swoop, and trying to find bearings again in another life and an entirely different world'.[2]

*

Flora Sandes was exceptional, but her moral response to combat was not aberrant. Although women were generally refused access to the frontlines of destruction, they were an integral part of the killing process in a number of other ways. They were much more than handmaidens, madonnas, and patriotic mothers. Women did not thrust bayonets into living flesh, but they imagined doing so. When they were refused arms, they fought back: demanding combat training and taking it upon themselves to learn how to fire weapons. Most importantly, they did this in the name of 'womanhood'. While the armed forces assured their male recruits that warfare would 'make men' of them, it tried to reassure women of the opposite: the military would *not* 'make men' of women but would enhance feminine stereotypes. Ironically, while the military could not fulfil their promise to male recruits (a large proportion of whom found themselves physically or psychologically dismembered by combat), they had more success with their female supporters. Thoughout the twentieth century, however, there was a problem: gender identities were much more uncertain, less contained, than was allowed for. While it was commonly agreed that there was no clearly distinguishable 'masculinity' which assured ideal combat behaviour, the military could fall back on traditional stereotypes that were familiar and allowed for a degree of diversity. When women were heard clamouring for a more active role, historical stereotypes were called into question and military institutions and masculine society felt threatened.

Femininity and pacifism

Battle narratives (as opposed to other kinds of war stories, most notably those concerned with the so-called 'home front') have tended to contrast feminine pacifism with a more bellicose masculinity, with one school of thought believing that such differences are biologically determined and another that men and women are moulded into these stereotypes by various social agents. Despite recent attempts to break down this dichotomy,[3]

discussions which strictly separate male and female 'spheres' in
wartime have prevailed. For instance, Alix Strachey's psychoanalytical study of war in 1957 insisted that the rise of women in the
public and political sphere would do much to eradicate war.[4] In
a feminist analysis of war and masculinity, Lynne B. Iglitzin
belted out a familiar refrain: while most people had the capacity
to act aggressively, 'it has been *men* who have fought in wars'. In
our society ('as in almost every patriarchal society', Iglitzin
reminded her readers), 'militarism and violence have been identified with manliness'.[5] Or, as Helen Caldicott asserted in *Missile
Envy* (1984), women have been more pacific than men because
they give birth.[6] Militarism would be stifled by the womb.

In the first half of this century, this identification of women
with a pacifist temperament was not so certain. Certainly, the
dominant rhetoric praised women's nurturing nature as either a
'natural' consequence of their potential to create life or as something learned on their mothers' laps. The immensely popular
pamphlet allegedly written by 'A Little Mother' (1916), which
sold 75,000 copies in less than a week, expressed this ideology in
no uncertain terms: 'women are created for the purpose of giving
life, and men to take it', it broadcast.[7] Journalistic responses to
frontline gender roles made similar observations. In a book on
English soldiers, called *Golden Lads* (1916), Arthur Gleason con-
trasted the boastful accounts by male soldiers about the number
of enemies they had 'potted' with the delicacy with which
Englishwomen cared for wounded Germans. While sharing
hardship alongside their menfolk, these women 'had no desire for
retaliation, no wish to wreak their will on human life', he
observed. Danger did not 'excite them to a nervous explosion
where they grab for a gun and shoot the other fellow'.[8] The feminist and pacifist Helen Mana Lucy Swanwick noted in 1915
that when women *did* seem to be supporting the war effort, this
was only due to their sense of familial loyalty and a refusal to
shun the sacrifices made by their menfolk.[9] More than twenty
years later, the Bloomsbury feminist Virginia Woolf reminded

her readers that 'scarcely a human being in the course of history has fallen to a woman's rifle' and the suffrage leader Carrie Chapman Catt rejoiced in affirming that women were 'devoid of the war spirit'.[10]

There have been powerful reasons why feminists (like Woolf) joined with conservatives (like 'A Little Mother') to embrace a rhetoric which stressed gender difference, whether this was constructed in biological or in cultural terms. For both groups, women's social influence and political advancement was at stake. The power of middle-class women as domestic and moral arbiters depended upon their separation (at the theoretical level more than in practice) from the sordid world of money-making and life-taking. On the other hand the positive relationship between bearing arms and citizenship (particularly the right to vote) was a pre-eminent concern for early-twentieth-century feminists.[11] While radical suffragettes effortlessly translated their aggressive oratory into recruitment speech, others attempted to garble any suggestion that there was an inverse relationship between fighting and suffrage. A First World War cartoon (published on the first page of the *Woman's Journal*) is an example of the latter efforts. It showed a woman holding a baby and saying 'Votes for Women'. Nearby, a heavily armed soldier declared that 'Women can't bear arms', to which a suffragist is replying, 'No! Women bear armies.'[12] Although radical in tone, such reminders were no repudiation of war but an assertion of separate spheres and a demand for equal, gendered recognition of *both* violence and sacrifice.

Opposed to such assertions of gender difference was a long line of women posing as men in order to go into battle. What was considered to be 'appropriate' female behaviour could shift dramatically within relatively short periods of time and by geographical location (the most startling reminder being the American frontier where hand-to-hand combat with Indians was considered to be behaviour wholly appropriate to the good American wife and mother).[13] There is also a formidable history of individual female combatants. 'Molly Pitcher' (1778), Lucy Brewer (war of 1812), Sarah Borginis (Mexican War), and Sarah Edwards (American Civil War) continue to generate exciting narratives. Whether we focus on great icons such as the Amazons, Boadicea, and Joan of Arc, or draw attention to the sedentary labours of women like Maria von Clausewitz who edited her husband's 700-page classic *On War* after his death, female bellicosity and the feminine warring imagination have a long and distinguished past.[14]

In the twentieth century, also, the idea that women were more peace-loving than men was being questioned. Early-twentieth-century feminism had infused a new uncertainty into assumptions about the gentleness and nurturing behaviour of women. During the major conflicts, most commentators noticed that women constructed elaborate and pleasurable fantasies around the murderous antics of their menfolk. Indeed, as we saw in Chapter 5, it was widely feared that women were taking more vicarious pleasure in the bloodshed than male combatants. Even the feminist and pacifist Helen Mana Lucy Swanwick ruefully admitted that although men made war, they could not have done so had women not been so adoring of their efforts.[15] Caroline Playne agreed, writing in the early 1930s that the 'souls of women were as much possessed by [military] passion as the souls of men'.[16]

Throughout the twentieth century, novels, short stories, magazine articles and autobiographies indulged the demand to hear more about female combatants. It was generally agreed that there seemed to be 'a great desire' on the part of women to 'take some active part on the battlefield'.[17] The popular press highlighted feminine fondness for the gun. During the Second World War, Miss Marjorie Stevens, a seventeen-year-old member of the Australian Women's Army Service (AWAS[18]), begged to be allowed to go overseas: 'I would just like to have a go at the enemy,' she pleaded. 'Give me a rifle and I would be satisfied if I only got one of them!' Similarly, Miss June Buckley of Kings

Cross (Sydney) argued, 'Why should women always be asked to be the cooks? I want to go overseas and take my part with the men. Girls in Russia have proved capable fighters and Australian girls could do the same as they have!'[19] Furthermore, the autobiographies of women such as Flora Sandes testified to the joy of being a warrior. Peggy Hill, a member of the Women's Royal Navy Service during the Second World War, never got to shoot at anyone, but would have. She volunteered to do rifle training: 'I was quite a good shot,' she recalled, 'It was like playing darts, I thought! And I didn't give it a thought that there might be a person at the other end.'[20] They could also be more warlike than their husbands. Joyce Carr worked on an AA gun site during the Second World War. She later admitted:

> I never worried about killing when I was on the guns: I
> wasn't actually killing the Germans, I was killing those
> that were flying with their bombs. I thought that was
> good, I really felt that. The only thing Tom [her husband,
> a bomber pilot] worried about later on was when he saw
> how much damage he'd done, and how many people were
> killed. But in war the innocent do suffer, don't they?[21]

Fictional accounts also glorified well-armed, spunky women (in the 1945 issue of the *Girls' Own Paper* a couple of recently demobbed officers from the Women's Auxiliary Air Force coolly threatened to shoot the men who refused to let them refuel their aeroplane).[22] In the writings of Tim O'Brien, the story was told of a woman (Mary Ann) who visited her lover in Vietnam. Here, women were permitted to bear arms for defensive purposes,[23] but Mary Ann went further than this: she eventually became a particularly aggressive combatant in the Special Forces. She explains to her former lover (who worked within a safe area) the pleasures of martial violence:

> You just don't *know* . . . You hide in this little fortress,

behind wire and sandbags, and you don't know what's out there or what it's all about or how it feels to really live in it. Sometimes I want to *eat* this place. Vietnam. I want to swallow the whole country – the dirt, the death – I just want to eat it and have it there inside me . . . When I'm out there at night, I feel close to my own body. I can feel my blood moving, my skin and my fingernails, everything, it's like I'm full of electricity and I'm glowing in the dark – I'm on fire almost – I'm burning away into nothing – but it doesn't matter because I know exactly who I am. You can't feel like that anywhere else.[24]

For women, as for men, killing could be a profoundly sexual and an exhilarating experience. In an article entitled 'The Ecstasy of the Fighter Pilot, October 1939', F. Tennyson Jesse enthused over the way a Spitfire pilot destroyed his enemy, explaining that his exploits 'made us all feel very elated'. She compared the act of killing in war with human desire. 'Love is more beautiful for the lust that is inextricably mingled with it', she noted, so too the 'high qualities of a fighting man' could not be separated from the 'ecstasy of the hunt and the kill, even when the quarry is another human being'.[25] F. Tennyson Jesse might have echoed the complaint of the heroine of Margaret W. Griffiths' novel, *Hazel in Uniform* (1945) when she gloomily explained to her brother that she was unhappy because 'Daddy's in the Army and *you* are in the Air Force, while I . . . I am nothing but a girl at home'[26] or she might have recited the lament of the poet Rose Macaulay:

> *Oh it's you have the luck,*
> *out there in blood and muck.*[27]

Women's role in wartime

Despite these desires for a more active role in battle, the behaviour of women in war was not particularly violent. But even so, traditional, non-combatant roles continued to meet with some opposition (especially from middle-class parents such as Vera Brittain's).[28] Of 706 women interviewed in the early 1990s about their military service during the Second World War, 41 per cent recalled opposition from close relatives and only half said that their closest male friend was supportive (in contrast to gaining the support of 80 per cent of their closest female friends).[29] Women's experience of war continued to be primarily concerned with persuading men that dishonour was worse than death, and then buckling on men's psychological (if not military) armour.[30] As Virginia Woolf put it in *A Room of One's Own* (1929), women served as magnifying mirrors 'reflecting the figure of man at twice its natural size. Without that power . . . the glories of all our wars would be unknown.' Such mirrors, she continued, were 'essential to all violent and heroic action'.[31]

Yet certain groups of women were expected to play a *larger* role in wartime than certain groups of men. Thus, an American public opinion poll carried out on 10 March 1944 revealed that three quarters of people who were asked whether they preferred the army to draft 300,000 young single women for non-combatant duties or to draft the same number of young fathers for the same work, preferred drafting single women.[32] Similarly, an American survey of nurses stationed in Saipan and Oahue revealed very high levels of agreement that female medical personnel had a duty to be posted to the frontlines.[33] As medical personnel, these women were exempted from bearing arms (unless directly attacked), in the same way as male medical personnel. Aside from nursing, what were these non-combatant duties? The range of roles open to women was immense: they worked in factories and farms, harvested crops and cared for troops on leave. They also enlisted in the armed forces. In the First World War, approximately 80,000 women served in the three British women's forces. In addition, over 25,000 American women saw overseas service. During the Second World War, British women became liable for the first time in British history for compulsory service in the armed forces or in other forms of national service. Consequently, the proportion of British women in the armed forces rose from 2.3 per cent (or 66,900 women) in December 1940 to a high of 9.2 per cent (449,100) by December 1943, before dropping to just over 8 per cent (415,800) by the end of the war.[34] Around 40,000 Australian women served in the forces during the Second World War. American estimates of the number of women in all branches of the military between 1939 and 1945 exceed 350,000. Between 7,500 and 15,000 American women served in Vietnam.[35] Of course, women were not spread evenly between the three branches of the armed services. There were proportionately more women serving in the air force, for instance, in part because of the 'moral' nature of aerial combat but largely as a result of the higher ratio of support to combatant personnel in this force.[36] With certain exceptions (such as in the Special Operations Executive (SOE) which sent fifty female agents to fight in occupied France), women were still not regarded as 'combatants'.[37]

In the context of this book, however, exhorting departing troops to 'kill the bastards'[38] and being exhilarated by gruesome accounts of the number of enemy slaughtered by their menfolk[39] was very different from actually desiring to kill. The poet Nora Bomford, who railed against 'sex' for excluding women from combat in the poem introducing this chapter, was contributing to a wider debate about the armed status of women working in combat zones. Should they be allowed to fire weapons? Although women have tended to be more dovish than men (opinion polls since the 1930s consistently show that women were less keen on violence in international relations than were men),[40] surveys asking whether or not women should be armed must be analysed cautiously. D'Ann Campbell surveyed 706 female veterans from the Second World War (221 of whom had been nurses, and the remainder had served in one of the other branches of the armed forces) and discovered that one quarter believed women should be allowed to serve in combat. One third of the nurses and one half of the non-nurses felt that women should *not* be allowed to serve in combat, even as volunteers, and one third of the nurses and one fifth of the non-nurses were undecided. It was not surprising that members of veterans' organizations were the most supportive of allowing women in combat units.[41] Given the fact that Campbell was interviewing women who had served in the forces nearly fifty years earlier, these statistics show remarkably high levels of approval for female combatants.

Other historians have generally used such evidence as proof of the strength of feeling *against* the breaking of gender norms. For instance, three researchers from the University of Maryland and the US Army Research Institute used a survey conducted during the Second World War and two surveys conducted in the 1970s to argue that the prohibition on arming women had met with widespread agreement. In this article, published in *Sex Roles* (1977), David R. Segal, Nora Scott Kinzer and John C. Woelfel examined a British poll which was carried out in November 1941 where respondents were asked whether they would approve or disapprove of women being allowed to become fighting members of the armed forces. Sixty-five per cent disapproved, 25 per cent approved, and 10 per cent were undecided. These statistics can, however, be turned around to read that over one third of respondents either approved of arming women or were sufficiently ambivalent about it to register an 'undecided' verdict. This is no clear-cut abhorrence of combatant women. The second survey these researchers cited was carried out thirty-two years later in the Detroit area of the United States, in which 576 adults were asked to state the degree to which they agreed with the statement that 'if anyone should bear arms, it should be men rather than women'. Three quarters of the civilian respondents agreed or agreed strongly with the statement. Again, this response could be turned on its head to show that one quarter of respondents did not regard it as obvious that men rather than women should bear arms. Furthermore, the response does not indicate how people might have felt about women bearing arms once the male potential was used up or in the case of invasion. The third survey cited by these three researchers involved 724 American army personnel in 1974. About 60 per cent felt that women should not serve on the frontlines; around half did not think that women would make good frontline soldiers, even if trained; and over half felt that if women were assigned to combat units, the army would become less effective. Three quarters of the sample registered the opinion that women should not become combat infantry personnel. Again, if we turn these statistics around, a very different image is created: about 40 per cent felt that women *should* serve on the frontline and around half stated that women would make good frontline soldiers and would not harm (and might even enhance) combat efficiency. Admittedly, three quarters thought that women should not be allowed to join combatant infantry units, but there is no indication here that they should be excluded on the grounds of feminine roles, rather than because of perceptions about women's physical strength and endurance levels in comparison with men.[42] However these surveys are interpreted, they do not support arguments that there should be an impenetrable barrier to women becoming combatants.

Female warriors?

When women *did* kill, their ability to do so was explained under one of two mutually exclusive headings: psychosexual confusion or maternal instincts. Flora Sandes would fall into the former category since the first sentence of her book proclaimed that she always wanted to be a boy.[43] Despite late-nineteenth-century pathologizing of lesbians and the burgeoning interest in sexology during the 1920s and 1930s (with its obsessive interest in sexual inversion), Sandes was completely unabashed by this admission. She described intimate friendships with men and women with equal fervour and felt no need to emphasize her eventual marriage to a Russian sergeant. She wanted to be a boy; she became a soldier; she loved and married a man: an uncomplicated trajectory, Sandes seemed to assume.

Other effective female warriors were painted with this 'mas-

culine brush'. Tim O'Brien, in the story of the Special Forces
female killer mentioned earlier, noted how Mary Ann's body
changed once she took up the rifle and grenade. Her body
became 'foreign somehow – too stiff in places, too firm where the
softness used to be . . . Her voice seemed to reorganize itself at a
lower pitch.'[44] The way in which combat effectiveness was somehow inscribed on the body and personality can also be seen in the
context of the Second World War. In 1945, Megan Llewellyn
McCamley (a keen member of the ATS, who described herself as
hating sewing) recalled a woman in her unit called Private Nancy
Brown. Private Brown was a single woman who hailed from a
Glasgow slum but no longer had any family alive. She was useless
in the conventionally 'female' military jobs since she knew nothing 'that smacked even faintly of domesticity'. As a strong,
clumsy woman, Private Brown eventually ended up heaving coal
and digging trenches. She 'loved to work with the men, not for
any other reason than that she fitted in with their kind of work.
She was "Matey" with them in a sexless sort of way,' McCamley
tells us. Private Brown soon tired of the spade and went Absent
Without Leave, not to be seen in person again. One day,
McCamley went to the cinema and – to her great surprise –
identified the 'daring guerrilla fighter' on the screen as Private
Brown. McCamley flushed with pride, recognizing that Private
Brown 'needed to live dangerously, beyond anything that we were
prepared, or needed to allow in the A.T.S.' She wanted to yell
aloud: 'That's one of my girls, one of the A.T.S.!'[45] McCamley's
breathless prose would have alienated many other commentators
who despised undomesticated women like Private Brown and
indeed, such 'masculine type' women were increasingly caricatured as freaks and lesbians or as 'hysterical creatures such as the
French Revolution produced'.[46]

Supporters of women's corps obviously appealed against such
crude stigmatization which endangered recruitment as much as
morale (in the ATS, for instance, such stereotypes damaged
combat effectiveness by causing women to become obsessed with

their femininity and fertility).[47] Spokeswomen for the female armed services repeatedly insisted upon their commitment to traditional female values: the military would not 'make men' out of these women but would reaffirm the feminine principle. The ousting (in 1941) of Dame Helen Gwynne Vaughan from her position as Chief Controller of the ATS was one example of the difficulties that too obvious identification with 'male' values could cause. Although Vaughan was ostensibly replaced on grounds of age (she was in her sixties), her successor told a different story. According to Dame Leslie Whateley, Vaughan had made herself unpopular by being 'so imbued with the military spirit that she was quite unable to see that women could not be treated as men'. Whateley adopted a different standpoint: 'Before all else we were women, and before anything else came our determination to remain so.' Indeed, Whateley took such a stance to an extreme: in her autobiography (published in 1949) she went out of her way to deny any knowledge whatsoever of rifles and to assure her readers of the fact that she was profoundly 'gun-shy'.[48]

The 'womanliness' of the female services was also stressed by other propagandists. Authors such as J. B. Priestley were employed to counter fears about the defeminization of service-women. In *British Women Go to War* (1943), Priestley placed great emphasis on the girlish charms of women attending to the guns. He informed his readers that women in the Women's Royal Naval Service (WRNS), for instance, donned 'very feminine saucy style' hats while servicing machine-guns on board motor torpedo boats. In his words:

It is part of the fascinating irony of our strange times that this fair and meditative maiden, who looks as though she might be dreamily contemplating a cluster of bluebells in a spring wood, is actually preparing for its dreadful task a thing that may soon blow a great hole in a ship's side and send hundreds of men to their death.[49]

As Robert Williamson succinctly put it in his defence of the Women's Volunteer Reserve: these women were 'taking domesticity with them into the firing line'. They were neither 'a pack of Amazons' nor were they 'cranky'.[50]

But there was another line of argument which the female armed services found much more congenial to their ethos. In contrast to insinuations that female combatants were 'not really' women, it was asserted that women killed because they were super-feminine. Although sometimes caricatured and often feared, this group of female combatants commanded a much higher social prestige. There was almost complete agreement that women would have little difficulty killing in defence of their husbands, lovers and children. Just as the pacifists mentioned earlier on contended that women could not kill because they were mothers, so, too, explanations for why women *did* kill were linked to their maternal nature.

The idea of female aggression in defence of the weak owed a lot to contemporary psychological thought, particularly instinct theory. According to this view, men were able to kill because they possessed the 'killer' or pugnacious instinct, evolved through centuries of combat. At one level, instinct theory provided commentators with a ready explanation for women's exclusion from combat: for evolutionary reasons, women simply did not possess this instinct, according to William McDougall.[51] Clyde B. Moore in 'Some Psychological Aspects of War' (1916) agreed, observing that throughout the ages it was men rather than women who hunted and fought. This was not only true of the human species, but of the 'higher animals' such as buffaloes, wild horses, deer, apes and monkeys. The absence of this instinct also resulted in women being physically smaller and weaker than men.[52] Paradoxically, though, while women lacked the 'killer instinct', they possessed another instinct which facilitated killing – that is, the maternal instinct. The military chaplain and amateur psychologist, W. N. Maxwell, writing immediately after the First World War, recognized that many women would have been prepared to fight 'under the sway of the maternal instinct, with its protective impulse and its tender emotion, which had been roused by the sight of the wounded or the stories of outrage'. The 'primitive cave-woman' would be aroused to fight and kill, he declaimed.[53] Dame Helen Gwynne Vaughan of the ATS strongly supported this view, arguing that although women were more likely than men to 'shrink from inflicting injury', such an inhibition would disappear if they had to defend themselves or if their children, or if weaker persons were in danger.[54]

Such bellicose femininity bestowed a distinctive 'style' of killing on female combatants. Literary accounts describe this most vividly. In Captain Frederick Sadleir Brereton's novel *With Rifle and Bayonet. A Story of the Boer War* (1900), the gallant male heroes and a modest young woman called Eileen were trapped inside a cottage under attack. Eileen was given a rifle and in no time a 'pile' of dead and wounded Boers 'blocked the entrance to the homely English flower-garden'. One of the heroes, Frank Russel, 'hoarsely' justified the slaughter by claiming that although it was 'awful to have to kill so many of them', it was 'their lives or ours, and besides, we've a glorious cause to fight for'. After a long, bloody fight, however, Eileen suddenly gave way under the strain and fainted, murmuring that she could no longer 'bear to hear their groans!' In the few minutes before she regained consciousness, the author subjected his readers to a short homily on the relationship between femininity and martial combat, reminding them that it was 'naturally' a 'sore trial' for a 'timid and gentle-mannered girl, to be called upon to use a rifle in earnest and deal mortal wounds'. Brereton pointed out that the

> need for strength, and the stern struggle in which she had so bravely borne a part, had, however, braced her for the work. But now, when it was all over . . . and she saw the wounded and heard their groans, the terrible sight and the unusual sounds unnerved her, and she was prostrate for a moment.

The men in this novel showed no such delicacy as they merrily massacred Boers. But feminine sensitivity did not mask cowardice. Later, encouraged by the 'absolute pluck' of her father and the two young Englishmen, Eileen chose to continue fighting, refusing the safe passage offered to her by the Boers (she consequently 'sent many a Boer to his last account').[55] In this way, Eileen epitomized English femininity: in mortal combat, she was both manfully courageous and as sensitive as any 'true' Englishwoman.

A further aspect of this theory was that while men fought for ideological reasons (the 'glorious cause' referred to by Frank Russel), female combatants were said to be compelled to bear arms because they could not contemplate being separated from their ideologically committed menfolk (Eileen would not leave her father's side).[56] Similarly, female fighters during the Spanish Civil War were portrayed as pawns to male ideologies; they were being used merely as an 'example to their troops in the field' and as a 'basis for propaganda, as a method for creating heroines, and for propagating a thirst for revenge' according to the pro-Franco writers, William Foss and Cecil Gerahty in *The Spanish Arena* (1938).[57] Even the anti-Franco feminist, Edith Shawcross, was convinced by this interpretation, insisting that commentators face up to feminine psychology. She mourned a friend who had been killed in Spain alongside her fiancé, and then questioned whether her friend would have volunteered 'on account of her principles' if her lover had stayed behind. Certainly not, Shawcross asserted, 'my friend only went because she could not bear to be left behind'.[58] As we shall see later in the chapter, it was precisely this belief that women would willingly engage in mortal combat for the sake of their lovers, husbands and children which rendered them the most terrifying of combatants.

Defending the 'Home Front'

In the twentieth century, the arming of women has only taken place during revolutions and large-scale invasions. Since neither

British, American nor Australian societies had experienced either of these traumas, women were not systematically taught to defend themselves or their country. Despite their absence from the killing fields, however, modern technology was diminishing the distance between the shell-holes and suburbia at an increasing rate, causing pacifists such as the Revd A. Belden to warn members of the Women's International Peace Crusade ominously in 1935 that (given the technological nature of modern military conflicts) women would inevitably have to be armed.[59] On the eve of the second major war of the century, the young Oxford feminist, Edith Shawcross, made a related point, arguing that the conflicts in Spain and China were ample proof that 'romantic nonsense' about protecting women and children had been thrown out the window. Women were already fighting alongside men in Spain and Russia and, once the enemy opted to utilize every able-bodied person irrespective of gender, the British government would immediately follow suit.[60]

Furthermore, at various stages in the world wars, invasion was a distinct possibility and if a case could be made for the arming of women, it was precisely in terms of *civil* defence. Many commentators recognized this, sometimes regretfully. In Australia, the popular patriot, Ion Idriess, reasoned in his *Guerrilla Tactics* (1942) that women combatants were essential in the event of an invasion and he reminded his readers that 'there is no need . . . to be a physical giant to be a guerrilla'.[61] In Britain, the Lord President of the Council, Sir John Anderson, summed up a general feeling in his Parliamentary address on 24 March 1942:

> If invasion is attempted, there will be one thought only in the mind of everyone in this country: to drive out or destroy the enemy. Everyone will want to do all he can to contribute to this end . . . the Government have always expected that the people of these Islands will offer a united opposition to an invader and that every citizen will regard it as his duty to hinder and frustrate the enemy by

every means which ingenuity can devise and common
sense suggest.[62]

Or, in the words of the Chief Controller of the ATS in 1941,
while it was men's job to fight, 'in their absence it is better that
women should fill the gap than that their cause should be lost'.[63]
These were the grounds upon which it was argued that
women needed to be armed – for home defence. Indeed it was
common practice in wartime for female heads of manor houses to
teach male and female employees on their estates to shoot.[64]
Numerous women's organizations now sprang up to meet the
need for a combat-ready 'home front'. In Australia during the
First World War, the Australian Local Volunteers set out to train
its members to become 'real rifle women and real soldiers' and
each state had its own legion of women preparing for a combatant role.[65] In Britain at this time, the powerful backing of Lord
Kitchener and Lord Roberts (president of the Women's Rifle
Association) enabled the establishment of the Women's Defence
Relief Corps. This consisted of two divisions: the Civil Section
which aimed to substitute women for men in employment in
order to free men for military service, and the 'Semi-Military or
good-citizen section' in which women actively recruited for the
armed forces underwent training in drill, marching, signalling
and scouting, and were instructed in the use of arms. Each
member of the later section was exhorted to defend not only
herself but also 'those dear to her'.[66] A similar example can be
found in the context of the Second World War when the
Women's Voluntary Reserve, commanded by the Marchioness of
Londonderry, proved popular. In this Reserve women were
encouraged (although not obliged) to practise marksmanship in
order to repel any invader.[67]
Women had less success within the established (male) home
defence forces. Prior to the Second World War, the Local
Defence Volunteers or LDV (the predecessors of the Home
Guards) employed women's talents in training men to shoot

while refusing to admit female members. By 1930, these women had had enough. Fifty women (including the 1930 winner of the Kings' Prize on the rifle range at Bisley, Marjorie E. Foster of the Women's Legion of Motor Drivers) established the Amazon Defence Corps dedicated to petitioning the LDV to admit female members. They insisted that women in each of the military services should receive training in small firearms and set themselves the task of nurturing in all women 'the spirit to resist the invader by all means available'.[68] The military establishment was not sympathetic so, in October 1941, when the War Office once again refused pleas to allow women to extend their role within the Home Guard beyond their unofficial status as cooks, nurses, drivers and telephone operators, the Women's Home Defence Corps was formed to teach women a wide range of military skills, including how to fire guns and throw bombs. In the words of one of the founders, the Labour MP, Dr Edith Summerskill,

> after the fall of France it seemed to me that it was rather incongruous that a woman like myself could have a husband in the army, children evacuated, but be entirely defenceless, if the moment arrived when instructions went out that everybody must resist to the uttermost.[69]

As an argument against giving women arms training, the excuse that weapons and ammunition were in too short supply was dismissed: after all, 'in the event of a male Home Guard being killed then his rifle should be used by whoever was near him at the time, irrespective of sex'. Similarly, the Women's Home Defence Corps held no illusions that Hitler would 'be scrupulous about the sex of his opponents' since Hitler 'did not differentiate in the gas ovens'.[70] Thousands of women were convinced. By February 1943, this corps had attracted 20,000 members in 250 units throughout the country.[71]

The resistance of the Home Guard to female members was weakened by the efforts of the Women's Home Defence Corps and the Royal United Service Institute, as well as by the persistent lobbying of women like Mrs Knox and Dr Summerskill, and one sector of the Guard at least were convinced by these women's arguments. *The Home Guard Encyclopedia* (1942) declared:

> Why not [train women]? They have never yet let down
> their country, and many have been tested in the fiercest
> fire . . . It is merely stupid to imagine that women cannot,
> or would not, fight. They have done so in Russia, Spain
> and elsewhere, and in a war such as this, they are entitled
> to do so. What is more, I am sure that many would, for
> British women are as spirited and courageous as any.

According to the *Encyclopedia*, the international prohibition against arming civilian women was formulated in the context of 'civilized wars' and therefore had no validity in a war against Nazism. Women could not (and would not) sit back to watch their homes ruined by a nasty invader. Both sexes would inevitably embrace the motto: 'Give no quarter and ask none.'[72]

Given such prominent support within the Home Guard, it is not surprising that in some branches women *were* given combat training.[73] On 14 December 1942, the *Daily Mail* broke the news that twenty female air raid wardens in Holywell, Flintshire, were being taught to throw hand grenades and to shoot with rifles and Stens (light-weight machine-guns).[74] This caused great consternation within the War Office: though they accepted women in an official capacity as clerks, telephonists, cooks, waitresses and drivers, they continued to stand firm on the prohibition on assigning women to combative duties. In an urgent memorandum issued by the War Office on 15 April 1943, it was forbidden to train women in the use of weapons.[75] As one military officer explained: women might 'degrease rifles or clean small arms, but not fire them'.[76] Furthermore, these women were to become 'registered', not 'enrolled' members of the Home Guard, were given no uniform save a plastic badge, and their numbers were limited to 5 per cent of the existing ceiling for men. In spite of these restrictions, within a year they had been given the title of Home Guard Auxiliaries and had attracted over 30,000 members.[77]

On the home front, there was one specialist form of active combat in which all three countries recognized the usefulness of women, that is, in manning Anti-Aircraft (AA) gun sites. AA batteries were involved in the long-range sighting, firing and destroying of enemy aeroplanes. In most countries, women were only engaged in the first of these activities: only women in the United States military were systematically taught to actually fire the guns. By the end of 1942, the American military had fully integrated female personnel in its Anti-Aircraft Artillery unit designed to protect Washington, DC, against air attack. The experiment (shielded from public attention lest Congress vetoed it) was regarded as a success in the sense that the unit had efficiently trained women to fire guns as well as to co-ordinate attacks.[78]

In Britain, women worked in 'mixed batteries' (that is, in teams consisting of both men and women) as range-finders and co-ordinators of attacks on enemy aircraft. They were 'the eyes of the guns' which men fired.[79] In this sense they were as much combatants as (male) range-finders in artillery units on the Western Front. In Britain, the possibility of using women in AA and searchlight units was first investigated in October 1939 when General Sir Frederick Pile was General of Command in Chief, AA Command, although it took another year and a half until final authorization was obtained from the War Office. The first mixed battery to open fire did so on 1 November 1941 and, a week later, female personnel for the first time aided in 'scoring a category 1 success'.[80] By 1942, there were more women serving in AA Command than men and, within a year, 60,000 British women had been assigned to these batteries.

Although mixed AA batteries were deemed a success, it took

some time for Regular army personnel to accept them. For instance, Major M. S. F. Millington from AA Command had expected women in AA units to be emotionally unstable and he recalled with surprise that 'practically the only tears shed by the A.T.S. in A.A. Command during the war were those of frustration when they failed to get "On Target" or were prevented for any reason from firing'.[81] Another commander of a battery in England confessed that when he was first assigned a mixed battery he 'loathed the idea'. Because he had previously served only with men and had 'loved it', he attempted (unsuccessfully) to have his orders changed. However, his experience of commanding a mixed AA battery convinced him of the short-sightedness of his prejudices and he subsequently warned his counterparts against the tendency to treat female AA personnel as 'musical comedy "girl soldiers"': if they were treated as 'women soldiers', they acted as such, he concluded. Ruefully he admitted that he had 'never been happier than I am now . . . My men and girls are great'.[82]

Coastal artillery – which brought with it a greater risk of having to physically ward off attack – was a different story. Australia's enormous length of coastline meant that there was very little resistance to utilizing women to defend the coastline and they were issued with .303 rifles and bayonets and encouraged to become markswomen.[83] In Britain, the authorities were less confident and, during the Second World War, women had been prevented from taking up posts in coastal artillery units. There were three reasons given for their exclusion: they would be at risk of actual combat if raided, there were an insufficient number of women to meet the requirements of AA Command, and there were difficulties about providing separate accommodation in the more remote batteries. Manpower shortages by the 1950s provided the incentive to overcome these problems. It was recognized that the coastal artillery duties upon which the Women's Royal Army Corps (Territorial Army), or WRAC (TA), could be employed were similar to those duties already performed by women in Area Command, Seaward Defence, Anti-Aircraft Operations and Mixed Anti-Aircraft Batteries. WRAC (TA) who were trained in coastal artillery were also allowed to work on anti-aircraft artillery since the training was almost identical. It was admitted that the administrative difficulties of mixed units had been exaggerated. Finally, the familiar argument about the requirements of total war were employed: all people, military and civilian, were now at risk of actual combat and this new fact of modern warfare had to be reflected in military policy. Rather than becoming a disincentive for recruitment, it was noted that recruitment in the Territorial Army was actually boosted when women served alongside men (it was not clear whether it was men or women, or both, who liked the idea).[84] Women were therefore granted permission for the first time to be employed as OsFC and Rangefinders in Coastal Artillery.

Combat overseas

AA Batteries and coastal artillery were exclusively concerned with the defence of the home front. The debate about women's right to bear arms was much more difficult to resolve within the regular military forces where there was a chance of being posted overseas. In America during the Second World War, the Marines were the only branch of the services which formally gave women arms training. Although members of the Women's Army Corps (WACs)[85] were non-combatants and were not officially given arms training until 1975, they were not prohibited from doing weapons training in private. The War Department was not unduly concerned with such activities until the press brought it to public attention, creating what WAC advisers called 'a serious public relations problem'. It led people to ask: were manpower shortages so severe that women had to be called to fight for their country? Were WACs insufficiently occupied if they were able to find time to practise on rifle ranges? Did their activities constitute a waste of military materials and, in consequence, show disrespect for the labour of munition workers? The Director of the WACs (the formidable Texas businesswoman, Colonel Oveta Culp Hobby) thus issued a directive prohibiting the carrying, training and use of any weapon (or replica of a weapon) by any member of the WAC. Immediately, this led to problems in the field: women instructors in the air force used gun-shaped training devices; it was essential that WAC officers in the finance department wore, or had available, revolvers; women engaged in signal and communications duties were (by regulation) required to keep a revolver in the code room; in overseas operations, it was forbidden to take out any vehicle without arms. Hobby had to back down, allowing arms to be carried 'if the assignment was otherwise suitable and non-combatant, and if the women had suitable training'. She was, however, overridden by the Director of Personnel to the Army Service Force who insisted that women should not receive weapons training. Hobby continued the fight to allow certain exemptions to the prohibition of women bearing arms after she was transferred to the War Department, General Staff (G-1 Division). This Division published a circular which allowed commanding generals to permit the bearing of arms by specific, named women. The official history of the WACs, however, noted that 'this authority was so extensively abused by field commanders that G-1 Division rescinded it' after accusing field commands of encouraging 'wholesale participation of W.A.C. personnel in familiarization courses in the use of weapons and arms'.[86] In 1948, the Women's Armed Services Integration Act excluded women from combat assignments in the military.

Similar tensions existed within the Australian armed forces. In 1942, the Directorate of Military Training informed the Australian Women's Army Service (AWAS) that they would not be allowed to carry out arms training. Only personnel who were allocated to units where attack or sabotage was a possibility would be taught to use weapons and then solely for the purposes of self-defence. Women on duty in anti-aircraft and searchlight units were allowed to be armed, but with the lighter .310 rather than .303 rifles. Women performing guard duty were closely watched and (in the words of one official report) 'members not temperamentally suited for such duty' were to be 'immediately relieved without any adverse effect on their service record'. Once the war was over, however, AWAS personnel were refused arms training and stripped of their weapons.[87]

As all these debates suggest, the problem was not the use of women in manning anti-aircraft guns or other forms of long-distance, remote-controlled instruments of war, but the use of personal weapons. In 1949, the War Office debated the issue anew and reported their discussions in a document entitled 'The Defensive Role in War of Women in the Army'. They agreed that the basic function of soldiers was to fight and that (with the exception of a small Non-Combatant Corps) all military person-nel were required to undergo weapons training. With the full incorporation of women into the armed services, weapons train-ing became an extremely sensitive issue. Should women in all three services be treated identically to the male soldiers? Contrary to popular belief, it was noted that there was no legal restraint to training women in the use of arms. In the past, the exclusion had been solely a matter of policy, not law.

Arguments against arming women were considered. For one thing, arming women soldiers was thought to be distasteful to the broader public. Male soldiers would find it 'incongruous to say the least' to have to deal with well-equipped women. It was emphasized that women should never be placed in a position where men 'could not afford them the necessary protection'. In a service suffering from severe manpower shortages, it was feared that the arming of women might hamper recruitment even fur-ther. Not only might existing and potential members of the women's corps object, their parents might also. Financial con-siderations were also discussed, with attention being drawn to the extremely low risk that women would have to use their weapons, as weighed against the financial and time costs of providing them with weapons training. Finally, there was apprehension that if women were taught to shoot to kill they might 'forfeit the measure of protection which, unarmed, their sex would demand and which our opponents might concede'.

There were a number of arguments in favour of arming women in the Women's Royal Army Corps. It was noted that at least one other nation (the Soviet Union) had armed their female soldiers. There were also the rights of women to self-defence: if the military assigned women to situations where they might be placed at risk, they also had to provide them with the training and weapons to protect themselves. This was regarded as particularly important in Asia where 'conventions and a natural respect for women' were liable to be ignored. In such situations 'most women would much prefer to be shot rather than raped', the 1949 report avouched. Finally, it was argued, the speed of modern warfare meant that no theatre of operations could be deemed immune from attack.

The document concluded that since the women's corps was part of the fighting service, its members could not be denied means of self-defence. In time of war, members of the WRAC would be required to carry out their normal duties until operational conditions or the approach of the enemy rendered it impossible. They would then be withdrawn. If withdrawal was not possible, all ranks would be permitted to defend themselves against attack and those who had volunteered to be trained in the use of arms would be issued with weapons. Unlike men in these corps, no woman was to be compelled to bear arms or receive arms training and women were only to be issued with personal arms suitable for defensive (rather than offensive) purposes.[88]

The story was slightly different in the Royal Army Medical Corps (RAMC). The non-combatant status of servicemen in this corps posed special problems in relation to the arming of women. The corps was concerned lest the training of women in the use of personal arms would strip them of their right to special treatment in the case of capture. The Director-General of the Army Medical

Services warned local commanders that, if in extremity they armed women, they would be responsible for increasing the women's likelihood of being shot, with or without prior rape. He advised that women in the RAMC, RADC and QURANC (or QARANC) should not be considered in the same way as the men in these corps so far as weapons were concerned. Since women in these corps would 'always be occupied in professional duties with patients' they should be viewed as coming 'under the same category as patients' and would therefore be protected by male medical personnel who were entitled in an extremity to carry firearms in the defence of themselves or their patients. The Director-General put it bluntly when he wrote that, since RAMC male personnel were permitted to use firearms in defence of their patients, 'it is safe to assume that in these [emergency] circumstances women personnel of the medical services working with them would come within the sphere of protection of patients'. There were therefore 'no circumstances in which a woman doctor or dentist should be required to carry or use firearms in war in the course of her hospital or professional duties'.[89]

Despite hopes that all three military services would adopt an identical policy regarding their female members, in reality there were significant differences. The Admiralty admitted that it had not thought about arming women but expected that in the future some form of compulsory training would have to be introduced. The Air Ministry also had not considered the matter but declared that they had no intention, at this date, of introducing any form of compulsory training in the use of arms within the Women's Royal Air Force (WRAF).

The patronizing tone of many of these pronouncements is striking. How was the army going to teach women to shoot 'without tears', asked the Director of Personnel Services?[90] Men mused about whether women really would prefer to be killed than be raped.[91] In a minute dated 14 June 1949, the Director of Staff Duties, R. A. Hull appealed to knightly honour, insisting that it would be psychologically unsound and an expensive waste of equipment, ammunition and training time to train women in the use of personal arms. The fact that 'Little Olga' is trained to kill and prides herself on the number of notches cut on her revolver butt is no reason why we, too, should cry 'Annie get your gun'. It is still the soldier's duty to protect his womenfolk whatever they are wearing. Even in these days when war means total war let us at least retain that degree of chivalry.[92]

The Director of Finances agreed, commenting that the idea of arming women was like teaching a dog to walk on its hind legs: 'It is not well done; but you are surprised to see it done at all.'[93] Sensitive to such attitudes, the Director of the WRAC was primarily concerned about the use of the term 'pistols' in the report on the arming of women, lest comedians and cartoonists would tease them with words like 'lay that pistol down baby, lay that pistol down' and call them 'pistol packing Momas'.[94]

Keeping women out of combat

Administrative, strategic and ideological arguments were put forward to keep women out of combat. The administrative reasons were the most regularly employed, and were the least convincing. It was alleged that having armed women in either the Home Guard or the Regular forces would complicate the organization and administration of the services.[95] Providing arms to women would also mean that there would be fewer weapons distributed to men, thus endangering the lives of 'real' combatants.[96] Duplicating accommodation and sanitary facilities (in order to maintain separate access for the two sexes) would be expensive and difficult to enforce in battle. The fact that there were no 'female latrines' was one of the most favourite excuses for excluding women from combat zones (causing one pragmatic female war correspondent to retort: 'there is no shortage of bushes in Korea!').[97] There were other 'practical' considerations.

In particular, it was said to be impossible to subject women to the strenuous physical and emotional training that male soldiers underwent.[98] In the context of the Vietnam War, General Westmoreland claimed that women in combat would have to be 'freaks': he could not imagine any woman capable of carrying a heavy pack, living in a foxhole, or going a week without bathing.[99] The physical, psychological and emotional inferiority of women would threaten military effectiveness. The parliamentarian E. N. Bennett commented in *The Times* on 19 January 1942 that although arming women might be good for sentimental propaganda, 'any experienced military commander' recognized that actually giving women a combatant role would be more of an encumbrance than an asset.[100] Physically, women lacked upper body strength and possessed less stamina and endurance – important considerations since even in modern, technologically driven warfare, it would always be important to have combat soldiers on the ground. Psychologically, women were said to lack the aggressiveness of men and to have lower fear thresholds. The dishonour of being captured by a woman would reduce the likelihood of the enemy surrendering. In addition, if women were captured, they would be raped and otherwise maltreated. Spouses of military men, particularly in the navy, were extremely opposed to the assignment of women to non-traditional roles.[101]

Administrative protests were linked to more strategic concerns. Rational allocation of manpower preoccupied military and political planners. In wartime, labour was at a high premium: its distribution had to be economical and effective. Even feminists and active military women admitted the strength of this argument. In 1920, for instance, women like Elisabeth Crosby of the Women's Reserve Ambulance conceded that although women were not yet required in the trenches, they were essential in helping the men who were in the trenches. These men 'want lots of things doing for them' which, she believed, only women could do.[102] Similarly, the formidable suffragist Millicent Garrett Fawcett emphasized that although society had evolved past the point at which women were required on the battlefield, their work on the home front was as vital for the welfare of the nation.[103] Of course, the military establishment itself was also supremely conscious of the need to manage labour effectively.

However, the primary arguments against arming women were explicitly political and ideological, though this varied from country to country. In America, for instance, George H. Quester argued that women were kept out of combat forces through pressure from Congress, rather than from the military. The professional military would not have immediately pushed women into combat had this pressure been absent, but they preferred to leave the issue in limbo, for use in emergency. The Congressional pressure against female combatants came most strongly from Congressmen in the South but the regional pattern should not be exaggerated. As Quester noted, 'southern congressmen tended by seniority to be powerful, but also to be more visible as spokesmen for their colleagues'. Incumbents were hardly likely to lose their seats based on this issue and opinion polls suggested that the wider public were much more open-minded about the issue than the politicians.[104] There is evidence that the parliaments of Britain and Australia may not have been quite as opposed to female combatants as their American counterparts, in part due to the greater fear of invasion.

Ideological responses can be divided into two categories: the urge for peace and the desire to retain traditional gender roles. To many people, it seemed important to emphasize women's abhorrence of killing in order to reduce the likelihood of war. This was the stance taken by certain feminists, but even arch-conservatives could see its political value. Thus, John Laffin argued that one of the greatest 'inducements to the end of a war' was

the intense desire of men to return home to women and bed. If a man is to have women at war with him, if he is to think of women as comrades-in-arms rather than as mistress-on-mattress, the inducement disappears.[105]

More to the point, it was feared that gender relations would inevitably be called into question by arming women. Masculine identity would be threatened by admitting women into the services on an entirely equal basis. Socially, the presence of women at the frontlines would be demoralizing for men: it would disrupt processes of bonding and destroy a self-consciously 'masculine' warrior ethic. Combat was the ultimate signifier of manliness: women would symbolically castrate the armed forces. In 1978, M. D. Feld argued that if a significant proportion of the armed forces consisted of women, this would damage the status of the soldier and destroy all distinctions between 'polite society' and the armed camp. Male non-combatants would be accused of effeminacy. At the same time, the military would become more representative and thus less elite. This would be extremely damaging since it was these 'elitist notions' which motivated 'effective combat performance' and justified 'the sacrifice it entails'.[106] To admit women to fighting corps was to undermine's men's privileged access to the rights of being a combatant and the knowledge that this entailed.

In addition, female soldiers threatened to devalue the sacrifice made by their male counterparts. The strength of this view is illustrated by the way in which female warriors faded out of popular literature in the interwar years. Indeed, some female combatants were erased from literary record altogether: in the 1900 and 1908 editions of John Finnemore's *Two Boys in War-Time*, the pretty Katrina and her aunt engage the enemy in bloody combat for conventional reasons, that is, for the protection of their families. Using a blunderbuss, they kill many black South Africans (described as 'mad for blood and plunder') who are attacking their home. By 1928, when a new edition of this novel was released, this particular chapter was omitted.[107] Furthermore, when female fighters were included in post-1918 novels, they were firmly located in 'savage lands' or on the American frontier (as in Terence T. Cuneo's 1943 story entitled 'The Trail of the Iron Horse' which described women taking their places 'beside their men-folk' in warding off an Indian assault on their train).[108]

The spectre of combatant women remained a frightening prospect for many men. It physically and militarily threatened their sense of their own virility. After all, if the sight of women wearing uniforms was thought to emasculate men, imagine the effect of women with guns![109] In the words of one unnamed Englishman on 25 May 1940: 'a few million women with rifles was the most frightening prospect a man could face'.[110] During the Vietnam War in particular, female combatants really came to epitomize all that was emasculating about women in war. In the film *Full Metal Jacket* (1987), one scrawny woman is able to decimate an entire unit, and causes comrades to scream in mortal agony. Military prowess was also threatened. In 1944, fantastic stories about female snipers diverting soldiers from the main battle and leading them on 'wild goosechases' led the War Office to issue instructions directing soldiers not to 'waste [their] time chasing women-snipers. They may exist, but probably do not.'[111]

Similarly, if female combatants could disrupt the concept of war as a masculine activity they could also pose a threat to the home front as a female domain. Like the women training at the WAAC centre at Daytona Beach, Florida, who (in 1940) were reported to be 'touring in groups, seizing and raping sailors and coast guardsmen', it was feared that once women were permitted to become combatants, they would no longer be content with their position in society and an uncontrollably aggressive feminist movement would flourish.[112] The race would also be imperilled.[113] John Laffin in *Women in Battle* (1967) agreed, declaring that a 'woman's place should be in the bed and not the battlefield, in crinoline or terylene rather than in battledress, wheeling a pram rather than driving a tank'. They should 'stop men from fighting' rather than joining in the fray.[114] Women would be 'the mothers of the children who will rebuild Australia', commented the Controller of the Australian Women's Army Service (AWAS); they 'must not have the death of another man's son on their hands'.[115]

The intense fears surrounding female combatants stemmed

largely from anxiety over women's sexual and reproductive powers. Although the killer instinct might lie at the heart of masculinity, violence was harboured in the maternal body. Even the anthropologist, Margaret Mead, who generally emphasized the cultural origins of aggression and war,[116] reverted to biological difference when (in 1968) she speculated on the basis of aggression in the two sexes. 'The female characteristically fights only for food or in defense of her young, and then fights to kill,' Mead noted. Why was this an important difference? Mead feared that it might mean that women lacked 'the built-in checks on conspecific murder that are either socially or biologically present in males'. Arming women might turn out to be a suicidal course, she warned.[117]

Because women soldiers either lacked femininity or possessed too much of the maternal impulse, they were uncontrollable, more ferocious, and more deceitful than their male counterparts, it was felt.[118] There were fewer restraints on their conduct, and they reverted quickly to a state of 'barbaric fury . . . bereft of reason or feeling'.[119] Again, popular fiction endorsed and encouraged this belief. John Finnemore's novels aimed at adolescent boys frequently portrayed terrifying female warriors. His novel *Foray and Fight* (1906), chronicling the 'Remarkable Adventures' of an Englishman and an American in Macedonia, invoked female killers. Finnemore bestowed on them an abundant share of femininity. In the climactic rescue scene when the young English hero, Maurice, is just about to be stabbed by a wild Kurd, the Kurd's 'gloating' scream of triumph was choked off by a 'strange combatant' whose 'gown' brushed against Maurice's face. This woman (Maurice's 'hostess of the last night he had spent in the village') seized the Kurd's head and, with a 'dexterous slash', passed a 'keen knife' across his throat, 'carving' a wound from which blood 'spouted'. Finnemore's youthful readers were warned not to 'make light of' these women: it was unwise to 'disregard the wild-cat fighting for her young', he reminded them. These were 'big-framed, powerful women, their muscles toughened and hardened by many years of heavy labour in the fields' and (more to the point), their strength was 'backed by a flaming fury beside which the ardour of the men showed as pale as a candle in the sun'. Armed with broad-blade knives, they were 'fighting to save their children from horrors unspeakable' and thus 'raged in the battle like tigresses'.[120] Maternal passions transformed women into formidable killers.

Biological urges, in other words, rendered women less sensitive to the so-called rules of chivalrous warfare. In Spain, female soldiers were said to be crueller to prisoners than their male counterparts.[121] During the Second World War, the woman in uniform was similarly castigated as being harsher than her male equivalent because she lacked 'a man's more impersonal sense of justice and tolerance'.[122] In Korea, women commandos were described as 'uncompromising' and 'as dangerous as their male counterparts, or even more so!'[123] It must be noted that there is no evidence that female combatants actually *were* more liable to play dirty. Indeed, Flora Sandes recalled at least one incident where she berated her male comrades for unsporting behaviour (they had challenged her to a shooting match in which the target was a wounded Bulgar) and, as she threw down her rifle, she sneered at them: 'What plucky chaps you are! Why don't you shoot at a man who can return your fire?'[124] Clearly, though, when women were 'unsporting', there was much greater consternation than when all-male units declared 'no holds barred'.[125]

Changing attitudes

It is only since the Vietnam War that the position of women within the military in these three countries has changed dramatically. Since the 1970s, the proportion of women in peacetime armies has grown dramatically. Between 1979 and 1990, the percentage of women in the Australian Defence Force increased from less than 6 per cent of the total to 11 per cent. By 1990, Australian women began to be allowed closer to the frontlines in 'combat related' positions although they continued to be excluded from positions which might involve fighting at close quarters (such as the corps of armour, artillery and infantry). In America, the proportion of women in the US military increased from less than 2 per cent in the first two years of the 1970s, to nearly 5 per cent by the middle of the decade, and 7 per cent by 1990. In Britain, 10 per cent of the forces were female by the end of the 1980s.[126] British women in the WRAC have been allocated weapons since 1981, although army women with 'genuine objections' were exempted from these provisions. Women in the air force were given the option of having such training, but those in the navy were not. British women are most likely to be active combatants in Northern Ireland where women (known as 'greenfinches' because of their high-pitched voices on the radio) have been formed into the Women's Ulster Defence Regiment to counter female terrorism.[127]

The increased acceptance of women in the military since the 1970s can be explained in a number of ways. Probably the most important reason is the shortage of males of military age to fill the required posts. The dramatic decline in the birth rate through the 1950s to 1970s severely restricted the pool of potential men. In the United Kingdom between 1982 and 1994, the number of young adults aged between fifteen and nineteen years of age declined by 30 per cent. This occurred at the same time as compulsory service had become unacceptable to most civilians. In America, the political crisis over conscription during the war in Vietnam effectively ruled it out in the near future. Personnel was urgently required in all three services: therefore women became a valuable labour pool. Indeed, it was precisely the hawkish policies of prominent military officers (such as General H. H. Arnold of the US Army Air Forces and Admiral Elmo Zumwalt, the chief of US Naval Operations) which made them more willing to admit groups which had formerly been marginalized by the military.[128]

Women's liberation, increased sexual freedom, and greater mingling of the sexes in the workplace and in leisure also had an impact. The legal legitimacy for discriminating against women had been progressively weakening and in 1975 the passing of the Sex Discrimination Act in Britain confirmed this. Although it excluded the military, a European Community directive a year later questioned the legal foundations for this exemption. From that time, the exemption of women from combat had to be made along the lines of combat effectiveness rather than gender and this was a progressively difficult argument to prove with technological shifts generally reducing the need for physical aggression in combat effectiveness.

In addition, many women became increasingly keen to participate in combat. Although there was a furore over Captain Linda L. Bray who led a military police unit into combat during the invasion of Panama, levels of approval for female combatants was high: according to the *New York Times* in January 1990 seven out of ten Americans favoured women in combat.[129] A lower – but still significant – favourable response was reported in New Zealand where an opinion poll in 1986 found that nearly half of the population believed that women should be allowed to take part in combat. Although the responses were similar for men and women, those under the age of thirty-five were more favourable to women in combat roles (59 per cent agreed) compared with those in the older age group (less than one third agreed).[130] Military women were even more keen. In Australia during the 1980s, one survey that questioned one fifth of all women in the Australian forces found that 87 per cent favoured allowing women to be trained for combat roles. Seventy-seven per cent thought that women who had been so trained should be allowed to serve in combat-related positions and 57 per cent thought they should be allowed to serve in combat. When asked whether they would be willing to serve, 61 per cent said they would serve in combat-related positions while 45 per cent were willing to serve in combat. Those who were more in favour of combat roles were younger women who saw the military as their profession.[131] It is widely believed that women employed within the armed forces (particularly officers)[132] are 'consistently eager' to widen their military roles in order to gain acceptance and status.[133]

Rather than being (as some historians assert[134]) the 'other' in war, women were an integral part of the slaughter of war and myths surrounding it. The military establishment was historically opposed to allowing women to carry arms, fearful of the chaos that might result from the disruption of traditional gender roles and anxious least the unleashing of female bellicosity would morally disenfranchise both sexes. In contrast, many women in this period began to realize that the rhetoric used by the established military forces was out of touch with both the nature of modern warfare (where the distinction between the 'home front' and the 'frontlines' was blurred) and the psychology of combatant troops (who were more liable to be motivated to kill in defence of their male comrades than in chivalrous protection of 'defenceless' womenfolk). While men found their corporeal and emotional manliness threatened in combat, women were able to refashion their gender role more creatively in wartime precisely by asserting their bellicosity. The pleasures of violence were shared by women but (since they were denied the experience of combat and consequently debarred from its realist literary representation), they responded by offering up the bodies of their sons, male lovers and husbands to the killing fields. Through this violence, they earned their right to grief.

Chapter 11

RETURN TO CIVILIAN LIFE

I'm afraid to hold a gun now.
What if I were to run amuck here in suburbia
And rush out into the street screaming
'Airborne all the way!'
And shoot the milkman.

Charles M. Purcell, '. . . In That Age When
We Were Young', 1972[1]

Throughout this book, men and women have been seen imposing upon the bloody fields of war their own individual, complex and, often, pleasurable landscapes. Even within inherently chaotic situations, these people attempted to create some form of order, while still insisting upon the authenticity of their own unique experiences. The issue of the inherent disorder attendant upon killing is equally problematical when examining the 'impact' of combat upon fighters, their families, and friends. Diaries, letters, memoirs, social statistics, and other constructions of the past can provide the historian with a snapshot or a 'still life' portrait of the impact of war. So, too, can fiction and drama.

Emily Mann's play *Still Life*, written and directed in 1981, claims to be 'shaped by the author from conversations with the people whose experience she sets forth'. This play is not only 'about' the struggle to transcend the chaos of combat, it represents the impact of war. The three characters who speak in this play have been familiar figures throughout this book: a veteran (Mark), his wife (Cheryl), and his lover (Nadine). Mark is portrayed as a typical, angry American, twenty-one years of age when he volunteered to serve in the Marine Corps in Vietnam. He thinks that combat will prove that he is 'a man'. Instead, it turns him, in his words, into 'an animal'. Like the combatants discussed in Chapter 5, 'Love and Hate', he does not hate the Communists. Indeed, it is his civilian parents who 'pushed him into going. They believed all those terrible clichés.' But after seeing his friends and comrades killed, castrated and otherwise mutilated, he turns himself into a hardened, enthusiastic killer. Mark collects souvenirs and even sends his mother a bone from a man he has killed. He would have been just as aggressive, he believes, if he had been fighting during the Second World War. Despite recognizing that it was easier to kill 'zips, or dinks, or gooks' than 'white men', he insists that he would not have had 'any trouble shooting anything'. Indeed, he enjoys the blood-letting, admitting to 'getting off' on the power of killing: 'it was so nice', he recalls, echoing Broyles in Chapter 1, 'The Pleasures of War': 'I had the power of life and death . . . It's like the best dope you ever had, the best sex you've ever had.'

As we saw earlier in this book, though, there was a distinction between legitimate slaughter and atrocity, and it was the massacre of three children and their parents 'in cold blood' that twists Mark's pleasure into agonizing pain. He knows that he could have avoided this senseless slaughter: 'I could have just said: I won't do it, I could have said: "I got a tooth-ache", gotten out of it.' But, orders were orders, and he shot. The murder was followed by a period of rationalization and, when this fails, all-consuming guilt. He observes that

all that a person can do is try and find words to try and excuse me, but I know it's the same damn thing as lining Jews up . . . I know that I'm not alone. I know that other people did it, too. More people went through more hell

than I did . . . but they didn't do this. I don't know . . . I
don't know . . . if it's a terrible flaw of *mine*, then I guess
deep down I'm just everything that's bad.

These conflicts are exacerbated when he returns to America,
wearing his medals. Despite his success as a 'warrior', there is no
ceremony when he arrives home. 'I came home from a war, walked
in the door, they don't say anything. I asked for a cup of coffee,
and my mother starts bitching at me about drinking coffee.'
Almost from the start, he begins picking fights with people in the
street and in bars. He violently assaults and rapes his wife and, in
his art, creates sadistic images using pornographic photographs of
his wife bound to a stake and surrounded with razor blades and
broken glass. Although sometimes he feels invincible, convinced
that even if he kills someone in the streets, no court will convict a
Vietnam hero, more frequently he is terrified that he, or his children, will be punished for his crimes in Vietnam:

I thought people were . . . uh . . . I mean I was kind of
paranoid. I thought everybody knew . . . I thought everybody knew what I did over there and that they were
against me. I was scared. I felt guilty.

This acknowledgement of guilt provides some kind of barrier
to self-destructive pain, however. Like the men discussed in
Chapter 7, 'The Burden of Guilt', Mark insists upon the importance of 'obeying orders' and is convinced that by confessing his
crimes to his lover he will be forever *un*able to 'wash my hands of
the guilt, because I did things over there'. Mark has been deeply
scarred by combat and, in turn, he hurts people when he returns,
but by retaining his guilt and embracing a mother–confessor, he
begins the process of healing.

Cheryl, Mark's wife, is also portrayed as a 'victim of war', and
although she has never killed anyone, the map she drafts of the
'killing fields' is much closer to home. Cheryl knows that violence

also exists outside the field of military combat. Her brother, an ordinary civilian, has nearly killed his wife a few times. His wife eventually 'snapped' and murdered her son. Unlike Mark, Cheryl's solution to the problem of violence is to shut her eyes to it. She pleads with Mark to 'forget' and when he tells her that he was a murderer, she merely reminds him that he 'can be a husband'. 'He blames it all on the war,' she observes, then turns to the audience and says: 'but I want to tell you . . . don't let him.' '*It takes two*,' she insists.

While Cheryl wants to forget the killing, Nadine (Mark's lover) looks straight-faced at violence, even embracing it. Mark's murderous impulses are simply 'dastardly deeds' and 'naughtiness' to her: his sadistic art is 'brilliant, humorous'. When Mark tells her that he 'felt good' about killing, she understands ('I didn't bat an eye,' she recalls). War is war; which means that 'everything Mark did was justified. We've all done it. Murdered someone we loved, or ourselves.' 'Sophistication' for Nadine is 'the inability to be surprised by anything'. For her, Mark is not unusual – yes, he is angry, but so are all of her friends: 'Mark's just been demonstrating it, by picking up weapons . . . Leading a whole group into group sex, vandalism, theft. That's not uncommon in our culture,' she explains. Like the women discussed in this book, she believes that violence is a part of a gendered world in which (man's) battle is simply three times worse than (woman's) parturition. For Nadine, the impact of war is uncomplicated: 'He calls himself a time-bomb. But so are you, aren't you?'

The fictional characters Mark, Cheryl and Nadine respond in different ways to combat: guilt, denial, acceptance. The two non-combatants are implicated in Mark's murderous aggression – whether expressed 'legitimately' (in the enthusiastic slaughter of other combatants) or 'illegitimately' (as in the atrocity). At the end of the play, the audience is directed to look at a photograph containing two grapefruit, an orange, a broken egg and some fresh bread. In the midst of this 'still life' lies a grenade.[2]

*

Images of violent veterans like Mark trouble our modern conscience, conjuring up frightening scenarios of 'the beast within' which refuses to be harnessed to more creative enterprises or eradicated altogether. The negative transformative power of spilling human blood creates a potent myth, which has been used to interpret the behaviour of servicemen and to add urgency to much anti-war rhetoric. After all, the argument goes, if bloody combat could be shown to have *no* long-term brutalizing impact upon the surviving participants, yet another restraint on armed conflict would be removed. But linked to that, the ideological message underlying the 'brutalizing thesis' is equally unsettling. Is the combatant *so* different from the non-combatant, ask Cheryl and Nadine? Certainly, the two groups are broadly distinguishable by gender, age and (at the start of each conflict) physical robustness, but is the gap between 'them' and 'us' really so wide? In the attempt to present a 'still life' of the impact of war, the stigma and fear attached to the combatant re-entering civilian society turns out to have more to do with our own hearts of darkness.

'The Beast Within'

Historians and other commentators have emphasized the way that combat brutalizes its participants: combatants pay an extremely high moral and psychic cost for their gruesome profession, which changes them into 'inferior' and degraded human beings, argued Alfredo Bonadeo in his aptly titled *Mark of the Beast* (1989).[3] During all three conflicts examined in this book, civilians were profoundly anxious about the ability of young men to cope with the need to kill, and risk being killed, for the sake of the polity. 'Boys' were sent into battle and some of them killed other 'boys'. When the survivors returned, they were 'men' faced with the task of re-establishing relationships with friends and family, sweethearts and spouses. No one expected this to be easy.

Men's experiences in combat could not simply be ignored on

their return to civilian life and in the process of retelling and making sense of their experiences these had to be shared with and also shaped by civilians. Their narratives provoked a range of responses – empathy, envy, pity – but also fear. How could anyone who spilt human blood remain untainted? Would the repressed aggression eventually 'leak' out of former combatants? Accusatory fingers were pointed at veterans. 'Murder, too, is there,' boomed Dr Francis Rowley in an SPCA magazine, 'his bloody beak betrays him. Killing has dulled all sensitiveness to the sacredness of life. These are some of the brood war hatches out and fosters under its foul wings.'[4] Was it any wonder, mused psychoanalysts, that men whose aggressive impulses had been nurtured within the military murdered their wives and girlfriends when these women refused to 'comply with their demands' or that they had fewer, if any, inhibitions against aggressive out-bursts?[5] The acute disillusionment of men who had enlisted with 'profound emotionalism' was said to have destroyed their capacity to feel sympathy upon returning to civilian society.[6] As one First World War medical orderly observed, human sensibilities were eradicated in the process of being 'turned into devilish machines to kill and to be killed'.[7] The transformation was even portrayed as being inscribed upon the body: men yowled the dog-language of militarism, their hands twisted into claws, and their brows sunk to resemble apes.[8] Veterans returned home and found that their families, friends and acquaintances looked at them warily: 'they seem to think you might go wild . . . Like you're a freak, likes to kill,' complained one soldier.[9] Vietnam veterans, in par-ticular, were forced to respond to hostile crowds calling them 'murderers' and 'butchers': 'Did you really enjoy killing babies and people?' snarled passers-by.[10]

In all three countries after the two world wars, civilians expounded frightening prophecies about the violence that would be wreaked upon peaceable societies once combatants returned home. Governments would tremble under the stern gaze of ex-servicemen, it was predicted. In *The ReMaking of a Mind* (1920),

Henry de Man issued a chilling warning to the state, declaring that:

> should conditions arise in the life of these masses that either make it in their interest to murder, or else create a common feeling in favour of class terrorism, they might remember how easy it is to take another man's life, and what a delight there is in doing it.[11]

Criminological studies predicted disaster. For more than four years, Clarence Darrow argued, 'most of the western world did nothing but kill'. Men were rewarded for proposing 'new and more efficient ways' to take men's lives and every school and church 'joined in the universal craze'. Was it any wonder that there would be an 'after-war harvest of crimes', asked Darrow?[12] Major-General G. B. Chisholm, the Deputy Minister of National Health and President of the National Committee for Mental Hygiene, Canada, echoed this warning, explaining in 1944 that the

> whole object of [the serviceman's] existence and the focus of all endeavour about him is killing. Effective and whole-sale killing for years has been given precedence as the highest moral value and the most admirable of virtues . . . Aggressive urges which have been carefully nurtured and developed over a period of years are supposed to disappear overnight, leaving a peaceful civilian with no such pressures and consequently no need of outlet . . . With the memory . . . of his friends or relatives who have been killed or maimed or even tortured by the enemy, fresh in his experience and kept alive as a spur to his aggressions, this changeover in attitude may be very difficult indeed.[13]

The threat these men posed to society and state might even warrant restrictions on their freedom. At the conclusion of the

Second World War, for instance, one prominent New York woman proposed that 'reorientation camps' be established (perhaps in the Panama Canal Zone) for repatriated soldiers. Even after veterans had been released into American society, she suggested, they should be required to wear an identification patch, such as a skull. In the bitter words of one Marine, this patch would warn civilians 'of our lethal instincts, sort of like a yellow star'.[14] Veteran benefits – described as a sort of bribe to safeguard civilians against the aggressive urges of servicemen[15] – could not be guaranteed to pacify resentful combat veterans. As August B. Hollingshaw drily commented in the *American Journal of Sociology* in 1946: 'It will be impossible for [veterans] to communicate their inner sense of accomplishment in the fine art of killing to civilians.'[16]

Such fears were not quelled in the aftermath of the Vietnam War. Indeed, they were considerably magnified. Again, it was noted that the skills developed by the 'trained killer' might endanger civil society. Boasting about their proficiency as 'sharpshooters' might look impressive on military records but, as a member of the US Civil Service Commission argued, it was 'no help at all on an application blank'.[17] Quite the opposite: these skills *dis*equipped men for life outside war zones. This worried Jerome Johns, an instructor at Phan Rang running a tough five-day indoctrination course for new troops arriving in Vietnam from the United States. 'Whenever I see something about a killing in the paper,' he admitted, 'I look to see if it was done by a Vietnam veteran . . . You remember how we had to *motivate* those kids to kill; we *programmed* them to kill, man . . . Well, nobody's *unprogramming* them.'[18] Such concerns were not to be lightly dismissed, as Mardi J. Horowitz and George F. Solomon reminded their readers in 1975 – after all, there *was* a distinction between combatants and obsessional neurotics who had thoughts of doing violence to others. In contrast to men who had not 'done the degree of harm that they imagine', veterans

will have witnessed such violence and may have participated in it. They know not only that such violence is really
possible since they have committed it, but also that it may
be pleasurable as well as guilt-provoking.

This 'shortening of the conceptual distance between impulse
and act, fantasy and reality' meant that 'conditioned inhibitions
to destructive behavior' were reduced and would be 'difficult to
reimpose'.[19]

These fears were particularly evident in relation to black
American soldiers. After the First World War, there was anxiety
that black ex-servicemen would demand full rights as citizens
and, if refused, would wreak havoc on the nation. In a letter
from Lieutenant Melville Hastings (an Englishman living in
Canada) in 1917, he lamented the fact that 'these blacks and yel-
lows will require hats two sizes larger when they return . . . the
dominant race is ominously unpedestralled [sic]. They may be
good trench diggers and rod repairers, but Britain, I fear, is
laying up . . . a toughish road for her own broad back.'[20] These
fears were again heightened during and after the Vietnam War.
After all, this was the first American conflict in the twentieth
century in which black and white troops had been completely
integrated and, in the field, black soldiers had generally shared a
greater degree of comradeship with white soldiers than they
could expect back home.

What will happen as the Negro veterans, filled with a new
sense of pride and equality, and thoroughly trained in all
the arts of killing and destruction, return to the ghettos to
confront the same squalor, joblessness and prejudice that
they knew when they left?

queried one journalist in *The Nation* (1968).[21] Black servicemen
had learnt many things, 'among them the skills of guerrilla warfare, of killing'.[22] Although these predictions about the transfer

of aggression from the Vietnamese towards American bigots were not fulfilled, black servicemen, with their newly acquired confidence and skills, *were* less willing to be treated as inferiors.[23] As David Parks wrote bitterly in his diary while serving in Vietnam, his 'white comrades' 'bug me more than Charlie [the VC]. I'm learning one hell of a lesson in here.'[24]

In all three conflicts, fear of returning veterans was exacerbated by the awareness that rifles, hand grenades and bombs were also finding their way back from the theatres of war and into civilian society, forcing governments to institute disarmament campaigns to encourage the handing over of illegal weapons.[25] Furthermore, violent veterans preoccupied film-makers who flooded the market with cinematic images of brutalized combatants.[26] No 'grade-B melodrama' based on the Vietnam War was complete without its 'standard vet – a psychotic, axe-wielding rapist'. As one commentator quipped:

> The demented-vet portrait has become so casual, so commonplace, that one pictures the children of Vietnam veterans shivering beneath their blankets and wondering if Daddy will come in with a goodnight kiss or a Black & Decker chain saw.[27]

Popular criminal reportage also exaggerated the extent of violence carried out by former servicemen. The press applied the label 'veteran' to any man who might have served and subsequently committed a violent crime, irrespective of the nature of that man's service or of any link between his particular action and his war service.[28] High-profile cases of murder, often associated with domestic violence and rape, dominated headlines after each conflict and yet were portrayed as the inevitable result of combat training and men's experiences of war. As the attorney defending one veteran accused of a particularly vicious rape and murder of a young Vietnamese woman in New York in 1977 argued, this former Marine had simply done what he had been 'trained to do', that is, 'kill women'. The attorney reminded the jury: 'what was so difficult about doing this again . . . kill one more Vietnamese girl?'[29]

As has been demonstrated throughout this book, 'effective combat behaviour' *did* require men to act in brutal, bloody ways. During battle, men typically lost their ability to feel shocked or disturbed and the observation that men who were initially sickened by warfare gradually 'got used to all these things . . . and rather liked the warfare' was often cited as proof of brutalization.[30] While the *first* 'kill' was often carefully recorded and reconsidered afterwards,[31] within a relatively short period slaughter became commonplace. The ability of gory blood-letting to appal civilians while leaving combatants unmoved was also an effective literary device employed by countless autobiographers, such as Richard Tregaskis in *Guadalcanal Diary* (1943). This 'diary' described in great detail the Japanese troops Tregaskis helped to kill:

Everywhere one turned there were piles of bodies; here one with a backbone visible from the front, and the rest of the flesh and bone peeled up over the man's head, like the leaf of an artichoke; there a charred head, hairless but still equipped with blackened eyeballs; pink, blue, yellow entrails drooping; a man with a red bullet-hole through his eye; a dead Jap private, wearing dark, tortoise-shell glasses, his buck teeth bared in a humorless grin, lying on his back with his chest a mess of ground meat.

Tregaskis then coolly reported that there was 'no horror to these things'. The first corpse was shocking: the rest were 'simply repetition'.[32] Or, in the words of Colonel David H. Hackworth, the most decorated officer in the American army when he retired in 1971, battle 'is like working in a slaughterhouse. At first the blood, the gore, gets to you. But after a while you don't see it, you don't smell it, you don't feel it.'[33]

But did this process of emotional numbing lead combatants to long-term brutalization? Certainly, military representatives were anxious to dispel such fears. Propagandists like William Ernest Hocking, who toured the battlefields in 1917, and representatives of the army medical services, were anxious to reassure readers that nothing that was 'a necessary duty' could ever be brutalizing.[34] Military films, posters and pamphlets all portrayed clean, sensible and 'civilized' warriors, and they rarely hinted at the filthy, bloodstained hands of men in battle. By the Vietnam War, even the established veteran organizations had done an about-turn: instead of demanding economic and social resources on the grounds that the particularly violent nature of combat in Vietnam had adversely affected these veterans, they strove to insist that their new members had suffered no more than earlier cohorts of veterans.[35]

Sociologists and historians were not always convinced and turned to statistical analyses in their attempts to quantify the long-term effects of combat on men once they returned to civilian society. The 'brutalization thesis' was tested in two ways: first, by examining whether veterans were disproportionately represented amongst men who committed violent crimes (the 'individual' approach), and secondly, whether there was a large increase in violent crimes after each major conflict (the 'group' or 'societal' approach). The surveys were complex, the interpretations confused, and the results contradictory. While a few studies found a correlation between crime rates and military participation,[36] most found no such relationship, or even a negative one (that is, crime rates *declined* after wars).[37] In the words of Commander Robert R. Strange and Captain Dudley E. Brown in 1970:

> combat zone experience does not eradicate controls of either internally or externally directed overtly aggressive behavior, although such controls may be temporarily over-come by group sanction, survival needs, and other factors

in combat . . . It is noteworthy that aggressive problems of
all types were slightly *less* frequent in the Viet Nam
returnees than in the noncombat group.[38]

Though statistical evidence neither proved nor disproved the
brutalization thesis, the weight of evidence found veterans
'innocent'.

Even in those cases where brutalization was admitted, the
trend was not consistent, in either the 'individual' approach or
the 'group' approach. Obviously, not every person who killed in
war suffered long-term psychological damage. The men who
participated in the My Lai massacre carried on both in Vietnam
and in America for a year before being brought to trial. During
that year, nothing in their behaviour differentiated them from
other men serving in Vietnam or from most other veterans in
America. Large-scale studies confirmed this observation. In the
1980s, two sociologists, Dane Archer and Rosemary Gartner,
carried out a huge statistical survey. Despite establishing a link
between war and crimes of violence, their evidence was
extremely contradictory: for example, homicide rates only
increased in Scotland and the USA after the First World War
and in Australia, England and Scotland after the Second World
War. They remained unchanged in England after the First World
War and they decreased in Australia and Canada after the First
World War and in Northern Ireland and the USA after the
Second World War.[39]

Moreover, the conceptual framework of many such studies
was contradictory. According to one theory, war would increase
crime rates 'because of the emotional instability in wartime'
while another theory predicted a decline in crime after war
because of 'an upsurge of national feeling': while one theory predicted an increase in crime because of 'the contagion of violence',
another predicted a decline because of 'the vicarious satiation of
the need for violence'.[40] Furthermore, the data employed was
unequal to the task. Researchers had no way of knowing what

levels of violence would have been if there had not been a war. Crime statistics did not distinguish between men who served in the armed forces and those who did not. In times of conscription, practically all fit men donned a uniform, and fit men might be more liable to commit violent crimes than unfit men. More to the point, any increase in violence after war might not have been due to the experience of killing. Unfortunately, in most of the large surveys, 'servicemen' could not be differentiated from '*combat* servicemen' yet the brutalization argument cannot be applied to servicemen who never left their offices or tents hundreds of miles from any battle zone.

The exact process whereby warfare might have led to an increase in crime was also difficult to define. Even if a statistical correlation was drawn between warfare and crime rates, the relationship might be traced to causes other than the actual experience of killing. For instance, in America, veterans were more liable to own guns, but this was because rural men (who owned more guns) were much more likely to join the armed forces. In other words, their gun ownership was due to early socialization into the use of arms rather than military training.[41] There were other 'pre-service' factors which were important. In an Australian study of Vietnam veterans, it was found that violent veterans were different from non-violent veterans. They had

> quite a number of pre-service and early background fac-
> tors predisposing them to their post-war aggression; these
> included an alcoholic family history, a poor relationship
> with their parents, loss of a parent early in life, and con-
> duct problems including violence in childhood or
> adolescence . . . Such violent beginnings continued in
> Vietnam but violence was often blamed on the war.[42]

In addition, combat fatigue, lack of appreciation for their 'sacrifices', and numerous other frustrations experienced upon re-entering civilian society often led veterans to act aggressively.

The breakdown of social controls and post-war insecurity could also be blamed for inciting violence. However, Archer and Gartner reached the conclusion in their survey that 'violent veterans' could *not* be responsible for increases in violent crimes after wars because the violence was carried out by people who could not have been combat veterans:

> During the ten-year period of the Vietnam War, for example, US arrests for homicide increased dramatically for both men and women – 101% and 59% respectively . . . Homicide arrests also increased for all age groups – including people over the age of 45. For the Vietnam War, therefore, the violent veteran model clearly is inadequate to explain the increase in the homicide rate.

This was also the case, they demonstrated, after the Second World War. Instead of the 'violent veteran model', Archer and Gartner emphasized the ways in which war generally legitimated violence: after war, this legitimation continued.

> War involves homicide legitimated by the highest auspices of the state. During many wars, the killing of enemy soldiers has been treated not merely as a regrettable and expedient measure but as praiseworthy and heroic . . . This legitimation is directed at both the nation's soldiers and the home front; but it may be more credible to civilians than to combat soldiers with direct experience of the realities of war.[43]

In other words, war made killing *seem* commonplace to combatants and civilians alike, but the actual *experience* of killing might dispel any romantic notions of its legitimacy.

The aftermath of the Vietnam War

Unlike previous conflicts when the disturbed veteran was marginalized, by the late 1960s popular culture had designated the resentful, angry and hurt Vietnam veteran as *the* archetypal combatant. Most of the histories of Vietnam veterans were heavily dependent upon accounts of hospitalized men, as though such men were representative. This 'disjuncture between myth and subjective experience' was particularly potent for Australian Vietnam veterans who *had* been welcomed home by large, cheering crowds. Peace protestors in Australia were disproportionately represented in the media, but in reality they were small in number and generally conducted their protests at times which did not coincide with scenes welcoming the veterans. Yet, the American myth of rejection followed by bitter fury has come to be 'the story' of the Vietnam experience for Australians as well.[44]

In America, on the other hand, the Vietnam War *did* see an increase in the proportion of disturbed, angry and aggressive veterans. For one thing, soldiers who participated in that war saw a good deal more combat than most soldiers in other twentieth-century conflicts. An American who enlisted after Pearl Harbor probably only saw a few weeks' combat duty and even in the Pacific theatre of conflict in the Second World War, Marines would have experienced just six weeks' combat. In contrast, men sent to Vietnam spent months in the field. Marines were committed to eighty days at a time and many served three or more rotations.[45] As we have seen, however, combat experience *per se* did not necessarily traumatize men. Hostile reactions were much more liable to stem from the feeling of having been 'fucked over' by military and civilian society on the return home rather than to any 'habit of violence' inculcated by military training or combat experiences.[46] Vietnam veterans were most susceptible to adverse effects because of their extreme youth, poor battlefield leadership, lack of unit cohesion, the guerrilla nature of the war, and the sense of purposelessness when they returned to the

US.[47] More than in any other conflict, Vietnam refused to allow men to behave like heroes; men went into battle with their minds stuffed with imaginative role models, none of which had the remotest chance of fulfilment.

But much more important in 'cementing' aggressive behaviour patterns was the process of returning to civilian life. For certain groups, such as American Indians, there were elaborate purification rites to ease the process and salve the consciences of men returning from battle.[48] In all the other conflicts, even that of Korea, the transition from combat zone to civilian society was typically a leisurely process, usually aboard a troopship, which provided men with time to readjust, share experiences and fears with fellow servicemen, and perform various rites of purification. This was not the case for American troops in Vietnam. Their shift from the field of war, through a processing centre at Long Binh or Cam Ranh Bay, to the army terminal in Oakland, California, and then to their own home often occurred within a couple of days. American troops arrived home alone, not with their unit. In other conflicts, ex-servicemen had been greeting by friendly, grateful crowds who confirmed the rightness of the slaughter, bestowed understanding and forgiveness upon uneasy consciences, and embraced lost sons as returning 'men'. In the case of Vietnam, returning Americans often faced a hostile reception and were liable to be reviled as lepers, child-murderers and idiots.[49]

Furthermore, during the Vietnam War, veterans were no longer allowed to claim exclusive rights to the stigma (and honour) of killing: civilians now insisted on claiming some degree of guilt, some responsibility, while simultaneously pushing away those who had actively participated. Civilians both shunned 'baby-killers' and expressed shame for their own part in prolonging the war. After the two world wars, there was no such reaction. By and large, the servicemen then retained their status as men who had sacrificed themselves for the polity and while in practice this did not entitle them to greater benefits in civilian life, it did confer upon the survivors a certain moral righteousness. After the

Vietnam War, incredulous combatants heard *civilians* confess to being brutalized by the war. For men whose own hands had been bloodied in the conflict, this created utter confusion about their own moral status.

Linked to this rejection was the difficulty of returning home to a country vastly different from the place the veterans had just left. Disorientation and disillusionment were common reactions. The Marine, Corporal Vito J. Lavacca, described his inability to comprehend the scale of the readjustment:

> I managed to survive Vietnam and got back to New York . . . The change was just shocking. It was totally devastating. Clothes had changed. People's attitudes had changed. Close friends that you had that had been clean-cut, athletic types when you last saw them now had hair down to their shoulders. They were wearin' big medallions around their necks. And . . . beards. Christ . . . earrings. And bell-bottomed pants and high heel boots. And . . . givin' peace signs.[50]

The values for which these young men had risked their lives had disappeared overnight. Consequently, the struggle to construct meaning out of the chaos was that much more difficult for Vietnam veterans. Don Browning, in an article entitled 'Psychiatry and Pastoral Counseling: Moral Content or Moral Vacuum?' (1974), elaborated on the problem.

> It is one thing to kill and be killed if one believes he is making a lasting contribution to the welfare of one's community, succeeding generations, and the future of the human species, [but] . . . it is another thing to kill or be killed when one is convinced that one is making no contribution at all except perhaps a negative one . . . there were no meaningful sacrifices in Vietnam.[51]

Browning exaggerated: for many Vietnam combatants, there was self-creation through violence and dreams of rebirth. But the bloodstained, rebirthed warrior was aborted on home soil.

Survivors

The stories told by combatants were central to their moral and emotional survival. Veterans were often anxious to reassure themselves and others that they had not been brutalized by combat. Admittedly, men as diverse as Lieutenant Frank Warren (1917), William Ernest Hocking (1918) and John B. Doyle (1944) confessed that in the short term they had become more 'stern', 'severe', or 'hardened', but this did not constitute brutalization.[52] In a typical fashion, Dixon Wecter, in his prize-winning book, *When Johnny Comes Marching Home* (1944), accused people who claimed that military training turned men into trained killers of speaking 'twaddle': 'teaching a soldier to shoot Japs or Germans under the rules of war does not mean that he will come home and shoot his neighbour'.[53]

Some combatants went even further, arguing that battle had made them less aggressive. Alfie Fowler, who served in Korea in 1952, claimed that the experience turned him into a 'much calmer person'.[54] Vietnam veterans interviewed by the psychiatrist Robert Jay Lifton often reported that the war had made them 'soft' or more willing to express their feelings.[55] Others believed that combat led to a 'deepening of tenderness': soldiers habitually experienced feelings of 'solemn gentleness like that sometimes attributed to the angel of death'.[56] The emotional distancing that a man cultivated in the face of his enemy could be compensated for by more tenderness to friends.[57] War might even make soldiers 'over-sensitive on behalf of sufferers'.[58] Indeed, the importance of gentleness in combatants was widely recognized. As Field-Marshal Sir William Slim, the Chief of the Imperial General Staff (1948–52), admitted in a BBC broadcast, gentleness was a 'soldierly quality'.[59] Temperate men were highly desired in combat. In the words of a Vietnam veteran: 'gentle people who somehow survive the brutality of war' were 'highly prized' in combat units because (even though they were 'highly efficient killers') they possessed 'the aura of priests'.[60]

This failure of combat to dent the moral and aesthetic sensibilities of most combatants is perhaps not surprising. After all, the assumption that men trained to kill in war would carry on killing after the war was based on a false notion of the 'killer personality', yet, as we have seen throughout this book, 'ordinary' people delighted in slaughter. Although some men became 'aberrant killers', there was no such thing as a 'typical' killer. 'Gentle' men who would burst into tears whenever they received a letter from their loved ones were capable of grotesque acts of brutality.[61] Yet, when these men returned home, they

> took off their uniforms, hung them in the closet, bought a new outfit or tried to get into the clothes they had when they volunteered, and went right out into life, getting up in the morning, going to work, coming home at night, some going to university, getting married, having fun, just as if there never had been a war on.[62]

When the conflict ended, most men were keen to return to their former lives and civilian sensibilities. In the words of Samuel A. Stouffer, referring to the period just after the end of the Second World War:

> in striking contrast to the expectations of those who anticipated deep-seated personal bitterness and disillusionment, manifesting itself either in widespread psychiatric breakdowns or in aggressive hostility against civilian society and institutions . . . by and large the evidence from men still in the Army pointed rather to an individualistic motivation to get back on the same civilian paths from which the war was a detour.[63]

Like Hitler's 'willing executioners',[64] allied troops in war were

> just American [or British or Australian] boys. They did
> not want that valley or any part of its jungle. They were
> ex-grocery clerks, ex-highway labourers, ex-bank clerks,
> ex-schoolboys, boys with a clean record and maybe a little
> extra restlessness, but not killers,

one commentator noted in 1943.[65] Enthusiastic killers arose out of a representative selection of the population.

As countless studies have shown, ordinary people placed in extraordinary situations act in ways they would not usually contemplate. The removal of external prohibitions against killing, and even encouragement to kill, did not nurture men who would kill in civilian contexts.[66] For most men (although not for those returning to certain inner-city slums), the circumstances of war had no counterpart in civilian society.[67] As we have seen throughout this book, combatants themselves drew a rigid distinction between legitimate killing (of enemy combatants) and illegitimate killing (of defenceless civilians). Even when immersed in the combat environment, civilian and martial contexts were differentiated. A striking – although not uncommon – example can be taken from the correspondence of Captain Alfred E. Bland. On 25 February 1916, he wrote to his wife:

> No longer do I expect to meet a German at every turn.
> But if I did I'd damned soon have a bullet through him. I
> have no compunction, no sympathy. If, as often happens,
> this man or set of men play the fool with a rifle or a
> grenade and get severely wounded – I don't care. I curse
> them for fools and ignore their pain. They deserve it. I
> can't be bothered to waste tears or feel fear.
>   The snow is beautiful, and the stars, oh! the stars. I
> came home last night over the top, stepping straight out
> from the support trench into the open and walked home

over the hill, with Sirius straight ahead of me due South
all the way. Bless the stars! I stand and watch them from
the narrow trench walls and think of you.[68]

Soldiers were obviously aware, too, of the different expectations
of their conduct in peace and in war. As an Irish soldier told
Michael MacDonagh in *The Irish on the Somme* (1917):

You do such things and get praise for them, such as
smashing a fellow's skull, or putting a bullet through him,
which if you were to do at home you'd soon be on the run,
with a hue and cry and all the police of the country at your
heels.[69]

Once the external props of the 'theatre of war' were removed,
only a tiny minority of men could continue exulting in the
slaughter. Indeed, as I argued in *Dismembering the Male: Men's
Bodies, Britain and the Great War* (1996), combat could make
men *more* aware of the love of women and the importance of
tender, domestic emotion.[70]

Combatants, then, were not passive moral subjects who, having
committed the ultimate transgression, must be permanently
scarred. Rather, they readily accepted their role as agents of war
and thus their moral natures could retain both creativity and
resilience. Most had not actively sought combat (although they
could generate pleasure from it), but in striving for their concept
of a 'better world', processes of brutalization were strenuously
resisted. This is not to deny that the logic of such resistance was
often highly fanciful. For instance, the commonplace distinction
made between short-term, contingent 'hardening' and more permanent processes of 'brutalization' depended upon metaphors of
'shedding outer skins' to enable the rebirthing of the 'real' man.[71]
The notion that killing in war was part of an inevitable cycle of
events by which 'boys' followed in their father's footsteps and
were 'tested', before being reintegrated into more mature society

as peaceable fathers preparing their own sons for the bloody ritual, represented a moral commitment to a warrior society. What's more, veterans were not the only people fascinated by blood and gore: civilians also were attracted, and repulsed, by rites of mass bloodshed.[72] How many civilians would ask returning combatants to tell them 'real life' tales of murder, with 'eyes glisten[ing] so eagerly'?[73] Even female nurses who had served in Vietnam were taunted with questions of how many babies they had killed.[74] As another veteran complained:

> We were depicted in the press as drug-addicted, chewiest [sic] killers and it was a stereotype that stuck . . . The gods have caused us to believe our own propaganda and so the reflection that Americans buy into is our own image. As a nation we are all crazed, drug-addicted killers. Because, if Joe-Blow working-class American can go over to Vietnam and come back that way, then we must all be that way.[75]

In this sense, soldiers were seen as reflecting civilian society.[76] As in Emily Mann's *Still Life*, killing was an essential part of civilian life as well – in fiction, art, films and (much more rarely) authorized by the state in rites of execution and enacted by people on the streets and in bedrooms in all three countries. Men who marched to war were 'brutalized' long before they donned the uniform.

Why has the brutalization thesis been so persuasive? The 'before-and-after' drama, with innocence shattered by a great trauma offers a certain narrative satisfaction and coherence which could be keenly embraced by combatants and civilians alike. There was also, to return to a theme in Chapter 1, the saturation of the martial imagination with combat literature and films. The journalist Michael Herr noted how young Marines suddenly became particularly aggressive fighters when television cameras approached because they began acting as though they were film heroes. Ironically, this callous refusal to regard combat as 'real' helped shield soldiers from grotesque brutalization. Thus, as Herr observed, the first few occasions he saw combat, 'nothing really happened':

> all the responses got locked in my head. It was the same familiar violence, only moved to another medium; same kind of jungle play with giant helicopters and fantastic special effects, actors lying out there in canvas body bags waiting for the scene to end so they could get up again and walk it off.[77]

Civilians continued to view killing in war as though it was just another cinematic image – combatants were the ones who were eventually forced to realize that there was 'no cutting it'. The brutalization thesis provided the victors with a scapegoat (the veteran) and laid the blame at the door of military training and combat experience rather than a more entrenched, yet nebulous sense of pervasive aggressiveness. Once again, 'the problem' became war (even better if it was located in a particularly foreign place), rather than troubled individual consciences and a society which promoted aggression.

EPILOGUE

Throughout this book, 'ordinary' men and women have been heard rejoicing as they committed grotesque acts of cruelty. The association of pleasure with killing and cruelty may be shocking, but it is familiar. The individuals whose stories introduce each chapter were passionately involved in creating desirable worlds and, through language, they attempted to achieve meaning in the midst of the chaos and terror. Undeniably, the stories they told failed to convey the 'reality' of combat and were often fantastical. This is not surprising: after all, there is no 'experience' independent of the ordering mechanisms of grammar, plot and genre and this is never more the case when attempting to 'speak' the ultimate transgression – killing another human being. They told their stories nonetheless.

Most refused to narrate their war-stories in self-destructive ways. Despite the confusing anonymity of battle, combatants insisted upon myths of agency. It was precisely this ability to assert their own individuality and sense of personal responsibility even within the disorder of combat that gave meaning to the warring enterprise and to their lives. This is not to argue that order and consistency could actually be achieved. It was not uncommon for people to hold contradictory beliefs simultaneously. As we have seen, for instance, declarations that they were 'only obeying orders' and that their murderous aggression was 'unnatural' could not be easily reconciled with the urge to assume responsibility for their violent impulses. But the acceptance of agency enabled combatants to take that one step towards the making of a bearable, and possibly entertaining war.

There were numerous ways in which exceptionally aggressive acts could be experienced as enjoyable. Many combatants in this book idealized themselves as warriors similar to heroic figures in combat literature and films. Of course, there was little that was 'really' heroic about their actions (indeed, actual acts of heroism were often regarded askance by many combatants who feared that such antics risked the survival of the group or were just plain silly). In the context of modern warfare, it was a feat of remarkable imagination that enabled combatants to construct their experiences in ways that emphasized skill, intimacy and chivalry, yet these three languages were used time and time again as a barrier against demoralization. It was these codes that enabled men to emphasize love as opposed to hatred, agency instead of chaos, martial pride as opposed to moral humiliation. Even a real hero like Roy Benavidez, who remained committed to mythical stories of gallantry despite subsequent disillusionment, strove throughout his life to reform military institutions in ways that would hold combatants in high esteem. Although conscious of the devaluation of the hero in modern warfare, his commitment to the memory of his dead comrades and to the sufferings of other veterans recalled a myth of 'clean kills', virile duellists and responsible citizens. For such men, dreams of fellowship gave strength to their violent hands and shrouded the horror of their actions with a mantle of honour.

There were simpler pleasures as well. As the historian Niall Ferguson argued in his momentous book *The Pity of War* (published as this book was being produced), most servicemen were not coerced into the firing line. Rather, for many, it was possible to become 'intoxicated' by 'violence for its own sake': 'fighting was fun'.[1] Combatants frequently dared to admit to orgasmic joy in unrestrained slaughter. The instruments and landscape of war could be seen as being aesthetically pleasing, and the potency of the weapons of death enchanting. When military technology meant that combatants could no longer 'see' the effect of their weapons, they conjured up face-to-face encounters. Furthermore, this process of personalizing the foe enabled them to kill. It validated combatants as moral men, innocent and protected from the threat of long-term brutalization. As the journalist and editor, William Broyles, admitted in the opening passage of the first chapter, the pleasures of killing were also linked to carnivalesque games (such as the humorous manipulation of corpses) and other rites of immorality (such as the taking and leaving of souvenirs). Through such grotesque acts, men were able to confront and even enjoy the horror.

These combatants returned to their friends and families after the war and proceeded to create more restful lives. Broyles raised a family and wrote books and screenplays, as did Ion Llewellyn Idriess (the Australian trooper during the First World War). The female combatant, Flora Sandes, married and continued with her busy life. All their attempts to impose meaning upon the chaos of combat (whilst inevitably jumbled and fictive, and necessitating a denial of the atrocious annihilation of fellow human beings) proved to be a personal bulwark against brutalization. In the long term, fears that combatants might become savage turned out to be illusory – indeed, it is more uncomfortable to hear civilians giggling over breakfast at descriptions of massacres or to read of 'peace' demonstrators who were still prepared to carry banners supporting (Communist) armed struggle. Face-to-face with the vicarious pleasure experienced by many men and women well behind the frontlines, combatants shrank away in despair.

Whatever their experiences, coming home from the battlefield was never easy, even for those men and women who told their combat stories in constructive ways. For instance, Stuart Smyth was one of many Vietnam veterans who came to realize that he was no longer able to live contentedly 'with the dwarfs/ of my hometown in quiet middle-age'. He wrote:

> *There have been too many new sensations,*
> *provocations to thought I might never have had,*
> *too many agitated tauntings of loves and*
> *fears I never knew existed.*[2]

For such men, it was precisely the knowledge of their appalling transgression which stunned them out of their complacency about life (and death). In the words of another veteran:

> I still take it right back to Vietnam, right after that first
> experience of looking at death. That's what pushed me
> into seeking. See, you kind of wonder what life is about
> when you see people die. The war, well, it made me have a
> clearer view of myself . . . It made me see what atrocities
> man can commit. I also got a lot of things off my chest
> just talking to people over there. So it showed me too the
> incredible love in the soldiers themselves, not the shit love
> on TV or the movies, real love that I haven't seen since.
> And just that experience alone gives me hope for a better
> world, helping each other with problems and things.[3]

For this minority, battle was employed to create positive imaginings: dreams of a war-free world, of love, of peace even. Wartime experiences could a man make vow never to touch a rifle again ('now I'm mightily afraid of guns', confessed the formerly keen hunter, Paul Sgroi)[4] or (as for the central character of *Still Life*) swear to break both his son's legs rather than let him go to war. For some men, like Father George Zabelka, forgiveness for their wartime bellicosity was sought by becoming active peace campaigners. An unknown number of men and women in war imagined a similar route towards peace. Others – like Hugh Thompson at My Lai – participated in combat but, when it became 'atrocious', protested according to their individual, moral logic. More typically, the desperate question asked by many servicemen about whether war could be about something other than mere slaughter was answered by William L. Calley: 'what the hell *else* is war than killing people'. Most combatants when faced with the demand to commit acts of violence that violated the laws of war merely withdrew passively from the scene rather than protesting actively. Their actions were those of the morally detached.

Combatants who could not maintain such a pose found themselves tortured with images of their dying prey. There was trauma and madness involved in killing. The guilt that tormented men like Arthur Hubbard (who served during the First World War), Bruce F. Anello (Vietnam soldier), and Dave Nelson (Vietnam sniper) was embraced as an alternative and appropriate moral response. It both confirmed their status as 'human' and enabled the slaughter to continue. The same combatants who admitted on one page in their diaries to feeling intense distress when killing another human being would confess, elsewhere, to feeling immensely happy while committing acts of murderous aggression. Contradictory emotions existed side by side, but historians have tended to examine only one half, assuming that the pleasure was 'sick' or 'abnormal' while the trauma was 'normal'. Yet it was through the language of trauma that people coped with the disturbing experience of killing. Most combat soldiers refused to countenance attempts to alleviate the emotional pain attendant upon sincere remorse. We do not know what happened to Hubbard after he returned to London and recovered from his nervous collapse. Anello's guilt was still-born – he was killed in action. Nelson, however, would continue to suffer, his guilt reinforced by a chorus of rebuke from his former comrades who did not share his version of the 'warrior ethos'. No amount of reassurance given by padres and (increasingly) psychologists could calm some consciences.

Of course, there were major differences between the conflicts, although this book has paid more attention to the similarities. For instance, it was crucial to the life stories of combatants during the Vietnam War that they were fighting an unpopular war. Their 'after' to combat was more traumatic than during other conflicts. The moral status of any particular conflict also influenced the bellicosity of padres and psychologists. Only during the Vietnam War was there widespread concern amongst both of these professional groups about the legitimacy of warfare, but these misgivings reflected wider social concerns about the legitimacy of this *particular* conflict rather than any reworking of specifically theological or psychological issues. Equally, killing Japanese or Vietnamese clearly meant something different to killing Germans – and it also mattered whether the man *doing* the killing was a black American, an Australian Aborigine, or an Irishman. The war in the Pacific (during the 1939–45 conflict) clearly had more in common with the Vietnam War than it did with either of the world wars being fought in Europe. There were also particular circumstances which facilitated or encouraged the most extreme acts of massacre: failures in military leadership, group pressure, ignorance about the legality of particular acts in wartime, and uncertainty about the limits to unquestioned obedience linked up with situational considerations (such as the inability to directly engage the enemy, a high death count amongst one's comrades, and racism) to render Vietnam a more atrocious war in terms of allied actions than the two world wars. The mass entrenched armies of the First World War were much more liable to engender feelings of respect and affection than in faster paced conflicts between 1939 and 1945 and in Vietnam. National traditions also affected battle narratives (for instance, Australian troops were probably more aggressive in combat when pitted next to English troops). Differences in education and expectations may have meant that the way some privates and officers attempted to find a language that would enable them to come to terms with killing had little in common.

There can be no unambiguous conclusion to a book about killing during wartime. In the final chapter, the impact of military experiences was examined, though millions of men did not live to see an 'after' to their war. We will never know whether Julian Francis Grenfell (who died of wounds in France on 26 May 1915) would have loved his German opponents during the Second World War as much as he did the 'Huns' in 1914–15 or whether the archetypal 'warrior', ace pilot Richard Hillary (Second World War), would always epitomize the 'beauty and courage and nobility of youth' because he never lived to see maturity. However, as this book has attempted to emphasize, warfare was as much about the business of sacrificing others as it was about *being* sacrificed. For many men and women, this was what made it 'a lovely war'.

NOTES

The following abbreviations are used in the Notes:
AWM: Australian War Memorial
CMAC: Contemporary Medical Archives Centre (Wellcome Institute for the History of Medicine)
HC: The House of Commons Parliamentary Papers
IWM: Imperial War Museum
PRO: Public Record Office (London)
PRONI: Public Records Office of Northern Ireland

Introduction

1 George A. Birmingham, *A Padre in France* (London, 1918), p. 64. My emphasis.

2 'Maxims for the Leader', *42nd East Lancashire Division, Handbook* (Aldershot, 1918), p. 8. Also see War Office, *Instructions for the Training of Platoons for Offensive Action 1917* (War Office, 1917), p. 14.

3 Graham H. Greenwell, *An Infant in Arms. War Letters of a Company Officer 1914–1918*, first published in 1935 (London, 1972), p. 142, letter to his mother, 9 October 1916; W. R. Kirby, 'The Battle of Cambrai, 1917', p. 21, IWM; 'Mark VII', *A Subaltern on the Somme in 1916* (London, 1927), p. 120.

4 Letter to the editor from Colonel George I. Forsythe (Infantry) and Lieutenant-Colonel Harold H. Dunwoody (Armor), 'Solidarity and the Mass Army', *The Army Combat Forces Journal*, 5.9 (April 1955), p. 5.

5 Letter by Private Wilson to Sal, quoted by Newham Nathaniel Davis, *Military Dialogues. On 'Active Service'* (London, 1900), p. 93.

6 Private Peter McGregor, 'Letters', letter to his wife, 7 July 1916, IWM and Sir Oliver J. Lodge, *Raymond or Life and Death with Examples of the*

Evidence for Survival of Memory and Affection After Death, 10th edition (London, 1918), p. 53, letter from Raymond Lodge to his family 28 June 1915.

7   William Carr, *A Time to Leave the Ploughshares. A Gunner Remembers 1917–1918* (London, 1985), pp. 47–8. Also see Kenneth T. Henderson, *Khaki and Cassock* (Melbourne, 1919), pp. 31–2; Cecil Sommers, *Temporary Heroes* (London, 1917), pp. 76–7; R. W. Thompson, *Men Under Fire* (London, 1946), p. 29.

8   Anthony S. Irwin, *Infantry Officer. A Personal Record* (London, 1943), p. 66.

9   For example, see Walter A. Briscoe and H. Russell Stannard, *Captain Ball, V.C. The Career of Flight Commander Ball, V.C., D.S.O.* (London, 1918), pp. 139 and 214.

10 Quoting a South African fighter pilot's 'Ten Commandments for Air Fighting', in Gavin Lyall (ed.), *The War in the Air, 1939–1945. An Anthology of Personal Experience* (London, 1968), p. 45. For descriptions of intimacy in aerial warfare, see Hector Bolitho, *Combat Report. The Story of a Fighter Pilot* (London, 1943), p. 54, quoting from a letter from pilot 'John', 10 April, probably 1940; Boyd Cable, *Air Men O' War* (London, 1918), p. 75; John J. Floherty, *The Courage and the Glory* (Philadelphia, 1942), p. 139; A. G. J. Whitehouse, *Hell in the Heavens. The Adventures of an Aerial Gunner in the Royal Flying Corps* (London, 1938), p. 44.

11 See the interview of K. O. Moore and Alec Gibb in Hector Bolitho, *Combat Report. The Story of a Fighter Pilot* (London, 1943), pp. 10–11.

12 John Slomm Riddell Hodgson, 'Letters of a Young Soldier', letters to parents, 28 March 1915 and 19 January 1916, Birmingham University Library Special Collection.

13 The most well-known examples include Hannah Arendt, *Eichmann in Jerusalem: A Report on the Banality of Evil* (New York, 1965); Zygmunt Bauman, *Modernity and the Holocaust* (Oxford, 1989); Christopher R. Browning, *Ordinary Men. Reserve Police Battalion 101 and the Final Solution in Poland* (New York, 1992); Daniel Jonah Goldhagen, *Hitler's Willing Executioners. Ordinary Germans and the Holocaust* (London, 1996); Rail Hilberg, *The Destruction of the European Jews* (London, 1961); Stanley Milgram, 'Behaviour Study of Obedience', in A. Etzioni and W. Wenglinsky (eds), *War and Its Prevention* (New York, 1970), part 5, pp. 245–59; Stanley Milgram, *Obedience to Authority: An Experimental View* (New York, 1974); Stanley Milgram, 'Some Conditions of Obedience and Disobedience to Authority', *Human Relations*, 18.1 (February 1965),

pp. 57–76; Ervin Staub, *The Roots of Evil. The Origins of Genocide and Other Group Violence* (Cambridge, 1989); Philip Zimbardo, Craig Haney, and Curtis Banks, 'Interpersonal Dynamics in a Simulated Prison', *International Journal of Criminology and Penology*, I (1973), pp. 89–90.

14 Edward C. McDonald, 'Social Adjustment to Militarism', *Sociology and Social Research*, 29.6 (July–August 1945), pp. 449–50.

15 Robert Jay Lifton, *Home From the War. Vietnam Veterans: Neither Victims Nor Executioners* (London, 1974), p. 350. Also see Shelford Bidwell, *Modern Warfare. A Study of Men, Weapons and Theories* (London, 1973), p. 62.

16 Richard A. Gabriel, *The Painful Field. The Psychiatric Dimension of Modern War* (New York, 1988), pp. 26–30.

17 Kenneth Macksey and William Woodhouse (eds), *The Penguin Encyclopedia of Modern Warfare* (London, 1991), p. 111.

18 Colonel A. G. Butler, *Official History of the Australian Army Medical Services 1914–1918. Volume II* (Canberra, 1940), p. 495 and Richard Holmes, *Firing Line* (London, 1985), p. 210. Also see Major-General J. F. C. Fuller, *Lectures on F.S.R. II* (London, 1931), pp. 14–15.

19 Sir Winston Churchill, quoted in Ronald W. Clark, *The Rise of the Boffins* (London, 1962), p. 17. 'Boffins' were scientists working in the military.

20 William J. Simon, 'My Country', in Larry Rottmann, Jan Barry, and Basil T. Paquet (eds), *Winning Hearts and Minds. War Poems by Vietnam Veterans* (New York, 1972), p. 42.

21 The most famous studies of this mode were carried out by Stanley Milgram, *Obedience to Authority: An Experimental View* (New York, 1974); 'Some Conditions of Obedience and Disobedience to Authority', *Human Relations*, 18.1 (February 1965), pp. 57–76; 'Behaviour Study of Obedience', in A. Etzioni and W. Wenglinsky (eds), *War and Its Prevention* (New York, 1970), part 5, pp. 245–59. Milgram's insights have been usefully employed by writers like Ervin Staub, *The Roots of Evil. The Origins of Genocide and Other Group Violence* (Cambridge, 1989), p. 43.

22 For a fascinating experiment into the relationship between the 'agentic mode' and assuming responsibility, see David Mark Mantell and Robert Panzarella, 'Obedience and Responsibility', *British Journal of Social and Clinical Psychology*, 15 (1976), pp. 239–45.

23 For instance, see Richard Kohn, 'Commentary', in Horst Boog (ed.), *The Conduct of the Air War in the Second World War. An International Comparison* (Oxford, 1992), pp. 410–11.

24 For instance, compare the stories of Captain John Long interviewed in Mary Penick Motley (ed.), *The Invisible Soldier. The Experience of the*

Black Soldier, World War II (Detroit, 1975), p. 152 and Ulysses Grant Lee, *United States Army in World War II. Special Studies. The Employment of Negro Troops* (Washington, DC, 1966).

25 Owen Spencer Watkins, *With French in France and Flanders* (London, 1915), p. 127.

26 Lieutenant-Colonel T. S. Wollocombe, 'Diary of the Great War', p. 34, IWM.

27 Edwin Weinstein, 'The Fifth U.S. Army Neuropsychiatric Centre – 601st', in Lieutenant-General Hal B. Jennings (ed.), *Neuropsychiatry in World War II. Volume II. Overseas Theatres* (Washington, DC, 1973), p. 137.

28 Richard J. Ford III interviewed in Wallace Terry, *Bloods. An Oral History of the Vietnam War by Black Veterans* (New York, 1984), p. 35.

29 T. W. Bacon, 'Letters Written to Mr. Bacon by his Nephews Ralph and Edmund Creyke and his Brother-in-Law, Sir Charles Barrington', letter from Ralph Creyke, 20 December 1914, IWM; Rear-Admiral Sir Douglas Brownrigg, *Indiscretions of the Naval Censor* (London, 1920), p. 188; Henry Gother Courtney, 'Letters', letter number 8/147, 26 November 1916, in Birmingham University Library Courtney Collection 8–10 and the letter by Captain Geoffrey Blemell Pollard, 19 October 1914, in Laurence Housman (ed.), *War Letters of Fallen Englishmen* (London, 1930), pp. 219–20; Major E. Henry E. Daniell, 'Letter to Mother', 26 September 1914, IWM; Lieutenant Rowland H. Owen, 'Letters Home', letter to parents, 30 September 1914, IWM.

30 William K. Willis, 'Letters to Miss Luttrell', 4 May 1917, AWM.

31 Tim O'Brien, *The Things They Carried* (New York, 1990), p. 21.

1 The Pleasures of War

1 R. L. Barth, 'A Letter to my Infant Son', in Barth, *A Soldier's Time. Vietnam War Poems* (Santa Barbara, 1987), pp. 38–9.

2 William Broyles, 'Why Men Love War', *Esquire*, November 1984, pp. 54–65. Ellipses in original.

3 Ken D. Jones and Arthur F. McClure, *Hollywood at War. The American Motion Picture and World War II* (South Brunswick, 1973), p. 16 and Peter A. Soderbergh, *Women Marines in the Korean War Era* (Westport, Connecticut, 1994), p. 71.

4 The best discussion of this remains Ted Bogacz, '"A Tyranny of Words": Language, Poetry, and Antimodernism in the First World War', *Journal of Modern History*, 58 (September 1986), pp. 643–68.

5 Quoted in Terry Christensen, *Reel Politics. American Political Movies from*

'The Birth of a Nation' to 'Platoon' (New York, 1987), p. 152.

6 Philip Caputo, A Rumor of War (London, 1977), p. 14.

7 For a discussion, see Nancy Anisfield, 'After the Apocalypse: Narrative Movement in Larry Heinemann's Paco's Story', in Owen W. Gilman, Jr., and Lorrie Smith (eds) America Rediscovered. Critical Essays on Literature and Film of the Vietnam War (New York, 1990), pp. 275–7 and Don Ringnaldo, Fighting and Writing the Vietnam War (Jackson, 1994), p. 176.

8 Larry Heinemann, Paco's Story (London, 1987), p. 4.

9 Tim O'Brien, interviewed by Eric James Schroeder, 'Two Interviews: Talks with Tim O'Brien and Robert Stone', Modern Fiction Studies, 30.1 (spring 1984), p. 146.

10 Vee Robinson, On Target (Wakefield, West Yorkshire, 1991), p. 3.

11 Jean Bethke Elshtain, Women and War (Chicago, 1995), p. 24.

12 Harold R. Peat, The Inexcusable Lie (New York, 1923), pp. 42–3.

13 Lieutenant-Colonel H. F. N. Jourdain, Ranging Memories (Oxford, 1934), pp. 15–16.

14 Eugene B. 'Sledgehammer' Sledge, With the Old Breed at Peleliu and Okinawa (New York, 1990), p. xii.

15 Audie Murphy, To Hell and Back (London, 1956), p. 16. For a Vietnam War example, see James Hebron, interviewed in Al Santoli, Everything We Had. An Oral History of the Vietnam War by Thirty-Three American Soldiers Who Fought It (New York, 1981), pp. 88 and 90.

16 Philip Caputo, A Rumor of War (London, 1977), p. 165 and Hugh Dundas, Flying Start. A Fighter Pilot's War Years (London, 1988), p. 3.

17 Allen Hunter, interviewed in Glen D. Edwards, Vietnam. The War Within (Salisbury, South Australia, 1992), p. 17. For similar examples, see the unnamed Vietnam veteran, interviewed by Mark Baker, Nam. The Vietnam War in the Words of the Men and Women Who Fought There (London, 1982), p. 17 and Sam Grashio, interviewed by Shirley Dicks, From Vietnam to Hell. Interviews with Victims of Post-Traumatic Stress Disorder (Jefferson, North Carolina, 1990), p. 118.

18 Ron Kovic, Born on the Fourth of July, first published in 1976 (Aylesbury, 1990), p. 42.

19 William D. Ehrhart, 'Why I Did It', The Virginia Quarterly Review, 56.1 (winter 1980), p. 26.

20 Donald Duncan, The New Legions (London, 1967), p. 199.

21 Ron Kovic, Born on the Fourth of July, first published in 1976 (Aylesbury, 1990), p. 43.

22 Dorothy Barclay, 'Behind all the Bang-Bang', The New York Times

Magazine, 22 July 1962, p. 47.

23 For a description, see Claudia Springer, 'Military Propaganda: Defense Department Films from World War II and Vietnam', *Cultural Critique*, 3 (spring 1986), p. 164.

24 Rowland Feilding, *War Letters to a Wife. France and Flanders, 1915–1919* (London, 1929), letter dated 5 September 1916, pp. 109-10.

25 Carl I. Hovland, Arthur A. Lumsdaine, and Fred D. Sheffield, *Experiments on Mass Communication. Volume III* (Princeton, 1949), p. 93.

26 Lieutenant-Commander Howard P. Rome, 'Motion Pictures as a Medium of Education', *Mental Hygiene*, 30 (January 1946), pp. 9–20. For more on the use of films in training, see 'The Work of Army Psychiatrists in Relation to Morale', January 1944, p. 3, PRO CAB 21/914.

27 Robert William MacKenna, *Through a Tent Door*, first published 1919 (London, 1930), pp. 87 and 89.

28 Told by John J. Floherty, *The Courage and the Glory* (Philadelphia, 1942), pp. 34–59.

29 Interview of Bill Stevens, by Mary Penick Motley (ed.), *The Invisible Soldier. The Experience of the Black Soldier, World War II* (Detroit, 1975), p. 78. For similar examples, see the interview in the same collection by Eddie Robinson, pp. 113–14 and Eugene B. 'Sledgehammer' Sledge, *With the Old Breed at Peleliu and Okinawa* (New York, 1990), p. 120.

30 For instance, see Mark Baker, *Nam: The Vietnam War in the Words of the Men and Women who Fought There* (London, 1982), pp. 12 and 17; Philip Caputo, *A Rumor of War* (London, 1977), p. 269; T. J. Kelly, interviewed in Otto J. Lehrack, *No Shining Armour. The Marines at War in Vietnam. An Oral History* (Lawrence, Kansas, 1992), p. 267; James Jones, *The Thin Red Line* (New York, 1962), p. 242; Ron Kovic, *Born on the Fourth of July*, first published in 1976 (Aylesbury, 1990), the epigraph and pp. 39, 42–3, 58, 67.

31 Allen Eyles, *John Wayne and the Movies* (London, 1976), p. 11.

32 Randy Cunningham, in Lou Drendel, . . . *And Kill Migs. Air to Air Combat in the Vietnam War* (Carrollton, 1974), p. 47.

33 Philip Caputo, *A Rumor of War* (London, 1977), p. 6.

34 Peter A. Soderbergh, *Women Marines in the Korean War Era* (Westport, Connecticut, 1994), pp. 30–1.

35 Gary McKay, *In Good Company. One Man's War in Vietnam* (Sydney, 1987), p. 162. For other disgruntled combatants, see the unnamed Vietnam veteran, interviewed by Mark Baker, *Nam. The Vietnam War in the Words of the Men and Women Who Fought There* (London, 1982), p. 50; Ron Kovic, *Born on the Fourth of July*, first published in 1976 (Aylesbury,

1990), pp. 148–9; Jon Oplinger, *Quang Tri.Cadence. Memoir of a Rifle Platoon Leader in the Mountains of Vietnam* (Jefferson, North Carolina, 1993), pp. 12–13.

36 Hugh Dundas, *Flying Start. A Fighter Pilot's War Years* (London, 1988), p. 3.

37 Interview of two pilots, in Squadron Leader Hector Bolitho, 'Two in Twenty-Two Minutes', in *Slipstream. A Royal Air Force Anthology* (London, 1946), p. 11.

38 Fighter pilot Doug Carter, interviewed in Rudy Tomedi, *No Bugles, No Drums. An Oral History of the Korean War* (New York, 1993), p. 171.

39 Geoffry R. Jones, 'A Short Memoir', 1986, p. 4, AWM. Also see Philip Caputo, *A Rumor of War* (London, 1977), p. 71 and Donald Gilchrist, *Castle Commando* (Edinburgh, 1960), pp. 5–6.

40 Tobias Wolff, *In Pharaoh's Army. Memories of a Lost War* (London, 1994), pp. 131–2. Also see Jon Oplinger, *Quang Tri Cadence. Memoir of a Rifle Platoon Leader in the Mountains of Vietnam* (Jefferson, North Carolina, 1993), p. 57.

41 Quoted by Revd E. J. Hardy, *The British Soldier. His Courage and Humour* (London, 1915), p. 37.

42 Sergeant Alexander Woollcott, 'With Pershing's A.E.F.', *Stars and Stripes*, 20 September 1918, cited in Herbert Mitgang (ed.), *Civilians Under Arms. The American Soldier – Civil War to Korea as He Revealed Himself in His Own Words in 'The Stars and Stripes', Army Newspaper* (Cleveland, Ohio, 1959), p. 110.

43 Unnamed Canadian informant, in Barry Broadfoot, *Six War Years 1939–1945. Memories of Canadians at Home and Abroad* (Don Mills, Ontario, 1974), p. 89. Also see Eugene B. 'Sledgehammer' Sledge, *With the Old Breed at Peleliu and Okinawa* (New York, 1990), pp. 56 and 79.

44 Mark Baker, *Nam. The Vietnam War in the Words of the Men and Women Who Fought There* (London, 1982), p. 58.

45 Philip Caputo, *A Rumor of War* (London, 1977), pp. 290 and 305–6.

46 William Manchester, *Goodbye Darkness. A Memoir of the Pacific War* (Boston, 1980), pp. 67–8.

47 Michael Herr, *Dispatches* (London, 1978), p. 169. Also see Hans Halberstadt, *Green Berets. Unconventional Warriors* (London, 1988), p. 133.

48 William Broyles, 'Why Men Love War', *Esquire* (November 1984), p. 56.

49 Josh Cruze, interviewed in Kim Willenson, *The Bad War. An Oral History of the Vietnam War* (New York, 1987), p. 61. Also see Dale Barnes and 'Nelson' interviewed by Shirley Dicks, *From Vietnam to Hell. Interviews*

with *Victims of Post-Traumatic Stress Disorder* (Jefferson, North Carolina, 1990), pp. 5 and 19.

50 Harold 'Light Bulb' Bryant, interviewed in Wallace Terry, *Bloods. An Oral History of the Vietnam War by Black Veterans* (New York, 1984), p. 25.

51 G. Belton Cobb, *Stand to Arms* (London, 1916), p. 4.

52 G. Belton Cobb, *Stand to Arms* (London, 1916), p. 5 and John Carver Edwards (ed.), 'Sergeant Jones Goes to War. Extracts from a U.S. Artilleryman's Diary, 1918', *The Army Quarterly and Defence Journal*, 104.1 (October 1973), pp. 62–3, diary for 4 October 1918.

53 Captain Alfred E. Bland, 'Letters to His Wife', 30 January 1916, IWM.

54 Unnamed Vietnam veteran, interviewed by Mark Baker, *Nam. The Vietnam War in the Words of the Men and Women Who Fought There* (London, 1982), p. 17.

55 Roy R. Grinker and John P. Spiegel, *Men Under Stress* (London, 1945), p. 44.

56 T. T. Paterson, *Morale in War and Work. An Experiment in the Management of Men* (London, 1955), pp. 88–9.

57 Samuel A. Stouffer et al., *The American Soldier: Combat and Its Aftermath. Volume II* (Princeton, 1949), p. 332.

58 Flight Lieutenant Richard Hillary, quoted by Gavin Lyall (ed.), *The War in the Air, 1939–1945. An Anthology of Personal Experience* (London, 1968), pp. 43–4. Also see Sam Grashio, interviewed by Shirley Dicks, *From Vietnam to Hell. Interviews with Victims of Post-Traumatic Stress Disorder* (Jefferson, North Carolina, 1990), p. 118.

59 Captain Wilfred Thomas Colyer, 'Memoirs', no page numbers, part 5, chapter 18 [labelled Chapter 2 but located between chapters 17 and 19], IWM. Admittedly, Colyer did not enjoy bayoneting: he was describing a comrade's feelings.

60 Letter by Second Lieutenant F. R. Darrow, in *Letters from the Front. Being a Record of the Part Played by Officers of the [Canadian] Bank [of Commerce] in the Great War 1914–1919*, vol. 1 (Toronto, 1920), p. 241.

61 'Anzac', *On the Anzac Trail. Being Extracts from the Diary of a New Zealand Sapper* (London, 1916), p. 121, entry for 28 April 1915.

62 Lieutenant-Colonel Neil Fraser Tytler, *Field Guns in France* (London, 1922), p. 35, letter to his father, 9 January 1916.

63 Henry de Man, *The Remaking of a Mind. A Soldier's Thoughts on War and Reconstruction* (London, 1920), pp. 198–9.

64 Philip Caputo, *A Rumor of War* (London, 1977), p. 81.

65 Gary McKay, *In Good Company. One Man's War in Vietnam* (Sydney,

1987), p. 162. Also see Lofty Large, *One Man's War in Korea* (Wellingborough, Northhamptonshire, 1988), p. 53.

66 Unnamed Vietnam veteran, quoted in Jonathan Shay, *Achilles in Vietnam. Combat Trauma and the Undoing of Character* (New York, 1994), p. 84. For an example from the Korean conflict, see Andy Barr, interviewed in Rudy Tomedi, *No Bugles, No Drums. An Oral History of the Korean War* (New York, 1993), p. 72.

67 James Hebron, interviewed in Al Santoli, *Everything We Had. An Oral History of the Vietnam War by Thirty-Three American Soldiers Who Fought It* (New York, 1981), pp. 98–9.

68 Michael Herr, *Dispatches* (London, 1979), p. 199.

69 Philip Caputo, *A Rumor of War* (London, 1977), p. 268 (and pp. xiii and xv) and James Jones, *The Thin Red Line* (New York, 1962), p. 197.

70 Mark Baker, *Nam. The Vietnam War in the Words of the Men and Women Who Fought There* (London, 1982), p. 51. Also see Richard Boyle, *The Flower of the Dragon. The Breakdown of the U.S. Army in Vietnam* (San Francisco, 1972), p. 70 and Arthur Brown, interviewed by Shirley Dicks, *From Vietnam to Hell. Interviews with Victims of Post-Traumatic Stress Disorder* (Jefferson, North Carolina, 1990), p. 132.

71 James Jones, *The Thin Red Line* (New York, 1962), pp. 197–8 (also see pp. 300, 356 and 438).

72 Major William Avery Bishop, *Winged Warfare. Hunting Huns in the Air* (London, 1918), p. 9.

73 James Byford McCudden, *Flying Fury*, first published in 1918 (London, 1930), p. 171.

74 'Bob', interviewed in Wing Commander Athol Forbes and Squadron Leader Hubert Allen, *The Fighter Boys* (London, 1942), p. 84.

75 See the interview of K. O. Moore and Alec Gibb, in Squadron Leader Hector Bolitho, 'Two in Twenty-Two Minutes', in *Slipstream. A Royal Air Force Anthology* (London, 1946), pp. 10–11.

76 Hector Bolitho, *Combat Report. The Story of a Fighter Pilot* (London, 1943), p. 54, quoting from a letter from pilot 'John', 10 April, probably 1940.

77 Spitfire pilot called 'Durex', interviewed by Wing Commander Athol Forbes and Squadron Leader Hubert Allen, *The Fighter Boys* (London, 1942), p. 41. Also see Flight-Lieutenant D. M. Crook, *Spitfire Pilot* (London, 1942), 28–9 and Kenneth Hemingway, *Wings Over Burma* (London, 1944), pp. 68–9.

78 Phil Buttigieg, interviewed in Gary McKay, *Vietnam Fragments. An Oral History of Australians at War* (St Leonards, New South Wales, 1992),

p. 99. Also see interview with unnamed veteran, in Mark Baker, *Nam. The Vietnam War in the Words of the Men and Women Who Fought There* (London, 1982), p. 56.

79 Private Peter Gates, 'Letters to His Family', 31 July 1967, p. 3, AWM.

80 Andrew Treffry, 'Letters to His Fiancé', letter dated 1 May 1969, AWM.

81 Quoted in Lieutenant-Colonel H. F. N. Jourdain, *Ranging Memories* (Oxford, 1934), p. 246. See also T. W. Bacon, 'Letters Written to Mr. Bacon by his Nephews Ralph and Edmund Creyke and his Brother-in-Law, Sir Charles Barrington', letter from Ralph Creyke, 20 December 1914, IWM; Rear-Admiral Sir Douglas Brownrigg, *Indiscretions of the Naval Censor* (London, 1920), p. 188; Henry Gother Courtney, 'Letters', letter number 8/147, 26 November 1916, in Birmingham University Library Courtney Collection; letter by Captain Geoffrey Blemell Pollard, 19 October 1914, in Laurence Housman (ed.), *War Letters of Fallen Englishmen* (London, 1930), pp. 219–20; Major E. Henry E. Daniell, 'Letter to Mother', 26 September 1914, IWM; James Jones, *The Thin Red Line* (New York, 1962), p. 300; Lieutenant Rowland H. Owen, 'Letters Home', letter to parents, 30 September 1914, IWM.

82 Lieutenant Rowland H. Owen, 'Letters Home', to his parents on 13 October 1914, p. 40, IWM.

83 Eugene B. 'Sledgehammer' Sledge, *With the Old Breed at Peleliu and Okinawa* (New York, 1990), p. 120 and Tobias Wolff, *In Pharaoh's Army. Memories of a Lost War* (London, 1994), pp. 15–16.

84 R. H. Kiernan, *Captain Albert Ball* (London, 1933), p. 45.

85 T. T. Paterson, *Morale in War and Work. An Experiment in the Management of Men* (London, 1955), p. 88.

86 William K. Willis, 'Letters to Miss Luttrell', 4 May 1917, AWM. It refers to the battle of Lagnicourt.

87 Sid T. Kemp, 'Remembrance. The 6th Royal West Kent Regiment 1914–1916', p. 44, IWM.

88 Interview with an unnamed soldier, in Barry Broadfoot, *Six War Years 1939–1945. Memories of Canadians at Home and Abroad* (Don Mills, Ontario, 1974), p. 395.

89 John J. Callan, *With Guns and Wagons. A Day in the Life of an Artillery Chaplain* (London, 1918), p. 21; Lofty Large, *One Man's War in Korea* (Wellingborough, Northamptonshire, 1988), p. 52; William E. Merritt, *Where the Rivers Ran Backwards* (Athens, 1989), p. 142.

90 Jack Strahan, 'Arclight Strike', in Jack Strahan, Peter Hollenbeck, and R. L. Barth, *Vietnam Literature Anthology: A Balanced Perspective* (Philadelphia, 1985), p. 16. Also see James N. Eastman, Walter Hanak,

and Lawrence J. Paszek (eds), *Aces and Aerial Victories. The United States Air Force in Southeast Asia 1965–1973* (Washington, DC, 1978), p. 66.

91 William E. Merritt, *Where the Rivers Ran Backwards* (Athens, 1989), pp. 102–3.

92 Norman Hampson, 'Corvette', in Brian Gardner, *The Terrible Rain. The War Poets 1939–1945* (London, 1966), p. 117.

93 Al Pinches, interviewed in Gary McKay, *Vietnam Fragments. An Oral History of Australians at War* (St Leonards, New South Wales, 1992), p. 169.

94 The Revd William Edward Drury, *Camp Follower. A Padre's Recollections of Nile, Somme and Tigris During the First World War* (Dublin, 1968), p. 101.

95 Sergeant John Henry Ewen, 'Bougainville Campaign', book 1, 1944, p. 68, AWM.

96 Jon Oplinger, *Quang Tri Cadence. Memoir of a Rifle Platoon Leader in the Mountains of Vietnam* (Jefferson, North Carolina, 1993), p. 169; Harold 'Light Bulb' Bryant and Arthur E. 'Gene' Woodley, Jr., interviewed in Wallace Terry, *Bloods. An Oral History of the Vietnam War by Black Veterans* (New York, 1984), pp. 26 and 251.

97 An unnamed Vietnam veteran, interviewed by Mark Baker, *Nam. The Vietnam War in the Words of the Men and Women Who Fought There* (London, 1982), p. 49; Philip Caputo, *A Rumor of War* (London, 1977), p. 67; Dennis Kitchin, *War in Aquarius. Memoir of an American Infantryman in Action Along the Cambodian Border During the Vietnam War* (Jefferson, North Carolina, 1994), p. 17; Wallace Terry, *Bloods. An Oral History of the Vietnam War by Black Veterans* (New York, 1984), p. 213; Kim Willenson, *The Bad War. An Oral History of the Vietnam War* (New York, 1987), p. 63.

98 James Jones, *The Thin Red Line* (New York, 1962), p. 71 and Ed Treratola, interviewed in Mark Lane, *Conversations with Americans* (New York, 1970), pp. 99–100.

99 John W. Dower in *War Without Mercy. Race and Power in the Pacific War* (London, 1986), p. 66 was only partially correct when he argued that it was 'virtually inconceivable . . . that teeth, ears, and skulls could have been collected from German or Italian war dead and publicized in the Anglo-American countries without provoking an uproar; and in this we have yet another inkling of the racial dimensions of the [Pacific] war.'

100 Captain D. W. J. Andrews, 'Diary', entry for 25 September 1915, CMAC RAMC 2021; Thomas Pincombe Bovingdon, 'Memoirs', p. 25, IWM; Alfred E. Bundy, 'War Diary', entry for 25 October 1916, IWM; Edward

Frederick Chapman, 'Letters from France', letters to his sister Hilda, 9 October 1916 and 2 November 1916, IWM; George Coppard, *With a Machine Gun to Cambrai* (London, 1980), pp. 88–9; Gerald V. Dennis, 'A Kitchener's Man's Bit (1916–1918)', p. 40, IWM; Sydney W. D. Lockwood, 'First World War Memories 1914–1918', p. 59, IWM; Sidney Rogerson, *Twelve Days*, first published in 1930 (Norwich, 1988), p. 57; Major O. P. Taylor, 'Diaries of a Sometime Trench Mortar Man', entry for 21 January [no year], IWM; *With the First Canadian Contingent* (Toronto, 1915), p. 63.

101 Letter from Wilfred Owen to Colin Owen, 9 April 1917, in Harold Owen and John Bell (eds), *Wilfred Owen. Collected Letters* (London, 1961), p. 451. He asked that the handkerchief be kept for him.

102 Edward Glover, *War, Sadism and Pacifism. Three Essays* (London, 1933), p. 17.

103 Interviews of Bill Stevens and Eddie Robinson, in Mary Penick Motley (ed.), *The Invisible Soldier. The Experience of the Black Soldier, World War II* (Detroit, 1975), pp. 78 and 113–14 and unnamed black Muslim Marine, interviewed in Mark Baker, *Nam. The Vietnam War in the Words of the Men and Women Who Fought There* (London, 1982), p. 50.

104 Interview of Eugene B. 'Sledgehammer' Sledge, in Studs Terkel, *'The Good War'. An Oral History of World War Two* (London, 1985), pp. 61–2. Also see Sledge, *With the Old Breed at Peleliu and Okinawa* (New York, 1990), pp. 64, 118, and 120.

105 Nearly every diary, series of letters, or autobiography from the Vietnam War contains examples.

106 George Coppard, *With a Machine Gun to Cambrai* (London, 1980), p. 88; James J. Fahey, *Pacific War Diary 1942–1945* (Boston, 1963), p. 192, entry for 18 July 1944; Lieutenant Hugh A. Munro, 'Diary', entry for 23 May 1915, p. 14, IWM.

107 George Coppard, *With a Machine Gun to Cambrai* (London, 1980), pp. 73 and 90. Also see James J. Fahey, *Pacific War Diary 1942–1945* (Boston, 1963), p. 192, diary for 18 July 1944.

108 Harold 'Light Bulb' Bryant, interviewed in Wallace Terry, *Bloods. An Oral History of the Vietnam War by Black Veterans* (New York, 1984), p. 26.

109 Sergeant Jarvi, quoted by George P. Hunt, *Coral Comes High* (New York, 1946), p. 62.

110 Arthur E. 'Gene' Woodley, Jr. , interviewed in Wallace Terry, *Bloods. An Oral History of the Vietnam War by Black Veterans* (New York, 1984), p. 251. Also see Alan Camden, interviewed in Mark Lane, *Conversations*

with Americans (New York, 1970).

111 Larry Heinemann, *Paco's Story* (London, 1987), p. 8.

112 Charles Baudelaire, 'The Essence of Laughter And, In General, On the Comic in the Plastic Arts', first published in 1855, in Baudelaire, *The Painter of Modern Life and Other Essays*, translated by Jonathan Mayne (London, 1964), p. 153.

113 Michael Herr, *Dispatches* (London, 1978), p. 161.

114 Allan Wright, interviewed in Mark Lane, *Conversations with Americans* (New York, 1970), p. 227 and Wallace Terry, *Bloods. An Oral History of the Vietnam War by Black Veterans* (New York, 1984), pp. 26 and 46.

115 James J. Fahey, *Pacific War Diary 1942–1945* (Boston, 1963), p. 231, diary for 27 November 1944.

116 Mikhail Bakhtin, *Rabelais and His World*, translated by H. Iswolsky (Bloomington, Indiana, 1985).

117 For a discussion, see Paul Fussell, *Wartime: Understanding and Behaviour in the Second World War* (Oxford, 1989), p. 117. Also see 'Proposed Scheme for a Raid', by the Commander of the 1st Battalion, the Royal Warwickshire Regiment, 24 June 1918, in Lieutenant P. L. Smith, 'Papers', IWM.

118 Cadet Ian Rashan, 'The Spirit of the Bayonet', in *The Blimp* (Cambridge, 1917), p. 43.

119 Interview with unnamed black Muslim Marine, in Mark Baker, *Nam. The Vietnam War in the Words of the Men and Women Who Fought There* (London, 1982), p. 50. Also see Peter V. Fossell, interviewed in Philip Scribner Balboni, 'Mylai Was Not an Isolated Incident. What Every Vietnam Veteran Knows', *The New Republic*, 19 December 1970, p. 13.

120 T. J. Kelly, interviewed in Otto J. Lehrack, *No Shining Armour. The Marines at War in Vietnam. An Oral History* (Lawrence, Kansas, 1992), p. 267.

2 The Warrior Myth

1 Keith Douglas, 'Sportsmen', written in 1943, in Douglas, *The Complete Poems*, edited by Ted Hughes (Oxford, 1987), p. 110.

2 Ion Llewellyn Idriess, 'Biographical Cuttings', in the Australian National Library; *The Desert Column. Leaves from the Diary of an Australian Trooper in Gallipoli, Sinai, and Palestine* (Sydney, 1932), p. 252, diary entry for 14 March 1917; 'Grim Diary of Lone Pine', *The Daily Telegraph Pictorial*, 12 November 1929, p. 5; *Guerrilla Tactics* (Sydney, 1942); *Lurking Death* (Sydney, 1942), pp. 82–3; *Must Australia Fight?* (Sydney, 1939); *Sniping. With an Episode from the Author's Experiences During the*

War of 1914–18 (Sydney, 1942), pp. 52, 97–8; Beverley Eley, *Ion Idriess* (Sydney, 1995), p. 89.

3 Lovat Dickson, *Richard Hillary* (London, 1950), pp. viii, 8, 11, 42 and 82–7; Richard Hillary, *The Last Enemy* (London, 1942), pp. 16, 120–2; Arthur Koestler, *The Yogi and the Commissar and Other Essays* (London, 1945), p. 49; Arthur Koestler, 'The Birth of a Myth. In Memory of Richard Hillary', *Horizon*, vii.40 (April 1943), pp. 227–43; Eric Linklater, *The Art of Adventure* (London, 1947), p. 73.

4 Dave Nelson, interviewed in J. T. Hansen, A. Susan Owen, and Michael Patrick Madden, *Parallels. The Soldiers' Knowledge and the Oral History of Contemporary Warfare* (New York, 1992), pp. 13–28.

5 Colonel A. G. Butler, *Official History of the Australian Army Medical Services 1914–1918. Vol. II* (Canberra, 1940), p. 495.

6 Robert W. Service in 'My Bay'nit', *The Rhymes of a Red-Cross Man* (London, 1916), pp. 87–8.

7 Ian Skennerton, *The British Sniper. British and Commonwealth Sniping and Equipments 1915–1983* (London, 1983), p. 16. Also see *H.C. Parliamentary Debates*, 19 May 1915, col. 2391–2, 5th series, vol. lxxi and 21 September 1915, col. 305, vol. lxxiv.

8 Lieutenant-Colonel N. A. D. Armstrong, *Fieldcraft, Sniping and Intelligence*, 5th edition (Aldershot, 1942), p. v.

9 Ian Skennerton, *The British Sniper. British and Commonwealth Sniping and Equipments 1915–1983* (London, 1983), pp. 16 and 136.

10 Captain R. J. Linwood, 'The Sniper – Part I,' *Australian Infantry*, xx.2 (December 1980), pp. 9–10.

11 Lieutenant-General Julian J. Ewell and Major-General Ira A. Hunt, *Sharpening the Combat Edge: The Use of Analysis to Reinforce Military Judgment* (Washington, DC, 1974), pp. 120–2.

12 Ion Llewellyn Idriess, *The Desert Column. Leaves from the Diary of an Australian Trooper in Gallipoli, Sinai, and Palestine* (Sydney, 1932), p. 45.

13 Bernard Adams, *Nothing of Importance. A Record of Eight Months at the Front with a Welsh Battalion. October 1915 to June 1916* (Stevenage, 1988), pp. 133 and 152.

14 Charles Bird, 'From Home to the Charge: A Psychological Study of the Soldier', *The American Journal of Psychology*, xxviii.3 (July 1917), p. 339; Albert N. Depew, *Gunner Depew* (London, 1918), p. 39; H. Hesketh-Prichard, *Sniping in France* (London, 1920), p. 120.

15 H. Hesketh-Prichard, *Sniping in France* (London, 1920), p. 40.

16 John Finnemore, *Two Boys in War-Time* (London, 1900 and 1928), pp. 216–17 and 200–1, 295–6 and 241–2, respectively, and *A Boy Scout with*

the Russians (London, 1915), p. 92.

17 Nat Gould, *Lost and Won. A Tale of Sport and War* (London, 1916), p. 244. Also see pp. 272–4.

18 Jack W. Bobbin, 'War to the Death', *The Boys' Journal*, iii.60 (7 November 1914), p. 196; Alan Deane, 'The Battle of Mons', *The Boys' Journal*, iii.58 (24 October 1914), p. 148; John Finnemore, *Foray and Fight. Being the Story of the Remarkable Adventures of an Englishman and an American in Macedonia* (London, 1906), p. 125; Eden Phillpotts, *The Human Boy and the War* (London, 1916), pp. 4–5. These quotations are not unusual: killing with the bayonet was a theme throughout boys' fiction.

19 Captain Frederick Sadleir Brereton, *With Rifle and Bayonet. A Story of the Boer War* (London, 1900), p. 271.

20 John Finnemore, *Two Boys in War-Time* (London, 1900 and 1928), pp. 296 and 242, respectively.

21 Revd E. J. Hardy, *The British Soldier. His Courage and Humour* (London, 1915), p. 42.

22 Patrick Mee, *Marine Gunner. Twenty-Two Years in the Royal Marine Artillery* (London, 1935), p. 190.

23 'An Officer', *Practical Bayonet-Fighting, with Service Rifle and Bayonet* (London, 1915), p. 14.

24 E. C. Palmer, 'It Happened Twice', p. 7, IWM; Wallace Reyburn, *Glorious Chapter. The Canadians at Dieppe* (London, 1943), p. 37; J. W. Stephenson, 'With the Dukes in Flanders', p. 2, IWM.

25 Battery 'C' 11th F.A., *From Arizona to the Huns* (Dijon, 1919), p. 38.

26 Sir Geoffrey Vickers, 'Papers', letter to his mother, 19 April 1915, Liddell Hart Centre.

27 Coningsby Dawson, *The Glory of the Trenches* (London, 1918), p. 74. Also see Stephen Graham, *A Private in the Guards* (London, 1919), p. 140.

28 Michael MacDonagh, *The Irish at the Front* (London, 1916), p. 80. Also see pp. 94–5.

29 T. W. Bacon, 'Letters', letter to Bacon from his nephew Ralph Creyke, IWM. Also see Rear-Admiral Sir Douglas Brownrigg, *Indiscretions of the Naval Censor* (London, 1920), p. 188. For examples from the Second World War, see Revd Thomas Cameron, *The Happy Warrior* (London, 1939), p. 23 and Wallace Reyburn, *Glorious Chapter. The Canadians at Dieppe* (London, 1943), p. 37.

30 For one example, see Michael MacDonagh, *The Irish at the Front* (London, 1916), pp. 17–18.

31 Robert William MacKenna, *Through a Tent Door*, first published 1919

(London, 1930), pp. 87 and 90–1. For an excruciating poem about a bay-onet fight, see his 'For Valour', in MacKenna, *Verses* (Edinburgh, 1897), p. 47.

32 See the correspondence and minutes from January and November 1936, in PRO WO32/4453.

33 *A Book of British Heroes* (London, 1914), p, 90, quoting from 'Heroes All', *Daily Express*, 27 August 1914. Also see Captain Leopold McLaglen, *Bayonet Fighting for War* (London, 1916), p. 4; *Pen Pictures of the War By Men at the Front. Vol. I. The Campaign in Natal to the Battle of Colenso* (London, 1900), p. 178 and throughout; Lieutenant-Colonel G. E. Thornton and Major H. de L. Walters, *Aids to Weapon Training. Some Ideas on Improvisation* (Aldershot, 1941), p. 17.

34 Sydney Duffield and Andrew G. Elliott, *Rough Stuff. For Home Guards and Members of H.M. Forces* (London, 1942), p. 10; Colonel G. W. Geddes, 'Papers', letter written in the Dardanelles, 30 April 1915, IWM; Ion Llewellyn Idriess, *The Desert Column. Leaves from the Diary of an Australian Trooper in Gallipoli, Sinai, and Palestine* (Sydney, 1932), p. 260; Private Frank E. Morley, 'War Diary', entry for 15 August 1915, CMAC.

35 'A Private', *The Private Life of a Private – Being Extracts from the Diary of a Soldier of Britain's New Army* (London, 1941), p. 14; Captain E. S. Archer, *The Warrior. A Military Drama Illustrative of the Great War* (Durban, 1917), p. 37; Major M. D. S. Armour, *Total War Training for Home Guard Officers and N.C.O.s* (London, 1942), p. 37; F. C. Bartlett, *Psychology and the Soldier* (Cambridge, 1927), p. 175; Alan Deane, 'The Battle of Mons', *The Boys' Journal*, iii.58 (24 October 1914), p. 148; Gerald V. Dennis, 'A Kitchener Man's Bit (1916–1918)', 1928, p. 67, IWM; Albert N. Depew, *Gunner Depew* (London, 1918), p. 63; Arthur Guy Empey, *First Call. Guide Posts to Berlin* (New York, 1918), p. 91; Collie Knox, *Heroes All* (London, 1941), p. 213; *Letters from the Front. Being a Record of the Part Played by Officers of the [Canadian] Bank [of Commerce] in the Great War 1914–1919*, vol. 1 (Toronto, 1920), pp. 17–18, letter from Private A. P. Glasgow; 'Secret Memo', 21 July 1941, PRO WO199/3249; Norman Shaw, 'Papers', letter dated 14 July 1916, IWM.

36 Sydney Duffield and Andrew G. Elliott, *Rough Stuff. For Home Guards and Members of H.M. Forces* (London, 1942), p. 10; 'Jungle Warfare Extracts', no date but Second World War, p. 2, AWM 54, 923/1/5; W. D. Jones, 'Precis on Jungle Fighting', 1942, p. 5, AWM.

37 Private First Class Rod Consalvo, in Otto J. Lehrack, *No Shining Armour. The Marines at War in Vietnam. An Oral History* (Lawrence, Kansas,

1992), p. 102.

38 David Lloyd George, 'A Nation's Thanks', 29 October 1917, in George, *The Great Crusade: Extracts from Speeches Delivered During the War* (London, 1918), p. 148. Also see C. G. Grey, *British Fighter Planes* (London, 1941), p. 3.

39 Kenneth Hemingway, *Wings over Burma* (London, 1944), p. 67.

40 Colin P. Sisson, *Wounded Warriors. The True Story of a Soldier in the Vietnam War and of the Emotional Wounds Inflicted* (Auckland, 1993), p. 32.

41 For a brilliant study, see George Mosse, 'The Knights of the Sky and the Myth of the War Experience', in Robert A. Hinde and Helen E. Watson (eds), *War: A Cruel Necessity? The Bases of Institutional Violence* (London, 1995), pp. 132–42.

42 Wing Commander Athol Forbes and Squadron Leader Hubert Allen, *The Fighter Boys* (London, 1942), pp. 110–14.

43 Samuel A. Stouffer et al., *The American Soldier: Combat and Its Aftermath. Volume II* (Princeton, 1949), pp. 333 and 335. The numbers involved in the questions to pilots were 351, 242, 200 and 654 for pilots of heavy bombers, medium bombers, light bombers and fighter planes, respectively.

44 Second Lieutenant Harold Warnica Price, 'Diary', in Brereton Greenhous (ed.), *A Rattle of Pebbles: The First World War Diaries of Two Canadian Airmen* (Ottawa, 1987), p. 250, diary for 17 December 1917.

45 James Byford McCudden, *Flying Fury*, first published in 1918 (London, 1930), p. 203. He also admitted, however, to feeling 'very sorry indeed' when seeing the dead pilot. He preferred shooting down planes in 'Hunland' where he could not see 'the results'.

46 Roderick Chrisholm, *Cover of Darkness* (London, 1953), p. 71.

47 Flight-Lieutenant D. M. Crook, *Spitfire Pilot* (London, 1942), pp. 28, 30–1 and 75.

48 Kenneth Hemingway, *Wings over Burma* (London, 1944), pp. 41–2 and 68–9.

49 William J. Simon, 'My Country', in Larry Rottmann, Jan Barry, and Basil T. Paquet (eds), *Winning Hearts and Minds. War Poems by Vietnam Veterans* (New York, 1972), p. 42.

50 Stephen Graham, *A Private in the Guards* (London, 1919), p. 78.

51 Harold Stainton, 'A Personal Narrative of the War', p. 22, IWM.

52 Right Revd John S. Vaughan, 'Thoughts on the Present War', *Irish Ecclesiastical Record*, 5th series, iv (July–December 1914), p. 562.

53 Captain R. J. Linwood, 'The Sniper – Part I', xx.2, *Australian Infantry*,

(December 1980), p. 9.

54 David Lloyd George, 'A Nation's Thanks', 29 October 1917, in George, *The Great Crusade. Extracts from Speeches Delivered During the War* (London, 1918), pp. 148–9. Also see Walter A. Briscoe, *The Boy Hero of the Air. From Schoolboy to V.C.* (London, 1921), p. 59; Walter A. Briscoe and H. Russell Stannard, *Captain Ball, V.C. The Career of Flight Commander Ball, V.C., D.C.O.* (London, 1918), p. 23; Major-General H. H. Eaker and Colonel Ira C. Eaker, *Winged Warfare* (New York, 1941), p. 36.

55 'Captain Albert Ball, V. C., D. S. O.', by W. N. Cobbold, *Poems on the War. March 21st to November 11th, 1918 and After the Armistice* (Cambridge, 1919), pp. 2–5.

56 'The Revival of Knighthood' by W. N. Cobbold, *Poems on the War. March 21st to November 11th, 1918 and After the Armistice* (Cambridge, 1919), pp. 10–11.

57 Cecil Lewis, *Sagittarius Rising* (London, 1936), p. 45. Also see Revd William Duncan Geare, *Letters of an Army Chaplain* (London, 1918), p. 77; Kenneth Hemingway, *Wings Over Burma* (London, 1944), p. 76; James Molony Spaight, *Air Power and War Rights* (London, 1924), p. 103.

58 Boyd Cable, *Air Men O' War* (London, 1918), pp. 139–40.

59 Major William Avery Bishop, *Winged Warfare. Hunting Huns in the Air* (London, 1918), pp. 1–5; Boyd Cable, *Air Men O' War* (London, 1918), pp. 139–40; Percy Raymond Hall, 'Recollections', np, IWM. In the context of the Second World War, identical sentiments were expressed by Roderick Chrisholm, *Cover of Darkness* (London, 1953), pp. 18–19. For other comments during the Second World War and the Vietnam War, see Jeanne Christie interviewed in Kathryn Marshall, *In the Combat Zone. An Oral History of American Women in Vietnam 1966–1975* (Boston, 1987), p. 181 and Samuel A. Stouffer et al., *The American Soldier: Combat and Its Aftermath. Volume II* (Princeton, 1949), pp. 340–58.

60 Captain E. S. Archer, *The Warrior. A Military Drama Illustrative of the Great War* (Durban, 1917), p. 37.

61 Stephen Graham, *A Private in the Guards* (London, 1919), p. 79.

62 Cardinal Hinsley, *The Bond of Peace and Other War-Time Addresses* (London, 1941), p. 7, broadcast on 10 December 1939.

63 George Coppard, *With a Machine Gun to Cambrai* (London, 1980), p. 42; Sir Henry C. Darlington, 'Letters from Helles', 5 June 1915, p. 41, Liddell Hart Centre; Coningsby Dawson, *The Glory of the Trenches* (London, 1918), p. 74; Stephen Graham, *A Private in the Guards*

(London, 1919), p. 78; Sid T. Kemp, 'Remembrance. The 6th Royal West Kent Regiment 1914–1916', p. 41, IWM; W. R. Kirby, 'The Battle of Cambrai 1917', p. 98, IWM; W. A. Quinton, 'Memoirs', p. 53, IWM; Dorothy Scholes, 'Papers', diary of E. Williams, 22 September 1914, Wigan Archives Service.

64 Captain M. D. Kennedy, 'Their Mercenary Calling', 1931, p. 96, IWM.

65 Charles Cecil Miller, 'A Letter from India to My Daughter in England', p. 20, IWM.

66 Captain Guy Warneford Nightingale, 'The 1915 Letters', letter to his sister, 4 May 1915 from Gallipoli, IWM.

67 For a description of looking a long-distance target 'in the eyes' before firing, see Victor G. Ricketts, 'Account of His Service', p. 34, IWM.

68 H. Hesketh-Prichard, *Sniping in France* (London, 1920), p. 63. Also see Captain E. A. Godson, 'Diaries', no. 5, p. 1, entry for 16 November 1916, IWM; Ion L. Idriess, *Must Australia Fight?* (Sydney, 1939), p. 147; Murdoch C. McDougall, *Swiftly They Struck. The Story of No. 4 Commando* (London, 1954), p. 124.

69 F. de Margry, 'In Memoriam', pp. 2–3, IWM.

70 Henry Seton-Karr, *The Call to Arms 1900–1901* (London, 1902), p. 31.

71 War Office, *Notes for Infantry Officers on Trench Warfare*, (London, 1917), p. 41 and War Office, *Sniping* (London, 1951), p. 1.

72 Lieutenant-Colonel N. A. D. Armstrong, *Fieldcraft, Sniping and Intelligence*, 5th edition (Aldershot, 1942), pp. 4 and 15; Major H. Hesketh-Prichard, *Sniping in France With Notes on the Scientific Training of Scouts, Observers, and Snipers* (London, 1920), p. 37; *Sniping, Scouting and Patrolling* (Aldershot, nd), p. 1.

73 Foreword by Colonel Lord Cottesloe, in Lieutenant-Colonel N. A. D. Armstrong, *Fieldcraft, Sniping, and Intelligence*, 5th edition (Aldershot, 1942), p. vii.

74 Max Plowman, *War and the Creative Impulse* (London, 1919), pp. 8–9.

75 Herbert W. McBride, *A Rifleman Went to War* (Marines, North Carolina, 1935), p. 30. Also see War Office, *Infantry Training. Volume I. Infantry Platoon Weapons Pamphlet No. 10. Sniping 1951* (London, 1951), p. 3.

76 See Walter A. Briscoe, *The Boy Hero of the Air. From Schoolboy to V.C.* (London, 1921), throughout; Flight-Lieutenant D. M. Crook, *Spitfire Pilot* (London, 1942), pp. 41–2; Wing Commander Athol Forbes and Squadron Leader Hubert Allen, *The Fighter Boys* (London, 1942), p. 97; The Revd E. J. Kennedy, *With the Immortal Seventh Division*, 2nd edition (London, 1916), p. 114; T. T. Paterson, *Morale in War and Work. An*

Experiment in the Management of Men (London, 1955), p. 88.

77 Quoting John Beede, in Gavin Lyall (ed.), *The War in the Air, 1939–1945. An Anthology of Personal Experience* (London, 1968), p. 278. Also see C. G. Grey, *British Fighter Planes* (London, 1941), pp. 3–4.

78 *With the First Canadian Contingent* (Toronto, 1915), pp. 76–7.

79 Letter by Lieutenant Leslie Yorath Sanders, 4 March 1917, in Laurence Housman (ed.), *War Letters of Fallen Englishmen* (London, 1930), p. 231. He was killed in action six days later.

80 Roderick Chrisholm, *Cover of Darkness* (London, 1953), p. 73. Also see p. 71.

81 Hector Bolitho, *Combat Report. The Story of a Fighter Pilot* (London, 1943), pp. 45 and 58 and Lou Drendel, . . . *And Kill Migs. Air to Air Combat in the Vietnam War* (Carrollton, 1974), p. 17.

82 Hector Bolitho, *Combat Report. The Story of a Fighter Pilot* (London, 1943), p. 133; Walter A. Briscoe, *The Boy Hero of the Air. From Schoolboy to V.C.* (London, 1921), p. 50; *Universal Military Training – No!* (np, 1919), p. 8.

83 For one version of this story, see 'F.B.M.' [Frank Bertrand Merryweather], *The Defiance of Death. Being Some Thoughts on the Death of a Brave Soldier* (London, 1918), pp. 14–16. This story – although recounted numerous times – was a fabrication. Walter A. Briscoe's biography of Ball confirmed that Ball was not the pilot who shot down Immelmann: *The Boy Hero of the Air. From Schoolboy to V.C.* (London, 1921), pp. 62–3.

84 Paul Richey, *Fighter Pilot. A Personal Record of the Campaign in France* (London, 1944), pp. 11 and 42.

85 The diary of Wing Commander Roland Beaumont, in Edward Lanchbery, *Against the Sun. The Story of Wing Commander Roland Beaumont, D.S.O., O.B.E., D.F.C. Pilot of the Canberra and the P.1* (London, 1955), pp. 56–7.

86 See throughout Lou Drendel, . . . *And Kill Migs. Air to Air Combat in the Vietnam War* (Carrollton, 1974). This process of personification also occurred in tank warfare. 'Taking out' a tank was usually described as though the tank was the human: see Anthony S. Irwin, *Infantry Officer. A Personal Record* (London, 1943), pp. 62–3 and the interview with unnamed soldier in Barry Broadfoot, *Six War Years 1939–1945. Memories of Canadians at Home and Abroad* (Don Mills, Ontario, 1974), p. 3.

87 Boyd Cable, *Air Men O' War* (London, 1918), p. 24.

88 Alan W. Mitchell, 'Sergeant-Pilot James Allen Ward, V.C.', in Derek Tangye (ed.), *Went the Day Well* (London, 1942), p. 11.

89 Wing Commander Athol Forbes and Squadron Leader Hubert Allen, *The Fighter Boys* (London, 1942), p. 105.

90 See the interview of K. O. Moore and Alec Gibb in Squadron Leader Hector Bolitho, 'Two in Twenty-Two Minutes', *Slipstream. A Royal Air Force Anthology* (London, 1946), pp. 10–11.

91 Stephen A. Garrett, *Ethics and Airpower in World War II. The British Bombing of German Cities* (New York, 1993), p. 84, quoting H. R. Allen, *The Legacy of Lord Trenchard* (London, 1972), p. ix.

92 E. W. Colebrook, 'Letters Home', letter to his father, 28 May 1915, IWM.

93 For example, see 'A Flying Corps Pilot', *Death in the Air. The War Diary and Photographs of a Flying Corps Pilot* (London, 1933), p. 75 and James Byford McCudden, *Flying Fury* (London, 1930), p. 170.

94 'A Flying Corps Pilot', *Death in the Air. The War Diary and Photographs of a Flying Corps Pilot* (London, 1933), pp. 97 and 99. Also see p. 57.

95 Commander B. W. Hogan, 'Psychiatric Observations of Senior Medical Officer on Board Aircraft Carrier U.S.S. *Wasp* During Action in Combat Areas, at Time of Torpedoing, and Survivors' Reaction', *The American Journal of Psychiatry*, 100 (1943–4), 91. He also noted, however, that it was not long before they took this form of killing in their stride.

96 Stanley Johnston, *The Grim Reapers* (London, 1945), p. 77.

97 John H. Morrow, 'Knights of the Sky. The Rise of Military Aviation', in Frans Coetzee and Marilyn Shevin-Coetzee (eds), *Authority, Identity and the Social History of the Great War* (Providence, 1995), p. 321.

98 Robert L. O'Connell, *Of Arms and Men. A History of War, Weapons, and Aggression* (New York, 1989), p. 263.

99 Stanley Johnston, *The Grim Reapers* (London, 1945), p. 39.

100 See the verses in Lieutenant-Colonel H. F. N. Jourdain, *Ranging Memories* (Oxford, 1934), p. 216.

101 Captain F. C. Hitchcock, *'Stand To'. A Diary of the Trenches 1915–1918* (London, 1937), p. 47.

102 Arthur Guy Emprey, *First Call. Guide Posts to Berlin* (New York, 1918), p. 222.

103 Letter from an unnamed British officer, dated 19 November 1914 after sniping at four men (and hitting two), in Amy Gordon Grant, *Letters from Armageddon. A Collection Made During the World War* (Boston, 1930), p. 37 and Victor G. Ricketts, 'Account of his Service', p. 34, IWM. Also see A. J. Turner, 'Zero Hour', p. 51, IWM.

104 Brigadier-General F. P. Crozier, *The Men I Killed* (London, 1937), pp. 101–4.

105 *Sniping, Scouting and Patrolling* (Aldershot, nd), pp. 10–11 and Lieutenant-Colonel N. A. D. Armstrong, *Fieldcraft, Sniping and Intelligence*, 5th ed. (Aldershot, 1942), p. xiii.

106 The 'dreadful romances' is from Lieutenant Rowland H. Owen, 'Letters Home', letter to his parents on 30 September 1914, IWM. Also see Henry Gother Courtney, 'Letters', letter to his sister from Salonica on 26 November 1916, letter no. 8/147, in Birmingham University Library Collection and the letter by Captain Geoffrey Blemell Pollard to Mrs Hughes on 19 October 1914, in Laurence Housman (ed.), *War Letters of Fallen Englishmen* (London, 1930), pp. 219–20.

3 Training Men to Kill

1 Shawn O'Leary, 'The Bayonet', in O'Leary, *Spikenard and Bayonet. Verse from the Front Line* (Melbourne, 1941), p. 21.

2 Private Richard E. Marks, *The Letters of Pfc. Richard E. Marks, USMC* (Philadelphia, 1967), pp. 7–13, 17, 22, 26, 29–30, 77, 85, 136, 183–4.

3 For discussions of the relationship of these sciences to war, see Everett Mendelsohn, Merritt Roe Smith, and Peter Weingart (eds), *Science, Technology, and the Military*, 2 vols. (Dordrecht, 1988) and Lawrence B. Radine, *The Taming of the Troops. Social Control in the United States Army* (Westport, Connecticut, 1977), pp. 89–90.

4 Lieutenant-Colonel W. Shirley, *Morale. The Most Important Factor in War* (London, 1916), p. 23.

5 'The Principles of Basic Training. A Training Directive Given by the Commander-in-Chief, Home Forces, to Commanders of the Reserve Divisions and Commanding Officers of Training Units', 25 May 1944, in PRO WO199/839.

6 'The Jungle Belongs to Us', *Australian Army Journal*, 85 (May 1956), p. 5, and Lieutenant-Colonel J. O. Langtry, 'Man-the-Weapon: Neglected Aspects of Leader Training', *Australian Army Journal*, 202 (March 1966), pp. 3–13. Also see Stephen Graham, *A Private in the Guards* (London, 1919), p. 6 and Lieutenant-Colonel W. Shirley, *Morale. The Most Important Factor in War* (London, 1916), p. 23.

7 G. Stanley Hall, 'Morale in War and After', *The Psychological Bulletin*, 15 (1918), p. 369; Arthur Guy Empey, *First Call. Guide Posts to Berlin* (New York, 1918), p. 91; John T. MacCurdy, *The Structure of Morale* (Cambridge, 1943), p. 39.

8 Robert Graves, *Goodbye to All That*, first published in 1929 (London, 1977), p. 187.

9 *Psychology for the Fighting Man*, 2nd edition (Washington, DC, 1944), p. 267.

10 Ibid., p. 267.

11 Henry W. Brosin, 'Panic States and Their Treatment', *American Journal of Psychiatry*, 100 (1943–4), p. 58. Also see R. H. Ahrenfeldt, 'Military Psychiatry', in Sir Arthur Salisbury MacNalty and W. Franklin Mellor (eds), *Medical Services in the War. The Principal Medical Lessons of the Second World War Based on the Official Medical Histories of the United Kingdom, Canada, Australia, New Zealand and India* (London, 1968), p. 190 and G. Stanley Hall, 'Morale in War and After', *The Psychological Bulletin*, 15 (1918), p. 369.

12 Major-General J. S. Nicols, 'State of Training of Reinforcements', 21 September 1942, in PRO WO201/2590. Similar observations were made during the Boer War when it was argued that fourteen out of every fifteen British soldiers were incapable of accurate fire: see Henry Seton-Karr, *The Call to Arms 1900–1901* (London, 1902), pp. 200–1.

13 A First Battalion Officer, 'Some Glimpses of the Battle of Givenchy', *The Braganza*, 1.1 (May 1916), p. 3.

14 Tony Ashworth, *Trench Warfare 1914–1918. The Live and Let Live System* (London, 1980). For some descriptions, see Alfred E. Bundy, 'War Diary', pp. 29–31 March 1917, IWM; Nobby Clarke, 'Sniper on the Western Front', p. 10, IWM; Charles K. McKerrow, 'Diaries and Letters', 25 May 1916, IWM.

15 For discussions of refusing to go over the top except at gunpoint, see Edwyn E. H. Bate, 'Memoirs – Vol. III. 1914–1918', p. 55, IWM; Brigadier-General F. P. Crozier, *The Men I Killed* (London, 1937), pp. 62–8; Stephen Graham, *A Private in the Guards* (London, 1919), p. 2; R. L. MacKay, 'Diary', 31 July and 1 August 1917, pp. 31 and 34, IWM; Revd J. E. Roscoe, *The Ethics of War, Spying and Compulsory Training* (London, nd, 1914–18), p. 31. See Joanna Bourke *Dismembering the Male: Men's Bodies, Britain, and the Great War* (London and Chicago, 1996), chapter 2.

16 The term is from Albert J. Glass, 'Preventative Psychiatry in the Combat Zone', *U. S. Armed Forces Medical Journal*, iv.1 (1953), p. 684.

17 Martin Bingham, 'Adjustment Problems of American Youth in Military Service', *Sociology and Social Research*, 31.1 (September–October 1946), p. 35.

18 '152nd Infantry Brigade. Discussion on Lessons Learned During the Year of Fighting from El Alamein to Messina', 1943, in PRO WO231/16.

19 Colonel S. L. A. Marshall, *Men Against Fire. The Problem of Battle Command in Future War* (New York, 1947), p. 72. Also see Lieutenant-

Colonel George Juskalian, 'Why Didn't They Shoot More?', *Army Combat Forces Journal*, 5.2 (September 1954), p. 35.

20 Captain Blair W. Sparks and Brigadier-General Oliver K. Neiss, 'Psychiatric Screening of Combat Pilots' *U.S. Armed Forces Medical Journal*, vii.6 (June 1956), p. 811.

21 Hugh Dundas, *Flying Start. A Fighter Pilot's War Years* (London, 1988), p. 2.

22 Captain W. R. Chamberlain, 'Training the Functional Rifleman', *Canadian Army Journal*, 4.9 (February 1951), p. 29.

23 Irvin L. Child, 'Morale: A Bibliographical Review', *Psychological Bulletin*, 38 (1941), p. 413.

24 Colonel S. L. A. Marshall, *Men Against Fire. The Problem of Battle Command in Future War* (New York, 1947), pp. 50, 54–9, and 65. For critiques of Marshall's statistics, see Donald E. Graves, '"Naked Truths for the Asking". Twentieth-Century Military Historians and the Battlefield Narrative', in David A. Chartes, Marc Milner, and J. Brent Wilson (eds), *Military History and the Military Profession* (Westport, Connecticut, 1992), p. 49 and Roger J. Spiller, 'S. L. A. Marshall and the Ratio of Fire', *R.U.S.I. Journal*, 133 (winter 1988). Marshall's exaggerated claims were not exposed until Spiller's article was published in 1988, and the famous statistics continue to be cited. Obviously the authoritative voice of statistics and social science is the central reason for this, coupled with the audacity of Marshall's claim, but another part of the answer lies in the fact that we *want* to believe that it is difficult to kill other men.

25 For instance, see the statement by Lieutenant-Colonel W. L. Raws, 'Discipline and *Moral*', *The Australian Military Journal*, v (April 1914), p. 274.

26 The problems this caused was discussed by Jack W. Dunlap, 'Psychologists and the Cold War', *The American Psychologist*, 10 (1955), pp. 108–9 and Captain J. K. Leggett, 'The Human Factor in Warfare', *Australian Army Journal*, 183 (1964), p. 24.

27 For commentary, see Commander Jack W. Dunlap, 'The Sensitive Adjustment of Men to Machines', *The Army Combat Forces Journal*, 5.3 (1954), p. 2; Lieutenant-Colonel A. Green, 'Revolution in the Military Profession', *Australian Army Journal*, 61 (June 1954), pp. 6–7; Werner S. Landecker, 'Sociological Research and the Defense Program', *Sociology and Social Research*, xxvi.2 (November–December 1941), pp. 103-4; Major J. O. Langtry, 'Tactical Implications of the Human Factors in Warfare', *Australian Army Journal*, 107 (April 1958), p. 6; Major-General S. F. Legge, 'Soldier, Scientist or Socialite', *Australian Army Journal*, 65

(October 1954), p. 6; Captain J. K. Leggett, 'The Human Factor in Warfare', *Australian Army Journal*, 183 (August 1964), p. 24; General Matthew B. Ridgway, 'Man – The Vital Weapon', *Australian Army Journal*, 79 (December 1955), p. 17; Erwin O. Smigel, 'The Place of Sociology in the Army', *Sociology and Social Research*, xxvi.6 (July–August 1942), p. 503.

28 Colonel S. L. A. Marshall, *Men Against Fire. The Problem of Battle Command in Future War* (New York, 1947), pp. 44–8.

29 Lieutenant-General L. E. Kiggell, *Training of Divisions for Offensive Action*, (London, 1916), p. 2.

30 The estimate was based on the analysis of 1,218 candidates accepted for training for commissions in July and August 1945: Lieutenant-Colonel J. H. A. Sparrow (compiler), *Morale* (London, 1949), p. 22.

31 Furthermore, this was *increasingly* the case as the educational qualifications of *all* people improved. For instance, during the First World War, 9 per cent of white American soldiers were high school graduates or college men, compared with 41 per cent during the Second World War: Samuel A. Stouffer et al., *The American Soldier: Adjustment During the Army Life. Volume I* (Princeton, 1949), p. 58.

32 G. A. Brett, 'Recollections', p. 46, IWM.

33 N. L. Macky, 'Weapon and Target', *The Army Quarterly*, xxxiii (January 1937), p. 313.

34 K. Garry, 'Letter to Mother from France', January 1916, p. 25, IWM.

35 See Chapter 5.

36 Field Marshal Earl Wavell, 'The Good Soldier', in his *The Good Soldier* (London, 1948), p. 43, originally published in *Sunday Times*, 19 August 1945.

37 For a few examples, see Christian G. Appy, *Working-Class War. American Combat Soldiers and Vietnam* (Chapel Hill, 1993), p. 98; George Barker, interviewed in Glen D. Edwards, *Vietnam. The War Within* (Salisbury, South Australia, 1992), p. 45; Philip Caputo, *A Rumor of War* (London, 1977), pp. 8–10; R. Wayne Eisenhart, 'You Can't Hack It Little Girl: A Discussion of the Covert Psychological Agenda of Modern Combat Training', *Journal of Social Issues*, 31.4 (1975), pp. 13–23; J. T. Hansen, A. Susan Owen, and Michael Patrick Madden, *Parallels. The Soldiers' Knowledge and the Oral History of Contemporary Warfare* (New York, 1992), pp. 44–5.

38 For the best descriptions, see Peter G. Bourne, 'Some Observations on the Psychosocial Phenomena Seen in Basic Training', *Psychiatry*, 30.2 (May 1967), pp. 187–96; Donald Duncan, *The New Legions* (London,

1967), pp. 97–8; Arthur J. Vidich and Maurice R. Stein, 'The Dissolved Identity in Military Life', in Stein, Vidich, and David Manning White (eds), *Identity and Anxiety. Survival of the Person in Mass Society* (Glencoe, Illinois, 1960), pp. 493–506.

39 For descriptions remarkably similar in nature (although by no means degree), see Mika Haritos-Fatouros, 'The Official Torturer: A Learning Model for Obedience to the Authority of Violence', *Journal of Applied Social Psychology*, 18.13 (1988), pp. 1107–20 and Ervin Staub, 'The Psychology and Culture of Torture and Torturers', in Peter Suedfeld (ed.), *Psychology and Torture* (New York, 1990), pp. 49–76.

40 Lieutenant William L. Calley, *Body Count* (London, 1971), p. 27.

41 Stephen Graham, *A Private in the Guards* (London, 1919), pp. 25 and 58.

42 'Allied Land Forces in S. W. Pacific Area. Training Ideas – No. 5', 1943, p. 2, AWM.

43 Lieutenant-Colonel H. F. Wood, 'The Case of the Fainting Soldier', *Canadian Army Journal*, xii.4 (October 1958), p. 31. Wood worked for the Army Council Secretariat, Army Headquarters, Ottawa. For a description of this happening during 'toughness training' in the First World War, see A. J. Turner, 'Zero Hour', pp. 5–6, IWM.

44 Michael Rosenfield, interviewed in Gerald R. Gioglio, *Days of Decision. An Oral History of Conscientious Objectors in the Military During the Vietnam War* (Trenton, New Jersey, 1989), p. 82. Also see the description by Dennis Kitchin, *War in Aquarius. Memoir of an American Infantryman in Action Along the Cambodian Border During the Vietnam War* (Jefferson, North Carolina, 1994), pp. 15–16.

45 Mark Gerzon, 'The Soldier', in Reese Williams (ed.), *Unwinding the Vietnam War. From War into Peace* (Seattle, 1987), p. 145 and Montgomery of Alamein, *Morale in Battle: Analysis* (np, 30 April 1946), 15.

46 Philip Caputo, *A Rumor of War* (London, 1977), p. 10.

47 For their theories, see John Dollard, Leonard W. Doob, Neal E. Miller, O. H. Mowrer and Robert R. Sears, *Frustration and Aggression* (New Haven, Connecticut, 1939) and Neal E. Miller, 'The Frustration–Aggression Hypothesis', *Psychological Review*, 48 (1941), pp. 337–42. Also see Leonard Berkowitz, 'The Frustration–Aggression Hypothesis Revisited', in Berkowitz (ed.), *Roots of Aggression* (New York, 1969); Leonard Berkowitz, 'Situational Influences on Aggression', in Jo Groebel and Robert A. Hinde (eds), *Aggression and War. Their Biological and Social Bases* (Cambridge, 1989), pp. 91–100; Mark A. May, *A Social Psychology of War and Peace* (New Haven, 1943); Neal E. Miller and Richard Bugelski, 'Minor Studies in Aggression: II. The Influence of Frustrations

Imposed by the In-Group on Attitudes Expressed Toward Out-Groups', *The Journal of Psychology*, 25 (1948); P. M. Turquet, 'Aggression in Nature and Society (II)', *British Journal of Medical Psychology*, xxii.3 (1949), pp. 157 and 160.

48 For an example of the latter, see the psychoanalytical ideas expounded by the psychiatrist, Dr Alvin F. Poussaint, quoted in David Llorens, 'Why Negroes Re-enlist', *Ebony*, 23 August 1968, p. 88.

49 For instance, see Lieutenant-Colonel C. W. T. Kyngdon, 'The AMF Gold Medal Prize Essay, 1948–49', *Australian Army Journal*, 10 (December–January 1949–50), pp. 6–22.

50 A. Kardiner, 'Forensic Issues in the Neuroses of War', *The American Journal of Psychiatry*, 99 (1942–43), p. 659. Also see Lieutenant-Colonel C. D. Daly, 'A Psychological Analysis of Military Morale', *The Army Quarterly*, xxxii (April 1936), p. 71 and Professor F. G. Bartlett, 'Psychological Questions Relating to AA Personnel', 1942, PRO WO222/66.

51 For descriptions, see *Time*, 14 July 1967, p. 16; Paul R. Bleda and Robert H. Sulzen, 'The Effects of Simulated Infantry Combat Training on Motivation and Satisfaction', *Armed Forces and Society*, 6.2 (winter 1980), pp. 202–14; Captain D. Collins, 'Quick Kill', *Infantry. The Magazine of the Royal Australian Infantry Corps* (January 1969), np; Thomas D. Scott, 'Tactical Training for Ground Combat Forces', *Armed Forces and Society*, 6.2 (winter 1980), pp. 215–31; U.S. Army, *Principles of Quick Kill. Training Text 23-71-1* (Fort Benning, Georgia, May 1967). The techniques employed in 'Quick Kill' were not unknown during the First World War: see Sydney W. D. Lockwood, 'First World War Memories 1914–1918', p. 100, IWM.

52 'Home Forces. Training. Address by the Commander-in-Chief, Home Forces (General Franklyn) to Commanders of Training Units', 7 December 1944, p. 3, in PRO WO199/840; Lieutenant-Colonel A. A. James, 'Emotional Adjustment and Morale in War', *Canadian Army Journal*, 2.5-8 (August–September 1948 and October–November 1948), pp. 28–31 and 24–5; Colonel William Line, 'Morale and Leadership', *Canadian Army Journal*, 6.1 (April 1952), pp. 46–7; Major-General Milija Stanishich, 'Command Initiative', *Canadian Army Journal*, xiv.2 (spring 1960), p. 81.

53 Dixon Wecter, *When Johnny Comes Marching Home* (Cambridge, Massachusetts, 1944), p. 482. For an insightful history, see S. P. Mackenzie, 'Morale and the Cause: The Campaign to Shape the Outlook of Soldiers in the British Expeditionary Force, 1914–1918', *Canadian*

Journal of History, 25 (1990), pp. 215–32.

54 Dixon Wecter, *When Johnny Comes Marching Home* (Cambridge, Massachusetts, 1944), p. 482.

55 Leonard R. Sillman, 'Morale', *War Medicine*, 3.5 (May 1943), pp. 498–502. See the similar statement by Captain Martin Stein, 'Neurosis and Group Motivation', *The Bulletin of the U.S. Army Medical Department*, vii.3 (March 1947), pp. 320–1.

56 Charles Moskos, *The American Enlisted Man* (New York, 1970), p. 147. Also see Roger W. Little, 'Buddy Relations and Combat Performance', in Morris Janowitz (ed.), *The New Military. Changing Patterns of Organization* (New York, 1964), pp. 204–5.

57 Colonel S. L. A. Marshall, *Men Against Fire. The Problem of Battle Command in Future War* (New York, 1947), p. 78. Also see Major H. H. Garner, 'Psychiatric Casualties in Combat', *War Medicine*, 8.5 (1945), p. 345; Samuel P. Huntington, *The Soldier and the State* (New York, 1967), p. 389; Alfred O. Ludwig, 'Neuroses Occurring in Soldiers After Prolonged Combat Exposure', *Bulletin of the Menninger Clinic*, 11.1 (January 1947), pp. 18–21; Samuel A. Stouffer et al., *The American Soldier: Adjustment During Army Life. Volume I* (Princeton, 1949), p. 437; Samuel A. Stouffer et al., *The American Soldier: Combat and Its Aftermath. Volume II* (Princeton, 1949), p. 150.

58 Fear was often described as a virus: for example, see Major W. E. Garber, 'Every Rifleman Must be an Aggressive Fighter', *Canadian Army Journal*, 6.6 (January 1953), pp. 22–3 and Albert J. Glass, 'Preventative Psychiatry in the Combat Zone', *United States Forces Medical Journal*, iv.1 (1953), pp. 684–5.

59 George Barker, interviewed in Glen D. Edwards, *Vietnam. The War Within* (Salisbury, South Australia, 1992), p. 45.

60 Colonel S. L. A. Marshall, *Men Against Fire. The Problem of Battle Command in Future War* (New York, 1947), pp. 40 and 60.

61 Brian Holden Reid, *J. F. C. Fuller: Military Thinker* (London, 1987), p. 1.

62 Captain J. F. C. Fuller, *Training Soldiers for War* (London, 1914), pp. 7–9, 11, 16, 29, 103–6 and 111–17.

63 Colonel S. L. A. Marshall, *Men Against Fire. The Problem of Battle Command in Future War* (New York, 1947), pp. 36, 40–2, 64–6, 71–2 and 75–6.

64 For descriptions, see Dennis Kitchin, *War in Aquarius. Memoir of an American Infantryman in Action Along the Cambodian Border During the Vietnam War* (Jefferson, North Carolina, 1994), p. 20; 'Realism in Training', *The Times*, 27 April 1942, p. 2; C. J. Twine, *Bayonet Battle*

Training. A Realistic and Practical Series of Exercises on the Use of the Training Stick and Dummy (Aldershot, 1942), pp. 38–44; Bell I. Wiley, 'The Building and Training of Infantry Divisions' and William R. Keasy, 'The Training of Enlisted Replacements', both in Robert R. Palmer, Bell I. Wiley, and William R. Keasy (eds), *United States Army in World War II. The Army Ground Forces. The Procurement and Training of Ground Combat Troops* (Washington, DC, 1948), pp. 387–9 and 448–51; 'War Cabinet Agendum. Report from G.O.C., A.I.F. (Middle East)', 1942, p. 3, AWM. More realistic training was adopted by other services too. For an airforce example, see U.S. Headquarters Army Air Force Training Division, *An Appraisal of Wartime Training of Individual Specialists in Army Air Forces* (Washington, DC, 1946), p. 38.

65 Christian G. Appy, *Working-Class War. American Combat Soldiers and Vietnam* (Chapel Hill, 1993), pp. 114–15 and Bell I. Wiley, 'The Building and Training of Infantry Divisions', in Robert R. Palmer, Wiley, and William R. Keasy (eds), *United States Army in World War II. The Army Ground Forces. The Procurement and Training of Ground Combat Troops* (Washington, DC, 1948), pp. 448–9 and 451.

66 'Training in Battle Conditions', Second World War, p. 1, AWM 54, 937/1/8. Also see Ruth Danenhower Wilson, *Jim Crow Joins Up. A Study of Negroes in the Armed Forces of the United States* (New York, 1945), p. 28.

67 The quotation is from 'Training in Battle Conditions', Second World War, p. 1, AWM 54, 937/1/8. Also see the letter to Lieutenant-Colonel Brittan, 'G' Training, Home Forces, from Major A. I. M. Wilson, War Office, 5 May 1942, and 'Psychological Aspects of Training', 4 June 1942, both in PRO WO199/799.

68 Lance Kent, 'Autobiography', pp. 167–8, AWM.

69 Samuel A. Stouffer et al., *The American Soldier: Combat and Its Aftermath. Volume II* (Princeton, 1949), pp. 228–9 and 231. Also see Lieutenant-Colonel K. E. Lindeman, 'The Psychology of Fear', *Canadian Army Journal*, xiv.5 (January 1960), p. 95.

70 R. G. Allanson-Winn and C. Phillips-Wolley, *Broadsword and Singlestick* (London, 1890), p. 87. Also see W. W. Greener, *Sharpshooting for Sport and War* (London, 1900), p. 19.

71 Fred Cederberg, *The Long Road Home. The Autobiography of a Canadian Soldier in Italy in World War Two* (Toronto, 1985), p. 146.

72 Percy Raymond Hall, 'Recollections', 1985, p. 24, IWM. Also see Major-General J. F. C. Fuller, *Lectures on F. S. R. II* (London, 1931), pp. 14–15; Colonel Sir John MacDonald, 'The Knife in Trench Warfare', *Journal of the Royal United Service Institution*, lxii (1917), pp. 65–6; A. J. Turner,

'Zero Hour', p. 53, IWM; Spenser Wilkinson, *First Lessons in War* (London, 1914), p. 64; R. J. Wilkinson-Lathan, *British Military Bayonets from 1700 to 1945* (London, 1967), pp. 65–6.

73 'Memorandum on Trench to Trench Attacks', 31 October 1916, p. 11, in PRO WO158/344.

74 *The Bayonet (Bayonet Fighting)* (Aldershot, 1940), pp. 1–2; Elbridge Colby, *The Profession of Arms* (New York, 1924), pp. 94–5; 'Lewis Gunner', *Tactical Handling of Lewis Guns with Notes on Instruction, Etc.* (London, 1918), p. 32; 'Memorandum on Trench to Trench Attacks', 31 October 1916, p. 11, in PRO WO158/344; 'XYZ', *A General's Letters to his Son on Minor Tactics* (London, 1918), p. 17. It was often argued that there would be close, bayonet fighting even in atomic war: see Major J. T. Ashenhurst, 'Barbs, Bullets and Bayonets', *Canadian Army Journal*, 4.5 (October 1950), p. 42; P. J. Dawson, 'Some Incidents in the Development of Armaments for World War II', p. 10, IWM; A. N. Hardin, *The American Bayonet 1776–1964* (Philadelphia, 1964), p. vii; Dr (General) Lothar Rendulic, 'The Change in Tactics Through Atomic Weapons', *Canadian Army Journal*, xi.1 (January 1957), p. 87; Major-General E. Wanty, 'The Offensive', *Canadian Army Review*, xii.4 (October 1958), p. 98.

75 War Office General Staff, *Bayonet Training, 1916* (London, 1916), pp. 6–7 and 20–1.

76 Lieutenant-Colonel G. E. Thornton and Major H. de L. Walters, *Aids to Weapon Training. Some Ideas on Improvisation* (Aldershot, 1941), p. 17.

77 Hal Lawrence, *A Bloody War. One Man's Memories of the Canadian Navy, 1939–1945* (Toronto, 1979), p. 17.

78 Eric M. Bergerud, *Red Thunder, Tropic Lightning. The World of a Combat Division in Vietnam* (St Leonards, New South Wales, 1993), p. 95 and William Nagle, 'Do You Remember When?', p. 3, AWM.

79 Squadron Leader R. A. Lidstone, *Bloody Bayonets. A Complete Guide to Bayonet Fighting* (Aldershot, 1942), p. vii.

80 Colonel C. T. Dawkins, *Night Operations for Infantry* (Aldershot, 1916), pp. 14, 25–7; Andrew Elliott, *Shooting to Kill. A Book Which May Save Your Life* (London, 1941), p. 52; 'G.G.A.', 'The Bayonet for Mounted Riflemen', *The Army Review*, v.1 (July 1913), p. 87.

81 Captain O. G. Body, 'Bush and Forest Fighting Against Modern Weapons', *The Army Quarterly*, viii (July 1924), p. 315 and P. J. Dawson, 'Some Incidents in the Development of Armaments for World War II', p. 10, IWM.

82 'G.G.A.', 'The Bayonet for Mounted Riflemen', *The Army Review*, v.1

(July 1913), p. 87. The comparison was with rifle fire.

83 Squadron Leader R. A. Lidstone, *Bloody Bayonets. A Complete Guide to Bayonet Fighting* (Aldershot, 1942), p. vii; 'Progress Report on Infantry Weapon Developments', 19 November 1952, p. 4, PRO WO32/15178; Spenser Wilkinson, *First Lessons in War* (London, 1914), pp. 64–6.

84 War Office, *Platoon Training* (London, 1919), p. 4. For similar statements, see Captain O. G. Body, 'Bush and Forest Fighting Against Modern Weapons', *The Army Quarterly*, viii (July 1924), p. 319 and Captain J. P. Villies-Stuart, 'The Bayonet', *The Army Review*, 1.2 (October 1911), p. 311.

85 War Office, *Notes from the Front. Part III* (London, 1915), p. 10. Also see the lecture on bayonet fighting, in Alfred Downes, 'Notes Taken at Southern Army School of Instruction, Brentwood. Course Commencing Oct. 16, 1916', unpaginated, Birmingham City Archives, punctuation and spelling as in original.

86 Alfred Downes, 'Lectures on Instruction: Bayonet Fighting', 16 October 1916, Birmingham City Archives; 'G.G.A.', 'The Bayonet for Mounted Riflemen', *The Army Review*, v.1 (July 1913), pp. 85–6; War Office, *Instructions for the Training of Platoons for Offensive Action 1917* (London, 1917), pp. 11–12; War Office, *Notes for Young Officers* (London, 1917), p. 22.

87 J. F. C. Fuller, 'The Foundations of the Science of War', *The Army Quarterly*, I (October 1920–January 1921), pp. 94–5.

88 For instance, see Major M. D. S. Armour, *Total War Training for Home Guard Officers and N.C.O.s* (London, 1942), p. 37 and Squadron Leader R. A. Lidstone, *Bloody Bayonets. A Complete Guide to Bayonet Fighting* (Aldershot, 1942), pp. 1–2.

89 'G.G.A.', 'The Bayonet for Mounted Riflemen', *The Army Review*, v.1 (July 1913), pp. 85–6; 'Home Guard. Instruction No. 58–1943. Miscellaneous Notes. Notes from Theatres of War', 1943, for issue to Platoon Commanders, p. 7, PRO WO199/872B; Squadron Leader R. A. Lidstone, *Bloody Bayonets. A Complete Guide to Bayonet Fighting* (Aldershot, 1942), p. 2; War Office, *Home Guard Training 1952* (London, 1952), p. 37.

90 Andrew Elliott, *Shooting to Kill. A Book Which May Save Your Life* (London, 1941), p. 52.

91 'Home Forces. Training. Address by the Commander-in-Chief, Home Forces, (General Franklyn) to Commanders of Training Units', 7 December 1944, p. 20, PRO WO199/840.

92 Squadron Leader R. A. Lidstone, *Bloody Bayonets. A Complete Guide to*

Bayonet Fighting (Aldershot, 1942), p. vii.

93 William E. Merritt, *Where the Rivers Ran Backwards* (Athens, 1989), p. 18.

94 'Courses in Psychology for the Students' Army Training Corps', *Psychological Bulletin*, 15 (1918), p. 130.

95 For instance, the account by James E. Driskell and Beckett Olmstead, 'Psychology and the Military. Recent Applications and Trends', *American Psychologist*, 44.1 (January 1989), pp. 43–54 does not mention it. One important exception is Peter Watson's *War on the Mind. The Military Uses and Abuses of Psychology* (London, 1978).

96 For the transfer of these knowledges from the military to civilian contexts, see Joanna Bourke, *Dismembering the Male: Men's Bodies, Britain, and the Great War* (London and Chicago, 1996) and C. M. Louttit, 'Psychology During the War and Afterwards', *Journal of Consulting Psychology*, viii.1 (January–February 1944), p. 1.

97 For summaries, see James Burk, 'Morris Janowitz and the Origins of Sociological Research on Armed Forces and Society', *Armed Forces and Society*, 19.2 (winter 1993), pp. 167–85; John Carson, 'Army Alpha, Army Brass, and the Search for Army Intelligence', *ISIS*, 84.2 (June 1993), pp. 278–309; Morris Janowitz, 'Consequences of Social Science Research on the U. S. Military', *Armed Forces and Society*, 8.4 (summer 1982), pp. 507–24.

98 E. F. M. Durbin and J. Bowlby, *Personal Aggressiveness and War* (London, 1939), p. v.

99 D. E. Rose, 'Psychology and the Armed Forces', *Australian Journal of Psychology*, 10.1 (June 1958), p. 43.

100 See Lawrence B. Radine, *The Taming of the Troops. Social Control in the United States Army* (Westport, Conn., 1977), p. 142.

101 A. S. Walker, *Australians in the War of 1939–1945. Series 5: Medical. Vol. I: Clinical Problems of the War* (Canberra, 1952), pp. 672 and 705 and Major J. W. Wishart, 'Experiences as a Psychiatrist with BNAF & CMF – Jan 1943 to Jan 1944 with Special Reference to Work in Forward Areas', 23 January 1944, p. 1, in Major-General Sir Ernest Cowell, 'Papers', in CMAC.

102 For example, see the letter from Sir James Grigg to Sir Edward Bridges, 28 October 1942, in PRO WO32/11972.

103 Major R. M. C. Cubis, 'An Academy of Military Art', *Australian Army Journal*, 162 (November 1962), p. 40.

104 Lieutenant-Colonel W. L. Raws, 'Discipline and *Moral*', *The Australian Military Journal*, v (April 1914), pp. 273–84.

105 Captain Blair W. Sparks and Brigadier-General Oliver K. Niess, 'Psychiatric Screening of Combat Pilots', *U. S. Armed Forces Medical Journal*, vii.6 (June 1956), pp. 811–12. They were arguing that psychiatrists were necessary *despite* imperfect predictive techniques.

106 This was regretted by Brigadier I. R. Campbell, 'Should We Study Military History?', *Army Training Memorandum. Australian Military Forces*, 49 (June–July 1947), pp. 49–51. He was the Director of Military Training, Australian Army Headquarters.

107 Field Marshal Earl Wavell, *The Good Soldier* (London, 1948), p. 103: amendments in 1932 to the 'Training and Manoeuvre Regulations'.

108 Lieutenant-Colonel C. D. Daly, 'A Psychological Analysis of Military Morale', *The Army Quarterly*, xxxii (April 1936), p. 60.

109 Field Marshal Earl Wavell, *The Good Soldier* (London, 1948), pp. 103–4, lecture in 1933 and W. C. Westmoreland's lecture to the 69th annual meeting of the Association of Military Surgeons of the United States, in 'Mental Health – An Aspect of Command', *Military Medicine* (March 1963), pp. 211–13. Also see Norman C. Meier, *Military Psychology* (New York, 1943), p. xix and J. A. Shimeld, *Hints on Military Instruction. A Concise Presentation of Valuable Instructional Knowledge for the Military Officer, Warrant Officer and N. C. O.* (Sydney, 1941), p. 27.

110 Donald G. Marquis, 'The Mobilization of Psychologists for War Service', *Psychological Bulletin*, 41 (1944), p. 470. By 1919, the psychologist, James R. Angell, had replaced the astronomer, G. E. Hale, as chairman of the National Research Council and another psychologist, Robert M. Yerkes, had been appointed chairman of the Research Information Service (an influential Division of the N.R.C.).

111 Argued by Franz Samelson, 'World War I Intelligence Testing and the Development of Psychology', *Journal of the History of the Behavioral Sciences*, 133 (1977), p. 276.

112 Lewis M. Terman, 'The Mental Test as a Psychological Method', *Psychological Review*, 31 (1924).

113 Horace B. English, 'How Psychology Can Facilitate Military Training – A Concrete Example', *The Journal of Applied Psychology*, xxvi (1942), pp. 3–7.

114 N. A. B. Wilson, 'Application of Psychology in the Defence Departments', in C. A. Mace and P. E. Vernon (eds), *Current Trends in British Psychology* (London, 1953), p. 28.

115 John Rawlings Rees, *The Shaping of Psychiatry by War* (London, 1945), p. 82; Commander Leon J. Saul, Commander Howard Rome, and Edwin Leuser, 'Desensitization of Combat Fatigue Patients', *The American*

Journal of Psychiatry, 102 (1945–46), pp. 476–78; Peter Watson, *War on the Mind. The Military Uses and Abuses of Psychology* (London, 1978), p. 38.

116 C. W. Valentine, *The Human Factor in the Army. Some Applications of Psychology to Training, Selection, Morale and Discipline* (Aldershot, 1943), pp. 27–38.

117 Walter V. Bingham, 'Psychological Services in the United States Army', *The Journal of Consulting Psychology*, 5.5 (September–October 1941), p. 221.

118 John Rawlings Rees, *The Shaping of Psychiatry by War* (London, 1945), pp. 79–80; C. W. Valentine, *Principles of Army Instruction With Special Reference to Elementary Weapon Training* (Aldershot, 1942), p. 13; Dael Wolfle, 'Military Training and the Useful Parts of Learning Theory', *Journal of Consulting Psychology*, x (1946), pp. 73–5.

119 Lieutenant-Colonel J. W. Gibb, *Training in the Army* (London, nd [end of the Second World War]), p. 23.

120 For a discussion, see the minute by the Director-General, Army Medical Services, PRO WO32/11974 and Colonel B. Ungerson, *Personnel Selection* (London, 1953), p. 1. Personnel selection was relatively unimportant during the 1914–18 war since a professional elite was not in place to administer a complex system and (until conscription was introduced in 1916) volunteers were able to choose their own corps and (it was considered) were sensitive to their own strengths and weaknesses. The screening procedures were generally regarded as having been a failure. Despite the fact that the army ended up rejecting men for mental and emotional reasons at a rate almost seven times that which prevailed during the First World War, the rate of separations for these causes was five times higher than it had been in the First World War.

121 See the discussion by Hamilton Cravens, *The Triumph of Evolution. American Scientists and the Heredity–Environment Controversy 1900–1941* (Pennsylvania, 1978), pp. 76–8. Also see Harry Campbell, 'The Biological Aspects of Warfare – I', *The Lancet*, 15 September 1917, pp. 433–5 (also see parts two and three on 22 September 1917, pp. 469–71 and 29 September 1917, pp. 505–8); Walter B. Cannon, *Bodily Changes in Pain, Hunger, Fear and Rage. An Account of Recent Researches into the Function of Emotional Excitement*, second edition, first published in 1915 (New York, 1929), p. 379; William James, *Principles of Psychology*, vol. 2 (1910); William McDougall, *An Introduction to Social Psychology* (London, 1908); Clyde B. Moore, 'Some Psychological Aspects of War', *The Pedagogical Seminary. A Quarterly*, xxiii (1916), pp. 367–86; J. Howard

Moore, *Savage Survivals* (London, 1916), pp. 137–46; Lloyd Morgan, *Habit and Instinct* (1896); William A. White, *Thoughts of a Psychiatrist on the War and After* (New York, 1919), p. 65.

122 Revd Stopford A. Brooke, *A Discourse on War*, first published in 1905 (London, 1916), p. 1.

123 William James, 'The Moral Equivalent of War', first published in 1910, in his *'The Moral Equivalent of War' and Other Essays and Selections from 'Some Problems of Philosophy'*, edited by John K. Roth (New York, 1971), p. 5.

124 Gustave Le Bon, *Psychologie des foules* (1985) and Wilfred Trotter, *Instincts of the Herd in Peace and War* (London, 1916). Also see Charles Bird, 'From Home to the Charge: A Psychological Study of the Soldier', *The American Journal of Psychology*, xxviii.3 (July 1917), pp. 331–3 and 341–2; Sir Martin Conway, *The Crowd in Peace and War* (London, 1915), pp. 305–6; LeRoy Eltinge, *Psychology of War* (London, 1918); G. Stanley Hall, *Morale. The Supreme Standard of Life and Conduct* (New York, 1920), pp. 116–17; W. N. Maxwell, *A Psychological Retrospect of the Great War* (London, 1923), pp. 46 and 57–8; Professor George Gilbert Aime Murray, *Herd Instinct: For Good and Evil* (London, 1940), pp. 13–14.

125 G. Stanley Hall, *Morale. The Supreme Standard of Life and Conduct* (New York, 1920), pp. 36–7. Also see Major C. Battye, 'Infantry Training', *Commonwealth Military Journal*, 4 (October 1913), pp. 679–84.

126 L. L. Bernard, 'Are There Any Instincts?', *Journal of Abnormal and Social Psychology*, 14 (1920), pp. 307–11; L. L. Bernard, *Instinct: A Study in Social Psychology* (New York, 1924); L. L. Bernard, 'The Misuse of Instinct in the Social Sciences', *Psychological Review*, 28 (1921), pp. 96–118; Ellsworth Faris, 'Are Instincts Data or Hypotheses?', *American Journal of Sociology*, 27 (1921–2), pp. 184–96; R. E. Money-Kyrle, 'The Development of War. A Psychological Approach', *British Journal of Medical Psychology*, xvi.3 (1937); W. J. Perry, *War and Civilisation* (Manchester, 1918); Franz Alexander, 'Aggressiveness – Individual and Collective', in *The March of Medicine. The New York Academy of Medicine. Lectures to the Laity 1943* (New York, 1943), pp. 83–99; Knight Dunlan, 'The Cause and Prevention of War', *The Journal of Abnormal and Social Psychology*, 35.4 (October 1940); Mark A. May, *A Social Psychology of War and Peace* (New Haven, Connecticut, 1943); T. H. Pear, *Are There Human Instincts?* (Manchester, 1943); and the contributors to *The British Journal of Educational Psychology*, November 1941, February, June and November 1942 and February 1943.

127 Captain P. P. Manzie, 'Philosophy, Psychology and the Army', *Australian*

Army Journal, 188 (January 1965), pp. 38–9.

128 Captain G. M. Orr, 'Some Moral Factors in War', *The Commonwealth Military Journal*, 1 (May 1911), p. 197 and Major C. B. B. White, 'The Study of War', *The Commonwealth Military Journal*, 1 (July 1911), p. 284. For a novelistic impression of the leader who 'sways' his men by force of personality to continue firing, see Willa Cather, *One of Ours* (New York, 1922), p. 452.

129 The best discussion can be found in Jon Balkind, 'A Critique of Military Sociology: Lessons from Vietnam', *Journal of Strategic Studies*, 1.3 (December 1978), pp. 241–2 and Anthony Kellett, 'Combat Motivation', in Gregory Belenky (ed.), *Contemporary Studies in Combat Psychiatry* (New York, 1987), p. 211.

130 Arthur L. Hessin, 'Neuropsychiatry in Airborne Divisions', in Lieutenant-General Hal B. Jennings (ed.), *Neuropsychiatry in World War II. Volume II. Overseas Theatres* (Washington, DC, 1973), p. 397, quoting from a 1951 report. Also see Roger W. Little, 'Buddy Relations and Combat Performance', in Morris Janowitz (ed.), *The New Military. Changing Patterns of Organization* (New York, 1964), pp. 218–19; Earle Silber, 'Adjustment to the Army. The Soldiers' Identification with the Group', *U.S. Armed Forces Medical Journal*, v.9 (September 1954), p. 1344; and (for an angry rebuttal of Little) see the letter to the editor from Colonel George I. Forsythe and Lieutenant-Colonel Harold H. Dunwood, 'Solidarity and the Mass Army', *The Army Combat Forces Journal*, 5.9 (April 1955), pp. 5–6.

131 Captain Herbert X. Spiegel, 'Psychiatric Observations in the Tunisian Campaign', *American Journal of Orthopsychiatry*, xiv (1944), p. 312 and his 'Psychiatry in an Infantry Battalion in North Africa', in Lieutenant-General Hal B. Jennings (ed.), *Neuropsychiatry in World War II. Volume II. Overseas Theatres* (Washington, DC,1973), p. 122. See Chapter 5, 'Love and Hate'.

132 Major J. O. Langtry, 'Tactical Implications of the Human Factors in Warfare', *Australian Army Journal*, 107 (April, 1958), p. 16. He was justifying the introduction of Standard Operational Procedures during battle which decreed that junior leaders continually crawled along deployment lines calling to their men.

133 Colonel Albert J. Glass, 'Preventive Psychiatry in the Combat Zone', *U.S. Armed Forces Medical Journal*, iv.1 (1953), p. 689.

134 John Broadus Watson, *Behaviour: An Introduction to Comparative Psychology* (New York, 1914) and *Behaviourism* (New York, 1925). Also see M. F. Ashley Montagu, 'Introduction', in his (ed.) *Man and Aggression*

(New York, 1968), pp. xi–xiii; J. P. Scott, 'Biology and Human Aggression', *American Journal of Orthopsychiatry*, 40.4 (July 1970), p. 570; J. P. Scott, 'Biology and the Control of Violence', *International Journal of Group Tensions*, 3.3–4 (1973), p. 7; J. P. Scott, 'Biological Basis of Human Warfare: An Interdisciplinary Problem', in Muzafer Sherif and Carolyn W. Sherif (eds), *Interdisciplinary Relationships in the Social Sciences* (Chicago, 1969), p. 131.

135 For a detailed description, see Lawrence B. Radine, *The Taming of the Troops. Social Control in the United States Army* (Westport, Connecticut, 1977), pp. 130–2.

136 Major Jules V. Coleman, 'The Group Factor in Military Psychiatry', *American Journal of Orthopsychiatry*, xvi (1946), pp. 222 and 224–25.

137 Also see Harry Trosman and I. Hyman Weiland, 'Application of Psychodynamic Principles to Psychotherapy in Military Service', *U.S. Armed Forces Medical Journal*, viii.9 (September 1957), pp. 1359–60.

138 For instance, see Harry Trosman and I. Hyman Weiland, 'Application of Psychodynamic Principles to Psychotherapy in Military Service', *U. S. Armed Forces Medical Journal*, viii.9 (September 1957), pp. 1359–60.

139 John T. MacCurdy, *The Structure of Morale* (Cambridge, 1943), p. 45.

140 For a description of some of the techniques used, see Joanna Bourke, *Dismembering the Male: Men's Bodies, Britain, and the Great War* (London and Chicago, 1996). For other examples, see 'Memorandum on Layforce: July 1940–July 1941', in Evelyn Waugh, *The Diaries of Evelyn Waugh* (London, 1976), p. 518; War Diaries of the Australian Directorate of Psychology, 'Discussion with DDGMP – Tasmania re. Psychological Selection for Military Service', 26 June 1945, AWM52 33/1/1; R. M. Yorkes and C. S. Yoakum, *Army Mental Tests* (New York, 1920), p. 10.

4 Anatomy of a Hero

1 Lily Doyle, 'O'Leary, V.C.', in Doyle, *Bound in Khaki* (London, 1916), pp. 23–4.

2 Master Sergeant Roy Benavidez with John R. Craig, *Medal of Honor. A Vietnam Warrior's Story* (Washington, DC, 1995), pp. 6, 9, 19, 21, 31, 74, 78–9, 85, 100, 139–49 and 159–72.

3 For a discussion of heroic myths, see Robert A. Segal (ed.), *In Quest of the Hero* (Princeton, 1990).

4 Harold R. Peat, *Private Peat* (Indianapolis, 1917), p. 104.

5 G. R. Peberdy, 'Moustaches', *Journal of Mental Science*, 107.446 (January 1961), pp. 40–7.

6 W. G. Burchett, *Wingate's Phantom Army* (Bombay, 1944), p. 36 and Jack

L. Mahan and George A. Clum, 'Longitudinal Prediction of Marine Combat Effectiveness', *The Journal of Social Psychology*, 83 (1971), pp. 45–54.

7 Dr Flanders Dunbar, 'Medical Aspects of Accidents and Mistakes in the Industrial Army and in the Armed Forces', *War Medicine*, 4.2 (August 1943), pp. 161–75 and Hilary St George Saunders, *The Green Beret. The Story of the Commandos 1940–1945* (London, 1949), p. 39.

8 W. G. Burchett, *Wingate's Phantom Army* (Bombay, 1944), p. 36. Also see Colonel John Howard McEniry, *A Marine Dive-Bomber Pilot at Guadalcanal* (Tuscaloosa, Alabama, 1987), p. 89. This idea was contradicted by the research of Samuel A. Stouffer who showed that married men made better combat soldiers than unmarried men: Stouffer et al., *The American Soldier: Combat and its Aftermath. Volume II* (Princeton, 1949), p. 35.

9 Captain David Fallon, *The Big Fight (Gallipoli to the Somme)* (London, 1918), pp. 81–2.

10 Roy R. Grinker and John P. Spiegel, *Men Under Stress* (London, 1945), pp. 41–2.

11 R. W. Thompson, *Men Under Fire* (London, 1946), p. 126. Also see Captain Charles K. McKerrow, 'Diaries and Letters', letter dated 11 November 1915, IWM.

12 Newsclipping from *The Morning Adviser*, 4 July 1936, in PRO PIN15/2503. The greater likelihood that students majoring in English, philosophy, and humanities would question orders and be less liable to use nuclear weapons was used by Peter Karsten to argue that they were *needed* in the military: Karsten, '"Professional" and "Citizen" Officers: A Comparison of Service Academy and ROTC Officer Candidates', in Charles C. Moskos (ed.), *Public Opinion and the Military Establishment* (Beverly Hills, 1971), p. 55.

13 Lieutenant-Colonel Matheson (AIF), 'Notes on Jungle Combat', 28 February 1943, p. 11, in AWM. Also see Henry Dearchs, interviewed in Eric Bergerud, *Touch with Fire. The Land War in the South Pacific* (New York, 1996), p. 219.

14 Colonel Laycock, 'Commando Training in the Middle East', 1942, AWM.

15 Robert William MacKenna, *Through a Tent Door*, first published 1919 (London, 1930), pp. 65–6 and Philip Caputo, *A Rumor of War* (London, 1977), p. 85.

16 Norris F. Hall, 'Science in War', in Hall, Zechariah Chafee, Jr., and Manley O. Hudson (eds), *The Next War* (Cambridge, 1925), p. 26.

17 Field Marshal Earl Wavell, 'The Good Soldier', *Sunday Times*, 19 August 1945, reproduced in his *The Good Soldier* (London, 1948), p. 43.

18 Firstbrooke and Grosvenor Clarke, 'Letters', letter from Firstbrooke Clarke, 12 June 1917, Liddle Collection.

19 Sir Sydney Frank Markham, 'Papers', letter from his mother, 10 March 1917, IWM.

20 Robert William MacKenna, *Through a Tent Door*, first published 1919 (London, 1930), pp. 64–5.

21 For a discussion, see Joanna Bourke, *Dismembering the Male: Men's Bodies, Britain, and the Great War* (London and Chicago, 1996), pp. 172–3.

22 Charles Bird, 'From Home to the Charge: A Psychological Study of the Soldier', *The American Journal of Psychology*, xxviii.3 (July 1917), pp. 323–4; A. J. M. A. Chaillou and Leon MacAuliffe, *Morphologie medicale; étude des quatre types humains* (Paris, 1912); G. Stanley Hall, 'Practical Applications of Psychology as Developed by the War', *The Pedagogical Seminary. A Quarterly*, xxvi (1919), pp. 83–4. These types paralleled the older divisions of phlegmatic, sanguine, bilious and nervous.

23 Lieutenant-Colonel A. D. Carbery, 'Some Medical Aspects of Recruiting for War', *Transactions of the Australasian Medical Congress (British Medical Association). Second Session: Dunedin, February 3 to 10, 1927*, 12 November 1927, p. 369.

24 Clark W. Heath, W. L. Woods, L. Brouha, C. C. Seltzer, and A. V. Bock, 'Personnel Selection: A Short Method for Selection of Combat Soldiers', *Annals of Internal Medicine*, 19 (1943), pp. 415–26. Also see Clark W. Heath, *What People Are. A Study of Normal Young Men* (Cambridge, Massachusetts, 1946).

25 Samuel A. Stouffer et al., *The American Soldier: Combat and its Aftermath. Volume II* (Princeton, 1949), pp. 34–5.

26 Captain Martin Stein, 'Neurosis and Group Motivation', *The Bulletin of the U.S. Army Medical Department*, vii.3 (March 1947), p. 318.

27 Jeannie Basinger, *The World War II Combat Film. Anatomy of a Genre* (New York, 1986), pp. 36–7.

28 Captain J. Fuller, *Hints on Training Territorial Infantry. From Recruit to Trained Soldier* (London, 1913), pp. 2–3.

29 Dr H. W. Wills, 'Footnote to Medical History . . . General Aspects. Shell Shock', item 1, p. 6, Liddle Collection.

30 John T. MacCurdy, *War Neuroses* (Cambridge, 1918), p. 129.

31 Lieutenant Uzal Ent, interviewed in Rudy Tomedi, *No Bugles, No Drums. An Oral History of the Korean War* (New York, 1993), p. 19; Mark Lane,

Conversations with Americans (New York, 1970), p. 140; Al Santoli, *Everything We Had. An Oral History of the Vietnam War by Thirty-Three American Soldiers Who Fought It* (New York, 1981), pp. 201–2; Hilary St George Saunders, *The Green Beret. The Story of the Commandos 1940–1945* (London, 1949), p. 39; Samuel A. Stouffer, et al., *The American Soldier: Adjustment During the Army Life. Volume I* (Princeton, 1949), p. 329.

32 Major E. S. C. Ford, 'Principles and Problems of Maintenance of Fighter-Bomber Pilots', *War Medicine*, 8.1 (July 1945), p. 30; Eli Ginzberg, John L. Herma, and Sol W. Ginsburg, *Psychiatry and Military Manpower Policy. A Reappraisal of the Experience in World War II* (New York, 1953), p. 18; Colonel Albert J. Glass and Lieutenant-Colonel Calvin S. Drayer, 'Italian Campaign (1 March 1944–2 May 1945), Psychiatry Established at Division Level', in Lieutenant-General Hal B. Jennings (ed.), *Neuropsychiatry in World War II. Volume II. Overseas Theatres* (Washington, DC, 1973), p. 106; Roy R. Grinker and John P. Spiegel, *Men Under Stress* (London, 1945), p. 12; Arthur L. Hessin, 'Neuropsychiatry in Airborne Divisions', in Lieutenant-General Hal B. Jennings (ed.), *Neuropsychiatry in World War II. Volume II. Overseas Theatres* (Washington, DC, 1973), pp. 396–7; William C. Porter, 'Military Psychiatry and the Selective Services', *War Medicine*, 1.3 (May 1941), p. 370; Edward A. Strecker and Kenneth E. Appel, *Psychiatry in Modern Warfare* (New York, 1945), p. 13.

33 Major Dallas Pratt and Abraham Neustadter, 'Combat Record of Psychoneurotic Patients', *The Bulletin of the U.S. Army Medical Department*, vii.9 (September 1947), pp. 809–11.

34 William C. Porter, 'Military Psychiatry and the Selective Services', *War Medicine*, 1.3 (May 1941), p. 370. Also see Albert J. Glass, Francis J. Ryan, Ardie Lubin, C. V. Ramana, and Anthony C. Tucker, 'Psychiatric Predictions and Military Effectiveness. Part II', *U.S. Armed Forces Medical Journal*, vii.11 (November 1956), pp. 1581–82.

35 For examples of the downside, see Major E. S. C. Ford, 'Principles and Problems of Maintenance of Fighter-Bomber Pilots', *War Medicine*, 8.1 (July 1945), p. 30; Major Arthur O. Hecker, First Lieutenant Marvin R. Plessett, and First Lieutenant Philip C. Grana, 'Psychiatry Problems in Military Service During the Training Period', *The American Journal of Psychiatry*, 99 (1942–43), pp. 38–9; William C. Porter, 'Military Psychiatry and the Selective Services', *War Medicine*, 1.3 (May 1941), p. 370.

36 S. Kirson Weinberg, 'The Combat Neuroses', *The American Journal of*

Sociology, li.5 (March 1946), p. 472.

37   Felix Deutsch, 'Civilian War Neuroses and Their Treatment', *The Psychoanalytical Quarterly*, 13 (1944), pp. 302–3; Eli Ginzberg, John L. Herma, and Sol W. Ginsburg, *Psychiatry and Military Manpower Policy. A Reappraisal of the Experience in World War II* (New York, 1953), p. 24; Blair W. Sparks and Oliver K. Niess, 'Psychiatric Screening of Combat Pilots', *U.S. Armed Forces Medical Journal*, vii.6 (June 1956), p. 815.

38   Major Arthur O. Hecker, First Lieutenant Marvin R. Plessett, and First Lieutenant Philip C. Grana, 'Psychiatry Problems in Military Service During the Training Period', *The American Journal of Psychiatry*, 99 (1942–43), pp. 38–9.

39   It was widely believed that men used the prohibition on homosexuality as an excuse to leave the armed services: for an example from the Vietnam War, see Major Franklin Del Jones, 'Experiences of a Division Psychiatrist in Vietnam', *Military Medicine*, 132.12 (December 1967), p. 1003. For men who considered using, or did use, this as a way to escape the armed forces, see the interview with John H. Abbott, in Studs Terkel, *'The Good War'. An Oral History of World War Two* (London, 1985), p. 167.

40   Ernest Jones, 'War and Individual Psychology', *The Sociological Review*, viii (1915), p. 177. Also see Pryns Charles Hopkins, *The Psychology of Social Movements. A Psycho-Analytic View of Society* (London, 1938), p. 119.

41   R. E. Money-Kyrle, 'The Development of War. A Psychological Approach', *British Journal of Medical Psychology*, xvi.3 (1937), p. 235.

42   Charles Berg, 'Clinical Notes on the Analysis of a War Neurosis', *British Journal of Medical Psychology*, xix.2 (1942), p. 185. This was also the view of Frederick H. Allen, 'Homosexuality in Relation to the Problem of Human Difference', *The American Journal of Orthopsychiatry*, x (1940), pp. 129–35.

43   Interview of Ted Allenby, in Studs Terkel, *'The Good War'. An Oral History of World War Two* (London, 1985), pp. 179–81.

44   Charles Anderson, 'On Certain Conscious and Unconscious Homosexual Responses to Warfare', *British Journal of Medical Psychology*, xx.2 (1945), pp. 162 and 172.

45   Eli Ginzberg, John L. Herma, and Sol W. Ginsburg, *Psychiatry and Military Manpower Policy. A Reappraisal of the Experience in World War II* (New York, 1953), p. 19 and Elizabeth Rosenberg, 'A Clinical Contribution to the Psychopathology of the War Neuroses', *The International Journal of Psycho-Analysis*, xxiv (1943), pp. 32–41.

46   Charles Anderson, 'On Certain Conscious and Unconscious Homosexual Responses to Warfare', *British Journal of Medical Psychology*, xx.2 (1945),

pp. 163 and 173–4.

47 David Hume, 'Of National Character' (1741–2), in *Essays: Moral, Political and Literary* (Oxford, 1963), p. 202.

48 Ernest Barker, *National Character and the Factors in Its Formation* (London, 1927), p. 140.

49 Michael MacDonagh, *The Irish on the Somme* (London, 1917), p. 57.

50 Michael MacDonagh, *The Irish at the Front* (London, 1916), p. 124.

51 'Some Principles of Psychological Warfare Policy', May 1944, p. 4, in PRO WO241/3.

52 For an example, see Robert Graves, *Goodbye to All That*, first published in 1929 (London, 1977), p. 164.

53 H. G. L. Priday, *The War from Coconut Square. The Story of the Defence of the Island Bases of the South Pacific* (Wellington, New Zealand, 1945), pp. 46 and 128–30.

54 Shelford Bidwell, *Modern Warfare. A Study of Men, Weapons and Theories* (London, 1973), p. 146. Also see Frederic Hillersdon Keeling, *Keeling Letters and Recollections* (London, 1918), p. 260, letter to Revd William Danks, 31 December 1915.

55 Escott Lynn, *In Khaki for the King. A Tale of the Great War* (London, 1915), p. 151.

56 Alexander Catto, *With the Scottish Troops in France* (Aberdeen, 1918), pp. 21 and 42–3.

57 Cynthia H. Enloe, *Ethnic Soldiers. State Security in Divided Societies* (Harmondsworth, 1980) discusses this in depth. For contemporary comment, see Second Lieutenant John Bernard Brody, 'Diary', in Brereton Greenhous (ed.), *A Rattle of Pebbles: The First World War Diaries of Two Canadian Airmen* (Ottawa, 1987), p. 13; Patrick Mee, *Marine Gunner. Twenty-Two Years in the Royal Marine Artillery* (London, 1935), pp. 198–9; Harold R. Peat, *Private Peat* (Indianapolis, 1917), pp. 11 and 111–15; William F. Pressey, 'All for a Shilling a Day', p. 25, IWM; 'Report of a Conference on Psychiatry in Forward Areas', 8–10 August 1944, pp. 13–4, in PRO WO32/11550; Lieutenant-Colonel Vivian Edgar Olmar Stevenson-Hamilton, 'Papers', Liddell Hart Centre; Edgar Wallace, *Heroes All. Gallant Deeds of the War* (London, 1914), pp. 198–206; A. Hyatt Williams, 'A Psychiatric Study of Indian Soldiers in the Arakan', *British Journal of Medical Psychology*, xxiii.3 (1950), pp. 131–2; Douglas P. Winnifrith, *The Church in the Fighting Line* (London, 1915), pp. 130–1.

58 Saint Nihal Singh, *India's Fighters: Their Mettle, History and Services to Britain* (London, 1914), pp. 61–2. Also see pp. 71–5 and Singh, *India's Fighting Troops* (London, 1914), p. 30.

59 Master Sergeant Roy Benavidez with John R. Craig, *Medal of Honor. A Vietnam Warrior's Story* (Washington, DC, 1995), p. 85; Tom Holm, 'American Indian Veterans and the Vietnam War', in Walter Capps (ed.), *The Vietnam Reader* (New York, 1991), p. 193; Thomas M. Holm, 'Forgotten Warriors: American Indian Servicemen in Vietnam', *The Vietnam Generation*, 1.2 (spring 1989), p. 63; Harold Ickes, 'Indians Have a Name for Hitler', *Collier's*, 113 (1944), p. 58.

60 See James J. Fisher, *The Immortal Deeds of Our Irish Regiments in Flanders and the Dardanelles*, no. 1 (Dublin, 1916); Denis Gwynn, *Redmond's Last Years* (London, 1919), p. 201; Joseph Keating, 'Tyneside Irish Brigade', in Felix Lavery (ed.), *Great Irishmen in War and Politics* (London, 1920), pp. 128–9; Revd G. A. Studdert Kennedy, *Rough Talks by a Padre* (London, 1918), pp. 35–6; Michael MacDonagh, *The Irish at the Front* (London, 1916), pp. 3 and 158; Michael MacDonagh, *The Irish on the Somme* (London, 1917), pp. 26–7 and 114; N. Marlowe, 'The Mood of Ireland', *British Review*, xi.1 (July 1915), pp. 4 and 9; Randall Parrish, *Shea of the Irish Brigade. A Soldier's Story* (London, 1914), frontispiece and p. 343.

61 Michael MacDonagh, *The Irish at the Front* (London, 1916), p. 111.

62 Captain Stephen L. Gwynn, 'Irish Regiments', in Felix Lavery (ed.), *Great Irishmen in War and Politics* (London, 1920), p. 149 and Harold Spender, 'Ireland and the War', *Contemporary Review*, cx (November 1916), p. 567.

63 Professor Alfred O'Rahilly, *Father William Doyle S.J.: A Spiritual Study* (London, 1925), p. 439; Denis Gwynn, *The Life of John Redmond* (London, 1932), p. 404; Henry Hanna, *The Pals at Suvla Bay. Being the Record of 'D' Company of the 7th Royal Dublin Fusiliers* (Dublin, 1916), p. 7; Joseph Keating, 'Tyneside Irish Brigade', in Felix Lavery (ed.), *Great Irishmen in War and Politics* (London, 1920), p. 145; S. Parnell Kerr, *What the Irish Regiments Have Done* (London, 1916), p. 135; Michael MacDonagh, *The Irish at the Front* (London, 1916), p. 124; J. S. MacKenzie, *Arrows of Desire. Essays on Our National Character and Outlook* (London, 1920), pp. 182 and 196–7; Harold Spender, 'Ireland and the War', *Contemporary Review*, cx (November 1916), pp. 566–7.

64 Alice M. P. Cooke, *Irish Heroes in Red War* (Dublin, 1915), p. 16. Also see pp. 19–20.

65 Harold Spender, 'Ireland and the War', *Contemporary Review*, cx (November 1916), p. 570.

66 'Outis', 'Has Recruiting in Ireland been Satisfactory?', *The United Service Magazine*, new series, li (April–September 1915), pp. 567 and 571.

67 Gordon Holman, *Commando Attack* (London, 1942), p. 53.

68 Shelford Bidwell, *Modern Warfare. A Study of Men, Weapons and Theories* (London, 1973), p. 146.

69 Sean McCann, *The Fighting Irish* (London, 1972), p. 9. Also see the foreword by General Sir John Hackett in Brigadier A. E. C. Bredin, *A History of the Irish Soldier* (Belfast, 1987), p. x.

70 Peter Karsten, 'Irish Soldiers in the British Army, 1792–1922: Suborned or Subordinate?', *Journal of Social History*, 17 (1983), pp. 40 and 59.

71 Letter from Private Robert McGregor to his father in 1915, quoted in Amy Gordon Grant, *Letters from Armageddon. A Collection Made During the World War* (Boston, 1930), pp. 91–2 and S. Parnell Kerr, *What the Irish Regiments Have Done* (London, 1916), pp. 47–8 and 103.

72 Professor Alfred O'Rahilly, *Father William Doyle S.J. A Spiritual Study* (London, 1925), p. 473.

73 S. Parnell Kerr, *What the Irish Regiments Have Done* (London, 1916), pp. 47–8.

74 For a summary, see Horace Mann Bond, 'The Negro in the Armed Forces of the United States Prior to World War One', *Journal of Negro Education*, xii (1943), pp. 268–87.

75 Charles H. Williams, *Sidelights on Negro Soldiers* (Boston, 1923), p. 27.

76 Samuel A. Stouffer et al., *The American Soldier: Adjustment During Army Life. Volume I* (Princeton, 1949), p. 494.

77 Ulysses Grant Lee, *The Employment of Negro Troops* (Washington, DC, 1966), p. 134.

78 John J. Niles, *Singing Soldiers* (New York, 1927), p. 48. Niles believed that black soldiers were good singers, not fighters (p. ix).

79 Ulysses Grant Lee, *The Employment of Negro Troops* (Washington, DC, 1966), pp. 453–4.

80 John Richards, 'Some Experiences with Colored Soldiers', *Atlantic Monthly*, August 1919, quoted by Jerome Dowd, *The Negro in American Life* (New York, 1926), pp. 233–6 and 239–40.

81 General Bullard's report, cited by Jerome Dowd, *The Negro in American Life* (New York, 1926), p. 226.

82 Dorothy W. Seago, 'Stereotypes: Before Pearl Harbour and After', *The Journal of Psychology*, 23 (1947), pp. 55–63. Also see Lynn L. Ralya, 'Some Surprising Beliefs Concerning Human Nature Among Pre-Medical Psychology Students', *The British Journal of Educational Psychology*, xv.ii (June 1945), pp. 72 and 141.

83 Dr. Walter L. Wright, Jr., quoted by Ulysses Grant Lee, *The Employment*

of Negro Troops (Washington, DC, 1966), pp. 704–5.

84 William A. Hunt, 'The Relative Incidence of Psychoneurosis Among Negroes', *Journal of Consulting Psychology*, xi (1947), pp. 134–5.

85 John Richards, 'Some Experiences with Colored Soldiers', *Atlantic Monthly*, August 1919, quoted by Jerome Dowd, *The Negro in American Life* (New York, 1926), p. 240.

86 John J. Niles, *Singing Soldiers* (New York, 1927), pp. 109–10.

87 Letter from T. Corder Catchpool, dated 15 November 1914, in Catchpool, *On Two Fronts* (London, 1918), p. 28.

88 David Fitzpatrick, 'Militarism in Ireland, 1900–1922', in Thomas Bartlett and Keith Jeffery (eds), *A Military History of Ireland* (Cambridge, 1996), p. 379.

89 William McDougall was not the first to argue that instincts were the primary basis of human conduct. He drew to a great extent on William James, *Principles of Psychology*, vol. 2 (London, 1910), pp. 383–441 and Lloyd Morgan, *Habit and Instinct* (London, 1896).

90 The letter was originally published in an unnamed Scottish newspaper. It was reprinted in *The Tablet* in January 1916 and in Michael MacDonagh, *The Irish at the Front* (London, 1916), p. 71.

91 Michael MacDonagh, *The Irish on the Somme* (London, 1917), p. 58.

92 Harold Spender, 'Ireland and the War', *Contemporary Review*, cx (November 1916), p. 567.

93 Cynthia H. Enloe, *Ethnic Soldiers. State Security in Divided Societies* (Harmondsworth, 1980), p. 26.

94 For a few examples, see Revd R. H. Bassett, 'The Chaplain with the West African Forces', *Journal of the Royal Army Chaplains' Department*, vii.49 (July 1950), p. 24; Edmund Candler, *The Year of Chivalry* (London, 1916), p. 57; Revd E. J. Kennedy, *With the Immortal Seventh Division*, second edition (London, 1916), p. 132–3; Captain H. T. A. McKeag, 'Memoirs', April 1919, p. 112, IWM; Lieutenant-Colonel J. W. B. Merewether and Sir Frederick Smith, *The Indian Corps in France* (London, 1917), pp. 8 and 471; Ruth Danenhower Wilson, *Jim Crow Joins Up. A Study of Negroes in the Armed Forces of the United States* (New York, 1945), pp. 44–5.

95 For a discussion, see Pierre Bovet, *The Fighting Instinct*, translated by J. Y. T. Greig (London, 1923), p. 149.

96 Michael MacDonagh, *The Irish on the Somme* (London, 1917), p. 111. Also see Revd R. H. Bassett, 'The Chaplain with the West African Forces', *Journal of the Royal Army Chaplains' Department*, vii.49 (July 1950), p. 24 and David Killingray, ' "The Rod of Empire" : The Debate

Over Corporal Punishment in the British African Colonial Forces, 1888–1946', *Journal of African History*, 35 (1994), 202 for West African soldiers. For Indian troops, see Edmund Candler, *The Year of Chivalry* (London, 1916), pp. 273–4 and Lieutenant-Colonel J. W. B. Merewether and Sir Frederick Smith, *The Indian Corps in France* (London, 1917), pp. 110–11.

97 Brigadier W. Carden Roe, 'Memoirs', pp. 8–9, IWM. Also see Lieutenant-Colonel H. F. N. Jourdain, *Ranging Memories* (Oxford, 1934), p. 305.

98 Rowland Feilding, *War Letters to a Wife. France and Flanders, 1915–1919* (London, 1929), p. 121, letters dated 25 September 1916 and 14 June 1917.

99 Diary of unnamed black soldier, quoted by Lieutenant-Colonel Herbert S. Ripley and Major Stewart Wolf, 'Mental Illness Among Negro Troops Overseas', *American Journal of Psychiatry*, 103 (1946–7), p. 510. Also see David Parks, *GI Diary* (New York, 1968), pp. 34–5 and 40–1, diary entries for 10 March and 12 June 1966.

100 Jerome Dowd, *The Negro in American Life* (New York, 1926), pp. 231–2.

101 Anson Phelps Stokes, 'American Race Relations in War Time', *The Journal of Negro Education*, xiv (1945), p. 549.

102 Eli Ginzberg, *Breakdown and Recovery* (New York, 1950), p. 103; Eli Ginzberg, John L. Herma and Sol W. Ginsburg, *Psychiatry and Military Manpower Policy. A Reappraisal of the Experience in World War II* (New York, 1953), p. 33; Robert W. Smuts, 'The Negro Community and the Development of Negro Potential', *The Journal of Negro Education*, xxvi.4 (fall 1957), pp. 461–2; Rutherford B. Stevens, 'Racial Aspects of Emotional Problems of Negro Soldiers', *The American Journal of Psychiatry*, 103 (1946–7), pp. 493–8.

103 Jerome Dowd, *The Negro in American Life* (New York, 1926), p. 235; David Parks, *GI Diary* (New York, 1968), p. 52, diary for 15 December 1966; Charles H. Williams, *Sidelights on Negro Soldiers* (Boston, 1923), p. 29.

104 Charles H. Williams, *Sidelights on Negro Soldiers* (Boston, 1923), pp. 28 and 61–2.

105 Samuel A. Stouffer et al., *The American Soldier: Adjustment During Army Life. Volume I* (Princeton, 1949), pp. 500–1.

106 Ibid., p. 502.

107 See Ruth Danenhower Wilson, *Jim Crow Joins Up. A Study of Negroes in the Armed Forces of the United States* (New York, 1945), pp. 98–9.

108 Lieutenant-General Hal B. Jennings, *Neuropsychiatry in World War II*.

Volume II. Overseas Theatres (Washington, DC, 1973), p. 453. Also see John Sibley Butler, 'The Military as a Vehicle for Social Integration. The Afro-American Experience as Data', in Henry Dietz, Jerrold Elkin and Maurice Roumani (eds), *Ethnicity, Integration and the Military* (Boulder, San Francisco, 1991), pp. 27–50.

109 Samuel A. Stouffer et al., *The American Soldier: Adjustment During Army Life. Volume I* (Princeton, 1949), p. 521. He interviewed 1,607 black soldiers and 6,749 white soldiers.

110 Ibid., pp. 522–6. He surveyed 3,000 black soldiers and 1,000 matched white soldiers.

111 Comment of the black Philadelphian truck driver, Harry Carpenter to a black soldier, cited in Roi Ottley, *'New World A-Comin': Inside Black America* (Boston, 1943), p. 307. He was held on charges of treason for this statement.

112 Both quotations in Samuel A. Stouffer et al., *The American Soldier: Adjustment During Army Life. Volume I* (Princeton, 1949), p. 533. Also see Mary Penick Motley (ed.), *The Invisible Soldier. The Experience of the Black Soldier, World War II* (Detroit, 1975), p. 170.

113 General Patton's speech, as recalled by Captain John Long, quoted in Mary Penick Motley (ed.), *The Invisible Soldier. The Experience of the Black Soldier, World War II* (Detroit, 1975), p. 152.

114 Charles A. Gates, interviewed by Studs Terkel, *'The Good War'. An Oral History of World War Two* (London, 1985), p. 266.

115 Captain Charles A. Hill, Jr., interviewed in Mary Penick Motley (ed.), *The Invisible Soldier. The Experience of the Black Soldier, World War II* (Detroit, 1975), pp. 240–1.

116 Serviceman known as 'T.I.V.', cited in Eli Ginzberg, *Breakdown and Recovery* (New York, 1950), p. 233.

117 This was repeated frequently. For example, see the interviews in Mary Penick Motley (ed.), *The Invisible Soldier. The Experience of the Black Soldier, World War II* (Detroit, 1975).

118 Roi Ottley, *'New World A-Comin': Inside Black America* (Boston, 1943), p. 317.

119 Rowland Feilding, *War Letters to a Wife. France and Flanders, 1915–1919* (London, 1929), p. 317, 17 September 1918.

120 Roy R. Grinker and John P. Spiegel, *Men Under Stress* (London, 1945), pp. 44-5.

121 E. A. Godson, *The Great War 1914–1918. Incidents, Experiences, Impressions and Comments of a Junior Officer* (Hertford, nd), p. 31. Also see Roy R. Grinker and John P. Spiegel, *Men Under Stress* (London, 1945), p. 45.

122 Leslie T. Wilkins, *The Social Survey. Prediction of the Demand for Campaign Stars and Medals* (London, 1949), pp. 3, 14, and 17–19. Of those interviewed, 560 did not collect their medals and 1149 did collect them.

123 Canadian War Records Office, *Thirty Canadian V.C.s. 23rd April 1915 to 30th March 1918* (London, 1918), p. 6.

124 'Memorandum by the Director of Infantry on Extra Pay for the Infantry During the Present War', 1945, p. 2, PRO WO32/10924.

125 'Notes on Infantry', by the Commander of the 14th Army, 1945, p. 3, in PRO WO32/10924.

126 For instance, see Michael Herr, *Dispatches* (London, 1978), p. 14 and Roger W. Little, 'Buddy Relations and Combat Performance', in Morris Janowitz (ed.), *The New Military. Changing Patterns of Organization* (New York, 1964), p. 205.

127 Unnamed Vietnam veteran, quoted in Jonathan Shay, *Achilles in Vietnam. Combat Trauma and the Undoing of Character* (New York, 1994), p. 83. Also see D. J. Dennis, *One Day at a Time. A Vietnam Diary* (St Lucia, 1992), p. 16 and Jimmy Roberson, interviewed in Mark Lane, *Conversations with Americans* (New York, 1970), p. 60.

128 Brian Sullivan, quoted in Bernard Edelman (ed.), *Dear America. Letters Home from Vietnam* (New York, 1985), p. 132, letter to his wife, 2 March 1969.

129 Chet Pedersen, 'Wastelands', in J. Topham (ed.), *Vietnam Literature Anthology*, revised and enlarged (Philadelphia, 1990), p. 100.

130 Eugene B. Sledge, *With the Old Breed at Peleliu and Okinawa* (New York, 1990), p. 96. Also see Bruce Anello, in Indochina Curriculum Group, *Front Lines. Soldiers' Writings from Vietnam* (Cambridge, Massachusetts, 1975), p. 7, diary entry for 15 February 1968.

131 Dennis Kitchin, *War in Aquarius. Memoir of an American Infantryman in Action Along the Cambodian Border During the Vietnam War* (Jefferson, North Carolina, 1994), p. 46 and unnamed soldier, interviewed in Barry Broadfoot, *Six War Years 1939–1945. Memories of Canadians at Home and Abroad* (Don Mills, Ontario, 1974), p. 327.

132 Charles C. Moskos, *The American Enlisted Man* (New York, 1970), pp. 154–5.

133 Roger W. Little, 'Buddy Relations and Combat Performance', in Morris Janowitz (ed.), *The New Military. Changing Patterns of Organization* (New York, 1964), p. 205.

134 Letter from the Director of Organization, Colonel E. R. H. Herbert to Commanding Officers and others, 23 April 1941, PRO WO199/1849.

135 For three examples taken from each of the three conflicts, see William Manchester, *Goodbye Darkness. A Memoir of the Pacific War* (Boston, 1980), p. 141; W. G. Shipway, 'My Memories of the First World War', pp. 23–4, IWM; Richard J. Ford III, interviewed in Wallace Terry, *Bloods. An Oral History of the Vietnam War by Black Veterans* (New York, 1984), pp. 42–3. Jokes of becoming a hero because of confusion and fear rather than valour were a staple of military humour: see U.S. Department of War, Army Service Forces, Special Service Division, *Soldier Shows, Staging Area and Transport Entertainment Guide Comprising Blackouts, Sketches, Quizzes, Parodies, and Games* (Washington, DC, 1944), p. 8.

136 Scott Higgins, quoted in Al Santoli, *Everything We Had. An Oral History of the Vietnam War by Thirty-Three American Soldiers Who Fought It* (New York, 1981), p. 95.

137 Letter from unnamed army specialist, published in *Sepia* and cited in William King, '"Our Men in Vietnam": Black Media as a Source of the Afro-American Experience in Southeast Asia', *The Vietnam Generation*, 1.2 (spring 1989), p. 104.

138 *The Australian*, 17 January 1966.

139 Lieutenant Graham, quoted by Edmund Candler, *The Year of Chivalry* (London, 1916), p. 265. Also see the letter from Francis Grenfell (VC) to his twin brother, cited in John Buchan, *Francis and Riversdale Grenfell* (London, 1920), pp. 200–1.

140 Rowland Feilding, *War Letters to a Wife. France and Flanders, 1915–1919* (London, 1929), p. 375, 18 February 1919.

141 Ibid., pp. 161 and 316, 27 February 1917 and 17 September 1918. Also see Guy Chapman, *A Passionate Prodigality. Fragments of an Autobiography*, first published in 1933 (New York, 1966), p. 266 and Robert Lekachman, interviewed in Studs Terkel, *'The Good War'. An Oral History of World War Two* (London, 1985), p. 67.

142 Robert E. Holcomb, quoted in Wallace Terry, *Bloods. An Oral History of the Vietnam War by Black Veterans* (New York, 1984), p. 222.

143 Sherman Pratt, interviewed in Rudy Tomedi, *No Bugles, No Drums. An Oral History of the Korean War* (New York, 1993), p. 145

144 Richard A. Kulka et al., *Trauma and the Vietnam War Generation. Report of Findings from the National Vietnam Veterans Readjustment Study* (New York, 1990), p. 26.

145 George Wilson, quoted in Henry Wysham Lanier, *The Book of Bravery. Being True Stories in an Ascending Scale of Courage* (London, 1918), pp. 264–5.

146 Norman Copeland, *Psychology and the Soldier* (London, 1942), pp. 59–60;
Terence Denman, *Ireland's Unknown Soldiers. The 16th (Irish) Division in
the Great War, 1914–1918* (Dublin, 1992), p. 75; Amy Gordon Grant,
Letters From Armageddon. A Collection Made During the World War
(Boston, 1930), pp. 91–2; Robert Graves, *Goodbye to All That* (London,
1977), p. 152; Captain F. C. Hitchcock, *'Stand To'. A Diary of the Trenches
1915–18* (London, 1937), p. 8; Frederic Hillersdon Keeling, *Keeling
Letters and Recollections* (London, 1918), p. 260; Tom M. Kettle, *The
Ways of War* (London, 1917), p. 169; Michael MacDonagh, *The Irish at
the Front* (London, 1916), pp. 30 and 99; Michael MacDonagh, *The Irish
on the Somme* (London, 1917), pp. 21, 59 and 63; Major Edward 'Mick'
Mannock, *The Personal Diary* (London, 1966), p. 196; Thomas J. Mullen,
'"Mick" Mannock. The Forgotten Ace', *The Irish Sword*, x.39 (winter
1971), p. 77; Cuthberg Spurling, 'The Secret of the English Character',
The Contemporary Review, cx (November 1916), p. 638; Stephen
Stapleton, 'The Relations Between the Trenches', *The Contemporary
Review*, cxi (January–June 1917), p. 639; Sir A. P. Wavell, *Allenby*
(London, 1940), p. 203.

147 Gustav Bychowski, 'Personality Changes Characterizing the Transition
from Civilian to Military Life', *The Journal of Nervous and Mental
Disease*, 100.3 (September 1944), p. 292.

148 Thomas Carlyle, *On Heroes, Hero Worship and the Heroic in History*
(London, 1832).

149 Unnamed Canadian soldier, interviewed in Barry Broadfoot, *Six War
Years 1939–1945. Memories of Canadians at Home and Abroad* (Don Mills,
Ontario, 1974), p. 327.

5 Love and Hate

1 Edward Thomas, 'This is No Case of Petty Right or Wrong', written in
1915, in Thomas, *Collected Poems* (London, 1920), p. 165.

2 Laurence Housman (ed.), *War Letters of Fallen Englishmen* (London,
1930), pp. 117–20; Viola Meynell, *Julian Grenfell* (London, 1918), pp. 12,
18–19, 22, 24–5; Nicholas Mosley, *Julian Grenfell. His Life and the Times
of His Death 1888–1915* (London, 1976), pp. 237–8, 243, 247 and 260.

3 As I argue in *Dismembering the Male: Men's Bodies, Britain, and the Great
War* (London and Chicago, 1996), chapter 3.

4 Samuel A. Stouffer et al., *The American Soldier: Combat and Its
Aftermath. Volume II* (Princeton, 1949), p. 109. Survey of 568 enlisted
infantrymen who had seen combat in Sicily and North Africa, questioned
in April 1944.

5 Herbert X. Spiegel, 'Psychiatric Observations in the Tunisian Campaign', *American Journal of Orthopsychiatry*, xiv (1944), p. 310. Also see Spiegel, 'Psychiatry in an Infantry Battalion in North Africa', in Lieutenant-General Hal B. Jennings (ed.), *Neuropsychiatry in World War II. Volume II. Overseas Theatres* (Washington, DC, 1973), p. 115.

6 Roy R. Grinker and John P. Spiegel, *Men Under Stress* (London, 1945), p. 45. Also see Colonel S. L. A. Marshall, *Men Against Fire. The Problem of Battle Command in Future War* (New York, 1947), p. 42 and Major Edwin A. Weinstein and Lieutenant-Colonel Calvin S. Drayer, 'A Dynamic Approach to the Problem of Combat-Induced Anxiety', *The Bulletin of the U.S. Army Medical Department*, ix, supplemental number (November 1949), p. 24.

7 Richard C. Foot, 'Once a Gunner', p. 125, IWM; Jack W. Mudd, 'Letters', 22 October 1917, IWM; 'Barnacle' [pseud. Revd Waldo Edward Lovell Smith], *Talks to the Troops* (Toronto, 1944), p. 13; Jonathan Shay, *Achilles in Vietnam. Combat Trauma and the Undoing of Character* (New York, 1994), p. 40.

8 Allen Hunter, interviewed in Glen D. Edwards, *Vietnam. The War Within* (Salisbury, South Australia, 1992), p. 18. See a similar statement by George Ryan, in Murray Polner, *No Victory Parades. The Return of the Vietnam Veteran* (London, 1971), p. 39.

9 William F. Crandell, 'What did America Learn from the Winter Soldier Investigation?', *Vietnam Generation Journal. Nobody Gets Off the Bus*, 5.1–4 (March 1994), p. 3, on the Internet.

10 Letter from 'Jack', 23 November 1917, in Miss D. Williams, 'Letters', IWM.

11 Jack Strahan, 'The Machine Gunner', in Jack Strahan, Peter Hollenbeck, and R. L. Barth (eds), *Vietnam Literature Anthology: A Balanced Perspective* (Philadelphia, 1985), p. 30.

12 'A Few Hints to All Ranks', in 'T.F.N.', *A Few Helpful Hints on Drill by the Adjutant, 1st B.N.V.R.* (Norwich, 1918), p. 19.

13 Captain C. W. Blackall, *Songs from the Trenches* (London, 1915), p. 16.

14 'Some Tips for Gunners', in A. Douglas Thorburn, *Amateur Gunners. The Adventures of an Amateur Soldier in France, Salonica and Palestine in The Royal Field Artillery* (Liverpool, 1933), pp. 152–3.

15 Robert William MacKenna, *Through a Tent Door*, first published in 1919 (London, 1930), p. 87 and Richard Tregaskis, *Guadalcanal Diary* (New York, 1943), p. 15.

16 Ron Kovic, *Born on the Fourth of July*, first published in 1976 (Aylesbury, 1990), p. 69; Stanley Kubrick, *Full Metal Jacket*, 1987, film; Eugene B.

Sledge, *With the Old Breed at Peleliu and Okinawa* (New York, 1990), p. 10.

17 Colonel Harry Summers, quoted in Richard A. Gabriel, *The Painful Field: The Psychiatric Dimension of Modern War* (New York, 1988), p. 162.

18 For instance, see George P. Hunt, *Coral Comes High* (New York, 1946), p. 30.

19 For instance, Cecil H. Cox, 'A Few Experiences of the First World War', p. 2, IWM.

20 Guy Chapman, *A Passionate Prodigality. Fragments of an Autobiography*, first published in 1933 (New York, 1966), p. 131 and Lieutenant-Colonel W. L. Raws, 'Home Training', *The Australian Military Journal*, 7 (January 1916), 248. Also see Lieutenant-Colonel J. W. B. Merewether and Sir Frederick Smith, *The Indian Corps in France* (London, 1917), p. 471.

21 'The O.C.' [Major L. Bird], 'Birth of a Battery', in *Tackle 'Em Low*, 1.1 (December 1939), p. 6, and Norman Copeland, *Psychology and the Soldier* (London, 1942), pp. 75–6. Also see Colonel William Line, 'Morale and Leadership', *Canadian Army Journal*, 6.1 (April 1952), p. 251.

22 Norman Copeland, *Psychology and the Soldier* (London, 1942), pp. 75–6.

23 Unnamed Vietnam veteran, quoted by Jonathan Shay, *Achilles in Vietnam. Combat Trauma and the Undoing of Character* (New York, 1994), p. 49.

24 Roy R. Grinker and John P. Spiegel, *Men Under Stress* (London, 1945), pp. 39–40, 46–8 and 122–6.

25 Unnamed American soldier, cited in Alfred O. Ludwig, 'Neuroses Occurring in Soldiers After Prolonged Combat Exposure', *Bulletin of the Menninger Clinic*, 11.1 (January 1947), p. 18.

26 Abel Chapman, *On Safari. Big-Game Hunting in British East Africa With Studies in Bird Life* (London, 1908), p. 302; W. R. D. Fairbairn, 'Arms and the Child', in Fairbairn, *From Instinct to Self: Selected Papers of W. R. D. Fairbairn. Volume II: Applications and Early Contributions* (Northvale, New Jersey, 1994), p. 329, first published in *The Liverpool Quarterly*, 5.1 (January 1937); Eli Sagan, *Cannibalism. Human Aggression and Cultural Form* (New York, 1974), pp. 68–9 and 80; Oscar Wilde, *The Ballad of Reading Gaol* (London, 1896). Also see R. E. Money-Kyrle, 'The Development of War. A Psychological Approach', *British Journal of Medical Psychology*, xvi.3 (1937), pp. 230–1.

27 For a discussion, see Chapter 9 'Priests and Padres'.

28 Unnamed American officer's diary, 18 September, no year, in Amy Gordon Grant, *Letters from Armageddon. A Collection Made During the World War* (Boston, 1930), p. 196.

29 Lieutenant-Colonel James Young, *With the 52nd (Lowland) Division in*

Three Continents (Edinburgh, 1920), p. 11, diary entry for 13 July 1915.

30 Herbert W. McBride, *A Rifleman Went to War* (Marines, North Carolina, 1935), p. 209. Also see Colonel Robert J. Blackam, *Scalpel, Sword and Stretcher. Forty Years of Work and Play* (London, 1931), p. 255.

31 Hector Bolitho, *Combat Report. The Story of a Fighter Pilot* (London, 1943), p. 47 and Coningsby Dawson, *The Glory of the Trenches* (London, 1918), p. 13.

32 W. R. D. Fairbairn, 'Is Aggression an Irreducible Factor?', *British Journal of Medical Psychology*, xviii.2 (1940), pp. 166–7.

33 Henry Gother Courtney, 'Letters', letter to his sister from Salonica, 15 October 1916, letter no. 8/139, Birmingham University Library.

34 Cecil Lewis, *Sagittarius Rising* (London, 1936), pp. 45–6.

35 Private Anthony Brennan, 'Diaries', entry for 1 July 1916, p. 10, IWM; letter from Captain Duncan, quoted in *Letters from the Front. Being a Record of the Part Played by Officers of the [Canadian] Bank [of Commerce] in the Great War 1914–1919*, vol. 1 (Toronto, 1920), p. 16; 'G. G. A.', 'The Bayonet for Mounted Riflemen', *The Army Review*, v.1 (July 1913), pp. 86–7; Ion L. Idriess, *The Desert Column. Leaves from the Diary of an Australian Trooper in Gallipoli, Sinai, and Palestine* (Sydney, 1932), p. 57, diary entry for September 1915; Sid T. Kemp, 'Remembrance. The 6th Royal West Kent Regiment 1914–1916', p. 31, IWM; Francis Law, *A Man at Arms. Memoirs of Two World Wars* (London, 1983), p. 58.

36 For other comments, see Captain H. Meredith Logan, 'Military Training To-Day', *The Army Quarterly*, vi (April 1923), p. 72 and Dixon Wecter, *When Johnny Comes Marching Home* (Cambridge, Massachusetts, 1944), p. 488.

37 For the best discussion of fraternization, see Tony Ashworth, *Trench Warfare 1914–1918. The Live and Let Live System* (London, 1980), pp. 24–47. For a contemporary history of fraternization, see Stephen Stapleton, 'The Relations Between the Trenches', *The Contemporary Review*, cvi (January–June 1917), pp. 636–44.

38 Gerald V. Dennis, 'A Kitchener Man's Bit (1916–1918)', 1928, p. 129, IWM. For examples of fraternization, also see Nobby Clarke, 'Sniper on the Western Front', p. 10, IWM and J. Davey, 'Letter to the Curator of the Imperial War Museum', 20 July 1969, IWM.

39 For instance, see Edwyn E. H. Bate, 'Memoirs – Vol. III. 1914–1918', 23–4, IWM.

40 Clifford Nixon, 'A Touch of Memory', p. 83, letter from Ron to his father, from France, 25 August 1917, IWM. Also see Revd M. A. Bere, 'Papers', letter dated 17 September 1917, IWM.

41 Cecil H. Cox, 'A Few Experiences of the First World War', p. 2, IWM.

42 W. Stekel, *Sadism and Masochism: The Psychology of Hatred and Cruelty*, vol. 1 (New York, 1953), pp. 12 and 31–2.

43 Derek Freeman, 'Human Aggression in Anthropological Perspective', in J. D. Carthy and F. J. Ebling (eds), *The Natural History of Aggression* (London, 1964), p. 113.

44 Ralph R. Greenson, 'The Fascination of Violence', in Greenson, *On Loving, Hating, and Living Well* (Madison, 1992), pp. 187–99, lecture given in 1968.

45 Adrian Caesar, *Taking it Like a Man. Suffering, Sexuality and the War Poets* (Manchester, 1993), p. 234.

46 Daniel Ford, *Incident at Muc Wa* (London, 1967), p. 146.

47 John Carroll, *Token Soldiers* (Boronia, Victoria, 1983), pp. 215–16.

48 Sydney Duffield and Andrew G. Elliott, *Rough Stuff. For Home Guards and Members of H.M. Forces* (London, 1942), pp. 9, 11 and 42.

49 Philip Caputo, *A Rumor of War* (New York, 1977), p. 254.

50 Anthony S. Irwin, *Infantry Officer. A Personal Record* (London, 1943), pp. 64–5. Irwin provided no explanation for his need of the tin helmet to sleep. Being unable to sleep unless their steel helmet was placed over their genitals was a common reaction to combat according to 'Summary of Lectures on Psychological Aspects of War . . . Parts of the Body with a Morale Significance', p. 14, in PRO CAB 21/914 (annex).

51 The definition is from George Allport's classic, *The Nature of Prejudice* (Reading, Massachusetts, 1979), p. 363, first published in 1954.

52 Robert W. Rieber (ed.), *The Psychology of War and Peace. The Image of the Enemy* (New York, 1991). A rare instance where this is questioned can be found in John A. Ballard and Alliecia J. McDowell, 'Hate and Combat Behavior', *Armed Forces and Society*, 17.2 (winter 1991), pp. 229–41.

53 Professor H. J. Laski, *The Germans – Are They Human? A Reply to Sir Robert Vansittart* (London, 1941), p. 2 and Lewis F. Richardson, 'The Persistence of National Hatred and the Changeability of Objects', *British Journal of Medical Psychology*, xxii.3 (1949), p. 167.

54 Revd E. L. Allen, *Pacifism as an Individual Duty* (London, 1946), p. 21.

55 For the need to teach men to hate, see Major Bertie A. Pond, 'Memoirs', np (p. 20), IWM.

56 For the clearest description of this, see Donald E. Core, 'The "Instinct-Distortion" or "War-Neurosis"', *The Lancet*, 10 August 1918, p. 168.

57 William Ernest Hocking, *Morale and Its Enemies* (New Haven, 1918), pp. 161–2; Major Raymond Sobel, 'Anxiety-Depressive Reactions After Prolonged Combat Experience – the "Old Sergeant" Syndrome', *The*

Bulletin of the U.S. Army Medical Department, ix, supplemental number (November 1949), pp. 143–4; Captain Martin Stein, 'Neurosis and Group Motivation', in *The Bulletin of the U.S. Army Medical Department*, vii.3 (March 1947), p. 317; War Office, *Psychiatric Disorders in Battle, 1951* (London, 1951), p. 7; Major Edwin A. Weinstein and Lieutenant-Colonel Calvin S. Drayer, 'A Dynamic Approach to the Problem of Combat-Induced Anxiety', in *The Bulletin of the U.S. Army Medical Department*, ix, supplemental number (November 1949), p. 24.

58 Gregory Zilboorg, 'Fear of Death', *Psychoanalytic Quarterly*, 12 (1943), p. 474.

59 Leonard R. Sillman, 'A Psychiatric Contribution to the Problem of Morale', *Journal of Nervous Mental Disorder*, 97 (1943), pp. 283–95 and 'Morale', *War Medicine*, 3.5 (May 1943), pp. 498–502. Also see C. H. Rogerson, 'Letter to Dr. D. Ewen Cameron of Albany, 18 July 1940', *American Journal of Psychiatry*, 97.2 (1941), p. 970.

60 John J. Marren, 'Psychiatric Problems in Troops in Korea During and Following Combat', *U.S. Armed Forces Medical Journal*, vii.5 (May 1956), p. 721.

61 Herbert X. Spiegel, 'Psychiatry in an Infantry Battalion in North Africa', in Lieutenant-General Hal B. Jennings (ed.), *Neuropsychiatry in World War II. Volume II. Overseas Theatres* (Washington, DC,1973), p. 121.

62 For a First World War discussion, see Captain H. Meredith Logan, 'Military Training To-Day', *The Army Quarterly*, vi (April 1923), p. 72.

63 Humphry Beevor, *Peace and Pacifism* (London, 1938), pp. 53–4.

64 'Realism in Army Training. The Spirit of Hate', undated newspaper clipping, in PRO WO199/799; 'Realism in Training', *The Times*, 27 April 1942, p. 2; 'The New Battle Drill', *The Times*, 25 November 1941, p. 5; J. W. Bellah and A. F. Clark, 'The Lunk Trainer', *Infantry Journal*, 52.3 (1943), pp. 72–5; 'Object of Battle Inoculation', undated but after 1 May 1942, p. 3, PRO WO199/799; John Rawlings Rees, *The Shaping of Psychiatry by War* (London, 1945), p. 80.

65 As reported in '"Hate" in Army Training', *The Times*, 25 May 1942, p. 2.

66 Mr. R. J. Davies of Westhoughton (Labour Minister), in the House of Commons, as reported in *The Times*, 14 May 1942, p. 8.

67 '"Hate" in Army Training', *The Times*, 25 May 1942, p. 2, quoting from the letter from General Sir Bernard Paget, Commander-in-Chief, Home Forces, to all army commanders; debate in the House of Commons, question asked by Mr Stokes of Ipswich (Labour Minister), as reported in *The Times*, 14 May 1942, p. 8, re. 13 May; 'Realism in Army Training. The Spirit of Hate', undated newspaper clipping, in PRO WO199/799; B. H.

McNeel, 'War Psychiatry in Retrospect', *American Journal of Psychiatry*, 102 (1945–6), p. 503.

68 Thomas Howard, 'Hate in Battle', *The Times*, 1 May 1942.

69 'Battle Inoculation', 1944, p. 2, in PRO CAB 21/914. Also see Ernest Jones, 'Psychology and War Conditions', *The Psychoanalytic Quarterly*, 14 (1945), p. 15.

70 John Rawlings Rees, *The Shaping of Psychiatry by War* (London, 1945), pp. 80–1.

71 Undated and unsigned report in PRO WO199/799.

72 Letter from A. I. M. Wilson (of the War Office) to Lieutenant-Colonel Briton ('G' Training, Home Forces), May 1942, in PRO WO199/799.

73 R. H. Ahrenfeldt, 'Military Psychiatry', in Sir Arthur Salisbury MacNalty and W. Franklin Mellor (eds), *Medical Services in the War. The Principal Medical Lessons of the Second World War Based on the Official Medical Histories of the United Kingdom, Canada, Australia, New Zealand, and India* (London, 1968), pp. 189–90.

74 Ibid., p. 190.

75 Directorate of Army Psychiatry, Technical Memorandum No. 2, 'Suppose You Were a Nazi Agent – or Fifth Column Work for Amateurs', June 1942, p. 7, PRO CAB21/914.

76 See Chapters 6 and 7.

77 Eugene B. Sledge, *With the Old Breed at Peleliu and Okinawa* (New York, 1990), pp. 33–4.

78 Samuel A. Stouffer et. al., *The American Soldier: Combat and Its Aftermath. Volume II* (Princeton, 1949), p. 158, based on a survey of 5,558 enlisted American infantrymen, 1943–4.

79 'Sniping', nd [WWII], p. 1, AWM.

80 Lieutenant Uzal Ent, interviewed in Rudy Tomedi, *No Bugles, No Drums. An Oral History of the Korean War* (New York, 1993), p. 18.

81 Philip Caputo, *A Rumor of War* (New York, 1977), p. 231.

82 Lieutenant Frank Warren, 'Journals and Letters', letter to 'Bun' on 20 October 1917, IWM.

83 Lieutenant-Colonel Kenneth H. Cousland, 'The Great War 1914–1918. A Former Gunner Looks Back', p. 61, Liddell Hart Centre and Second Lieutenant Alfred Richard Williams, 'Letters from the West [sic] Front', 16 October 1916, p. 90, IWM.

84 Gary McKay, *In Good Company. One Man's War in Vietnam* (Sydney, 1987), p. 44.

85 Diary of Bruce Anello, 19 February 1968, in Indochina Curriculum Group, *Front Lines. Soldiers' Writings from Vietnam* (Cambridge,

Massachusetts, 1975), p. 8.

86	Both cited in Georgie Anne Geyer, 'Our New GI: He Asks Why', *Chicago Daily News*, 16 January 1969, in Virginia Elwood-Akers, *Women War Correspondents in the Vietnam War, 1961–1975* (Metuchen, New Jersey, 1988), pp. 174–5.

87	Harold Strickland Constable, *Something About Horses, Sport and War* (London, 1891), pp. 167–8.

88	D. W. Hastings, D. G. Wright, and B. C. Glueck, *Psychiatric Experiences of the Eighth Air Force* (New York, 1944), pp. 137–9.

89	John Helmer, *Bringing the War Home. The American Soldier in Vietnam and After* (New York, 1974), p. 169. Seven per cent did not reply. There were major political differences in men's responses: veterans who eventually joined the Vietnam Veterans Against the War movement were most likely to say that combat led them to respect the enemy (93 per cent gave this response compared with only 13 per cent of veterans who belonged to the more conservative organisation, Veterans of Foreign Wars).

90	Samuel A. Stouffer et al., *The American Soldier: Combat and Its Aftermath. Volume II* (Princeton, 1949), pp. 158 and 164–5. The first sample involved 4,064 enlisted infantrymen in the Pacific, 591 in the Mediterranean theatre, and 1,766 in Italy. The second sample involved 4,064 enlisted infantrymen in the Pacific, 1,022 in Europe, and 472 in the United States.

91	For instance, see the letter dated 18 July 1915, in Mark Servern, *The Gambordier. Giving Some Account of the Heavy and Siege Artillery in France 1914–1918* (London, 1930), p. 63.

92	G. Stanley Hall, 'Morale in War and After', *The Psychological Bulletin*, 15 (1918), p. 282. Also see Charles Bird, 'From Home to the Charge: A Psychological Study of the Soldier', *The American Journal of Psychology*, xxviii.3 (July 1917), p. 339 and G. Stanley Hall, *Morale. The Supreme Standard of Life and Conduct* (New York, 1920), p. 66.

93	Stewart H. Holbrook, *None More Courageous: American War Heroes of Today* (New York, 1942), p. 90.

94	Huntly Gordon, *The Unreturning Army. A Field-Gunner in Flanders, 1917–18* (London, 1967), p. 74.

95	Norman Copeland, *Psychology and the Soldier* (London, 1942), p. 58.

96	Lieutenant-Colonel Kenneth H. Cousland, 'The Great War 1914–1918. A Former Gunner Looks Back', nd, 61, Liddell Hart Centre. For an example from the Second World War, see Mary Penick Motley (ed.), *The Invisible Soldier. The Experience of the Black Soldier, World War II* (Detroit, 1975), p. 179.

97	A. G. J. Whitehouse, *Hell in Heavens. The Adventures of an Aerial Gunner in the Royal Flying Corps* (London, 1938), p. 43.

98 Oliver Elton, *C. E. Montague. A Memoir* (London, 1929), p. 197; G. C. Field, *Pacifism and Conscientious Objection* (Cambridge, 1945), p. 67; Frederic Hillersdon Keeling, *Keeling Letters and Recollections* (London, 1918), p. 285.

99 James Molony Spaight, *Air Power and War Rights* (London, 1924), p. 319; *The Weekly Dispatch*, 10 September 1916; *Daily Mail*, 4 and 7 May 1918.

100 R. H. Kiernan, *Captain Albert Ball* (London, 1933), p. 60.

101 J. H. Early, 'War Diary 1914–1918', letter for 4 May 1915, Cambridgeshire County Record Office B1/HF/J/2.

102 Cited in Stephen A. Garrett, *Ethics and Airpower in World War II. The British Bombing of German Cities* (New York, 1993), p. 95.

103 Daniel Ford, *Incident at Muc Wa* (London, 1967), p. 23.

104 Maurice Wright, 'Effect of War on Civilian Populations', *The Lancet*, 28 January 1939, pp. 189–90. Clifford Allen in 'Emotional Changes of War-Time', *The Lancet*, 11 May 1940, p. 901 thought that there was *less* sadism in the Second World War compared with the 1914–18 conflict and identified older men (and those behind the lines) as the ones most liable to be sadistic towards the enemy.

105 J. Glenn Gray, *The Warriors: Reflections on Men in Battle* (New York, 1959), p. 135 and Frederic Hillersdon Keeling, *Keeling Letters and Recollections* (London, 1918), pp. 259–60, letter to R. C. K. Ensor, 23 December 1915.

106 The Regular armed forces were anxious to limit the role of the Home Guard, particularly to prevent them from turning themselves into blood-thirsty guerrillas: see 'Lecture Given at Home Guard Company Commander's Course, Aldeburgh, 29 Aug. 42. Operational Role of the Home Guard', p. 4, PRO WO199/2487 and Major G. E. O. Walker, 'Guerilla Warfare and the Home Guard', 30 April 1942, in PRO WO199/387.

107 Tom Wintringham, *The Home Guard Can Fight. A Summary of Lectures Given at the Osterley Park Training School for Home Guard* (London, 1941), pp. 3–4 and 27.

108 Commander-in-Chief, Home Forces, 'Home Guard. Instruction No. 41–1942. Misc. Notes', p. 3, PRO WO199/872B.

109 Norman Demuth, *Harrying the Hun. A Handbook of Scouting, Stalking and Camouflage* (London, 1941), pp. 64 and 84.

110 Major M. D. S. Armour, *Total War Training for Home Guard Officers and N.C.O.s* (London, 1942), p. 46.

111 Lieutenant E. Hartley Leather, *Combat Without Weapons* (Aldershot, 1942), p. 7.

112 Sydney Duffield and Andrew G. Elliott, *Rough Stuff. For Home Guards and Members of H.M. Forces* (London, 1942), pp. 9, 11 and 42.

113 Andrew G. Elliott, 'J. B.', and 'Scientist', *The Home Guard Encyclopedia* (London, 1942), pp. 28 and 65.

114 Mark A. May, *A Social Psychology of War and Peace* (New Haven, Connecticut, 1943), p. 66. Also see Felix Deutch, 'Civilian War Neuroses and Their Treatment', *The Psychoanalytic Quarterly*, 13 (1944), p. 303.

115 Gregory Zilboorg, 'Fear of Death', *The Psychoanalytic Quarterly*, 12 (1943), p. 472.

116 F. M. Packham, 'Memoirs of an Old Contemptible 1912–1920', p. 3, IWM.

117 Robert Graves, *Goodbye to All That*, first published 1929 (London, 1977), p. 162.

118 Hiram Sturdy, 'Illustrated Account of His Service on the Western Front With the Royal Regiment of Artillery', pp. 51–2, IWM.

119 Sidney Rogerson, *Twelve Days*, first published in 1930 (Norwich, 1988), pp. 124–5.

120 Edward Glover, *The Psychology of Fear and Courage* (Harmondsworth, 1940), p. 84; Master Sergeant Floyd Jones, interviewed in Mary Penick Motley (ed.), *The Invisible Soldier. The Experience of the Black Soldier, World War II* (Detroit, 1975), p. 177; Ernie Pyle, *Here is Your War* (New York, 1943), pp. 241–2; Paul Richey, *Fighter Pilot. A Personal Record of the Campaign in France* (London, 1944), p. 71.

121 John J. Floherty, *The Courage and the Glory* (Philadelphia, 1942).

122 Sergeant Ehlers, quoted by Merle Miller, 'Introduction', in Don Congdon (ed.), *Combat: War in the Pacific* (London, 1958), p. 9. He had won the Congressional Medal of Honor, a Silver Star, and a Bronze Star with clusters.

123 Paul Thomas, interviewed in Glen D. Edwards, *Vietnam. The War Within* (Salisbury, South Australia, 1992), p. 67.

124 Trooper William Clarke, 'Random Recollections of "14/18"', p. 6, IWM. Similar statements were also made by Frederick Hunt, 'And Truly Serve', 1980, p. 25, IWM and Revd T. W. Pym and Revd Geoffrey Gordon, *Papers from Picardy by Two Chaplains* (London, 1917), p. 28.

125 Frederic Hillersdon Keeling, *Keeling Letters and Recollections* (London, 1918), pp. 253–4, letter to Mrs. Green, 13 November 1915.

126 Hamer, 'Peace Terms', *Aussie*, 6, August 1918, p. 11.

127 Samuel A. Stouffer et al., *The American Soldier: Combat and Its*

Aftermath. Volume II (Princeton, 1949), p. 161. In contrast, seeing Japanese prisoners made 42 per cent of American combatants feel *more* like killing. This was probably due to the particularly brutal nature of that war, where it seemed almost impossible to contact the enemy.

128 W. N. Maxwell, *A Psychological Retrospect of the Great War* (London, 1923), p. 84. Also see Stanley Diamond, 'War and Disassociated Personality', in Morton Fried, Marvin Harris, and Robert Murphy (eds), *War. The Anthropology of Armed Conflict and Aggression* (New York, 1968), p. 187.

129 *Psychology for the Fighting Man*, 2nd edition (Washington, DC,1944), p. 267.

130 Anthony F. C. Wallace, 'Psychological Preparations for War', in Morton Fried, Marvin Harris, and Robert Murphy (eds), *War. The Anthropology of Armed Conflict and Aggression* (New York, 1968), p. 178.

131 G. Stanley Hall, 'Morale in War and After', *The Psychological Bulletin*, 15 (1918), pp. 382–3. For explicit discussion of the experiences of killing and the absence of hatred, see Edward Frederick Chapman, 'Letters from France', letter to his sister Hilda, 2 November 1916, IWM; Trooper William Clarke, 'Random Recollections of "14/18" ', 16, IWM; letters by Sergeant-Major Frederic Hillersdon Keeling on 23 December 1915 and Captain Arthur Graeme West on 12 February 1916, both quoted in Laurence Housman (ed.), *War Letters of Fallen Englishmen* (London, 1930), pp. 161 and 290; Private Lionel Leslie Heming, 'When Down Came a Dragoon', p. 112, AWM; draft article by B. L. Lawrence, 'Sense or Sentiment', pp. 4–5, in C. K. Ogden, 'Papers', IWM; Second Lieutenant Alfred Richard Williams, 'Letters from the West Front', p. 90, letter dated 16 October 1916, IWM.

132 G. W. Colebrook, 'Letters Home', 28 May 1915, IWM and Ulysses Grant Lee, *United States Army in World War II. Special Studies. The Employment of Negro Troops* (Washington, DC, 1966), p. 698.

133 Captain H. Meredith Logan, 'Military Training To-Day', *The Army Quarterly*, vi (April 1923), p. 72.

134 Ray R. Grinkler and John P. Spiegel, *Men Under Stress* (London, 1945), p. 43.

135 Ray R. Grinkler and John P. Spiegel, *Men Under Stress* (London, 1945), p. 43. Also see E. F. M. Durbin and John Bowlby, *Personal Aggressiveness and War* (London, 1939), p. 74.

136 Major Edwin Weinstein, 'The Fifth U.S. Army Neuropsychiatric Centre – "601st"', in Lieutenant-General Hal B. Jennings (ed.), *Neuropsychiatry in World War II. Volume II. Overseas Theatres* (Washington,

DC, 1973) and Major Raymond Sobel, 'Anxiety-Depressive Reactions After Prolonged Combat Experience – the "Old Sergeant" Syndrome', in *The Bulletin of the U. S. Army Medical Department*, ix, supplemental number (November 1949), pp. 143–4.

137 Revd Harry W. Blackburne, *This Also Happened On the Western Front. The Padre's Story* (London, 1932), pp. 108–9.

138 Edward Glover, *The Psychology of Fear and Courage* (Harmondsworth, 1940).

139 Therese Benedek, *Insight and Personality Adjustment. A Study of the Psychological Effects of War* (New York, 1946), p. 54.

140 William Ernest Hocking, *Morale and Its Enemies* (New Haven, Connecticut, 1918), pp. 25–6.

141 Lord Moran, *The Anatomy of Courage* (London, 1945), pp. 56–7.

142 Revd T. W. Pym and Revd Geoffrey Gordon, *Papers from Picardy by Two Chaplains* (London, 1917), pp. 23–4.

143 Captain Herbert W. McBride, *A Rifleman Went to War* (Marines, North Carolina, 1935), p. 105 and Robert William MacKenna, *Through a Tent Door*, first published in 1919 (London, 1930), p. 108.

144 Revd Harry W. Blackburne, *This Also Happened On the Western Front. The Padre's Story* (London, 1932), pp. 108–9 and G. Stanley Hall, *Morale. The Supreme Standard of Life and Conduct* (New York, 1920), pp. 67–8.

145 Captain Herbert X. Spiegel, 'Psychiatric Observations in the Tunisian Campaign', *American Journal of Orthopsychiatry*, xiv (1944), p. 382 and Herbert X. Spiegel, 'Psychiatry in an Infantry Battalion in North Africa', in Lieutenant-General Hal B. Jennings (ed.), *Neuropsychiatry in World War II. Volume II. Overseas Theatres* (Washington, DC,1973), p. 122.

6 War Crimes

1 Walter MacDonald, 'Interview With a Guy Named Fawkes, U.S. Army', in William D. Ehrhart (ed.), *Unaccustomed Mercy. Soldier-Poets of the Vietnam War* (Lubbock, Texas, 1989) p. 102.

2 Lieutenant William L. Calley, *Body Count* (London, 1971), pp. 108–9.

3 Joseph Goldstein, Burke Marshall, and Jack Schwartz (eds), *The Peers Commission Report* (New York, 1976), pp. 89–90.

4 Lieutenant William L. Calley, *Body Count* (London, 1971), pp. 8 and 84; Michael Bilton and Kevin Sims, *Four Hours in My Lai. A War Crime and its Aftermath* (London, 1992); Joseph Goldstein, Burke Marshall, and Jack Schwartz (eds), *The Peers Commission Report* (New York, 1976); Richard Hammer, *The Court-Martial of Lt. Calley* (New York, 1971), p. 161; Joseph Lelyveld, 'The Story of a Soldier Who Refused to Fire at

Songmy', *The New York Times Magazine*, 14 December 1969, p. 116; Lieutenant-General William R. Peers, *The My Lai Inquiry* (New York, 1979); 'Who is Responsible for My Lai', *Time*, 8 March 1971, p. 19.

5 For instance, Martin Gershen, *Destroy or Die: The True Story of My Lai* (New York, 1971).

6 The most helpful include Geoffrey Best, *Humanity in Warfare. The Modern History of the International Law of Armed Conflict* (London, 1980); Geoffrey Best, *War and Law Since 1945* (Oxford, 1995); Richard L. Holmes, *On War and Morality* (Princeton, 1989); Michael Walzer, *Just and Unjust Wars* (New York, 1977). The Vietnam War was problematical because war had not been formally declared between sovereign states and the conflict did not comprise two or more identifiable military forces. From 1965, however, the US government adjudged the hostilities to be an armed international conflict with the North Vietnam Army as the enemy and the Viet Cong as agents of the belligerent government of North Vietnam. Therefore the laws were binding.

7 Leon Friedman (ed.), *The Law of War. A Documentary History*, vols 1 and 2 (New York, 1972), pp. 225, 229, 314, 318 and 526 and 'The Nuremberg Principles', in Richard A. Falk, Gabriel Kolko, and Robert Jay Lifton (eds), *Crimes of War* (New York, 1971), p. 107.

8 Major-General George S. Prugh, *Law at War: Vietnam 1964–1973* (Washington, DC, 1975), p. 76.

9 For an exhaustive review, see Major William H. Parks, 'Command Responsibility for War Crimes', *Military Law Review*, 62 (1973), pp. 1–104.

10 Lassa Oppenheim, *International Law. A Treatise*, vol. II, p. 264 and British *Manual of Military Law*, 1914 and 1944, ch. xiv, para. 433.

11 The best discussion of these shifts can be found in Guenter Lewy, 'Superior Orders, Nuclear Warfare, and the Dictates of Conscience', in Richard A. Wasserstrom (ed.), *War and Morality* (Belmont, California, 1970), pp. 115–34.

12 *The U.S. Army Field Manual*, 'The Law of Land Warfare', 1956, para. 509.

13 US *Manual for Courts Martial* (1969), para. 197.

14 First and fourth principles, 'The Nuremberg Principles', in Richard A. Falk, Gabriel Kolko, and Robert Jay Lifton (eds), *Crimes of War* (New York, 1971), p. 107.

15 For instance, see the interview with Captain Joseph B. Anderson, in Wallace Terry, *Bloods. An Oral History of the Vietnam War by Black Veterans* (New York, 1984), p. 233; Kenneth Maddock, 'Going Over the

Limit? – The Question of Australian Atrocities', in Maddock (ed.), *Memories of Vietnam* (Sydney, 1991), pp. 151–63; George L. Mosse, *Fallen Soldiers. Reshaping the Memory of the World Wars* (New York, 1990), p. 110.

16 Rifleman Barry Kavanagh, quoted by Sally Wilkins, 'A Glance Behind', *The Age*, 10 May 1975, p. 11. For more about atrocities committed by Australians, see Pat Burgess, 'The Village of Hidden Hate', *The Age*, 21 August 1976, p. 16; Ian McKay, 'Truths and Untruths of War', *The Advertiser*, 5 August 1976, p. 5; Alan Ramsey, 'How to Stop Worrying and Love the VC', *Nation Review*, 6–12 August 1976, p. 1046.

17 Australian warrant officer, cited in Gerald L. Stone, *War Without Heroes* (Melbourne, 1966), p. 131. For other Australian examples, see Alex Carey, *Australian Atrocities in Vietnam* (Sydney, 1968).

18 Arthur E. 'Gene' Woodley, Jr., in Wallace Terry, *Bloods. An Oral History of the Vietnam War by Black Veterans* (New York, 1984), p. 261.

19 Craig Weeden, 'My Lie', in Jan Barry and W. D. Ehrhart (eds), *Demilitarized Zones. Veterans After Vietnam* (Perkasie, Pa., 1976), p. 24.

20 For instance, Bill Adler (ed.), *Letters from Vietnam* (New York, 1967), p. 42; Mark Baker, *Nam. The Vietnam War in the Words of the Men and Women Who Fought There* (London, 1982), p. 50; Philip Scribner Balboni, 'Mylai was Not an Isolated Incident. What Every Vietnam Veteran Knows', *The New Republic*, 19 December 1970, pp. 13–15; Richard Boyle, *The Flower of the Dragon. The Breakdown of the U.S. Army in Vietnam* (San Francisco, 1972), pp. 137–9; Donald Duncan, *A 'Green Beret' Blasts the War* (London, 1966), p. 9; Mark Lane, *Conversations with Americans* (New York, 1970); Eric Norden, *American Atrocities in Vietnam* (Sydney, 1966); 'The Case of the Green Berets', *Newsweek*, 25 August 1969, pp. 26–33.

21 Lieutenant William Crandell of Americal Division, in 'Vietnam Veterans Against the War', *The Winter Soldier Investigation. An Inquiry into American War Crimes* (Boston, 1972), pp. 2–3. Also see Herbert C. Kelman, 'War Criminals and War Resisters', *Society*, 12.4 (May–June 1975), pp. 18–19; 'The Interrogation of Captain Howard Turner at the Trial of Lieutenant James Duffer, 1970', in Richard A. Falk, Gabriel Kolko, and Robert Jay Lifton (eds), *Crimes of War* (New York, 1971), pp. 246–7 and the letter from Captain H. Miller (company commander, 25th US Army Infantry Division) to his parents on the 17 February 1970, reproduced in Richard A. Falk, Gabriel Kolko, and Robert Jay Lifton (eds), *Crimes of War* (New York, 1971), pp. 395–6.

22 Studies discussed in Richard Strayer and Lewis Ellenhorn, 'Vietnam

Veterans: A Study Exploring Adjustment Patterns and Attitudes', *Journal of Social Issues*, 31.4 (1975), p. 90.

23 Respondents were asked their opinion prior to the My Lai atrocity: Douglas Kinnard, 'Vietnam Reconsidered: An Attitudinal Survey of U.S. Army General Officers', *The Public Opinion Quarterly*, xxxix.4 (1975–6), p. 451. Sixty-one per cent said the rules were 'fairly well adhered to'.

24 John Helmer, *Bringing the War Home. The American Soldier in Vietnam and After* (New York, 1974), pp. 202–3.

25 This point is convincingly argued by Cornelius A. Cronin, 'Line of Departure. The Atrocity in Vietnam War Literature', in Philip K. Jason (ed.), *Fourteen Landing Zones. Approaches to Vietnam War Literature* (Iowa City, 1991), pp. 208–15.

26 See Eric Carlton, *Massacres. An Historical Perspective* (Aldershot, 1994), pp. 137–45. The International War Crimes Tribunal found the United States guilty of war crimes for the intentional bombing of civilian targets with no justified military objective (thus violating the 1923 Convention of the Hague, the 1949 Geneva Convention, and the findings of the Nuremberg Trials): John Duffett, *Against the Crime of Silence. Proceedings of the Russell International War Crimes Tribunal* (New York, 1968) and Gabriel Kolko, 'War Crimes and the Nature of the Vietnam War', *Journal of Contemporary Asia*, 1.1 (autumn 1970), pp. 5–14.

27 General Bruce Palmer, *The 25-Year War. America's Military Role in Vietnam* (New York, 1984), p. 85. He was attempting to portray the My Lai atrocity as an aberration, but his statement can be read in the opposite way (as I have done here).

28 Stephen Graham, *A Private in the Guards* (London, 1919), p. 217.

29 Lieutenant S. H. Monard, 'Fuel and Ashes', *The Incinerator*, 1.2 (June 1916), p. 18. Advice given by Major Campbell.

30 Captain Guy Warneford Nightingale, 'The 1915 Letters', p. 8, letter to his sister, 4 May 1915, IWM. For other examples, see Ralph Coburn Carson, 'Recollections of the Front', in James Carson (compiler), *The Carsons of Monanton, Ballybay, Co. Monaghan, Ireland* (Lisburn, 1931), p. 46; Thomas P. Dooley, *Irishmen or English Soldiers? The Times and World of a Southern Catholic Irish Man (1876-1916) Enlisted with the British Army During the First World War* (Liverpool, 1995), p. 134, quoting an Irish soldier in the *Waterford News*, 1 January 1915; Private W. Harley's entry, dated 25 May 1915, in Clifford Nixon, 'A Touch of Memory', 1915, p. 57, IWM 2508.

31 Robert Graves, *Goodbye to All That*, first published in 1929 (London, 1977), p. 163.

32 H. S. Clapman, *Mud and Khaki. Memories of an Incomplete Soldier* (London, 1930), pp. 151–2, diary for 19 June 1915; George Coppard, *With a Machine Gun to Cambrai* (London, 1980), pp. 70–1; F. P. Crozier, *A Brass Hat in No Man's Land* (London, 1930), p. 228; Arthur Guy Empey, *First Call. Guide Posts to Berlin* (New York, 1918), p. 91; Sydney W. D. Lockwood, 'First World War Memories 1914–1918', p. 106, IWM 90/21/1.

33 A. Ashurt Morris, 'Diaries', pp. 69–70, entry for 16 June 1915, IWM.

34 James J. Weingartner, 'Massacre at Biscari: Patton and an American War Crime', *The Historian*, lii.1 (November 1989), p. 30. Compton was acquitted. The other man accused of the crime was sentenced to life imprisonment. His defence had been weaker, and had claimed temporary insanity. The differences between the two sentences disturbed military commanders and the second prisoner was released within six months on condition that the case was given no publicity.

35 Other examples given in Eric Bergerud, *Touch with Fire. The Land War in the South Pacific* (New York, 1996), p. 423; Barry Broadfoot, *Six War Years, 1939–1945. Memories of Canadians at Home and Abroad* (Toronto, 1974), pp. 152–3; Lieutenant William L. Calley, *Body Count* (London, 1971), pp. 10–12; John Miller, *Guadalcanal: The First Offensive* (Washington, DC, 1949), p. 310; Rowland Walker, *Commando Captain* (London, 1942), pp. 54–5; Kim Willensen, *The Bad War. An Oral History of the Vietnam War* (New York, 1987), p. 62. For an interesting example drawn from the Korean War, see Colonel David H. Hackworth and Julie Sharman, *About Face* (Sydney, 1989), pp. 66–7.

36 Captain S. J. Cuthbert, *'We Shall Fight in the Streets!' Guide to Street Fighting* (Aldershot, 1941), p. 23. This was republished three times in 1942 and again in 1950.

37 Colonel Ted B. Borek, 'Legal Services During the War', *Military Law Review*, 120 (spring 1988), pp. 34–5.

38 A British lieutenant during the Second World War, cited in Andrew Wilson, 'The War in Vietnam', published in both *The Observer*, 30 November 1969, p. 9 and *Current*, January 1970, p. 3.

39 Sergeant John Henry Ewen, 'Diaries', book I, 1944, p. 60, AWM PR89/190.

40 Major General Raymond Hufft, quoted in 'Judgment at Fort Benning', *Newsweek*, 12 April 1971, p. 28.

41 Ex-Marine, Master Sergeant Stanley Gertner, cited in Tom Tiede, *Calley: Soldier or Killer?* (New York, 1971), p. 16. For a similar confession, see the comment by Second World War veteran Carl E. Savard,

cited by Kenneth Auchincloss, 'Who Else is Guilty?', *Newsweek*, 12 April 1971, p. 30.

42 Clayton D. Laurie, 'The Ultimate Dilemma of Psychological Warfare in the Pacific: Enemies Who Don't Surrender, and GIs Who Don't Take Prisoners', *War and Society*, 14.1 (May 1996), p. 117.

43 Arnold Krammer, 'Japanese Prisoners of War in America', *Pacific Historical Review*, 52 (1983), p. 69.

44 George MacDonald Fraser, *Quartered Safe Out Here. A Recollection of the War in Burma* (London, 1992), pp. xvi, 26, 73, 83, 87, 118, 125–6, 191–2.

45 William F. Crandell, 'What Did America Learn from the Winter Soldier Investigation?', *Vietnam Generation Journal. Nobody Gets Off the Bus*, 5 (March 1994), on the Internet, p. 1.

46 Captain W. C. Rowe, 'Ethics of Surrender', *Army Training Memorandum. Australian Military Forces*, 44 (October, 1946), p. 23.

47 Drill Sergeant Kenneth Hodges, quoted in Michael Bilton and Kevin Sims, *Four Hours in My Lai. A War Crime and its Aftermath* (London, 1992), p. 55.

48 Major W. Hays Parks, 'Crimes in Hostilities. Part I', *Marine Corps Gazette* (August 1976), p. 21.

49 Unnamed colonel, quoted by Revd T. W. Pym and Revd Geoffrey Gordon, *Papers from Picardy by Two Chaplains* (London, 1917), pp. 29–30.

50 Patton, cited in James J. Weingartner, 'Massacre at Biscari: Patton and an American War Crime', *The Historian*, lii.1 (November 1989), p. 37.

51 Billy Conway, interviewed in Mark Lane, *Conversations with Americans* (New York, 1970), p. 183.

52 Ed Treratola, interviewed in Mark Lane, *Conversations with Americans* (New York, 1970), p. 96.

53 Unnamed army officer, interviewed in December 1969, cited in Edward M. Opton, 'It Never Happened and Besides They Deserved It', in Nevitt Sanford and Craig Comstock (eds), *Sanctions for Evil* (San Francisco, 1971), p. 65.

54 James Jones, *The Thin Red Line* (London, 1962), p. 302.

55 Letter by Captain John Eugene Crombie to an unnamed friend, 3 March 1917, in Laurence Housman (ed.), *War Letters of Fallen Englishmen* (London, 1930), pp. 82–3.

56 George Barker, interviewed in Glen D. Edwards, *Vietnam. The War Within* (Salisbury, South Australia, 1992), p. 47.

57 Sam Damon, quoted in Anton Myrer, *Once an Eagle* (New York, 1968), pp. 787–8.

58 Donald Duncan, *The New Legions* (London, 1967), pp. 160–1.

59 James Adams, interviewed in Mark Lane, *Conversations with Americans* (New York, 1970), pp. 131–2.

60 For example, see the testimonies of Chuck Onan and Jimmy Roberson, interviewed in Mark Lane, *Conversations with Americans* (New York, 1970), pp. 28 and 60 and Daniel Lang, *Casualties of War* (New York, 1969), pp. 25–6.

61 Jimmy Roberson, interviewed in Mark Lane, *Conversations with Americans* (New York, 1970), p. 59.

62 The highest status was given to 'double veterans' (rape followed by murder), followed by babies, adult women, older men, and finally prisoners of war. Killing the victims closest to the perpetrators (POWs) brought little thrill: it was the breaking of a law which was always being broken.

63 Sergeant Michael McCuster, in 'Vietnam Veterans Against the War', *The Winter Soldier Investigation. An Inquiry into American War Crimes* (Boston, 1972), p. 29.

64 Chuck Onan, interviewed in Mark Lane, *Conversations with Americans* (New York, 1970), p. 30.

65 Lieutenant William L. Calley, *Body Count* (London, 1971), pp. 10–12 and Tom Tiede, *Calley: Soldier or Killer?* (New York, 1971), p. 130.

66 Herbert Rainwater, cited in 'Judgment at Fort Benning', *Newsweek*, 12 April 1971, p. 28.

67 For a discussion, see Guenter Lewy, *America in Vietnam* (New York, 1978), pp. 239–40 and 347. For Westmoreland's defence, see W. C. Westmoreland, *A Soldier Reports* (New York, 1976), pp. 378–9.

68 Michael Uhl, *Vietnam. A Soldier's View* (Wellington, New Zealand, 1971), p. 9.

69 Edward M. Opton, 'It Never Happened and Besides They Deserved It', in Nevitt Sanford and Craig Comstock (eds), *Sanctions for Evil* (San Francisco, 1971), p. 62.

70 1,608 people were questioned: 'The War: New Support for Nixon', *Time*, 12 January 1970, pp. 10–11.

71 Kenneth Maddock, 'Going Over the Limit? – The Question of Australian Atrocities', in Maddock (ed.), *Memories of Vietnam* (Sydney, 1991), p. 163.

72 'The Battle Hymn of Lieutenant Calley' can be heard on the recording held in the Australian National Film and Sound Archives.

73 For civilian attitudes towards the conviction of Calley, also see Wayne Greenhaw, *The Making of a Hero. The Story of Lieut. William Calley, Jr.* (Louisville, Kentucky, 1971), p. 191; Robert D. Heinl, 'My Lai in Perspective: The Court-Martial of William L. Calley', *Armed Forces*

Journal, 21 December 1970, pp. 38–41; Tom Tiede, *Calley: Soldier or Killer?* (New York, 1971), p. 16; Kenrick S. Thompson, Alfred C. Clarke and Simon Dinitz, 'Reactions to My-Lai: A Visual-Verbal Comparison', *Sociology and Social Research*, 58.2 (January 1974), pp. 122–9.

74 A Gallup poll based on telephone interviews of a representative cross section of 522 Americans: 'Judgment at Fort Benning', *Newsweek*, 12 April 1971, p. 28.

75 Herbert C. Kelman and Lee H. Lawrence, 'Assignment of Responsibility in the Case of Lt. Calley: Preliminary Report on a National Survey', *Journal of Social Issues*, 28.1 (1972), pp. 177–212.

76 Leon Mann, 'Attitudes Towards My Lai and Obedience to Orders: An Australian Survey', *Australian Journal of Psychology*, 25.1 (1973), pp. 11–21.

77 Herbert C. Kelman and Lee H. Lawrence, 'Assignment of Responsibility in the Case of Lt. Calley: Preliminary Report on a National Survey', *Journal of Social Issues*, 28.1 (1972), pp. 177–212.

78 Leon Mann, 'Attitudes Towards My Lai and Obedience to Orders: An Australian Survey', *Australian Journal of Psychology*, 25.1 (1973), pp. 11–21.

79 Edward M. Opton, 'It Never Happened and Besides They Deserved It', in Nevitt Sanford and Craig Comstock (eds), *Sanctions for Evil* (San Francisco, 1971), pp. 63–4.

80 Randy Olley and Herbert H. Kraus, 'Variables Which May Influence the Decision to Fire in Combat', *The Journal of Social Psychology*, 92 (1974), pp. 151–2.

81 Stanley Milgram, *Obedience to Authority: An Experimental View* (New York, 1974); Milgram, 'Some Conditions of Obedience and Disobedience to Authority', *Human Relations*, 18.1 (February 1965), pp. 57–76; Milgram, 'Behaviour Study of Obedience', in A. Etzioni and W. Wenglinsky (eds), *War and Its Prevention* (New York, 1970), part 5, pp. 245–59. Also see Philip Zimbardo, Craig Haney, and Curtis Banks, 'Interpersonal Dynamics in a Simulated Prison', *International Journal of Criminology and Penology*, I (1973), pp. 69–97.

82 Joseph Goldstein, Burke Marshall, and Jack Schwartz (eds), *The Peers Commission Report* (New York, 1976), pp. 83–4, 197–201.

83 W. C. Westmoreland, *A Soldier Reports* (New York, 1976), pp. 378–9.

84 Loren Baritz, *Backfire: A History of How American Culture Led Us Into Vietnam and Made Us Fight the Way We Did* (New York, 1985), pp. 314–15 and Richard A. Gabriel, *To Serve With Honor. A Treatise on Military Ethics and the Way of the Soldier* (Westport, Connecticut, 1982), p. 4.

85 Joseph Goldstein, Burke Marshall, and Jack Schwartz (eds), *The Peers Commission Report* (New York, 1976), p. 82.

86 Richard Tregaskis, *Guadalcanal Diary* (New York, 1943), p. 16. For other threats, see Bill Crooks, interviewed in Eric Bergerud, *Touch with Fire. The Land War in the South Pacific* (New York, 1996), pp. 423–4.

87 Daniel Lang, *Casualties of War* (New York, 1969), p. 35. Also see George Ryan, in Murray Polner, *No Victory Parades. The Return of the Vietnam Veteran* (London, 1971), p. 40.

88 Ronald Haeberle, quoted in Michael Bilton and Kevin Sims, *Four Hours in My Lai. A War Crime and its Aftermath* (London, 1992), p. 183.

89 Michael Bernhardt, quoted in a letter by Ron Ridenhour, in Joseph Goldstein, Burke Marshall, and Jack Schwartz (eds), *The Peers Commission Report* (New York, 1976), p. 37.

90 Greg Olsen, quoted in Michael Bilton and Kevin Sims, *Four Hours in My Lai. A War Crime and its Aftermath* (London, 1992), p. 82.

91 George Ryan (pseudonym), interviewed in Murray Polner, *No Victory Parades. The Return of the Vietnam Veteran* (London, 1971), p. 40.

92 William Calley, cited in Arthur Everett, Kathryn Johnson, and Harry F. Rosenthal, *Calley* (New York, 1971), p. 25.

93 Cited in Robert Jay Lifton, *Home from the War. Vietnam Veterans: Neither Victims Nor Executioners* (London, 1974), p. 119.

94 A long-serving Pathfinder navigator, quoted by Martin Middlebrook, *The Battle of Hamburg. Allied Bomber Forces Against a German City in 1943* (London, 1980), p. 349.

95 Lieutenant William L. Calley, *Body Count* (London, 1971), p. 47. Also see 'Lieutenant Duffy's Statement', in Richard A. Falk, Gabriel Kolko and Robert Jay Lifton (eds), *Crimes of War* (New York, 1971), p. 253

96 Edward Smithies (compiler), *War in the Air. Men and Women Who Built, Serviced, and Flew War Planes Remember the Second World War* (London, 1990), pp. 125–6.

97 Joseph Ellis and Robert Moore, *School for Soldiers. West Point and the Profession of Arms* (New York, 1974), p. 167. During the 1939–45 conflict, air force personnel classified LMF (or Lack of Moral Fibre) could have their commissions terminated, be demoted, refused subsequent lucrative positions in civil aviation, and stripped of badges and medals. For a superb analysis, see John McCarthy, 'Aircrew and "Lack of Moral Fibre" in the Second World War', *War and Society*, 2.2 (September 1984), pp. 88–9.

98 Jeff Needle in a pamphlet entitled 'Please Read This', 1970, cited in

Robert Jay Lifton, *Home from the War. Vietnam Veterans: Neither Victims Nor Executioners* (London, 1974), p. 313.

99 Rifleman Barry Kavanagh, quoted by Sally Wilkins, 'A Glance Behind', *The Age*, 10 May 1975, p. 11.

100 Rick Springman, interviewed in Indochina Curriculum Group, *Front Lines. Soldiers' Writings from Vietnam* (Cambridge, Massachusetts, 1975), p. 23.

101 L. C. Green, *Essays on the Modern Law of War* (New York, 1985), p. 27.

102 Westmoreland interview, in Michael Charlton and Anthony Moncrieff, *Many Reasons Why* (London, 1978), p. 148 quoted in Loren Baritz, *Backfire: A History of How American Culture Led Us into Vietnam and Made Us Fight the Way We Did* (New York, 1985), p. 294.

103 Loren Baritz, *Backfire: A History of How American Culture Led Us into Vietnam and Made Us Fight the Way We Did* (New York, 1985), p. 294.

104 The only exception was in the air force because pilots and men in fire control were examined on the rules prior to operational duty, and every three months thereafter: Guenter Lewy, *America in Vietnam* (New York, 1978), p. 234. Also see 'Lieutenant Duffy's Statement', in Richard A. Falk, Gabriel Kolko, and Robert Jay Lifton (eds), *Crimes of War* (New York, 1971), p. 253 and General Bruce Palmer, *The 25-Year War. America's Military Role in Vietnam* (New York, 1984), pp. 170–1.

105 Douglas Kinnard, 'Vietnam Reconsidered: An Attitudinal Survey of U.S. Army General Officers', *The Public Opinion Quarterly*, xxxix.4 (winter 1975–6), p. 451. Also see Guenter Lewy, *America in Vietnam* (New York, 1978), p. 234.

106 According to Michael Bernhardt, cited in Joseph Lelyveld, 'The Story of a Soldier Who Refused to Fire at Songmy', *The New York Times Magazine*, 14 December 1969, p. 101.

107 Lieutenant William L. Calley, *Body Count* (London, 1971), pp. 28–9.

108 William Calley, cited in Arthur Everett, Kathryn Johnson, and Harry F. Rosenthal, *Calley* (New York, 1971), p. 25. Paul Meadlo also claimed never to have heard of the possibility of an 'illegal order': Richard Hammer, *The Court-Martial of Lt. Calley* (New York, 1971), p. 158.

109 Richard Hammer, *The Court-Martial of Lt. Calley* (New York, 1971), p. 91.

110 Joseph Goldstein, Burke Marshall, and Jack Schwartz (eds), *The Peers Commission Report* (New York, 1976), pp. 204–5. Also see p. 83.

111 Joseph Goldstein, Burke Marshall, and Jack Schwartz (eds), *The Peers Commission Report* (New York, 1976), pp. 210–11. This was increased to two hours after the My Lai atrocity.

112 J. Glenn Gray, *The Warriors: Reflections on Men in Battle*, second edition (New York, 1970), p. 102.

113 Lieutenant-Colonel J. O. Langtry, 'Man-the-Weapon: Neglected Aspects of Leader Training', *Australian Army Journal*, 202 (March 1966), p. 12 and Major-General W. H. S. Macklin, 'Military Law', *Canadian Army Journal*, viii.1 (January 1954), pp. 31–2.

114 Philip Caputo, *A Rumor of War* (New York, 1977), pp. 229–30. Also see Seymour M. Hersh, 'A Reporter at Large. The Reprimand', *The New Yorker*, 9 October 1971, p. 114; Richard Holmes, *Firing Line* (London, 1985), p. 367; Guenter Lewy, *America in Vietnam* (New York, 1978), p. 235.

115 Based on a report carried out by the US Defence Department in mid-1967: the ratio did change in the Americans' favour later in the war. See D. Michael Shafer, 'The Vietnam Combat Experience: The Human Legacy', in Shafer (ed.), *The Legacy. The Vietnam War in the American Imagination* (Boston, 1990), p. 84. For a similar estimate, see John Helmer, *Bringing the War Home. The American Soldier in Vietnam and After* (New York, 1974), p. 166.

116 Joseph Goldstein, Burke Marshall, and Jack Schwartz (eds), *The Peers Commission Report* (New York, 1976), pp. 195–6.

117 Robert Jay Lifton, 'Existential Evil', in Nevitt Sanford and Craig Comstock (eds), *Sanctions for Evil* (San Francisco, 1971), p. 45.

118 Philip Caputo, *A Rumor of War* (New York, 1977), p. 229.

119 David E. Wilson, in Gerald R. Gioglio, *Days of Decision. An Oral History of Conscientious Objectors in the Military During the Vietnam War* (Trenton, New Jersey, 1989), p. 123. For similar accounts from all three conflicts, see Micheal [sic] Clodfelter, 'Snipers', in J. Topham, *Vietnam Literature Anthology*, revised and enlarged (Philadelphia, 1990), p. 87; Richard Elliott in a letter to his parents on 31 March 1966, in Bill Adler (ed.), *Letters from Vietnam* (New York, 1967), p. 22; Sergeant John Henry Ewen, 'Bougainville Campaign', book I, 1944, p. 60, AWM; 'Lieutenant Duffy's Statement', in Richard A. Falk, Gabriel Kolko, and Robert Jay Lifton (eds), *Crimes of War* (New York, 1971), p. 248; Rifleman John Muir, interviewed in Al Santoli, *Everything We Had. An Oral History of the Vietnam War by Thirty-Three American Soldiers Who Fought It* (New York, 1981), p. 25; Eric Norden, *American Atrocities in Vietnam* (Sydney, 1966), p. 5; Captain D. B. Reardon, 'Diary', entry for 4 August 1915, IWM.

120 Joseph Goldstein, Burke Marshall, and Jack Schwartz (eds), *The Peers Commission Report* (New York, 1976), pp. 99–100.

121 Eugene B. ('Sledgehammer') Sledge, interviewed in Studs Terkel, *'The*

Good War'. An Oral History of World War Two (London, 1985), p. 62.

122 Lieutenant William L. Calley, *Body Count* (London, 1971), p. 31. Also see Joseph Goldstein, Burke Marshall, and Jack Schwartz (eds), *The Peers Commission Report* (New York, 1976), p. 194.

123 Joseph Lelyveld, 'The Story of a Soldier Who Refused to Fire at Songmy', *The New York Times Magazine*, 14 December 1969, pp. 101 and 103.

124 Sergeant Scott Camil, in Vietnam Veterans Against the War, *The Winter Soldier Investigation. An Inquiry into American War Crimes* (Boston, 1972), p. 14.

125 John W. Dower, *War Without Mercy: Race and Power in the Pacific War* (New York, 1986), p. 147. Also see Seymour Leventman and Paul Comacho, 'The "Gook" Syndrome: The Vietnam War as a Racial Encounter', in Charles R. Figley and Seymour Leventman (eds), *Strangers at Home. Vietnam Veterans Since the War* (New York, 1990), pp. 55–70 and James J. Weingartner, 'Trophies of War: U.S. Troops and the Mutilation of Japanese War Dead, 1941–1945', *Pacific Historical Review*, lxi.1 (February 1992), pp. 53–67.

126 Michael Bilton and Kevin Sims, *Four Hours in My Lai. A War Crime and its Aftermath* (London, 1992), p. 52 and Joseph Goldstein, Burke Marshall, and Jack Schwartz (eds), *The Peers Commission Report* (New York, 1976), pp. 83–4.

127 Michael Bilton and Kevin Sims, *Four Hours in My Lai. A War Crime and its Aftermath* (London, 1992), p. 37; Joseph Goldstein, Burke Marshall, and Jack Schwartz (eds), *The Peers Commission Report* (New York, 1976), p. 220; Major-General George S. Prugh, *Law at War: Vietnam 1964–1973* (Washington, DC, 1975), p. 76.

128 Joseph Goldstein, Burke Marshall, and Jack Schwartz (eds), *The Peers Commission Report* (New York, 1976), p. 211 and Major-General George S. Prugh, *Law at War: Vietnam 1964–1973* (Washington, DC, 1975), pp. 74–5.

129 William Calley, cited in Arthur Everett, Kathryn Johnson, and Harry F. Rosenthal, *Calley* (New York, 1971), p. 25.

130 Sergeant Michael McCuster, in Vietnam Veterans Against the War, *The Winter Soldier Investigation. An Inquiry into American War Crimes* (Boston, 1972), p. 29. Also see the interview with Jerry Samuels, in Philip Scribner Balboni, 'Mylai was Not an Isolated Incident. What Every Vietnam Veteran Knows', *The New Republic*, 19 December 1970, p. 15.

131 A former senior army lawyer, quoted in Seymour M. Hersh, 'A Reporter

at Large. The Reprimand', *The New Yorker*, 9 October 1971, p. 119. For an interesting Korean War example, see Colonel David H. Hackworth and Julie Sharman, *About Face* (Sydney, 1989), pp. 66–7.

132 Richard Hammer, *The Court-Martial of Lt. Calley* (New York, 1971), pp. 18–21.

133 Michael Bernhardt quoted in a letter by Ron Ridenhour, in Joseph Goldstein, Burke Marshall, and Jack Schwartz (eds), *The Peers Commission Report* (New York, 1976), p. 37.

134 For an insider's account of the cover-up, see Bruce E. Jones, *War Without Windows. A True Account of a Young Army Officer Trapped in an Intelligence Cover-Up in Vietnam* (New York, 1987), pp. 221–5 and 257–63. Also see Seymour M. Hersh, *Cover-Up. The Army's Secret Investigation of the Massacre at My Lai 4* (New York, 1972) and Lieutenant-General William R. Peers, *The My Lai Inquiry* (New York, 1979), pp. 199–209.

135 Major-General George S. Prugh, *Law at War: Vietnam 1964–1973* (Washington, DC, 1975), p. 74. In total, there were 241 allegations of war crimes committed by American troops but two thirds were determined to be unsubstantiated. Of the thirty-six court martials, 56 per cent resulted in a conviction.

136 Major W. Hays Parks, 'Crimes in Hostilities. Part I', *Marine Corps Gazette* (August 1976), p. 18. His statistics were slightly different to the ones quoted above. The total number of court martials in his table was 259 for the US Army, Navy, Marine Corps and Air Force.

137 For their experiences, see Howard Levy and David Miller, *Going to Jail. The Political Prisoner* (New York, 1970).

138 My Lai survivors talking to Keith Suter, in 'My Lai: Everyone Wins Except the Dead Vietnamese', *The National Times*, 18–23 November 1974, p. 33.

139 John DiFusco, interviewed in Kim Willenson, *The Bad War. An Oral History of the Vietnam War* (New York, 1987), p. 58. Also see Private Anthony R. Brennan, 'Diaries', p. 21, entries for January to June 1917, IWM; Bill Crooks, interviewed in Eric Bergerud, *Touch with Fire. The Land War in the South Pacific* (New York, 1996), pp. 423–4; and Roy R. Grinker and John P. Spiegel, *Men Under Stress* (London, 1945), p. 307.

140 Charles J. Levy, 'A.R.V.N. as Faggots: Inverted Warfare in Vietnam', *Trans-Action* (October 1971), pp. 18–27 and Ian McNeill, 'Australian Army Advisers. Perceptions of Enemies and Allies', in Kenneth Maddock and Barry Wright (eds), *War. Australia and Vietnam* (Sydney, 1987), pp. 41–2.

141 Herbert C. Kelman and Lee H. Lawrence, 'Assignment of Responsibility in the Case of Lt. Calley: Preliminary Report on a National Survey', *Journal of Social Issues*, 28.1 (1972), p. 196. Also see Gabriel Kolko, 'War Crimes and the American Conscience', in Jay W. Baird (ed.), *From Nuremberg to My Lai* (Lexington, Massachusetts, 1972), p. 238.

142 James Joseph Dursi, testifying in Richard Hammer, *The Court-Martial of Lt. Calley* (New York, 1971), p. 143.

143 Michael Bilton and Kevin Sims, *Four Hours in My Lai. A War Crime and its Aftermath* (London, 1992), p. 19.

144 William E. Merritt, *Where the Rivers Ran Backwards* (Athens, 1989), pp. 101–11 and John Tuma of US Military Intelligence, interviewed in Willa Seidenberg and William Short (eds), *A Matter of Conscience. GI Resistance During the Vietnam War* (Andover, Massachusetts, 1991), p. 24.

145 Private First Class Reginald 'Malik' Edwards, interviewed in Wallace Terry, *Bloods. An Oral History of the Vietnam War by Black Veterans* (New York, 1984), pp. 16–17.

146 Thomas John Kinch, testifying in Richard Hammer, *The Court-Martial of Lt. Calley* (New York, 1971), p. 197.

147 Rick Springman, interviewed in the Indochina Curriculum Group (eds.), *Front Lines. Soldiers' Writings from Vietnam* (Cambridge, Massachusetts, 1975), p. 23. Also see Robert S. Laufer, M. S. Gallops, and Ellen Frey-Wouters, 'War Stress and Trauma: The Vietnam Veteran Experience', *Journal of Health and Social Behavior*, 25 (1984), p. 78, and Keith William Nolan, *Sappers in the Wire. The Life and Death of Firebase Mary Ann* (College Station, 1975), pp. 94–5.

148 Robert Rasmus, interviewed in Studs Terkel, *'The Good War': An Oral History of World War Two* (New York, 1985), p. 44.

149 Lieutenant-Colonel David Lloyd Owen, *The Desert My Dwelling Place* (London, 1957), pp. 129–30.

150 Charles Stewart Alexander, 'Letters to His Cousin Amy Reid 1916–1918', 3 November 1917, Auckland Institute and Museum Library MSS. 92/70.

151 Tony Ashworth, *Trench Warfare 1914–1918* (London, 1980).

152 Richard Pendleton, quoted in Martin Gershen, *Destroy or Die: The True Story of My Lai* (New York, 1971), p. 31.

153 Philip Scribner Balboni, 'Mylai was Not an Isolated Incident. What Every Vietnam Veteran Knows', *The New Republic*, 19 December 1970, p. 15.

154 Richard Boyle, *The Flower of the Dragon. The Breakdown of the U.S. Army in Vietnam* (San Francisco, 1972), p. 22.

155 Joseph Lelyveld, 'The Story of a Soldier Who Refused to Fire at Songmy', *The New York Times Magazine*, 14 December 1969, pp. 32–33, 101–3 and 113–16.

156 Sergeant Michael McCuster, in Vietnam Veterans Against the War, *The Winter Soldier Investigation. An Inquiry into American War Crimes* (Boston, 1972), p. 29. Simply walking a short distance away was also the response of another serviceman who refused to rape and murder a woman: Daniel Lang, *Casualties of War* (New York, 1969), p. 36.

157 Sergeant Ed Murphy, in Vietnam Veterans Against the War, *The Winter Soldier Investigation. An Inquiry into American War Crimes* (Boston, 1972), p. 48.

158 Terry Whitmore, interviewed in Mark Lane, *Conversations with Americans* (New York, 1970), p. 76.

159 Mary McCarthy, *Medina* (New York, 1972), p. 74.

7 The Burden of Guilt

1 Wilfred Owen, 'Apologia Pro Poemate Meo', in Jon Stallworthy (ed.), *Wilfred Owen: The Complete Poems and Fragments*, vol. 1 (London, 1983), p. 124.

2 The diary of Sergeant Bruce F. Anello and an interview with his closest friend, Dave Lang, in Indochina Curriculum Group (eds), *Front Lines. Soldiers' Writings from Vietnam* (Cambridge, Massachusetts, 1975), pp. 1–19. First published in *WIN Magazine*, 15 May 1971.

3 George M. Kren, 'The Holocaust: Moral Theory and Immoral Acts', in Alen Rosenberg and Gerald E. Myers (eds), *Echoes from the Holocaust. Philosophical Reflections on a Dark Time* (Philadelphia, 1988), p. 255.

4 R. D. Gillespie, *Psychological Effects of War on Citizen and Soldier* (London, 1942), p. 180; Edward A. Strecker and Kenneth E. Appel, *Psychiatry in Modern Warfare* (New York, 1945), pp. 24–5; Major Edwin Weinstein, 'The Fifth U. S. Army Neuropsychiatric Centre – "601st"', in Lieutenant-General Hal B. Jennings (ed.), *Neuropsychiatry in World War II. Volume II. Overseas Theatres* (Washington, DC, 1973), p. 134.

5 Robert William MacKenna, *Through a Tent Door*, first published in 1919 (London, 1930), p. 103.

6 Charles Bird, 'From Home to the Charge: A Psychological Study of the Soldier', *American Journal of Psychology*, xxviii.3 (July 1917), p. 343.

7 Irving N. Berlin, 'Guilt as an Etiologic Factor in War Neuroses', *Journal of Nervous and Mental Disease*, 111 (January–June 1950), pp. 239–45. Also see Therese Benedek, *Insight and Personality Adjustment. A Study of*

the Psychological Effects of War (New York, 1946), p. 56.

8 Major Jules V. Coleman, 'The Group Factor in Military Psychiatry', *American Journal of Orthopsychiatry*, xvi (1946), p. 224.

9 Private John Doran, 'Letters', letter to his mother, 3 July 1916, IWM.

10 Sir Ivone Kirkpatrick, 'Memoirs', p. 29, IWM. Also see Philip Caputo, *A Rumor of War* (London, 1977), pp. 305–6.

11 Michael MacDonagh, *The Irish on the Somme* (London, 1917), p. 76.

12 Fred Branfman, 'The Era of the Blue Machine: Laos', *Washington Monthly*, July 1971, cited in Robert Jay Lifton, *Home from the War. Vietnam Veterans: Neither Victims Nor Executioners* (London, 1974), p. 349.

13 J. Glenn Gray, *The Warriors: Reflections on Men in Battle* (New York, 1959), p. 173. Also see the similar comments by Therese Benedek, *Insight and Personality Adjustment. A Study of the Psychological Effects of War* (New York, 1946), pp. 54–5; Irvin L. Child, 'Morale: A Bibliographical Review', *Psychological Bulletin*, 38 (1941), p. 411; Robert L. Garrard, 'Combat Guilt Reactions', *North Carolina Medical Journal*, 10.9 (September 1949), p. 489.

14 Ted Van Kirk in *Newsweek*, July 1985, p. 44, quoted by Bernard J. Verkamp, *The Moral Treatment of Returning Warriors in Early Medieval and Modern Times* (Scranton, 1993), p. 151.

15 Frank Elkins's diary for 1 July 1966, in Indochina Curriculum Group, *Front Lines. Soldiers' Writings from Vietnam* (Cambridge, Massachusetts, 1975), p. 101.

16 Philip Caputo, *A Rumor of War* (London, 1977), p. 124.

17 Poem by M. Grover, in Malvern Van Wyk Smith, *Drummer Hodge. The Poetry of the Anglo-Boer War, 1899–1902* (Oxford, 1978), p. 152.

18 R. H. Stewart, quoted by Philip Orr, *The Road to the Somme. Men of the Ulster Division Tell Their Story* (Belfast, 1987), p. 155.

19 Albert N. Depew, *Gunner Depew* (London, 1918), p. 61.

20 Major Edwin A. Weinstein and Lieutenant-Colonel Calvin S. Drayer, 'A Dynamic Approach to the Problem of Combat-Induced Anxiety', *The Bulletin of the U. S. Army Medical Department*, ix, supplemental number (November 1949), p. 16.

21 'Bogles', quoted in Wing Commander Athol Forbes and Squadron Leader Hubert Allen, *The Fighter Boys* (London, 1942), p. 70. Also see A. G. J. Whitehouse, *Hell in the Heavens. The Adventures of an Aerial Gunner in the Royal Flying Corps* (London, 1938), p. 43; Roy R. Grinker and John P. Spiegel, *Men Under Stress* (London, 1945), pp. 132–3; an unnamed aircrew, quoted by Martin Middlebrook, *The Battle of*

Hamburg. Allied Bomber Forces Against a German City in 1943 (London, 1980), p. 349.

22 Jerry Samuels, in Philip Scribner Balboni, 'Mylai was Not an Isolated Incident. What Every Vietnam Veteran Knows', *The New Republic*, 19 December 1970, p. 15. Also see Private First Class Reginald 'Malik' Edwards and Arthur E. 'Gene' Woodley, Jr., interviewed in Wallace Terry, *Bloods. An Oral History of the Vietnam War by Black Veterans* (New York, 1984), pp. 12 and 243–4; Joe Walker, *Parsons and War and Other Essays in War Time* (Bradford, 1917), p. 8.

23 Robert L. Garrard, 'Combat Guilt Reactions', *North Carolina Medical Journal*, 10.9 (September 1949), p. 489.

24 Naomi Breslau and Glenn C. Davis, 'Posttraumatic Stress Disorder: The Etiologic Specificity of Wartime Stressors', *American Journal of Psychiatry*, 144.5 (May 1987), pp. 578–83; Richard Strayer and Lewis Ellenhorn, 'Vietnam Veterans: A Study Exploring Adjustment Patterns and Attitudes', *Journal of Social Issues*, 31.4 (1975), pp. 85 and 87; Rachel Yehuda, Steven M. Southwick, and Earl L. Giller, 'Exposure to Atrocities and Severity of Chronic Posttraumatic Stress Disorder in Vietnam Combat Veterans', *American Journal of Psychiatry*, 149.3 (March 1992), pp. 333–6.

25 'Fred', interviewed in Shirley Dicks, *From Vietnam to Hell. Interviews with Victims of Post-Traumatic Stress Disorder* (Jefferson, North Carolina, 1990), p. 30. This volume gives numerous examples. Also see the interview with nurse Eunice Splawn, in Kathryn Marshall, *In the Combat Zone. An Oral History of American Women in Vietnam 1966–1975* (Boston, 1987), p. 95.

26 John Garcia, interviewed in Studs Terkel, *'The Good War'. An Oral History of World War Two* (London, 1985), pp. 23–4. Also see Richard A. Newman, 'Combat Fatigue: A Review to the Korean Conflict', *Military Medicine*, 129.10 (October 1964), p. 924.

27 Samuel Futterman and Eugene Pumpion-Mindlin, 'Traumatic War Neuroses Five Years Later', *American Journal of Psychiatry*, 108 (December 1951), p. 401; Richard A. Newman, 'Combat Fatigue: A Review to the Korean Conflict', *Military Medicine*, 129.10 (October 1964), p. 924; John Rawlings Rees, *The Shaping of Psychiatry by War* (London, 1945), p. 15; Charles O. Sturdevant, 'Residuals of Combat Induced Anxiety', *The American Journal of Psychiatry*, 103 (1946–7), p. 58; Eric Wittkower and J. P. Spillane, 'A Survey of the Literature of Neuroses in War', in Emanuel Miller (ed.), *The Neuroses in War* (London, 1940), p. 6.

28 Major Jules V. Coleman, 'The Group Factor in Military Psychiatry',

American Journal of Orthopsychiatry, xvi (1946), p. 222.

29 Stephen Graham, *A Private in the Guards* (London, 1919), p. 3.

30 Irvin L. Child, 'Morale: A Bibliographical Review', *Psychological Bulletin*, 38 (1941), p. 411.

31 Colonel Albert J. Glass and Lieutenant-Colonel Calvin S. Drayer, 'Italian Campaign (1 March 1944–2 May 1945), Psychiatry Established at Division Level', in Lieutenant-General Hal B. Jennings (ed.), *Neuropsychiatry in World War II. Volume II. Overseas Theatres* (Washington, DC,1973), p. 105. These cases were not due to the desire to avoid battle (since the battle was over) and the men did not seem to be depressed, suicidal or fearful of returning home. Instead, Glass and Drayer were forced to speculate that the wounds must express some underlying guilt relating to combat experiences.

32 *Psychology for the Fighting Man. Prepared for the Fighting Man Himself by a Committee of the National Research Council with the Collaboration of Science Service as a Contribution to the War Effort*, second edition (Washington, DC, 1944), pp. 287–8.

33 Steve Harper (pseudonym), interviewed by Murray Polner, *No Victory Parades. The Return of the Vietnam Veteran* (London, 1971), p. 24 and William Kerr, 'Canadian Soldier', p. 109, IWM.

34 Lieutenant William L. Calley, *Body Count* (London, 1971), p. 106 and Margaret Garnegie and Frank Shields, *In Search of Breaker Morant. Balladist and Bushveldt Carbineer* (Armadale, 1979), p. 5.

35 Stephen Graham, *A Private in the Guards* (London, 1919), p. 3.

36 Stephen Graham, *The Challenge of the Dead* (London, 1921), p. 78.

37 That is, except (as we have just seen) in the case of atrocities.

38 'Bogles', interviewed in Wing Commander Athol Forbes and Squadron Leader Hubert Allen, *The Fighter Boys* (London, 1942), p. 72. Bogles was later killed in a flying accident.

39 Alex Cumstie, 'Diary', 29 November 1915, IWM and W. R. Kirby, 'The Battle of Cambrai, 1917', p. 104, IWM. Also see Hector Bolitho, *Combat Report. The Story of a Fighter Pilot* (London, 1943), p. 61; John M. Cordy, 'My Memories of the First World War', p. 18, IWM; F. de Margry, 'From Saddle-Bag to Infantry Pack', pp. 13–14, IWM; Donald Gilchrist, *Castle Commando* (Edinburgh, 1960), p. 7; Roy R. Grinker and John P. Spiegel, *Men Under Stress* (London, 1945), p. 43; Sydney W. D. Lockwood, 'First World War Memories 1914–1918', pp. 47 and 85, IWM; Charles K. McKerrow, 'Diaries and Letters', 21 March 1916, IWM; R. G. Plint, 'Memoirs', p. 17, IWM; Siegfried Sassoon, *Diaries 1915–1918*, edited by Rupert Hart-Davis (London, 1983), pp. 52–3,

entries for 1 and 4 April 1916; Dixon Scott, 'Letters', letter to A. N. Monkhouse, 6 January 1915, Manchester City Council Local Studies Unit; A. J. Turner, 'Zero Hour', pp. 48–9, IWM.

40 Thomas Alexander Ervine, 'Undiminished Memories', in Michael Hall (ed.), *Sacrifice on the Somme* (Newtownabbey, 1993), p. 23.

41 Unnamed Vietnam veteran, quoted by Jonathan Shay, *Achilles in Vietnam. Combat Trauma and the Undoing of Character* (New York, 1994), pp. 78–9. Also see Philip Caputo, *A Rumor of War* (London, 1977), xvii and 231 and Private First Class Chuck Fink, interviewed in Otto J. Lehrack, *No Shining Armour. The Marines at War in Vietnam. An Oral History* (Lawrence, Kansas, 1992), p. 44.

42 Ibid., p. 89. Also see p. 96. Similar words were recited by the First World War American, Albert N. Depew, *Gunner Depew* (London, 1918), p. 145 and by John Lohman, interviewed in Shirley Dicks, *From Vietnam to Hell. Interviews with Victims of Post-Traumatic Stress Disorder* (Jefferson, North Carolina, 1990), p. 34.

43 Robert Jay Lifton, 'The Postwar World', *Journal of Social Issues*, 31.4 (1975), pp. 181–2.

44 Neil J. Smelser, 'Some Determinants of Destructive Behavior', in Nevitt Sanford and Craig Comstock (eds), *Sanctions for Evil* (San Francisco, 1971), p. 17.

45 For examples from all three conflicts, see Barry Broadfoot, *Six War Years 1939–1945. Memories of Canadians at Home and Abroad* (Don Mills, 1974), p. 240; Nobby Clarke, 'Sniper on the Western Front', p. 7, IWM; Coningsby Dawson, *The Glory of the Trenches* (London, 1918), pp. 74 and 134; Harold Dearden, *Medicine and Duty. A War Diary* (London, 1928), pp. 66–7; 'Nelson', interviewed in Shirley Dicks, *From Vietnam to Hell. Interviews with Victims of Post-Traumatic Stress Disorder* (Jefferson, North Carolina, 1990), p. 19; Steve Harper (pseudonym), quoted by Murray Polner, *No Victory Parades. The Return of the Vietnam Veteran* (London, 1971), p. 22; F. de Margry, 'From Saddle-Bag to Infantry Pack', pp. 13–14, IWM; Gerald V. Dennis, 'A Kitchener Man's Bit (1916–1918)', p. 67, IWM; Edward Foukes, 'A V. C. Episode', p. 3, IWM; Sir Philip Gibbs, *The War Dispatches* (Isle of Man, 1964), p. 131; Captain M. D. Kennedy, 'Their Mercenary Calling', 1932, p. 152, IWM; W. R. Kirby, 'The Battle of Cambrai, 1917', p. 98, IWM 78/51/1; Charles K. McKerrow, 'Diaries and Letters', 14 April 1916, IWM; Revd J. E. Roscoe, *The Ethics of War, Spying and Compulsory Training* (London, nd [during the war]), p. 31; Revd Harold Augustine Thomas, 'A Parson', p. 43, IWM; War Office General Staff, *Bayonet Training, 1916* (London, 1916), p. 5.

46 Sydney W. D. Lockwood, 'No Man's Land', p. 10, IWM.

47 Victor G. Ricketts, 'Account of his Service', p. 34, IWM.

48 Private Jeff Porter, quoted by Alan Ramsey, 'How I Shot Two of the Vietcong', in an unidentified newscutting for the 13 August 1965, in Mrs Barbara Arnison, 'Collection of Newspaper Cuttings', AWM. Also see W. D. Ehrhart, 'Full Moon', in Larry Rottmann, Jan Barry and Basil T. Paquet (eds), *Winning Hearts and Minds. War Poems by Vietnam Veterans* (New York, 1972), p. 14.

49 Four sergeants in the *San Francisco Chronicle*, 31 December 1969, cited in Edward M. Opton, 'It Never Happened and Besides They Deserved It', in Nevitt Sanford and Craig Comstock (eds), *Sanctions for Evil* (San Francisco, 1971), p. 65.

50 James Morgan Read, *Atrocity Propaganda 1914–1919* (New Haven, Connecticut, 1941), p. 188.

51 Sir Robert Vansittart, *Black Record: Germans Past and Present* (London, 1941), pp. 4, 9, and 19.

52 *Vietcong Atrocities and Sabotage in South Vietnam* (np, 1966).

53 For examples, see Albert N. Depew, *Gunner Depew* (London, 1918), throughout; Captain David Fallon, *The Big Fight (Gallipoli to the Somme)* (London, 1918), pp. 107–8 and 154–5; Gerald French, *The Life of Field Marshal Sir John French* (London, 1931), p. 304; Margaret Garnegie and Frank Shields, *In Search of Breaker Morant. Balladist and Bushveldt Carbineer* (Armadale, 1979), p. 76; Stephen Graham, *A Private in the Guards* (New York, 1919), p. 220; William Ernest Hocking, *Morale and Its Enemies* (New Haven, Connecticut, 1918), p. 59; W. W. Holdsworth, 'Impressions of a Hospital Chaplain', *The Contemporary Review*, cix (May 1916), p. 638; Harold D. Lasswell, *Propaganda Technique in the World War* (London, 1927), pp. 62 and 81; Captain J. E. H. Neville, *The War Letters of a Light Infantryman* (London, 1930), pp. 51–2; Major John Leslie, 'Diaries', entry for 8 March 1915, AWM; Harold R. Peat, *Private Peat* (Indianapolis, 1917), pp. 45–6; James Morgan Read, *Atrocity Propaganda 1914-1919* (New Haven, Connecticut, 1941), p. 3; Samuel A. Stouffer et al., *The American Soldier: Combat and Its Aftermath. Volume II* (Princeton, 1949), pp. 156–7; Second Lieutenant Alfred Richard Williams, 'Letters from the West [sic] Front', letter home, 8 February 1916, p. 49, IWM.

54 Harold R. Peat, *The Inexcusable Lie* (New York, 1923), pp. 87–8. Ellipses in the original.

55 Captain John Long of the 761st (Tank Division), interviewed in Mary

Penick Motley (ed.), *The Invisible Soldier. The Experience of the Black Soldier, World War II* (Detroit, 1975), p. 155. For a similar account see Interview with Timuel Black, in Studs Terkel, *'The Good War'. An Oral History of World War Two* (London, 1985), p. 281.

56 John Bennetts, 'A Weird War', *The Age*, 25 January 1968, p. 5.

57 Testimony of Scott Camil, in Frances FitzGerald, *Fire in the Lake. The Vietnamese and the Americans in Vietnam* (Boston, 1972), p. 371. Also see Lieutenant William L. Calley, *Body Count* (London, 1971), p. 106.

58 James Morgan Read, *Atrocity Propaganda 1914–1919* (New Haven, Connecticut, 1941), pp. 141–2 and Harold R. Peat, *The Inexcusable Lie* (New York, 1923), pp. 154–5.

59 Donald Hankey, *A Student in Arms. Second Series* (London, 1917), pp. 65–6. Also see Sir Philip Gibbs, *The War Dispatches* (Isle of Man, 1964), p. 131 and Captain J. E. H. Neville, *The War Letters of a Light Infantryman* (London, 1930), p. 27.

60 Hugh W. Spurrell, 'My Impressions of Life in the Great War', p. 64, IWM. Also see Captain Charles K. McKerrow, 'Diaries and Letters', 15 April 1916, IWM.

61 Captain M. D. Kennedy, 'Their Mercenary Calling', pp. 153–4, IWM. Also see Sid T. Kemp, 'Remembrance', pp. 19–20, IWM.

62 Quoted in Millais Culpin, *Psychoneuroses of War and Peace* (Cambridge, 1920), p. 75.

63 Charles Stewart Alexander, 'Letters to his Cousin', 3 November 1917, Auckland Institute and Museum Library MSS 92/70.

64 John T. MacCurdy, *War Neuroses* (Cambridge, 1918), p. 35.

65 William A. Bacher (ed.), *The Treasury Star Parade* (New York, 1942), p. 359; James J. Fahey, *Pacific War Diary 1942–1945* (Boston, 1963), p. 178, entry for 27 June 1944; John J. Floherty, *The Courage and the Glory* (Philadelphia, 1942), p. 94; Henry ('Jo') Gullett, *Not as a Duty Only. An Infantryman's War* (Melbourne, 1976), p. 127; John Hersey, *Into the Valley. A Skirmish of the Marines* (London, 1943), pp. 39–40; George P. Hunt, *Coral Comes High* (New York, 1946), pp. 59 and 82; Neville Jason, 'Letters', letter to his sister Roz from Vietnam, 11 December 1965, AWM; George H. Johnston, *The Toughest Fighting in the World* (New York, 1944), p. 207; *Pen Pictures of the War by Men at the Front. Vol. 1. The Campaign in Natal to the Battle of Colenso* (London, 1900), p. 60; Colonel R. G. Pollard, '6th. Aust. Div. Training Instruction, No. 1. Jungle Warfare', 27 March 1943, p. 1, in Lieutenant-General Sir F. H. Berryman, 'Papers', AWM; Frederick Treves, *The Tale of a Field Hospital* (London, 1901), p. 12; Wade Williams, *Infantry Attack* (Sydney, 1955),

pp. 52 and 93–5.

66 Captain Charles James Lodge Patch, 'Memoir', pp. 37–8, IWM; Lieutenant-Colonel Kenneth H. Cousland, 'The Great War 1914–18. A Former Gunner Looks Back', nd, p. 61, Liddell Hart Centre for Military Archives.

67 E. Gardiner Williams, 'The Last Chapter', p. 8, Liverpool Records Office. Also see Lieutenant-Colonel Kenneth H. Cousland, 'The Great War 1914–18. A Former Gunner Looks Back', p. 61, Liddell Hart Centre for Military Archives.

68 Barry Broadfoot, *Six War Years 1939–1945. Memories of Canadians at Home and Abroad* (Don Mills, Ontario, 1974), p. 240; Lieutenant William L. Calley, *Body Count* (London, 1971), pp. 101 and 104–5; Studs Terkel, *'The Good War'. An Oral History of World War Two* (London, 1985), p. 259.

69 Trooper Simon Cole, quoted in Sally Wilkins, 'A Glance Behind', *The Age*, 10 May 1975, p. 11.

70 Harry O'Connor (pseudonym), in Murray Polner, *No Victory Parades. The Return of the Vietnam Veteran* (London, 1971), p. 71.

71 Neville Jason, 'Letters', letter to his sister Roz, 13 December 1965, AWM. Also see Philip Caputo, *A Rumor of War* (London, 1977), p. xvii.

72 For instance, Lieutenant William L. Calley recalled being asked after one fight how many were killed. He replied 'six to nine', but it was reported as being sixty-nine: Calley, *Body Count* (London, 1971), p. 101.

73 Lieutenant-General Julian J. Ewell and Major-General Ira A. Hunt, *Sharpening the Combat Edge: The Use of Analysis to Reinforce Military Judgment* (Washington, DC, 1974), pp. 227–8.

74 Reg Saunders, quoted in Robert A. Hall, *Fighters from the Fringe: Aborigines and Torres Strait Islanders Recall the Second World War* (Canberra, 1995), pp. 72–3. Also see W. R. H. Brown, 'The Great War. Descriptive Diary', pp. 14–15, IWM; Lieutenant-General Kenneth H. Cousland, 'The Great War 1914–18. A Former Gunner Looks Back', p. 61, Liddell Hart Centre and E. Gardiner Williams, 'The Last Chapter', p. 8, Liverpool Records Office; Frederick Hunt, 'And Truly Serve', 1980, p. 22, IWM; Charles James Lodge Patch, 'Memoir', pp. 37–8, IWM; William F. Pressey, 'All for a Shilling a Day', pp. 132–3, IWM; Victor G. Ricketts, 'Account of His Service', p. 34, IWM; Dixon Scott, 'Letters', letter to A. N. Monkhouse, 6 January 1915, in Manchester City Council Local Studies Unit; Sir Geoffrey Vickers, 'Papers', diary of his brother, Willie Vickers, 24 November 1916, in Liddell Hart Centre; Second Lieutenant Alfred Richard Williams, 'Letters from the West [sic] Front',

letter to his brother, 10 July 1916, pp. 68–9, IWM.

75 W. R. H. Brown, 'The Great War. Descriptive Diary', pp. 14–15, IWM; Alexander Catto, *With the Scottish Troops in France* (Aberdeen, 1918), p. 35; Gerald V. Dennis, 'A Kitchener Man's Bit (1916–1918)', p. 67, IWM; Coningsby Dawson, *The Glory of the Trenches* (London, 1918), p. 12; H. Hesketh-Prichard, *Sniping in France* (London, 1920), p. 37; Sir Edward Hulse, 'Letters', letter to Uncle Mi, 2 February 1915, p. 3, IWM; *Sniping, Scouting and Patrolling* (Aldershot, nd), p. 1; Charles Gordon Templer, 'Autobiography of an Old Soldier', p. 25, IWM; A. Douglas Thorburn, *Amateur Gunners. The Adventures of an Amateur Soldier in France, Salonica and Palestine in the Royal Field Artillery* (Liverpool, 1933), pp. 155–7; Lieutenant-Colonel Neil Fraser Tytler, *Field Guns in France* (London, 1922), pp. 90, 131–2 letters on 17 July 1916, 10 November 1916 and 8 May 1917.

76 Henry S. Salt, *The Creed of Kinship* (London, 1935), p. 34.

77 Letter by Sergeant J. A. Caw, August 1916, in *Letters from the Front. Being a Record of the Part Played by Officers of the [Canadian] Bank [of Commerce] in the Great War, 1914–1919*, vol. 1 (Toronto, 1920), p. 143.

78 Major Gordon Casserly, *The Training of the Volunteers for War* (London, 1915), p. 75; A. T. Walker, *Field Craft for the Home Guard* (Glasgow, 1940), p. 13; General Sir Archibald Wavell, 'Rules and Stratagems of War', in Wavell, *Speaking Generally. Broadcasts, Orders and Addresses in Time of War (1939–43)* (London, 1946), p. 80, issued July 1942.

79 Colonel Jack Houghton-Brown, 'Farmer-Soldier', Liddell Hart Centre, p. 154.

80 Ibid., p. 168.

81 John DiFusco, Vincent Caristi, Richard Chaves, Eric E. Emerson, Rick Gallavan, Merlin Marston, Harry Stephens, with Sheldon Lettich, *Tracers* (New York, 1986), pp. 55-6, cited in David J. DeRose, 'A Dual Perspective: First-Person Narrative in Vietnam Film and Drama', in Owen W. Gilman, Jr., and Lorrie Smith (eds), *America Rediscovered. Critical Essays on Literature and Film of the Vietnam War* (New York, 1990), pp. 114–15.

82 Major Tony Mellor, *Machine-Gunner* (London, 1944), p. 70. Also see Major-General Arthur Smith, (Deputy Chief, General Staff), 'Middle East Training Memorandum No. 10. Destruction of Enemy Tanks', 19 April 1941, p. 1, PRO WO201/2588 and 'M. E. Training Memorandum No. 3. Tank-Hunting Platoons', 8 October 1940, at the same location.

83 Robert Hughes, *Through the Waters. A Gunnery Officer in H.M.S. Scylla 1942–43* (London, 1956), p. 119.

84 Edwyn E. H. Bate, 'Memoirs – Vol. III. 1914–1918', pp. 23–4, IWM; Captain Ralph H. Coventon, 'Fifty Odd Years of Memoirs', pp. 28–9, Liddell Hart Centre; Brigadier W. Carden Roe, 'Memoirs', chapter x, p. 1, IWM.

85 Poem by M. Grover, quoted in Malvern Van Wyk Smith, *Drummer Hodge. The Poetry of the Anglo-Boer War 1899–1902* (Oxford, 1978), p. 152.

86 George MacDonald Fraser, *Quartered Safe Out Here. A Recollection of the War in Burma* (London, 1992), p. 173.

87 J. B. S. Haldane, *Callinicus. A Defence of Chemical Warfare* (London, 1925), pp. 29–30.

88 Ernie Pyle, *Here is Your War* (New York, 1943), p. 141.

89 Max Plowman, *War and the Creative Impulse* (London, 1919), p. 8.

90 Letters of Private Daniel John Sweeney to his fiancé, Ivy Williams, beginning of November 1916, quoted in Michael Moynihan (ed.), *Greater Love. Letters Home 1914–1918* (London, 1980), pp. 84–5. Punctuation as in the original.

91 Foreword by Louis Heren, in Lieutenant William L. Calley, *Body Count* (London, 1971). Also see Stephen Graham, *A Private in the Guards* (London, 1919), pp. 3–4.

92 Samuel A. Stouffer et al., *The American Soldier: Combat and Its Aftermath. Volume II* (Princeton, 1949), p. 162, a survey of 4,495 American infantrymen in 1943–4.

93 Robert William Fairfield Johnston, 'The British Army Officer and the Great War', p. 11, IWM.

94 Bob Swanson, in J. T. Hansen, A. Susan Owen, and Michael Patrick Madden, *Parallels. The Soldiers' Knowledge and the Oral History of Contemporary Warfare* (New York, 1992), p. 123.

95 George Coppard, *With a Machine Gun to Cambrai* (London, 1980), p. 116.

96 William Ernest Hocking, *Morale and Its Enemies* (New Haven, Connecticut, 1918), pp. 56–8. Also see Professor H. J. Laski, *The Germans – Are They Human? A Reply to Sir Robert Vansittart* (London, 1941), pp. 3–6.

97 'Morale (Far Eastern) Inter-Services Committee. Interim Report: Second Draft', 1944, p. 4, in PRO WO32/11195. Also see 'Morale (Far Eastern) Inter-Services Committee. Minutes of the Eighth Meeting Held in Room 433, Hobart House, on Wednesday, 16th August, 1944', pp. 4 and 7, PRO WO32/11195.

98 Philip Caputo, *A Rumor of War* (London, 1977), p. 124. Also see p. 109.

99 Robert Jay Lifton, *Home from the War. Vietnam Veterans: Neither Victims*

Nor Executioners (London, 1974), pp. 52–3.

100 J. Glenn Gray, *The Warriors: Reflections on Men in Battle* (New York, 1959), pp. 152–3.

101 John Joseph Conray describing the reaction of a young Marine, in James D. Horan and Gerald Frank, *Out in the Boondocks. Marines in Action in the Pacific. 21 U. S. Marines Tell Their Story* (New York, 1943), p. 79.

102 Edward Glover, *War, Sadism and Pacifism. Further Essays on Group Psychology and War* (London, 1947), pp. 251–2.

103 Darren Gates, interviewed in J. T. Hansen, A. Susan Owen, and Michael Patrick Madden, *Parallels. The Soldiers' Knowledge and the Oral History of Contemporary Warfare* (New York, 1992), pp. 125–6.

104 Corporal Tom Michlovic in a letter to the editor of the *Pittsburgh Post Gazette*, 7 April 1971, p. 10, cited by Peter Karsten, *Law, Soldiers and Combat* (Westport, Connecticut, 1978), p. 54.

105 Unnamed veteran, quoted by Arthur Egendorf, 'Vietnam Veteran Rap Groups and Themes of Postwar Life', *Journal of Social Issues*, 31.4 (1975), p. 121.

106 Peter Marin, 'Living with Moral Pain', *Psychology Today*, 15.11 (November 1981), p. 71.

107 Ibid., p. 72.

108 Bernard J. Verkamp, *The Moral Treatment of Returning Warriors in Early Medieval and Modern Times* (Scranton, 1993), p. 105.

109 William P. Mahedy, *Out of the Night. The Spiritual Journey of Vietnam Vets* (New York, 1986), p. 105.

110 Peter Marin, 'Living with Moral Pain', *Psychology Today*, 15.11 (November 1981), pp. 71 and 74. Also see Bernard J. Verkamp, *The Moral Treatment of Returning Warriors in Early Medieval and Modern Times* (Scranton, 1993), pp. 79–80.

111 Chaim F. Shatan, 'The Grief of Soldiers: Vietnam Combat Veterans' Self-Help Movement', *American Journal of Orthopsychiatry*, 43.4 (July 1973), pp. 648–9. There were numerous ways of doing this: for example, see Robert Davis, interviewed in Shirley Dicks, *From Vietnam to Hell. Interviews with Victims of Post-Traumatic Stress Disorder* (Jefferson, North Carolina, 1990), p. 27. Also see Arthur Egendorf, 'Vietnam Veteran Rap Groups and Themes of Postwar Life', *Journal of Social Issues*, 31.4 (1975), pp. 111–24; Robert Jay Lifton, *Home from the War. Vietnam Veterans: Neither Victims Nor Executioners* (London, 1974); Peter Marin, 'Living with Moral Pain', *Psychology Today*, 15.11 (November 1981), pp. 68–80.

8 Medics and the Military

1 Shawn O'Leary, 'Shell Shock', in *Spikenard and Bayonet. Verse from the Front Line* (Melbourne, 1941), p. 20.

2 Private Arthur H. Hubbard, 'Letters Written May–November 1916', IWM: letter to 'Mother and All', 7 July 1916; 20 May 1916; to 'Mother and All', 17 June 1916; to his brother Fred, 13 June 1916; to 'Mother and All', 17 June 1916; to his sisters Nellie and Ivy, 29 June 1916; to Nellie and Ivy, 29 June 1916.

3 Albert J. Glass, 'Introduction', in Peter G. Bourne (ed.), *The Psychology and Physiology of Stress with Reference to Special Studies of the Vietnam War* (New York, 1969), p. xxi.

4 Eli Ginzberg, John L. Herma, and Sol W. Ginsburg, *Psychiatry and Military Manpower Policy. A Reappraisal of the Experience in World War II* (New York, 1953), p. 31 and Albert J. Glass, Introduction, in Peter G. Bourne (ed.), *The Psychology and Physiology of Stress with Reference to Special Studies of the Vietnam War* (New York, 1969), pp. xx–xxi.

5 Eli Ginzberg, *Patterns of Performance* (New York, 1959), p. 50.

6 Sara McVicker, quoted in Keith Walker, *A Piece of My Heart. The Stories of 26 American Women who Served in Vietnam* (Novato, CA, 1985), p. 114.

7 Meyer Maskin, *Psychiatry*, 9 (May 1946), pp. 133–41.

8 Isidore S. Edelman, *Diseases of the Nervous System*, 8 (June 1947), pp. 171–4.

9 William Needles, 'The Regression of Psychiatry in the Army', *Psychiatry*, 9.3 (August 1946), p. 176.

10 'Psychiatry – Arakan Campaign', nd, p. 3, PRO WO222/1571 and Anne-Marie Conde, '"The Ordeal of Adjustment": Australian Psychiatric Casualties of the Second World War', *War and Society*, 15.2 (October 1997), pp. 61–74.

11 Richard A. Gabriel, *The Painful Field: The Psychiatric Dimension of Modern War* (New York, 1988), p. 29.

12 Richard A. Kulka et al. (eds), *Trauma and the Vietnam War Generation. Report of Findings from the National Vietnam Veterans Readjustment Study* (New York, 1990), p. v.

13 Arthur Egendorf et al., *Legacies of Vietnam* (New York, 1981), p. 52.

14 Thomas W. Salmon, *The Care and Treatment of Mental Diseases and War Neuroses ('Shell Shock') in the British Army* (New York, 1917), pp. 23–4.

15 E. E. Southard, *Shell-Shock and Other Neuro-Psychiatric Problems Presented in Five Hundred and Eighty Nine Case Histories from the War Literature, 1914–1918* (Boston, 1919), p. 408.

16 Dr G. Roussy and J. Lhermitte, *The Psychoneuroses of War* (London,

1917), p. 105.

17 'Soldiers' Dreams', *The Lancet*, 23 January 1915, p. 210.

18 Quoted by Harvey Cushing, *From a Surgeon's Journal 1915–1918* (London, 1936), p. 489. Also see the Vietnam war poem by John McAfee, 'Open Season', in J. Topham (ed.), *Vietnam Literature Anthology*, revised and enlarged (Philadelphia, 1990), p. 80.

19 For some examples, see W. R. H. Brown, 'The Great War. Descriptive Diary', p. 13, IWM; Robert A. Clark, 'Aggressiveness and Military Training', *American Journal of Sociology*, 51.5 (March 1946), p. 423; Millais Culpin, *Psychoneuroses of War and Peace* (Cambridge, 1920), p. 63; Revd C. I. O. Hood, 'Diary. Gallipoli 1915', entry for 9 October 1915, IWM; Robert William Fairfield Johnston, 'Some Experiences in the Great War of 1914–1918', p. 17, IWM; F. W. Mott, 'Two Addresses on War Psycho-Neurosis (I) Neurasthenia: The Disorders and Disabilities of Fear', *The Lancet*, 26 January 1918, pp. 127–9; F. W. Mott, *War Neuroses and Shell Shock* (London, 1919), pp. 114–23; Captain E. D. Ridley, 'Diary and Letters', letter to his mother 19 November 1914, Cambridge University Library, Manuscripts Department, Acc. 7065; E. E. Southard, *Shell-Shock and Other Neuro-Psychiatric Problems Presented in Five Hundred and Eighty-Nine Case Histories from the War Literature, 1914–1918* (Boston, 1919), throughout.

20 Rowland Myrddyn Luther, 'The Poppies are Blood Red', p. 37, IWM.

21 Quoted in Edwin Weinstein, 'The Fifth U. S. Army Neuropsychiatric Centre – "601st"', in Lieutenant-General Hal B. Jennings (ed.), *Neuropsychiatry in World War II. Volume II. Overseas Theatres* (Washington, DC, 1973), p. 135. Also see p. 134.

22 William Manchester, *Goodbye Darkness. A Memoir of the Pacific War* (Boston, 1980), p. 7.

23 O. P. Napier Pearn, 'Psychoses in the Expeditionary Forces', *The Journal of Mental Science*, lxv (April 1919), p. 101.

24 War Cabinet, 'Ministerial Committee on the Work of Psychologists and Psychiatrists in the Services. Report by the Expert Committee', 31 January 1945, p. 8, PRO CAB21/915. In 10 per cent of cases, the 'cause' was categorized under 'miscellaneous'. For an Australian discussion, see Anne-Marie Conde, '"The Ordeal of Adjustment": Australian Psychiatric Casualties of the Second World War', *War and Society*, 15.2 (October 1997), pp. 61–74.

25 A. J. M. Sinclair, 'Psychiatric Aspects of the Present War', *Medical Journal of Australia*, 3 June 1944, p. 508 and Sinclair, 'Psychological Reactions of Soldiers', *Medical Journal of Australia*, 25 August 1945, p. 229.

26 R. D. Gillespie, *Psychological Effects of War on Citizen and Soldier* (New York, 1942), p. 180.

27 John T. MacCurdy, *War Neuroses* (Cambridge, 1918), p. 11.

28 Ibid., pp. 14 and 111. Similar arguments were made by G. Elliott Smith and T. H. Pear, *Shell Shock and Its Lessons* (Manchester, 1919), pp. 9–10.

29 Lieutenant-Commander R. A. Cohen and Lieutenant J. G. Delano, 'Subacute Emotional Disturbances Induced by Combat', *War Medicine*, 7.5 (May 1945), p. 285; Eli Ginzberg, John L. Herma, and Sol W. Ginsberg, *Psychiatry and Military Manpower Policy. A Reappraisal of the Experience in World War II* (New York, 1953), p. 27; Maurice Silverman, 'Causes of Neurotic Breakdown in British Service Personnel Stationed in the Far East in Peacetime', *The Journal of Mental Science*, xcvi.403 (April 1950), p. 497; 'The Third Meeting of Command Specialists in Psychological Medicine', 21 September 1940, p. 3, PRO WO222/1584; Major Edwin A. Weinstein and Lieutenant-Colonel Calvin S. Drayer, 'A Dynamic Approach to the Problem of Combat-Induced Anxiety', *The Bulletin of the U.S. Army Medical Department*, ix, supplemental number (November 1949), 1949, pp. 15–16; Lieutenant-Commander Meyer A. Zeligs, 'War Neurosis. Psychiatric Experiences and Management on a Pacific Island', *War Medicine*, 6.3 (September 1944), p. 168.

30 Major Marvin F. Greiber, 'Narcosynthesis in the Treatment of the Noncombatant Psychiatric Casualty Overseas', *War Medicine*, 8.2 (August 1945), p. 85.

31 Colonel Rees, 'Memorandum. Psychoneurosis in the Army – September 1939 to June 1940', p. 1, PRO WO222/1584.

32 Lieutenant-Commander R. A. Cohen and Lieutenant J. G. Delano, 'Subacute Emotional Disturbances Induced by Combat', *War Medicine*, 7.5 (May 1945), p. 285.

33 'Psychiatry – Arakan Campaign', nd, p. 6, in PRO WO222/1571.

34 Major Edwin A. Weinstein and Captain Martin H. Stein, 'Psychogenic Disorders of the Upper Gastrointestinal Tract in Combat Personnel', *War Medicine*, 8.5 (November–December 1945), p. 367.

35 Donald P. Hall, 'Stress, Suicide, and Military Service During Operation Uphold Democracy', *Military Medicine*, 161.3 (March 1996), p. 161. Also see J. R. Rundell, R. J. Ursano, H. C. Holloway, and D. R. Jones, 'Combat Stress Disorders and the U. S. Air Force', *Military Medicine*, 155 (1990), pp. 515–18.

36 Ernest Jones, 'War Shock and Freud's Theory of the Neuroses', in S. Ferenczi, Karl Abraham, Ernest Simmel, and Ernest Jones (eds), *Psycho-*

Analysis and the War Neuroses (London, 1921), pp. 47–8.

37 John T. MacCurdy, *War Neuroses* (Cambridge, 1918), p. 11.

38 Commander Francis J. Braceland and Lieutenant-Commander Howard P. Rome, 'Problems of Naval Psychiatry', *War Medicine*, 6.4 (October 1944), p. 219.

39 John T. MacCurdy, *The Psychology of War* (London, 1917), pp. 53–4.

40 Dr Edward W. Lazell, 'Psychology of War and Schizophrenia', *The Psychoanalytic Review*, vii (1920), pp. 227–8. For a Second World War example of this argument, see Edward A. Strecker and Kenneth E. Appel, *Psychiatry in Modern Warfare* (New York, 1945), p. 24.

41 Joseph D. Teicher, '"Combat Fatigue" or Death Anxiety Neurosis', *The Journal of Nervous and Mental Disease*, 117 (January–June 1953), pp. 240–1 and Gregory Zilboorg, 'Fear of Death', *The Psychoanalytic Quarterly*, 12 (1943), pp. 465–75.

42 Circular to all medical officers, 'Morale, Discipline and Mental Fitness', nd, Second World War, p. 4, PRO WO222/218; Eli Ginzberg, John L. Herma, and Sol W. Ginsburg, *Psychiatry and Military Manpower Policy. A Reappraisal of the Experience in World War II* (New York, 1953), pp. 17–18 and 31; Albert J. Glass, Introduction, in Peter G. Bourne (ed.), *The Psychology and Physiology of Stress with Reference to Special Studies of the Vietnam War* (New York, 1969), pp. xx–xxi; Major W. G. Lindsell, 'War Exhaustion', *Journal of the Royal Artillery*, liii (1926–7), p. 462; Charles S. Myers, *Shell Shock in France 1914–18. Based on a War Diary* (Cambridge, 1940), pp. 38–9; 'The Third Meeting of Command Specialists in Psychological Breakdown', 21 September 1940, p. 3, PRO WO222/1584; Major Edwin A. Weinstein and Lieutenant-Colonel Calvin S. Drayer, 'A Dynamic Approach to the Problem of Combat-Induced Anxiety', *The Bulletin of the U.S. Army Medical Department*, ix, supplemental number (November 1949), p. 12.

43 Samuel A. Stouffer et al., *The American Soldier: Combat and Its Aftermath. Volume II* (Princeton, 1949), p. 79, based on a survey of 1,766 soldiers in rifle and heavy weapons companies. Also see Eli Ginzberg, *Patterns of Performance* (New York, 1959), p. 38.

44 'Preventative Psychiatry', *The Bulletin of the U.S. Army Medical Department*, vi.5 (November 1946), p. 493.

45 Major Charles W. Miller, 'Delayed Combat Reactions in Air Force Personnel', *War Medicine*, 8.4 (October 1945), p. 256.

46 Naomi Breslau and Glenn C. Davis, 'Posttraumatic Stress Disorder: The Etiologic Specificity of Wartime Stressors', *American Journal of Psychiatry*, 144.5 (May 1987), pp. 578–83 and Rachel Yehuda, Steven M.

Southwick, and Earl L. Giller, 'Exposure to Atrocities and Severity of Chronic Posttraumatic Stress Disorder in Vietnam Combat Veterans', *American Journal of Psychiatry*, 149.3 (March 1992), pp. 333–6.

47 Amalgamated Union of Building Trade Workers, *The Case of Sapper Hasties* (London, 1943), p. 3; 'Malingering', *British Medical Journal*, 28 July 1917, p. 117; 'Minister's Meeting with Representatives of War Pensions Committees. Extract from the Summary of Proceedings of the Meeting Held at Bristol on 25th April 1930', PRO PIN15/2946; Charles S. Myers, *Shell Shock in France 1914–18. Based on a War Diary* (Cambridge, 1940), pp. 51–3; Thomas W. Salmon, *The Care and Treatment of Mental Diseases and War Neuroses ('Shell Shock') in the British Army* (New York, 1917); *Mental Hygiene*, i.4 (October 1917), p. 516; R. T. Williamson, 'The Treatment of Neurasthenia and Psychosthenia Following Shell Shock', *British Medical Journal*, 1 December 1917, p. 714.

48 'Ex-Private X', *War is War* (London, 1930), pp. 97–8; J. Hutchinson, 'The Early Reminiscence of a Royal Irish Rifleman, 1917–1919', 1982, p. 33, in PRONI D3804; *Report of the War Office Committee of Enquiry into 'Shell-Shock'* [Cmd 1734], H. C. 1922, xii, p. 43, evidence by William Brown; E. E. Southard, *Shell-Shock and Other Neuro-Psychiatric Problems Presented in Five Hundred and Eighty-Nine Case Histories from the War Literature, 1914–1918* (Boston, 1919), pp. 106 and 643. Note that, by some definitions, any man who deliberately set out to fake madness, *was* mad: De Witt MacKenzie, *Men Without Guns* (Philadelphia, 1945), p. 30.

49 Edward Casey, 'The Misfit Soldier', p. 33, IWM. See my edition of this typescript (with introduction), Joanna Bourke, *The Misfit Soldier* (Cork, 1999).

50 Haywood T. 'The Kid' Kirkland, interviewed in Wallace Terry, *Bloods. An Oral History of the Vietnam War by Black Veterans* (New York, 1984), p. 104.

51 Lord Moran, *The Anatomy of Courage* (London, 1945), p. 186.

52 Dr. H. W. Wills, 'Footnote to Medical History . . . General Aspects: Shell Shock', nd, p. 2, Liddle Collection.

53 See 'Report of a Conference on Psychiatry in Forward Areas', Calcutta, 8–10 August 1944, p. 2, address by Major-General Cantlie, in PRO WO32/11550.

54 Surgeon Captain D. Curran, 'Functional Nervous States in Relation to Service in the Royal Navy', in Major-General Sir Henry Letheby Tidy (ed.), *Inter-Allied Conferences on War Medicine 1942–1945* (London, 1947), p. 220. Also see 'Survey of Conditions by DDMS 5 Corps', 9 March to 31 May 1943, p. 8, CMAC.

55 For a further discussion, see Joanna Bourke, *Dismembering the Male:*

Men's Bodies, Britain, and the Great War (London and Chicago, 1996), chapter two.

56 John T. MacCurdy, *War Neuroses* (Cambridge, 1918), pp. 7–8.

57 Philip S. Wagner, 'Psychiatric Activities During the Normandy Offensive, June 20–August 20, 1944', *Psychiatry*, 9.4 (November 1946), p. 356.

58 For example, see Colonel Francis W. Pruitt, 'Doctor–Patient Relationships in the Army', *U.S. Armed Forces Medical Journal*, v.2 (February 1954), p. 204. See also Chapter 9 in this book.

59 *Psychology for the Armed Forces* (Washington, DC, 1945).

60 Colonel Francis W. Pruitt, 'Doctor–Patient Relationships in the Army', *U.S. Armed Forces Medical Journal*, v.2 (February 1954), p. 204.

61 Scott H. Nelson and E. Fuller Torrey, 'The Religious Functions of Psychiatry', *American Journal of Orthopsychiatry*, 43.3 (April 1973), p. 362.

62 Quoted in Peter Marin, 'Living with Moral Pain', *Psychology Today*, 15.11 (November 1981), p. 68.

63 Letter to the editor, from Francis J. Johnston, 'Solidarity – Sociologist Comments', *The Army Combat Forces Journal*, 5.10 (May 1955), p. 7.

64 Report on the speech by Dr C. S. Myers at the annual medical conference of the Ex-Services Welfare Society, 'War Neuroses', *The Lancet*, 15 July 1939, 153. The argument that pensions should not be given to men with psychological problems as a result of their war service was widespread. For some Australian examples, see R. S. Ellery, 'A Psychiatric Programme for Peace', *Medical Journal of Australia*, 1.14 (6 April 1946), p. 459; C. M. McCarthy, 'The Rehabilitation of War Neurotics', *Medical Journal of Australia*, 1.26 (29 June 1946), p. 911; H. Hastings Willis, 'The Rehabilitation of War Neurotics', *Medical Journal of Australia*, 1.26 (29 June 1946), p. 915.

65 'Summary of Lectures on Psychological Aspects of War', lecture entitled 'The Reaction to Killing', p. 5, PRO CAB 21/914 (Annex).

66 For a discussion, see Chapter 9 in this book.

67 Minute from the Secretary of State, 30 June 1915, in PRO WO32/11550.

68 A study of 731 discharges from the Red Cross Military Hospital at Mughum in the year ending 30 June 1917, in Thomas W. Salmon, *The Care and Treatment of Mental Diseases and War Neuroses ('Shell Shock') in the British Army* (New York, 1917); *Mental Hygiene*, i.4 (October 1917), p. 525.

69 For a history, see Robert H. Ahrenfeldt, *Psychiatry in the British Army in the Second World War* (London, 1958), p. 17. Also see Eli Ginzberg,

Patterns of Performance (New York, 1959), p. 13. Note, however, that during the Vietnam War, the Australian forces did not employ any psychiatrists on a regular basis.

70 'Memorandum on Conference of Neurological Deputy Commissioners of Medical Services Held at Headquarters on Friday, June 17th, 1921', 1921, p. 1, PRO PIN15/56.

71 George Rutherford Jeffery, 'Some Points of Interest in Connection with the Psychoneurosis of War', *The Journal of Mental Science*, lxvi.273 (April 1920), p. 140.

72 J. H. Hebb, 'Hardening of Neurasthenics', 6 December 1923, PRO PIN15/2946. Also see PRO PIN15/57, minute dated 20 June 1922.

73 Lieutenant-Colonel J. H. A. Sparrow (compiler), *Morale* (London, 1949), p. 13 and Major C. Kenton, 'Medical History of the War . . . Report for Quarter Ending 30 June 1944', in PRO WO222/1458.

74 Philip S. Wagner, 'Psychiatric Activities During the Normandy Offensive, June 20–August 20, 1944', *Psychiatry*, 9.4 (November 1946), p. 343.

75 Major Alfred O. Ludwig, 'Malingering in Combat Soldiers', *The Bulletin of the U.S. Army Medical Department*, ix, supplemental number (November 1949), pp. 30 and 97.

76 Philip S. Wagner, 'Psychiatric Activities During the Normandy Offensive, June 20–August 20, 1944', *Psychiatry*, 9.4 (November 1946), pp. 348–9.

77 Private Frederick Bratten, 'Conscript RAMC', pp. 29–30, CMAC.

78 Lieutenant-Colonel Louis L. Tureen and Major Martin Stein, 'The Base Section Psychiatric Hospital', *The Bulletin of the U.S. Army Medical Department*, ix, supplemental number (November 1949), p. 127.

79 Jeffrey Streimer and Christopher Tennant, 'Psychiatric Aspects of the Vietnam War. The Effect on Combatants', in Kenneth Maddock and Barry Wright (eds), *War. Australia and Vietnam* (Sydney, 1987), p. 236.

80 F. W. Mott, 'Two Addresses on War Psycho-Neurosis. (I) Neurasthenia: The Disorder and Disabilities of Fear', *The Lancet*, 26 January 1918, p. 129 and Philip S. Wagner, 'Psychiatric Activities During the Normandy Offensive, June 20–August 20, 1944', *Psychiatry*, 9.4 (November 1946), p. 345.

81 A. Jack Abraham, '1914–1918: Memoirs of a Non-Hero', c.1973, p. 84, IWM and Hiram Sturdy, 'Illustrated Account of His Service on the Western Front with the Royal Regiment of Artillery', pp. 25–6, IWM.

82 James Norman Hall, *Kitchener's Mob. The Adventures of an American in the British Army* (London, 1916), p. 184. For a Second World War exam-

ple, see Audie Murphy, *To Hell and Back* (London, 1956), p. 24.

83 Philip Caputo, *A Rumor of War* (London, 1977), p. 289.

84 Undated and untitled typescript by W. D. Esplin, first paragraph, in PRO PIN15/2502.

85 Robert A. Clark, 'Aggressiveness and Military Training', *American Journal of Sociology*, 51.5 (March 1946), p. 423.

86 Captain James Henry Dible, 'Diary and Account', entry for 24 August 1914, p. 29, IWM.

87 War Office, *Army Medical Department Bulletin No. 36* (London, June 1944), p. 5, quoting from *The Bulletin of the U.S. Army Medical Department*, no. 71, December 1943, p. 31.

88 Unnamed Vietnam veteran, cited by Robert Jay Lifton, *Home from the War. Vietnam Veterans: Neither Victims Nor Executioners* (London, 1974), p. 165, ellipses in the original.

89 G. Elliott Smith and T. H. Pear, *Shell Shock and its Lessons* (Manchester, 1919), p. 2.

90 'Report of a Conference on Psychiatry in Forward Areas', 8–10 August 1944, p. 13, PRO WO32/11550.

91 Major J. O. Langtry, 'Tactical Implications of the Human Factors in Warfare', *Australian Army Journal*, 107 (April 1958), p. 14 and Major H. A. Palmer, 'The Problem of the P & N Casualty – A Study of 12,000 Cases', 1944, p. 3, in CMAC.

92 Sir Andrew MacPhail, *Official History of the Canadian Forces in the Great War 1914–19. The Medical Services* (Ottawa, 1925), p. 278 and Philip S. Wagner, 'Psychiatric Activities During the Normandy Offensive, June 20–August 20, 1944', *Psychiatry*, 9.4 (November 1946), p. 356.

93 Franz Alexander, 'The Psychiatric Aspects of War and Peace', *The American Journal of Sociology*, xlvi.4 (January 1941), p. 505.

94 John T. MacCurdy, *The Structure of Morale* (Cambridge, 1943), pp. 44–5.

95 Major H. A. Palmer, 'The Problem of the P & N Casualty – A Study of 12,000 Cases', 1944, p. 11, in CMAC.

96 Captain Eugene R. Hering, 'The Medical Officer's Responsibility to Command', *U. S. Armed Forces Medical Journal*, 4.2 (1953), pp. 1496–1500.

97 Lieutenant-Colonel Robert W. Augustine, 'The Relation of Physician and Physical Therapist', *U.S. Armed Forces Medical Journal*, 4.2 (1953), p. 1515.

98 Philip S. Wagner, 'Psychiatric Activities During the Normandy Offensive, June 20–August 20, 1944', *Psychiatry*, 9.4 (November 1946), p. 358.

99 Lieutenant-General Sir Neil Cantlie, 'Papers', p. 1, his reply to a War Office questionnaire, 1946, CMAC; Marc G. Cloutier, 'Medical Care Behind Enemy Lines: A Historical Examination of Clandestine Hospitals', *Military Medicine*, 158.12 (December 1993), p. 817; George S. Goldman, 'The Psychiatrist's Job in War and Peace', *Psychiatry*, 9.3 (August 1946), p. 265; John Rawlings Rees, *The Shaping of Psychiatry by War* (London, 1945), p. 19.

100 Captain J. C. Dunn, *The War the Infantry Knew 1914–1919*, first published 1938 (London, 1987), p. 585.

101 Philip S. Wagner, 'Psychiatric Activities During the Normandy Offensive, June 20–August 20, 1944', *Psychiatry*, 9.4 (November 1946), p. 348.

102 Eli Ginzberg, *Breakdown and Recovery* (New York, 1950), p. 86. Also see Harry Trosman and I. Hyman Weiland, 'Application of Psychodynamic Principles to Psychotherapy in Military Service', *U.S. Armed Forces Medical Journal*, viii.9 (September 1957), p. 1359.

103 Meredith P. Crawford, 'Military Psychology and General Psychology', *American Psychologist*, 25 (1970), p. 335.

104 Robin M. Williams, 'Some Observations on Sociological Research in Government During World War II', *American Sociological Review*, 11.5 (October 1946), p. 573.

105 Major J. W. Wishart, 'Experiences as a Psychiatrist with BNAF and CMF – Jan 1943 to Jan 1944 With Special Reference to Work in the Forward Areas', 23 January 1944, p. 7, in Major-General Sir Ernest Cowell, 'Papers', CMAC.

106 James Burk, 'Morris Janowitz and the Origins of Sociological Research on Armed Forces and Society', *Armed Forces and Society*, 19.2 (winter 1993), p. 179.

107 'Shell Shock', July 1915, in Sir Charles Burtchaell, 'Papers and Reports', CMAC.

108 Quoted in Donald B. Peterson, 'Discussion', *The American Journal of Psychiatry*, 123.7 (January 1967), p. 819.

109 Ibid., p. 819.

110 L. Kirshner, 'Counter-Transference Issues in the Treatment of the Military Dissenter', *American Journal of Orthopsychiatry*, 43 (1973), p. 657.

111 James E. Driskell and Beckett Olmstead, 'Psychology and the Military. Recent Applications and Trends', *American Psychologist*, 44.1 (January 1989), p. 45.

112 For instance, see Colonel Roy E. Clausen and Arlene K. Daniels, 'Role

Conflicts and their Ideological Resolution in Military Psychiatric Practice', *American Journal of Psychiatry*, 123.3 (September 1966), p. 281; Arlene K. Daniels, 'The Captive Professional: Bureaucratic Limitations in the Practice of Military Psychiatry', *Journal of Health and Social Behavior*, 10.4 (December 1969), p. 257; Arlene K. Daniels, 'Military Psychiatry. The Emergence of a Subspecialty', in Eliot Freidson and Judith Lorber (eds), *Medical Men and their Work* (Chicago, 1972), p. 160.

113 *Red Devils. A Parachute Field Ambulance in Normandy*, 1944, p. 45, in CMAC.

114 Dr Gordon S. Livingstone, 'Healing in Vietnam', in Richard A. Falk, Gabriel Kolko, and Robert Jay Lifton (eds), *Crimes of War* (New York, 1971), pp. 436–7 and Rose Sandecki, interviewed in Keith Walker, *A Piece of My Heart. The Stories of 26 American Women who Served in Vietnam* (Novato, CA, 1985), p. 9.

115 Private Frederick Bratten, 'Conscript RAMC', p. 9, CMAC.

116 *44 Indian Airborne Division. Training Pamphlet No. 1. The Medical Services in the Division* (np, 1944), pp. 8 and 14. Also see A. S. Walker, *Medical Services of the R.A.N. and R.A.A.F.. With a Section on Women in the Army Medical Services* (Canberra, 1961), p. 512, for difficulties distinguishing between 'front' and 'behind-lines' which might encourage the arming of medical personnel.

117 Circular to all medical officers, 'Morale, Discipline and Mental Fitness', undated but Second World War, in PRO WO222/218. Also see John Rawlings Rees, *The Shaping of Psychiatry by War* (London, 1945), pp. 18–19.

118 Jack Strahan, 'The Medic', in J. Topham (ed.), *Vietnam Literature Anthology*, revised and enlarged (Philadelphia, 1990), p. 37. Also see the statement by the medic Lee Reynolds, in Eric M. Bergerud, *Red Thunder, Tropic Lightning. The World of a Combat Division in Vietnam* (St Leonards, New South Wales, 1993), p. 55.

119 M. Ralph Kaufman and Lindsay E. Beaton, 'South Pacific Area', in Lieutenant-General Hal B. Jennings (ed.), *Neuropsychiatry in World War II. Volume II. Overseas Theatres* (Washington, DC, 1973), p. 461.

120 J. Geiger, 'Hidden Professional Roles: The Physician as Reactionary, Reformer, Revolutionary', *Social Policy*, 1 (March–April 1974), p. 28.

121 Gary Gianninoto, interviewed in Mark Lane, *Conversations with Americans* (New York, 1970), p. 214.

122 For instance, see H. R. L. Sheppard, *We Say 'No'. The Plain Man's Guide to Pacifism* (London, 1935), pp. 68–9.

123 Sarah Haley, cited in Wilbur J. Scott, *The Politics of Readjustment. Vietnam Veterans Since the War* (New York, 1993), p. 5.

124 William Needles, 'The Regression of Psychiatry in the Army', *Psychiatry*, 9.3 (August 1946), pp. 167–85.

125 R. Dale Givens and Martin A. Nettleship (eds), *Discussions on War and Human Aggression* (Paris, 1976), p. 8.

126 See the excellent article by Arthur Egendorf, 'Vietnam Veteran Rap Groups and Themes of Postwar Life', *Journal of Social Issues*, 31.4 (1975), pp. 111–24.

127 For instance, see Norman I. Barr and Leonard M. Zunin, 'Clarification of the Psychiatrist's Dilemma While in Military Service', *American Journal of Orthopsychiatry*, 41.4 (July 1971), pp. 672–4; A. Daniels, 'The Captive Professional: Bureaucratic Limitations in the Practice of Military Psychiatry', *Journal of Health and Social Behavior*, 10.4 (December 1969), pp. 255–65; L. Kirshner, 'Counter-Transference Issues in the Treatment of the Military Dissenter', *American Journal of Orthopsychiatry*, 43 (1973), pp. 654–9; Marc Pilisuk, 'The Legacy of the Vietnam Veteran', *Journal of Social Issues*, 31.4 (1975), pp. 6–7. For a history of the anti-psychiatry movement, see Norman Dain, 'Critics and Dissenters: Reflections on "Anti-Psychiatry" in the United States', *Journal of the History of the Behavioral Sciences*, 25 (January 1989), pp. 3–25.

9 Priests and Padres

1 Ian Serraillier, 'The New Learning', in Brian Gardner, *The Terrible Rain. The War Poets 1939–1945* (London, 1966), pp. 30–1.

2 Father George Zabelka, interviewed in Studs Terkel, *'The Good War'. An Oral History of World War Two* (London, 1985), pp. 531–6.

3 Joe Walker, *Parsons and War and Other Essays in War Time* (Bradford, 1917), pp. 3–4.

4 'A Flying Corps Pilot', *Death in the Air. The War Diary and Photographs of a Flying Corps Pilot* (London, 1933), p. 69 and Charles E. Raven, *Musings and Memories* (London, 1931), pp. 166–7.

5 Edith Shawcross, 'The Hand that Rocks the Cradle', in Keith Briant and Lyall Wilkes (eds), *Would I Fight?* (Oxford, 1938), pp. 99–100.

6 Lactantius, *Divine Institutes*, vi, p. 20, cited in Albert Marrin, *The Last Crusade. The Church of England in the First World War* (Durham, North Carolina, 1974), p. 120.

7 *The Military Chaplaincy. A Report to the President by the [U.S.A.] President's Committee on Religion and Welfare in the Armed Forces*

(Washington, DC, 1 October 1950), p. 7 and War Office, *Statistics of the British Empire During the Great War 1914–1920* (London, 1922), p. 91. For histories of the churches during war, see Donald F. Crosby, *Battlefield Chaplains. Catholic Priests in World War II* (Lawrence, Kansas, 1994); Tom Johnstone and James Hagerty, *The Cross on the Sword. Catholic Chaplains in the Forces* (London, 1996); Michael McKernan, *Australian Churches at War. Attitudes and Activities of the Major Churches 1914–1918* (Sydney, 1980); Albert Marrin, *The Last Crusade. The Church of England in the First World War* (Durham, North Carolina, 1974); Brigadier The Rt. Hon. Sir John Smyth, *In This Sign Conquer. The Story of the Army Chaplains* (London, 1968); Alan Wilkinson, *The Church of England and the First World War* (London, 1978); Alan Wilkinson, *Dissent or Conform? War, Peace, and the English Churches 1900–1945* (London, 1986).

8 Revd J. Crookston, 'The Role of Chaplains', p. 5, AWM and Right Revd Gilbert White, 'The Doctrine of Non-Resistance', *The Commonwealth Military Journal*, 4 (October 1913), p. 723.

9 For many examples, see George Bedborough, *Arms and the Clergy (1914–1918)* (London, 1934).

10 Quoted in George M. Marsden, *Fundamentalism and American Culture. The Shaping of Twentieth Century Evangelicalism 1870–1925* (New York, 1980), p. 142.

11 Frank Ballard, *Mistakes of Pacifism or Why a Christian Can have Anything to Do with War* (London, 1915), p. 27 and Captain E. J. Solano, *The Pacifist Lie: A Book for Sailors and Soldiers* (London, 1918), pp. 21–3.

12 Henry Hallam Tweedy, 'The Ministry and the War', in E. Hershey Sneath (ed.), *Religion and the War* (New Haven, Connecticut, 1918), p. 85. Also see William Temple (the Archbishop of York), *Thoughts in War-Time* (London, 1940), pp. 34–5.

13 Rolland W. Schloerb, 'An Uneasy Conscience About Killing', in Dewitte Holland (ed.), *Sermons in American History. Selected Issues in the American Pulpit 1630–1967* (Nashville, 1971), p. 316, reprinted from *The Pulpit*, September 1943.

14 Henry Scott Holland, *So As By Fire. Notes on the War* (London, 1915), p. 24. He was Regius Professor of Divinity in the University of Oxford and Canon of Christ Church.

15 Paul B. Bull, *Peace and War. Notes of Sermons and Addresses* (London, 1917), pp. 33–4.

16 Henry Wace, *The Christian Sanction of War. An Address at the Service of Intercession for the King's Naval and Military Forces, Held in Canterbury Cathedral: August 9th. 1914* (London, 1914), pp. 7–8.

17 Revd A. C. Buckell, *The Greatest War. Six Addresses* (London, 1915), pp. 55–6.

18 George W. Downs speaking at the Asbury Methodist Episcopalian Church in Pittsburgh in November 1917, quoted in Ray H. Abrams, *Preachers Present Arms* (Philadelphia, 1933), p. 67.

19 For his popularity during the Second World War: see Joan Clifford, *Thank You, Padre. Memories of World War II* (London, 1989) and Revd Canon John Wallis, *With God's Blessing and a Green Beret. A Pilgrimage* (Poole, Dorset, 1994), p. 16.

20 William Purcell, *Woodbine Willie. An Anglican Incident. Being Some Account of the Life and Times of Geoffrey Anketell Studdert Kennedy, Poet, Prophet, Seeker After Truth, 1883–1929* (London, 1962), p. 109.

21 Revd G. A. Studdert Kennedy, *Rough Talks by a Padre. Delivered to Officers and Men of the B. E. F.* (London, 1918), pp. 25–6 and 29–30.

22 Interview with an unnamed Canadian soldier, in Barry Broadfoot, *Six War Years 1939–1945. Memories of Canadians at Home and Abroad* (Don Mills, Ontario, 1974), p. 233; A. J. Hoover, *God, Germans, and Britain in the Great War. A Study in Clerical Nationalism* (New York, 1989), p. 6.

23 For instance, see Paul B. Bull, *Peace and War. Notes of Sermons and Addresses* (London, 1917), pp. 34–5 and Thomas Henry Sprott, *Christianity and War. Five Addresses Delivered to the Student Christian Movement Conference. Te Awamutu. January, 1916* (Wellington, 1916), p. 7.

24 'Ourselves and Our Enemies', *Capellanus*, November 1944, p. 2.

25 Revd J. E. Roscoe, *The Ethics of War, Spying and Compulsory Training* (London, nd), p. 26.

26 Harry Emerson Fosdick, *The Challenge of the Present Crisis* (New York, 1917), pp. 38–9.

27 William Temple, *A Conditional Justification of War* (London, 1940), pp. 29 and 32–4.

28 Revd Robert Forman Horton, *The War. Its Cause and Its Conduct* (London, 1914), pp. 31–2. Also see Charles Plater, *A Primer of Peace and War. The Principles of International Morality*, edited for the Catholic Social Guild (London, 1915), p. 86 and Right Revd Arthur F. Winnington Ingram, *The Church in Time of War* (London, 1915), pp. 33–4.

29 Father Grayson, quoted in Jon Oplinger, *Quang Tri Cadence. Memoir of a Rifle Platoon Leader in the Mountains of Vietnam* (Jefferson, North Carolina, 1993), p. 91.

30 Edward Increase Bosworth, *The Christian Witness in War* (New York,

1918), pp. 8–10. Also see Marshall Broomhall, *'Mine Own Vineyard'. Personal Religion and War* (London, 1916), pp. 45–6; Charles Plater, *A Primer of Peace and War. The Principles of International Morality*, edited for the Catholic Social Guild (London, 1915), p. 87; Henry Wace (Dean of Canterbury), *The Christian Sanction of War. An Address at the Service of Intercession for the King's Naval and Military Forces, Held in Canterbury Cathedral: August 9th 1914* (London, 1914), pp. 8–9; Clement Webb, *In Time of War. Addresses Upon Several Occasions* (Oxford, 1918), p. 21; Right Revd Gilbert White, 'The Doctrine of Non-Resistance', *The Commonwealth Military Journal*, 4 (October 1913), p. 722.

31 E. Griffith Jones, *The Challenge of Christianity to a World at War* (London, 1915), p. 186. Also see Revd J. M. Wilson, 'Christ's Sanction as Well as Condemnation of War', *Hibbert Journal*, xiii.4 (July 1915), pp. 839–58.

32 An unnamed 'influential clergyman', quoted by Robert Coope, *Shall I Fight? An Essay on War, Peace, and the Individual* (London, 1935), p. 16 and Marshall Broomhall, *'Mine Own Vineyard'. Personal Religion and War* (London, 1916), pp. 45–6. For protests against sermons exhorting soldiers to love the German soldier while 'thrust[ing] his bayonet into his abdomen', see G. Stanley Hall, 'Morale in War and After', *The Psychological Bulletin*, 15 (1918), p. 384 and Morris N. Kertzer, *With an H on my Dog Tag* (New York, 1947), p. 44.

33 For instance, Fenner Brockway, *Bermondsey Story. The Life of Alfred Salter* (London, 1949), p. 59 and Henry William Pinkham, *Was Christ a Pacifist?* (Melrose, Massachusetts, 1917), p. 14. This question was not only asked by pacifists: see Philip Gibbs, *Realities of War* (London, 1920), p. 82.

34 George Stewart and Henry B. Knight, *The Practice of Friendship* (New York, 1918), pp. 22–4. Also see Sydney E. Ahlstrom, *A Religious History of the American People* (New Haven, Connecticut, 1972), p. 885.

35 Martin Ceadel, 'Christian Pacifism in the Era of Two World Wars', in W. J. Sheils (ed.), *The Church and War* (London, 1983), p. 404.

36 For this fascinating story, see Guenter Lewy, *Peace and Revolution. The Moral Crisis of American Pacifism* (Grand Rapids, Michigan, 1988).

37 G. K. A. Bell, *The Church and Humanity (1939–1946)* (London, 1946), p. 129; G. K. A. Bell, *Christianity and World Order* (Harmondsworth, 1940), pp. 81–2; letter from Dean Thicknesse of St Albans, *The Times*, 16 August 1945; Gordon C. Zahn, *Chaplains in the R.A.F. A Study in Role Tension* (Manchester, 1969), p. 118.

38 Clarence L. Abercrombie, *The Military Chaplain* (Beverly Hills, 1977), p.

142. Prior to 1968, Jewish leaders 'drafted' rabbis into the armed forces, thus bypassing the normal self-selection process and ensuring an opposing culture.

39 Alan Wilkinson, *Dissent or Conform? War, Peace, and the English Churches 1900–1945* (London, 1986), p. 53.

40 Frederick C. Spurr, *Some Chaplains in Khaki. An Account of the Work of Chaplains of the United Navy and Army Board* (London, 1916), p. 52. Also see Charles T. Bateman, 'The Salvation Army and the War', *The Contemporary Review*, cix (January–June 1916), p. 106.

41 Revd Reginald John Campbell, *With Our Troops in France* (London, 1916), p. 12.

42 Henry Hallam Tweedy, 'The Ministry and the War', in E. Hershey Sneath (ed.), *Religion and the War* (New Haven, Connecticut, 1918), pp. 86–7.

43 Tom M. Kettle, *The Ways of War* (London, 1917), pp. 188–205; Major William Redmond, *Trench Pictures from France* (London, 1917), pp. 128–9; 'Junius Redivivus', *The Holy War. Diabolus Extremes: Generosity and Avarice. A Satire* (London, 1915), p. 3; Edw. Winton, 'The Clergy and Military Service', *The Contemporary Review*, cix (January–June 1916), pp. 153–4.

44 J. G. Simpson, 'The Witness of the Church in Time of War', in G. K. A. Bell (ed.), *The War and the Kingdom of God* (London, 1915), pp. 104–9.

45 Revd Arthur Foley Winnington Ingram, *Church Times*, 4 June 1915, cited in Albert Marrin, *The Last Crusade. The Church of England in the First World War* (Durham, North Carolina, 1974), p. 189.

46 Charles Gore, *The War and the Church and Other Addresses* (London, 1914), pp. 60–5.

47 Revd John Sinker, *The War, Its Deeds and Lessons. Addresses Delivered in Lytham Parish Church* (London, 1916), p. 104.

48 Edw. Winton, 'The Clergy and Military Service', *The Contemporary Review*, cix (January–June 1916), pp. 153–60. For the importance of their responsibilities on the home front, see Revd Arthur F. Winnington Ingram, *A Message for the Supreme Moment Delivered to the Assembled Clergy of the London Dioceses by the Bishop of London, November 23rd, 1915, at St Martin's-in-the-Fields* (London, 1915), p. 10.

49 See the *Guardian*, 14 September 1916.

50 Only Canadian clergymen were called by their rank (ie. Captain, Major, Colonel). In the other armies, they possessed officer ranks but were simply called 'padres' by the men. See Revd Harry W. Blackburne, *This*

Also Happened On the Western Front. The Padre's Story (London, 1932), pp. 103–4.

51 J. Fraser McLuskey, *Parachute Padre* (London, 1951), p. 136.

52 For instance, see Ernest Northcote Merrington's statement, in Michael McKernan, *Padre. Australian Chaplains in Gallipoli and France* (Sydney, 1986), p. 82.

53 John J. Callan, *With Guns and Wagons. A Day in the Life of an Artillery Chaplain* (London, 1918), p. 7; Michael McKernan, *Australian Churches at War. Attitudes and Activities of the Major Churches 1914–1918* (Sydney, 1980), p. 54; Revd Dennis Jones, *The Diary of a Padre at Suvla Bay* (London, 1916), p. 17. The stories that Fahey led a charge, although widely rumoured, were untrue according to McKernan. Also see 'Temporary Chaplain', *The Padre* (London, 1916), p. 9.

54 J. Fraser McLuskey, *Parachute Padre* (London, 1951), p. 16.

55 Revd R. L. Barnes, *A War-Time Chaplaincy* (London, 1939), p. 69.

56 Joseph McCulloch, *We Have Our Orders* (London, 1944), p. 176.

57 Barry Broadfoot, *Six War Years 1939–1945. Memories of Canadians at Home and Abroad* (Don Mills, Ontario, 1974), p. 203; Donald F. Crosby, *Battlefield Chaplains. Catholic Priests in World War II* (Lawrence, Kansas, 1995), pp. xx–xxi, 48, and 189; George MacDonald Fraser, *Quartered Safe Out Here. A Recollection of the War in Burma* (London, 1992), p. 110.

58 Quoted in Donald F. Crosby, *Battlefield Chaplains. Catholic Priests in World War II* (Lawrence, Kansas, 1995), p. 13. The priest was said to be William A. Maguire, although he denied it.

59 Larry Rottmann, 'Man of God', in Larry Rottmann, Jan Barry, and Basil T. Paquet (eds), *Winning Hearts and Minds. War Poems by Vietnam Veterans* (New York, 1972), p. 24.

60 Brigadier-General F. P. Crozier, *The Men I Killed* (London, 1937), pp. 76–7.

61 Revd Arthur F. Winnington Ingram, *A Message for the Supreme Moment Delivered to the Assembled Clergy of the London Dioceses by the Bishop of London. November 23rd, 1915, at St Martin's-in-the-Fields* (London, 1915), pp. 9–10; C. E. Montague, *Disenchantment* (London, 1922), pp. 67–8; Edw. Winton, 'The Clergy and Military Service', *The Contemporary Review*, cix (January–June 1916), p. 153.

62 Roy J. Honeywell, *Chaplains of the United States Army* (Washington, DC, 1958), p. 253.

63 Letter by Lieutenant Robert Furley Calloway of the Sherwood Foresters, to his wife on 2 September 1916, quoted in Laurence Housman (ed.), *War Letters of Fallen Englishmen* (London, 1930), pp. 58–9.

64 Revd P. Middleton Brumwell, *The Army Chaplain. The Royal Army Chaplains' Department. The Duties of Chaplains and Morale* (London, 1943), p. 38 and Ronald Selby Wright, *The Padre Presents. Discussions About Life in the Forces* (Edinburgh, 1944), p. 19.

65 Albert Marrin, *The Last Crusade. The Church of England in the First World War* (Durham, North Carolina, 1974), p. 188.

66 Michael McKernan, *Australian Churches at War. Attitudes and Activities of the Major Churches 1914–1918* (Sydney, 1980), p. 95.

67 'Training of Theological Students in War Time', 18 April 1950, PRO DEFE7/46.

68 Charles Edward Montague, *Disenchantment* (London, 1922), pp. 67–8.

69 Canon J. E. Gethyn-Jones, *A Territorial Army Chaplain in Peace and War. A Country Cleric in Khaki 1938–1961* (East Wittering, West Surrey, 1988), pp. 32–3.

70 Noted by Revd R. J. Campbell, *With Our Troops in France* (London, 1916), p. 10.

71 Spencer Leigh Hughes, *Things That Don't Count* (London, 1916), pp. 34–5 and Revd Arthur F. Winnington Ingram, *A Message for the Supreme Moment Delivered to the Assembled Clergy of the London Dioceses by the Bishop of London. November 23rd, 1915, at St Martin's-in-the-Fields* (London, 1915), pp. 10–11.

72 Revd R. L. Barnes, *A War-Time Chaplaincy* (London, 1939), p. 69.

73 John A. Boullier, *Jotting by a Gunner and Chaplain* (London, 1917), p. 13.

74 Revd R. M. Hickey, *The Scarlet Dawn* (Campbellton, N.B., 1949), p. 52 and Brigadier The Rt. Hon. Sir John Smyth, *In This Sign Conquer. The Story of the Army Chaplains* (London, 1968), p. 245.

75 Interview with the chaplain, David Knight, in Shirley Dicks, *From Vietnam to Hell. Interviews with Victims of Post-Traumatic Stress Disorder* (Jefferson, North Carolina, 1990), p. 70.

76 Revd Mervyn S. Evers, 'Memoirs', p. 66, IWM; Roy J. Honeywell, *Chaplains of the U.S. Army* (Washington, DC, 1958), p. 193; letter from Revd John Fahey to the Archbishop of Perth, Patrick Clune, published in *The Advocate* in 1915 and quoted by Michael McKernan, *Padre. Australian Chaplains in Gallipoli and France* (Sydney, 1986), pp. 49–50.

77 Revd P. Middleton Brumwell, *The Army Chaplain. The Royal Army Chaplains' Department. The Duties of Chaplains and Morale* (London, 1943), p. 38. For Boer War example, see Revd J. L. Findlay, *Fighting Padre. (Memoirs of an Army Chaplain)* (London, 1941), p. 26.

78 Brigadier-General F. P. Crozier, *The Men I Killed* (London, 1937), pp. 76–7.

79 Interview with an unnamed Canadian officer, quoted in Barry Broadfoot, *Six War Years 1939–1945. Memories of Canadians at Home and Abroad* (Don Mills, Ontario, 1974), p. 328.

80 Roy J. Honeywell, *Chaplains of the United States Army* (Washington, DC, 1958), p. 253 and Canon William Rook, 'Interview', 21 April 1989, pp. 14–15, AWM.

81 'Even the Padre Has Strapped on a Revolver', unidentified newspaper cutting from the Vietnam War, in Mrs Barbara Arnison, 'Collection of Newspaper Cuttings', AWM.

82 Bryan Cox, *Too Young to Die. The Story of a New Zealand Fighter Pilot in the Pacific War* (Ames, Iowa, 1989), p. 157. Also see Monsignor McCosker, 'Interview', pp. 33–4, AWM.

83 Chaplain Ernest N. Merrington, 'With the Anzacs, 1914–1915', 1922, p. 1, AWM.

84 Professor Alfred O'Rahilly, *Father William Doyle S.J. A Spiritual Study* (London, 1925), throughout.

85 Revd J. Smith, 'The Black Dragoon', nd, np, AWM.

86 For a discussion, see Albert Marrin, *The Last Crusade. The Church of England in the First World War* (Durham, North Carolina, 1974), p. 153.

87 Revd Arthur Herbert Gray, *The Captain* (London, 1921), p. 15 and 'Sunday, May 14th, 1922', *The Mill Hill Magazine*, June 1922, p. 6.

88 Revd G. C. Breach, *'So Fight I'. Thoughts Upon the Warfare in which Every Soul is Engaged and in which There Can Be No Neutrals* (London, 1917), p. 17.

89 'Junius Redivivus', *The Holy War. Diabolus Extremes: Generosity and Avarice. A Satire* (London, 1915), pp. 4–5.

90 Spencer Leigh Hughes, *Things That Don't Count* (London, 1916), p. 40.

91 Spencer Leigh Hughes, *Things That Don't Count* (London, 1916), p. 40; 'Junius Redivivus', *The Holy War. Diabolus Extremes: Generosity and Avarice. A Satire* (London, 1915), pp. 5–6; Edw. Winton, 'The Clergy and Military Service', *The Contemporary Review*, cix (January–June 1916), pp. 153 and 158.

92 Revd J. Smith, 'The Black Dragoon', nd, np, AWM. Also see Edw. Winton, 'The Clergy and Military Service', *The Contemporary Review*, cix (January–June 1916), p. 154.

93 Professor H. M. B. Reid, *Theology After the War* (Glasgow, 1916), pp. 4–6.

94 John Smith, quoted by Michael McKernan, *Australian Churches at War. Attitudes and Activities of the Major Churches 1914–1918* (Sydney, 1980), pp. 132–3.

95 See the praise given to that most famous of combatant padres (Revd E. J. Kennedy) in the preface by the Right Revd The Lord Bishop of Winchester to Kennedy's *With the Immortal Seventh Division*, 2nd edition (London, 1916), p. v. Also see Donald F. Crosby, *Battlefield Chaplains. Catholic Priests in World War II* (Lawrence, Kansas, 1994), p. 48.

96 See the debates in George Bassett, *This Also Happened* (London, 1947), pp. 23–4; Henry Scott Holland, *So As By Fire. Notes on the War. Second Series* (London, 1916), pp. 23–4; Michael MacDonagh, *The Irish at the Front* (London, 1916), p. 104.

97 Joan Clifford, *Thank You, Padre. Memories of World War II* (London, 1989), pp. 90–1.

98 F. P. Crozier, 'Army Chaplains' Usefulness', *Daily Mirror*, 25 April 1930.

99 S. P. MacKenzie, 'Morale and the Cause: The Campaign to Shape the Outlook of Soldiers in the British Expeditionary Force, 1914–1918', *Canadian Journal of History*, 25 (1990), pp. 215–32.

100 Noted in Albert Marrin, *The Last Crusade. The Church of England in the First World War* (Durham, North Carolina, 1974), p. 181.

101 Brigadier-General F. P. Crozier, *A Brass Hat in No Man's Land* (London, 1930), p. 43; Father Joseph T. O'Callahan, *I Was Chaplain on the 'Franklin'* (New York, 1961), p. 40; W. E. Sellers, *From Aldershot to Pretoria. A Story of Christian Work Among Our Troops in South Africa* (London, 1900), p. 113.

102 Quoted in the book by the Methodist chaplain, John A. Boullier, *Jotting by a Gunner and Chaplain* (London, 1917), p. 78.

103 *The Military Chaplaincy. A Report to the President by the [U.S.A.] President's Committee on Religion and Welfare in the Armed Forces* (Washington, DC, 1 October 1950), p. 18. Also see the comments by General Pershing and General Brehan Somervell on p. 13.

104 Revd G. A. Studdert Kennedy, *Rough Talks by a Padre. Delivered to Officers and Men of the B.E.F.* (London, 1918), p. 31.

105 Revd P. Middleton Brumwell, *The Army Chaplain. The Royal Army Chaplains' Department. The Duties of Chaplains and Morale* (London, 1943), p. 51.

106 'F. Ll. H.', 'The Chaplain's Duty in Battle', *The Chaplains' Magazine. Middle East*, 1.2 (Easter 1943), pp. 65–6. The article was followed by a statement by General B. L. Montgomery recommending the advice to all chaplains.

107 For a Vietnam War example, see Haywood T. 'The Kid' Kirkland, interviewed in Wallace Terry, *Bloods. An Oral History of the Vietnam War by Black Veterans* (New York, 1984), pp. 94–5.

108 Ronald Selby Wright, *The Padre Presents. Discussions About Life in the Forces* (Edinburgh, 1944), p. 33, radio broadcast. I have only found one example where it was declared that soldiers never asked padres about the legitimacy of killing: Robert William MacKenna, *Through a Tent Door*, first published 1919 (London, 1930), p. 103.

109 Revd J. Smith, 'The Black Dragoon', nd, np, AWM.

110 Revd A. Irving Davidson, 'A Padre's Reminiscences', p. 12, AWM.

111 Gordon C. Zahn, *Chaplains in the R.A.F. A Study in Role Tension* (Manchester, 1969), p. 114. For a Second World War example, see Monsignor McCosker, 'Interview', June 1989, p. 20, AWM.

112 Revd G. A. Studdert Kennedy, *Rough Talks by a Padre. Delivered to Officers and Men of the B.E.F.* (London, 1918), p. 26. Also see S. A. Alexander, 'Peace and War', *Guardian*, 11 February 1915.

113 Revd William Archibald Moore, 'Experiences of a Chaplain in the A.I.F.', p. 10, AWM.

114 Michael McKernan, *Australian Churches at War. Attitudes and Activities of the Major Churches 1914–1918* (Sydney, 1980), p. 2.

115 Joseph McCulloch, *We Have Our Orders* (London, 1944), p. 171.

116 Lionel Leslie Heming, 'When Down Came a Dragoon', p. 22, AWM.

117 Sermon by Revd E. W. Brereton, in *John Bull*, 10 July 1915.

118 William Temple, *A Conditional Justification of War* (London, 1940), p. 34.

119 Revd H. D. A. Major, 'Sentimentalists and Casuists', *Modern Churchman* (August 1917), pp. 212–13.

120 His book was widely reviewed in the religious press. For instance, see *Intercom. A Journal for Australian Defence Force Chaplains*, 2 and 3 (November 1972 and March 1973), pp. 14–17 and 5–9, respectively.

121 Although Clarence L. Abercrombie, *The Military Chaplain* (Beverly Hills, 1977), p. 97 gives different results, his failure to ask specifically *what* action the padre would take seriously weakens his argument against Zahn.

122 Gordon C. Zahn, *Chaplains in the R.A.F. A Study in Role Tension* (Manchester, 1969), pp. 139, 144–5 and 199–200. For a study which agrees with Zahn, see Waldo W. Burchard, 'Role Conflicts of Military Chaplains', *American Sociological Review*, 19.5 (October 1954), p. 531.

123 Joseph Goldstein, Burke Marshall, and Jack Schwartz (eds), *The Peers Commission Report* (New York, 1976), pp. 266–8.

124 David Griffith, 'There was a Man Sent from God', *Intercom. A Journal for Australian Defence Force Chaplains*, 33 (October 1986), p. 14.

125 John Guest, *Broken Images. A Journal* (London, 1949), p. 214.

126 Robert L. Garrard, 'Combat Guilt Reactions', *North Carolina Medical*

Journal, 10.9 (September 1949), p. 491.

127 Andrew Chandler, 'The Church of England and the Obliteration Bombing of Germans in the Second World War', *English Historical Review* (October 1993), p. 922 and Albert Marrin, *The Last Crusade. The Church of England in the First World War* (Durham, North Carolina, 1974), p. 16.

128 For Bell's story, see Alan Wilkinson, *Dissent or Conform? War, Peace, and the English Churches 1900–1945* (London, 1986). Also see Gordon C. Zahn, *Chaplains in the R.A.F. A Study in Role Tension* (Manchester, 1969), p. 118.

129 Albert Marrin, *The Last Crusade. The Church of England in the First World War* (Durham, North Carolina, 1974), p. 245.

130 Andrew Chandler, 'The Church of England and the Obliteration Bombing of Germans in the Second World War', *English Historical Review* (October 1993), pp. 920–4.

131 For instance, see Revd R. L. Barnes, *A War-Time Chaplaincy* (London, 1939), p. 7; *The Military Chaplaincy. A Report to the President by the [U.S.A.] President's Committee on Religion and Welfare in the Armed Forces* (Washington, DC, 1 October 1950), p. 11; Chaplain George L. Waring, *Chaplain's Duties and How Best to Accomplish His Work* (Washington, DC, 1912), pp. 2–3.

132 Revd Harry W. Blackburne, *This Also Happened On the Western Front. The Padre's Story* (London, 1932), p. 87; Joan Clifford, *Thank You, Padre. Memories of World War II* (London, 1989), pp. 17–19; Tom Johnstone and James Hagerty, *The Cross on the Sword. Catholic Chaplains in the Forces* (London, 1996), p. 101; Revd Leslie Skinner, *'The Man Who Worked on Sundays'. Personal War Diary* (Stoneleigh, Surrey, nd), p. 5.

133 Letter from Rogers to the Director of the Australian War Memorial, 5 December 1921, in Revd Canon E. Jellicoe Rogers, 'Letter', AWM.

134 Revd R. L. Barnes, *A War-Time Chaplaincy* (London, 1939), pp. 17–18.

135 Everard Digby, *Tips for Padres. A Handbook for Chaplains* (London, 1917), pp. 15–17.

136 Bernard J. Verkamp, *The Moral Treatment of Returning Warriors in Early Medieval and Modern Times* (Scranton, 1993), p. 80 and William P. Mahedy, *Out of the Night. The Spiritual Journey of Vietnam Vets* (New York, 1986), p. 7.

137 Albert Marrin, *The Last Crusade. The Church of England in the First World War* (Durham, North Carolina, 1974), pp. 98–103, 140–1, 245, 251–3.

138 Also see Father Joseph T. O'Callahan, *I Was Chaplain on the 'Franklin'* (New York, 1961), p. 40.

139 Waldo W. Burchard, 'Role Conflicts of Military Chaplains', *American Sociological Review*, 19.5 (October 1954), pp. 528, 531, 533–4.

140 Clarence L. Abercrombie, *The Military Chaplain* (Beverly Hills, 1977), p. 52.

141 Albert Marrin, *The Last Crusade. The Church of England in the First World War* (Durham, North Carolina, 1974), p. 187.

142 This point is made most strongly by Tom Johnstone and James Hagerty in *The Cross on the Sword. Catholic Chaplains in the Forces* (London, 1996), p. 103.

143 Clarence L. Abercrombie, *The Military Chaplain* (Beverly Hills, 1977), p. 126.

144 T. A. Bickerton, 'The Wartime Experiences of an Ordinary "Tommy"', August 1964, p. 15, IWM; Lieutenant Frank Warren, 'Journals and Letters', diary for 11 November 1916, IWM; Dennis Kitchin, *War in Aquarius. Memoir of an American Infantryman in Action Along the Cambodian Border During the Vietnam War* (Jefferson, North Carolina, 1994), p. 132; Private Peter McGregor, 'Letters', 7 July 1916, IWM; Andrew Treffry, 'Letters to His Fiancé', letter dated 12 February 1968, AWM. Not everyone had such problems reconciling religious services with bellicosity: see Edward Frederick Chapman, 'Letters from France', letter to his mother, 24 September 1916, IWM.

145 Letter from Wilfred Owen to his mother, 16 May 1917, in Harold Owen and John Bell (eds), *Wilfred Owen. Collected Letters* (London, 1961), p. 461.

146 For one accusation, see the letter from S. L. Edwards (Private Secretary to Sir Edward Griggs), to the MP Cecil H. Wilson, 30 May 1941, in PRO WO32/15272.

147 Rolland W. Schloerb, 'An Uneasy Conscience About Killing', in Dewitte Holland (ed.), *Sermons in American History. Selected Issues in the American Pulpit 1630–1967* (Nashville, 1971), p. 314, reprinted from *The Pulpit*, September 1943. Also see William P. Mahedy, *Out of the Night. The Spiritual Journey of Vietnam Vets* (New York, 1986), p. 7.

148 Rolland W. Schloerb, 'An Uneasy Conscience About Killing', in Dewitte Holland (ed.), *Sermons in American History. Selected Issues in the American Pulpit 1630–1967* (Nashville, 1971), p. 314, reprinted from *The Pulpit*, September 1943.

149 Letter from Montague to his wife, 26 November 1917, and diary entry for 'late 1917', both quoted in Oliver Elton, *C. E. Montague. A Memoir* (London, 1929), pp. 167 and 169.

150 An unnamed radio operator who was killed on his next operational flight, quoted by Canon L. Collins, *Faith Under Fire* (London, 1965), pp. 85–6.

151 Mike Pearson, in Murray Polner, *No Victory Parades. The Return of the Vietnam Veteran* (London, 1971), p. 82.

152 Haywood T. 'The Kid' Kirkland, interviewed in Wallace Terry, *Bloods. An Oral History of the Vietnam War by Black Veterans* (New York, 1984), p. 95. Also see Dennis Kitchin, *War in Aquarius. Memoir of an American Infantryman in Action Along the Cambodian Border During the Vietnam War* (Jefferson, North Carolina, 1994), p. 132.

153 Andrew Treffry, 'Letters to his Fiancé', 12 February 1968, AWM.

154 Robert Jay Lifton, 'The Postwar World', *Journal of Social Issues*, 31.4 (1975), p. 186.

155 Unnamed chaplain, cited in Robert Jay Lifton, *Home From the War. Vietnam Veterans: Neither Victims Nor Executioners* (London, 1974), p. 336.

10 Women Go to War

1 Nora Bomford, 'Drafts', in Catherine Reilly (ed.), *Scars Upon my Heart* (London, 1990), p. 12.

2 Flora Sandes, *An English Woman-Sergeant in the Serbian Army* (London, 1916), pp. 17–18, 34, 72, 115–16, 131, 139, 140, 154 and 220 and Flora Sandes, *The Autobiography of a Woman Soldier. A Brief Record of Adventure with the Serbian Army, 1916–1919* (London, 1927), pp. 9, 12–14, 16, 18–19, 23, 25, 30–1, 72, 82–3, 100, 156 and 220. Also see Alan Burgess, *The Lovely Sergeant* (London, 1963).

3 Helen Cooper, Adrienne Auslander Munich, and Susan Merrill Squier, Introduction, in their (eds), *Arms and the Woman. War, Gender, and Literary Representation* (Chapel Hill, 1989), p. xiii; Jean Bethke Elshtain, *Women and War* (Chicago, 1995), p. 10; Nancy Huston, 'Tales of War and Tears of Women', *Women's Studies International Forum*, 5 (1982), p. 280; C. Tavris, 'The Mismeasure of Women', in J. D. Goodchilds (ed.), *Psychological Perspectives on Human Diversity in America* (Washington, DC, 1991).

4 Alix Strachey, *The Unconscious Motives of War. A Psycho-Analytical Contribution* (London, 1957), p. 266.

5 Lynne B. Iglitzin, 'War, Sex, Sports, and Masculinity', in L. L. Farrar, Jr. (ed.), *War. A Historical, Political, and Social Study* (Santa Barbara, 1978), p. 63.

6 Helen Caldicott, *Missile Envy. The Arms Race and Nuclear War* (New York, 1984), p. 296. Also see John Archer (ed.), *Male Violence* (London, 1994); Alfredo Bonadeo, *Mark of the Beast. Death and Degradation in the Literature of the Great War* (Lexington, 1989); Marion Bromley,

'Feminism and Non-Violent Revolution', in Pam McAllister (ed.), *Reweaving the Web of Life. Feminism and Nonviolence* (Philadelphia, 1982), p. 154; Adrian Caesar, *Taking It Like a Man. Suffering, Sexuality, and the War Poets* (Manchester, 1993); Alice Echols, 'New Feminism of Yin and Yang', in Ann Snitow et al. (eds), *Powers of Desire* (New York, 1983); Klaus Theweleit, *Male Fantasies*, 2 vols. (Minneapolis, 1987).

7 'A Little Mother', 'A Mother's Answer to "A Common Soldier"', first published in *The Morning Post* and cited by Robert Graves, *Goodbye to All That*, first published 1929 (London, 1977), pp. 203–4.

8 Arthur Gleason and Helen Hayes Gleason, *Golden Lads* (New York, 1916), p. 175. This chapter was written by Arthur Gleason.

9 Helen Mana Lucy Swanwick, *Women and War* (London, 1915), pp. 10–11.

10 Mary Beard, *Women as a Force in History* (New York, 1962), p. 48 and Virginia Woolf, *Three Guineas* (London, 1938), p. 13.

11 For a modern discussion, see David R. Segal, Nora Scott Kinzer, and John C. Woelfel, 'The Concept of Citizenship and Attitudes Toward Women in Combat', *Sex Roles*, 3.5 (1977), pp. 469–77. Histories of the early feminist movement summarize the contemporary debates and the way these arguments divided activists.

12 May Wright Sewall, *Women, World War and Permanent Peace* (San Francisco, 1915), p. 172.

13 Linda Grant De Pauw, 'Women in Combat. The Revolutionary War Experience', *Armed Forces and Society*, 7.2 (winter 1981), pp. 222–3.

14 Vern L. Bullough and Bonnie Bullough, *Cross Dressing, Sex, and Gender* (Philadelphia, 1993), pp. 99–102, 134 and 157–64; Karl von Clausewitz, *On War*, edited and translated by Michael Howard and Peter Paret (London, 1993), pp. 73–6; Courtney Ryley Cooper, *Annie Oakley. Woman at Arms* (London, 1928); Dianne Dugaw, *Warrior Women and Popular Balladry 1650–1850* (Cambridge, 1989); Francis Gribble, *Women in War* (London, 1916); Reginald Hargreaves, *Women-at-Arms. Their Famous Exploits Throughout the Ages* (London, 1930); John Laffin, *Women in Battle* (London, 1967); C. J. S. Thompson, *Ladies or Gentlemen? Women Who Posed as Men and Men Who Impersonated Women* (New York, 1993). Warring women remain an important part of certain military traditions: see Miriam Cooke, *Women and the War Story* (Berkeley, 1996).

15 Helen Mana Lucy Swanwick, *I Have Been Young* (London, 1935), p. 246.

16 Caroline Playne, *Society at War 1914–1916* (London, 1931), p. 143.

17 Quoting the Principal Medical Officer of the Commonwealth Forces in Victoria (Australia), Colonel Charles Ryan, in F. Sternberg, 'Women's

Part in the War Time', *Every Lady's Journal*, 6 September 1916, p. 555, cited by Carmel Shute, 'Heroines and Heroes: Sexual Mythology in Australia 1914–1918', in Joy Damousi and Marilyn Lake (eds), *Gender and War. Australians at War in the Twentieth Century* (Cambridge, 1995), p. 32.

18 The AWAS was called the Australian Army Women's Service (AAWS) prior to 1941.

19 Both cited by Ann Howard, *You'll Be Sorry!* (Sydney, 1990), p. 155. Also see Patricia Pitman, 'Private Hatred', in Mavis Nicholson (ed.), *What Did You Do in the War, Mummy? Women in World War II* (London, 1995), p. 172 and Vee Robinson, *On Target* (Wakefield, West Yorkshire, 1991), p. 44, pp. 40 and 44.

20 Peggy Hill, 'Wartime Bride', in Mavis Nicholson (ed.), *What Did You Do in the War, Mummy? Women in World War II* (London, 1995), p. 131.

21 Joyce Carr, 'Just Like William', in Mavis Nicholson (ed.) , *What Did You Do in the War, Mummy? Women in World War II* (London, 1995), p. 112.

22 Captain W. E. Johns, 'Worrals in the Wilds', *Girls' Own Paper* (November 1945), p. 12.

23 For examples, see Kathryn Marshall, *In the Combat Zone. An Oral History of American Women in Vietnam, 1966–1975* (Boston, 1987).

24 Tim O'Brien, *The Things They Carried* (New York, 1990), p. 121.

25 F. Tennyson Jesse, 'The Ecstasy of the Fighter Pilot, October 1939', in her *London Front. Letters Written to America, August 1939–July 1940* (London, 1940), cited by Jenny Hartley (ed.), *Hearts Undefeated. Women's Writing of the Second World War* (London, 1995), p. 218.

26 Margaret W. Griffiths, *Hazel in Uniform* (London, 1945), p. 8. The ellipses in original.

27 Rose Macaulay, 'Many Sisters to Many Brothers'.

28 Vera Brittain, *Testament of Youth*, first published 1930 (London, 1960), p. 156.

29 D'Ann Campbell, 'Servicewomen of World War II', *Armed Forces and Society*, 16.2 (winter 1990), p. 254.

30 Thomas Tiplady, *The Kitten in the Crater and Other Fragments from the Front* (London, 1917), p. 63.

31 Virginia Woolf, *A Room of One's Own*, first published in 1929 (Harmondsworth, 1945), p. 31.

32 George H. Gallup, *The Gallup Poll*, vol. 1 (New York, 1972), p. 435.

33 Desmond P. Wilson and Jessie Horack, 'Military Experience as a Determinant of Veterans' Attitudes', in *Studies Prepared for the President's Commission on an All-Volunteer Armed Force* (Washington, DC, 1970),

vol. 2, part 3, study 7, cited by D'Ann Campbell, 'Servicewomen of World War II', *Armed Forces and Society*, 16.2 (winter 1990), p. 264.

34 Nancy Loring Goldman and Richard Stites, 'Great Britain and the World Wars', in Goldman (ed.), *Female Soldiers – Combatants or Noncombatants? Historical and Contemporary Perspectives* (Westport, Connecticut, 1982), pp. 26 and 31. The numbers include WAAC, WRNS, and WRAF for the First World War and the WRNS, WAAF, and ATS for the Second World War.

35 Lynda Van Devanter and Joan A. Furey (eds), *Visions of War, Dreams of Peace. Writings of Women in the Vietnam War* (New York, 1991), p. xvii; Kathryn Marshall, *In the Combat Zone. An Oral History of American Women in Vietnam, 1966–1975* (Boston, 1987), p. 4; Carol Lynn Mithers, 'Missing in Action: Women Warriors in Vietnam', *Cultural Critique*, 3 (spring 1986), p. 79.

36 According to one estimate, while 76 per cent of jobs in the air force were classified as 'non-combatant', the respective ratios for the army and navy were 50 and 38 per cent: George H. Quester, 'The Problems', in Nancy Loring Goldman (ed.), *Female Soldiers – Combatants or Noncombatants? Historical and Contemporary Perspectives* (Westport, Connecticut, 1982), p. 222. Also see M. J. Castle, 'The Role of Women in the Armed Forces', *The Australian Journal of Defence Studies*, 2.2 (November 1978), p. 116.

37 This veils difficulties in defining what constituted a 'combatant' role. For instance, there were chaplains and medical personnel who wore uniforms yet were legally exempted from combat; there were personnel such as AA operators who attacked the enemy, but were under little or no risk themselves; and there were personnel who attacked the enemy and might be directly attacked themselves either at a distance (as in the artillery and many aerial battles) or at closer quarters (as in the infantry). For a useful discussion see Hugh Smith, 'Conscientious Objection to Particular Wars: Australia's Experience During the Vietnam War, 1965–1972', *War and Society*, 8.1 (May 1990), pp. 135–6.

38 As did the women described by John Mercer, *Mike Target* (Lewes, Sussex, 1990), p. 13 and Edward R. Murrow, *This is London* (London, 1941), p. 116.

39 Siegfried Sassoon, *Diaries 1915–1918*, edited by Rupert Hart-Davis (London, 1983), p. 175, entry for 19 June 1917.

40 D'Ann Campbell, 'Servicewomen of World War II', *Armed Forces and Society*, 16.2 (winter 1990), p. 264 and Nancy W. Gallagher, 'The Gender Gap in Popular Attitudes Toward the Use of Force', in Ruth H. Howes

and Michael R. Stevenson (eds), *Women and the Use of Military Force* (Boulder, 1993), pp. 23–37.

41 D'Ann Campbell, 'Servicewomen of World War II', *Armed Forces and Society*, 16.2 (winter 1990), p. 264.

42 David R. Segal, Nora Scott Kinzer, and John C. Woelfel, 'The Concept of Citizenship and Attitudes Toward Women in Combat', *Sex Roles*, 3.5 (1977), pp. 471–3.

43 Flora Sandes, *The Autobiography of a Woman Soldier. A Brief Record of Adventure with the Serbian Army, 1916–1919* (London, 1927), p. 9.

44 Tim O'Brien, *The Things They Carried* (New York, 1990), p. 101.

45 Megan Llewellyn McCamley, 'Absent Without Leave', in *'Off Duty'. An Anthology. Verse and Short Stories by Men and Women of the Forces – Members of the War Service Women's Club (English Speaking Union) at the Stanley Palace, Chester* (Chester, 1945), pp. 44–6.

46 William Foss and Cecil Gerahty, *The Spanish Arena* (London, 1938), p. 304. Also see Margaret W. Griffiths, *Hazel in Uniform* (London, 1945), p. 95 and Edith Shawcross, 'The Hand that Rocks the Cradle', in Keith Briant and Lyall Wilkes (eds), *Would I Fight?* (Oxford, 1938), p. 94.

47 John Rawlings Rees, *The Shaping of Psychiatry by War* (London, 1945), pp. 94–5. He advised the Service to introduce elementary hygiene lectures where servicewomen could be reassured and told about others who had become pregnant while in military employment ('to indicate that sterility was not a necessary complement of operational service').

48 Dame Leslie Whateley, *As Thoughts Survive* (London, 1949), p. 20.

49 J. B. Priestley, *British Women Go to War* (London, 1943), p. 24.

50 Robert Williamson, 'Women Mobilise', nd [WWII]. It refers to the Women's Volunteer Reserve.

51 William McDougall, *An Introduction to Social Psychology*, 9th edition (London, 1915), p. 59.

52 Clyde B. Moore, 'Some Psychological Aspects of War', *The Pedagogical Seminary. A Quarterly*, xxiii (1916), pp. 141–2.

53 W. N. Maxwell, *A Psychological Retrospect of the Great War* (London, 1923), p. 49.

54 Dame Helen Gwynne Vaughan, *Service with the Army* (London, 1941), p. 138.

55 Captain Frederick Sadleir Brereton, *With Rifle and Bayonet. A Story of the Boer War* (London, 1900), pp. 196–7 and 206.

56 Margaret W. Griffiths, *Hazel in Uniform* (London, 1945), p. 94.

57 William Foss and Cecil Gerahty, *The Spanish Arena* (London, 1938), p. 304.

58 Edith Shawcross, 'The Hand that Rocks the Cradle', in Keith Briant and Lyall Wilkes (eds), *Would I Fight?* (Oxford, 1938), p. 94.

59 H. R. L. Sheppard, *We Say 'No'. The Plain Man's Guide to Pacifism* (London, 1935), p. 81.

60 Edith Shawcross, 'The Hand that Rocks the Cradle', in Keith Briant and Lyall Wilkes (eds), *Would I Fight?* (Oxford, 1938), p. 84.

61 Ion L. Idriess, *Guerrilla Tactics* (Sydney, 1942), p. 14.

62 Sir John Anderson (Lord President of the Council), *Hansard*, 24 March 1942, 5th series, vol. 378, col. 1812–13.

63 Dame Helen Gwynne Vaughan, *Service with the Army* (London, 1941), p. 138.

64 David Lampe, *The Last Ditch* (London, 1968), p. 6.

65 Carmel Shute, 'Heroines and Heroes: Sexual Mythology in Australia 1914–1918', in Joy Damousi and Marilyn Lake (eds), *Gender and War. Australians at War in the Twentieth Century* (Cambridge, 1995), p. 32 and Clare Stevenson and Honor Darling (eds), *The W.A.A.F. Book* (Sydney, 1984), p. 83. For an example from Toronto, see Barbara M. Wilson (ed.), *Ontario and the First World War 1914–1918. A Collection of Documents* (Toronto, 1977), pp. lxxxvi–lxxxxviii and the *Globe*, 30 July 1915, 2 and 27 August 1915, p. 6.

66 Mrs Dawson Scott (ed.), *The Official Handbook of the Women's Defence Relief Corps*, third (enlarged) edition (Southall, Middlesex, 1915), pp. 11 and 15.

67 Mrs Charles Henderson, 'The Women's Voluntary Reserve', nd, single paged leaflet. She was the Country Branch Secretary. Also see Robert Williamson, 'Women Mobilise', undated leaflet, Second World War.

68 Quoted in David Lampe, *The Last Ditch* (London, 1968), pp. 5–6.

69 Edith Summerskill, *A Woman's World* (London, 1967), p. 73.

70 Ibid., p. 73.

71 Vera Douie, *Daughters of Britain* (Oxford, 1949), p. 47 and Sir Frank Bernard Sanderson, in *Hansard*, 5th series, vol. 383, 13 October 1942, col. 1467.

72 Andrew G. Elliot, 'J.B.', and 'Scientist', *The Home Guard Encyclopedia* (London, 1942), pp. 71–3.

73 Letter from Major M. B. Jenkins, HQ Western Command, Chester, to Major G. E. O. Walker, GHQ Home Forces, 17 December 1942, in PRO WO199/401.

74 *Daily Mail*, 14 December 1942.

75 The War Office, 'Urgent Memorandum: Employment of Women to Assist the Home Guard', 15 April 1943, p. 1, PRO WO199/401.

76 Vera Douie, *Daughters of Britain* (Oxford, 1949), p. 47.

77 Ibid., p. 47.

78 George H. Quester, 'The Problems', in Nancy Loring Goldman (ed.), *Female Soldiers – Combatants or Noncombatants? Historical and Contemporary Perspectives* (Westport, Connecticut, 1982), pp. 220–1. It was, of course, not tested in actual combat since North America was not attacked. Also see Mattie E. Treadwell, *United States Army in World War II. Special Studies. The Women's Army Corps* (Washington, DC, 1954), pp. 301–3.

79 Vera Douie, *Daughters of Britain* (Oxford, 1949), p. 41.

80 Major M. S. F. Millington, 'The Women Who Served the Guns', in *On Target. The Great Story of Ack-Ack Command. Official Souvenir* (London, 1955), pp. 41–2. The press incorrectly reported that a mixed battery had been in action as early as September 1941.

81 Ibid., p. 42. Also see Vee Robinson, *On Target* (Wakefield, West Yorkshire, 1991), pp. 2 and 29.

82 'J. W. N.', 'Mixed Batteries', 1942, pp. 3 and 7–8, AWM. Also see Vera Douie, *Daughters of Britain* (Oxford, 1949), p. 41; Edith Summerskill, *A Woman's World* (London, 1967), p. 76; Mattie E. Treadwell, *United States Army in World War II. Special Studies. The Women's Army Corps* (Washington, DC, 1954), p. 191.

83 Ann Howard, *You'll Be Sorry!* (Sydney, 1990), pp. 133–4.

84 'Executive Committee of the Army Council. Employment of WRAC (TA) in Coast Artillery', 5 October 1950, pp. 1–2, in PRO WO32/14100.

85 WAC became the Women's Auxiliary Army Corps (WAAC) in September 1943.

86 Mattie E. Treadwell, *United States Army in World War II. Special Studies. The Women's Army Corps* (Washington, DC, 1954), pp. 337 and 552–3.

87 'History of Matters Affecting Army Women's Services', compiled by Assistant Adjutant-General (WS), August 1947, p. 293 and 'History of the Australia Women's Army Service', 7 January 1947, p. 76, in AWM 54, 88/1/1[2].

88 'The Defensive Role in War of Women in the Army', 1949, pp. 1–3, PRO WO32/13689.

89 Minute signed by the Director-General Army Medical Services, 5 October 1949, in PRO WO32/13689.

90 Minute dated 8 June 1949 from the Director of Personnel Services, in PRO WO32/13689.

91 Summary of Director's Comments on AG16/BM/5007, paragraph 6, in PRO WO32/13689.

92 Minute by R. A. Hull, DSD, 14 June 1949, in PRO WO32/13689.

93 Minute by McGregor, DF, 26 July 1949, in PRO WO32/13689.

94 Comments by the DWRAC, 26 July 1949, in PRO WO32/13689.

95 Memorandum, 15 January 1941, PRO WO32/9423.

96 J. R. Eastwood, 'Employment of Women in the Home Guards', 27 December 1940, in PRO WO32/9423. Also see the comments by the Secretary of State for War (Sir James Grigg) in *Hansard*, 5th series, vol. 383, 29 September 1942, cols. 690–1.

97 Marguerite Higgins, *War in Korea* (New York, 1951), p. 100. Also see Dickey Chapelle, *What's a Woman Doing Here?* (New York, 1962), p. 53; Virginia Elwood-Akers, *Women War Correspondents in the Vietnam War, 1961–1975* (Metuchen, New Jersey, 1988), pp. 3–4; Gloria Emerson, 'Hey, Lady, What are You Doing Here?', *McCalls*, 98 (August 1971), p. 108.

98 Letter to the editor from E. N. Bennett (House of Commons), 'Role of the A.T.S.', *The Times*, 19 January 1942, p. 5.

99 Quoted in Virginia Elwood-Akers, *Women War Correspondents in the Vietnam War, 1961–1975* (Metuchen, New Jersey, 1988), p. 3.

100 Letter to the editor from E. N. Bennett (House of Commons), 'Role of the A.T.S.', *The Times*, 19 January 1942, p. 5.

101 M. J. Castle, 'The Role of Women in the Armed Forces', *The Australian Journal of Defence Studies*, 2.2 (November 1978), p. 126 and Clare Stevenson and Honor Darling (eds), *The W.A.A.F. Book* (Sydney, 1984), p. 219.

102 Elisabeth Cosby, *Meditations of a Khaki Woman* (Amersham, Buckinghamshire, 1920), pp. 3–4.

103 Millicent Garrett Fawcett, 'Women's Work in War Time', *The Contemporary Review*, cvi (June–December 1914), p. 775. For reference to the Second World War, see Memorandum, 15 January 1941, PRO WO32/9423.

104 George H. Quester, 'The Problems', in Nancy Loring Goldman (ed.), *Female Soldiers – Combatants or Noncombatants? Historical and Contemporary Perspectives* (Westport, Connecticut, 1982), pp. 219–20.

105 John Laffin, *Women in Battle* (London, 1967), p. 185.

106 M. D. Feld, 'Arms and the Woman. Some General Considerations', *Armed Forces and Society. An Interdisciplinary Journal*, 4.4 (summer 1978), p. 565.

107 John Finnemore, *Two Boys in War-Time* (London, 1908), pp. 245–8. The 1908 edition is the same as the 1900 edition: the 1928 edition omits this chapter.

108 Terence T. Cuneo, 'The Trail of the Iron Horse', in *Brave and Bold. Stories for Boys* (Oxford, 1943), p. 12.

109 Therese Benedek, *Insight and Personality Adjustment. A Study of the Psychological Effects of War* (New York, 1946), pp. 270–1.

110 Edward R. Murrow, *This is London* (New York, 1941), p. 118.

111 'Notes for Reinforcement Groups and Residues', July 1944, p. 1, PRO WO199/3092.

112 Mairin Mitchell, *Storm Over Spain* (London, 1937), p. 35 and Mattie E. Treadwell, *United States Army in World War II. Special Studies. The Women's Army Corps* (Washington, DC, 1954), p. 209.

113 Letter to the editor from E. N. Bennett (House of Commons), 'Role of the A.T.S.', *The Times*, 19 January 1942, p. 5.

114 John Laffin, *Women in Battle* (London, 1967), p. 185.

115 Colonel Dorie Irving, Controller of the AWAS, quoted in Ann Howard, *You'll Be Sorry!* (Sydney, 1990), p. 20.

116 For instance, see Margaret Mead, 'Warfare is Only an Invention – Not a Biological Necessity', in Leon Bramson and George W. Goethals (eds), *War: Studies from Psychology, Sociology, Anthropology* (New York, 1964), pp. 269–74.

117 Margaret Mead, 'Alternatives to War', in Morton Fried, Marvin Harris, and Robert Murphy (eds), *War. The Anthropology of Armed Conflict and Aggression* (New York, 1968), p. 220.

118 This was not unique to war. Jean Bethke Elshtain in her controversial book *Women and War* (Chicago, 1995) noted that the image of women's violence being formless, savage and uncontrollable was commonplace.

119 Caroline Playne, *Society at War 1914–1918* (London, 1931), p. 140.

120 John Finnemore, *Foray and Fight. Being the Story of the Remarkable Adventures of an Englishman and an American in Macedonia* (London, 1906), pp. 123–4.

121 Peter Kemp, *Mine Were of Trouble* (London, 1957), p. 23. For a fascinating discussion of female combatants in the Spanish Civil War, see Mary Nash, '"Milicianas" and Homefront Heroines: Images of Women in Revolutionary Spain (1936–1939)', *History of European Ideas*, 11 (1989), pp. 235–44.

122 Collie Knox, *Heroes All* (London, 1941), p. 156. Also see W. N. Maxwell, *A Psychological Retrospect of the Great War* (London, 1923), pp. 48–9.

123 Lionel Heming, 'When Down Came a Dragoon', p. 151, AWM.

124 Flora Sandes, *The Autobiography of a Woman Soldier. A Brief Record of Adventure with the Serbian Army, 1916–1919* (London, 1927), p. 156.

125 For an example, see William Foss's and Cecil Gerahty's account of the

trick played by the Women Militia in *The Spanish Arena* (London, 1938), p. 304.

126 Elisabetta Addis, 'Women and the Economic Consequences of Being a Soldier', in Addis and Valeria E. Russo (eds), *Women Soldiers. Images and Realities* (Basingstoke, 1994), pp. 7–10; George H. Quester, 'The Problems', in Nancy Loring Goldman (ed.), *Female Soldiers – Combatants or Noncombatants? Historical and Contemporary Perspectives* (Westport, Connecticut, 1982), p. 232; Hugh Smith, 'Conscientious Objection to Particular Wars: Australia's Experience During the Vietnam War, 1965–1972', *War and Society*, 8.1 (May 1990), pp. 125 and 131.

127 Nancy Loring Goldman and Richard Stites, 'Great Britain and the World Wars', in Goldman (ed.), *Female Soldiers – Combatants or Noncombatants? Historical and Contemporary Perspectives* (Westport, Connecticut, 1982), pp. 38 and 45.

128 Sandra Carson Stanley and Mady Wechsler Segal, 'Military Women in NATO: An Update', *Armed Forces and Society*, 14.4 (summer 1980), p. 560 and George H. Quester, 'The Problems', in Nancy Loring Goldman (ed.), *Female Soldiers – Combatants or Noncombatants? Historical and Contemporary Perspectives* (Westport, Connecticut, 1982), pp. 230–1.

129 *The New York Times*, 25 January 1990, p. A1. This was a *New York Times* and CBS News Poll.

130 *Defence and Security: What New Zealanders Want* (Wellington, NZ, 1986), p. 58.

131 Major Kathryn Quinn, 'Women and the Military Profession', in Hugh Smith (ed.), *The Military Profession in Australia* (Canberra, 1988), pp. 63–6.

132 Charles Moskos, 'Army Women', *The Atlantic Monthly* (August 1990), on the Internet. Three quarters of enlisted women felt that women should *not* be allowed in combat units and about a quarter said that physically capable women should be allowed to *volunteer* for combat roles. None of the enlisted women supported compulsory assignment of women into combat units. Amongst female officers, however, three quarters believed that qualified women should be allowed to volunteer for combat and one quarter even approved of women being compulsorily assigned to combat units.

133 George H. Quester, 'The Problems', in Nancy Loring Goldman (ed.), *Female Soldiers – Combatants or Noncombatants? Historical and Contemporary Perspectives* (Westport, Connecticut, 1982), p. 223.

134 For historians from each of the three countries examined in this book who make this argument, see Miriam Cooke, *Women and the War Story*

(Berkeley, 1996), p. 113; Adrian Howe, 'Anzac Mythology and the Feminist Challenge', in Joy Damousi and Marilyn Lake (eds), *Gender and War. Australians at War in the Twentieth Century* (Cambridge, 1995), p. 305; Sharon Ouditt, *Fighting Forces, Writing Women. Identity and Ideology in the First World War* (London, 1994), p. 217.

11 Return to Civilian Life

1 Charles M. Purcell, holder of the Vietnamese Cross of Gallantry, '. . . In That Age When We Were Young', in Larry Rottmann, Jan Barry, and Basil T. Paquet (eds), *Winning Hearts and Minds. War Poems by Vietnam Veterans* (New York, 1972), p. 109.

2 Emily Mann, *Still Life* (New York, 1982).

3 Alfredo Bonadeo, *Mark of the Beast. Death and Degradation in the Literature of the Great War* (Lexington, 1989), p. 2. Also see Maria Tatar, *Lustmord. Sexual Murder in Weimar Germany* (Princeton, 1995) and Klaus Theweleit, *Male Fantasies*, 2 vols. (Minneapolis, 1987).

4 Dr Francis Rowley in an unnamed and undated SPCA (Society for the Protection of Children and Animals) magazine, quoted by Dixon Wecter, *When Johnny Comes Marching Home* (Cambridge, Massachusetts, 1944), p. 414.

5 Therese Benedek, *Insight and Personality Adjustment. A Study of the Psychological Effects of War* (New York, 1946), p. 90. Also see Ray R. Grinker and John P. Spiegel, *Men Under Stress* (London, 1945), pp. 308 and 362; G. C. Field, *Pacifism and Conscientious Objection* (Cambridge, 1945), p. 23; William Ernest Hocking, *Morale and Its Enemies* (New Haven, Connecticut, 1918), p. 113.

6 Raymond English, *The Pursuit of Purpose. An Essay on Social Morale* (London, 1947), pp. 16–17.

7 Ernest Sheard, 'Manuscript Memoirs', p. 436, IWM.

8 Stephen Graham, *The Challenge of the Dead* (London, 1921), p. 121.

9 Unnamed veteran, quoted by Haynes Johnson, 'The Veterans – Aliens in Their Land', *The Washington Post*, 27 January 1971, p. A12. Also see Jon Nordheimer, 'Postwar Shock Besets Veterans of Vietnam', *The New York Times*, 28 August 1972, p. 20.

10 Harry A. Wilmer, 'The Healing Nightmare: A Study of the War Dreams of Vietnam Combat Veterans', in Reese Williams (ed.), *Unwinding the Vietnam War. From War into Peace* (Seattle, 1987), pp. 75–6.

11 Henry de Man, *The ReMaking of a Mind. A Soldier's Thoughts on War and Reconstruction* (London, 1920), p. 200.

12 Clarence Darrow, *Crime: Its Cause and Treatment*, first published 1922

(Montclair, 1972), pp. 215–16.

13 Major-General G. B. Chisholm, 'Emotional Problems of Demobilization', in *Military Neuropsychiatry. Proceedings of the Association [for Research in Nervous and Mental Disease], December 15 and 16, 1944, New York* (Baltimore, 1946), pp. 63–4.

14 William Manchester, *Goodbye Darkness. A Memoir of the Pacific War* (Boston, 1980), p. 273.

15 Morris H. Adler and Phillip H. Gates, 'Veteran Status Complicating Psychotherapy', *Journal of Nervous and Mental Disease*, 119 (January–June 1954), p. 58.

16 August B. Hollingshaw, 'Adjustment to Military Life', *American Journal of Sociology*, li.5 (March 1946), p. 446.

17 Commissioner L. J. Andolsek, 'Home and Hoping . . . The Vietnam Era Veteran', *Civil Service Journal*, 8.4 (April–June 1968), p. 12.

18 Jerome Johns, interviewed in Peter T. Chews, 'The Forgotten Soldiers, Black Veterans Say They're Ignored', *The National Observer*, 10 March 1973, p. 15.

19 Mardi J. Horowitz and George F. Solomon, 'A Prediction of Delayed Stress Syndromes in Vietnam Veterans', *Journal of Social Issues*, 31.4 (1975), p. 73.

20 Letter by Lieutenant Melville Hastings to the headmaster of Wycliffe College, Autumn 1917, quoted in Laurence Housman (ed.), *War Letters of Fallen Englishmen* (London, 1930), pp. 125–6. Also see Charles H. Williams, *Sidelights on Negro Soldiers* (Boston, 1923), pp. 17–18.

21 Sandy Goodman, 'The Invisible Veterans', *The Nation*, 3 June 1968. Also see Jon Castelli, 'Society and the Vietnam Veteran', *Catholic World*, January 1971; Sol Stern, 'When the Black GI Comes Back from Vietnam', *The New York Times Magazine*, 24 March 1968, p. 190; 'The Vietnam Vet: No One Gives a Damn', in *Newsweek*, 29 March 1971.

22 Whitney M. Young, 'When the Negroes in Vietnam Come Home', *Harper's Magazine*, June 1967, p. 65.

23 'The Vietnam Vet: No One Gives a Damn', *Newsweek*, 29 March 1971, pp. 27–30.

24 David Parks, *GI Diary* (New York, 1968), p. 87, diary for 2 February 1967.

25 For a description of these campaigns during the Second World War, see PRO WO32/11675.

26 For descriptions of hundreds of films with this theme, see Jean-Jacques Malo and Tony Williams (eds), *Vietnam War Films* (Jefferson, North Carolina, 1994) and Michael Fleming and Roger Manwell, *Images of*

Madness. The Portrayal of Insanity in the Feature Film (London, 1985).

27 George Swiers, '"Demented Vets" and Other Myths. The Moral Obligation of Veterans', in Harrison E. Salisbury (ed.), *Vietnam Reconsidered. Lessons from a War* (New York, 1985), p. 198.

28 Morris H. Adler and Philip H. Gates, 'Veteran Status Complicating Psychotherapy', *Journal of Nervous and Mental Disease*, 119 (January–June 1954), p. 58.

29 The veteran was acquitted on grounds that he lacked the ability to appreciate the wrongfulness of his conduct: Jacqueline E. Lawson, '"She's a Pretty Woman . . . For a Gook". The Misogyny of the Vietnam War', in Philip K. Jason (ed.), *Fourteen Landing Zones. Approaches to Vietnam War Literature* (Iowa City, 1991), pp. 15–16.

30 A young private in France, discussed by John T. MacCurdy, *War Neuroses* (Cambridge, 1918), p. 99.

31 Ernest Raymond, *Tell England. A Study in a Generation* (London, 1922), pp. 265–6.

32 Richard Tregaskis, *Guadalcanal Diary* (New York, 1943), p. 148.

33 Colonel David H. Hackworth and Julie Sharman, *About Face* (Sydney, 1989), p. 63.

34 For examples from the two world wars, see William Ernest Hocking, *Morale and Its Enemies* (New Haven, Connecticut, 1918), p. 113 and Captain Martin Stein, 'Neurosis and Group Motivation', *The Bulletin of the U.S. Army Medical Department*, vii.3 (March 1947), p. 319.

35 For instance, see the statements by Charles Stenger, a Veterans' Administration psychologist, in B. Drummond Ayres, 'The Vietnam Veteran: Silent, Perplexed, Unnoticed', *The New York Times*, 8 November 1970, p. 32; Jon Castelli, 'Society and the Vietnam Veteran', *Catholic World*, January 1971.

36 For instance, Richard A. Kulka et al., *Trauma and the Vietnam War Generation. Report of Findings from the National Vietnam Veterans Readjustment Study* (New York, 1990), pp. 180–6.

37 David Lester, 'The Association Between Involvement in War and Rates of Suicide and Homicide', *The Journal of Social Psychology*, 131.6 (1991), pp. 893–5; David Lester, 'War and Personal Violence', in G. Ausenda (ed.), *Effects of War on Society* (San Marino, 1992), p. 213; Colonel John J. Marren, 'Psychiatric Problems in Troops in Korea During and Following Combat', *U. S. Armed Forces Medical Journal*, vii.5 (May 1956), pp. 725–6; Jeffrey Streimer and Christopher Tennant, 'Psychiatric Aspects of the Vietnam War. The Effect on Combatants', in Kenneth Maddock and Barry Wright (eds), *War. Australia and Vietnam* (Sydney,

1987), pp. 230–61; James Webb, President Reagan's Assistant Secretary of Defense, Reserve Affairs, quoted by Timothy J. Lomperis, *'Reading the Wind'. The Literature of the Vietnam War. An Interpretive Critique* (Durham, North Carolina, 1987), p. 17.

38 Commander Robert E. Strange and Captain Dudley E. Brown, 'Home from the War: A Study of Psychiatric Problems in Viet Nam Returnees', *American Journal of Psychiatry*, 127.4 (October 1970), p. 492. My emphasis.

39 Dane Archer and Rosemary Gartner, *Violence and Crime in Cross-National Perspective* (New Haven, Connecticut, 1984), p. 79.

40 For a discussion, see Edwin H. Sutherland, 'Wartime Crime', in Sutherland, *On Analyzing Crime*, first published in 1943 (Chicago, 1973), pp. 120–1.

41 Alan J. Lizotte and David J. Bordua, 'Military Socialization, Childhood Socialization, and Current Situation: Veterans' Firearms Ownership', *Journal of Political and Military Socialisation*, 8.2 (fall 1980), pp. 243–56.

42 Jeffrey Streimer and Christopher Tennant, 'Psychiatric Aspects of the Vietnam War. The Effect on Combatants', in Kenneth Maddock and Barry Wright (eds), *War. Australia and Vietnam* (Sydney, 1987), p. 249.

43 Dane Archer and Rosemary Gartner, *Violence and Crime in Cross-National Perspective* (New Haven, Connecticut, 1984), p. 92 and their 'Violent Acts and Violent Times: A Comparative Approach to Postwar Homicide Rates', *The American Sociological Review*, 41.6 (1976), p. 958.

44 Stephen Garton, *The Cost of War. Australians Return* (Melbourne, 1996), pp. 230–1.

45 D. Michael Shafer, 'The Vietnam Combat Experience: The Human Legacy', in Shafer (ed.), *The Legacy. The Vietnam War in the American Imagination* (Boston, 1990), p. 86.

46 R. Wayne Eisenhart, 'You Can't Hack It Little Girl: A Discussion of the Covert Psychological Agenda of Modern Combat Training', *Journal of Social Issues*, 31.4 (1975), p. 21.

47 D. Michael Shafer, 'The Vietnam Combat Experience: The Human Legacy', in Shafer (ed.), *The Legacy. The Vietnam War in the American Imagination* (Boston, 1990), pp. 92–4. As argued in the chapter on atrocities, we must be careful not to 'over-determine' the Vietnam situation: soldiers in general were young, battlefield leadership was often poor, unit cohesion frequently was unstable, guerrilla wars had been fought before, and postwar disillusionment with the values for which the wars were fought was a typical reaction in all conflicts.

48 For a discussion, see Thomas M. Holm, 'American Indian Veterans and the

Vietnam War', in Walter Capps (ed.), *The Vietnam Reader* (New York, 1991), pp. 191–204 and Holm, 'Forgotten Warriors: American Indian Servicemen in Vietnam', *The Vietnam Generation*, 1.2 (spring 1989), pp. 56–68.

49 Australian veterans rotated as a unit, spent time in troopships on the way home, and were more liable to be treated as war heroes when they returned: see Jane Ross, 'Veterans in Australia: The Search for Integration', in Jeff Doyle and Jeffery Grey (eds), *Australia R&R. Representations and Reinterpretations of Australia's War in Vietnam* (Chevy Chase, MD, 1991), pp. 50–73.

50 Corporal Vito J. Lavacca, interviewed in Otto J. Lehrack, *No Shining Armour. The Marines at War in Vietnam. An Oral History* (Lawrence, Kansas, 1992), p. 1.

51 Don Browning, 'Psychiatry and Pastoral Counseling: Moral Content or Moral Vacuum?', *The Christian Century*, 6 February 1974, p. 161.

52 Marine John B. Doyle, letter to his family quoted by Dixon Wecter, *When Johnny Comes Marching Home* (Cambridge, Massachusetts, 1944), pp. 493–4; William Ernest Hocking, *Morale and Its Enemies* (New Haven, Connecticut, 1918), p. 113; Lieutenant Frank Warren, 'Journals and Letters', letter to 'Bun' on 20 October 1917, IWM. Also see A. Douglas Thorburn, *Amateur Gunners. The Adventures of an Amateur Soldier in France, Salonica and Palestine in The Royal Field Artillery* (Liverpool, 1933), pp. 154–7.

53 Dixon Wecter, *When Johnny Comes Marching Home* (Cambridge, Massachusetts, 1944), p. 413.

54 Alfie Fowler, interviewed in Adrian Walker, *A Barren Place. National Servicemen in Korea, 1950–1954* (London, 1994), pp. 69–70.

55 Robert Jay Lifton, *Home from the War. Vietnam Veterans: Neither Victims nor Executioners* (London, 1974), p. 255. Also see George Bassett, *This Also Happened* (London, 1947), p. 16.

56 William Ernest Hocking, *Morale and Its Enemies* (New Haven, Connecticut, 1918), pp. 113 and 192 and G. Stanley Hall, *Morale. The Supreme Standard of Life and Conduct* (New York, 1920), p. 145.

57 For one discussion, see G. Stanley Hall, *Morale. The Supreme Standard of Life and Conduct* (New York, 1920), p. 145. Also see Joanna Bourke *Dismembering the Male: Men's Bodies, Britain, and the First World War* (London and Chicago, 1996).

58 H. E. Harvey, *Battle-Line Narratives 1915–1918* (London, 1928), p. 14.

59 BBC broadcast by Field-Marshal Sir William Slim, *Courage and Other Broadcasts* (London, 1957), p. 51. Also see Major-General J. M. Devine,

'What is a Tough Soldier?', *Canadian Army Journal*, 3.6 (September 1949), p. 1.

60 Unnamed Vietnam veteran, quoted in Jonathan Shay, *Achilles in Vietnam. Combat Trauma and the Undoing of Character* (New York, 1994), p. 44.

61 Dick Baxter described by Reg Saunders, in United Telecasters Sydney, 'Interview Typescript for the Television Programme "Australians at War"', nd, np.

62 An unnamed soldier, interviewed in Barry Broadfoot, *Six War Years 1939–1945. Memories of Canadians at Home and Abroad* (Don Mills, Ontario, 1974), p. 413.

63 Samuel A. Stouffer et al., *The American Soldier: Combat and its Aftermath. Volume II* (Princeton, 1949), p. 597.

64 Christopher R. Browning, *Ordinary Men. Reserve Police Battalion 101 and the Final Solution in Poland* (New York, 1992) and Daniel Jonah Goldhagen, *Hitler's Willing Executioners. Ordinary Germans and the Holocaust* (London, 1996).

65 John Hersey, *Into the Valley. A Skirmish of the Marines* (London, 1943), pp. 30–1.

66 Ray R. Grinker and John P. Spiegel, *Men Under Stress* (London, 1945), p. 308.

67 B. H. McNeel, 'War Psychiatry in Retrospect', *The American Journal of Psychiatry*, 102 (1945–6), p. 503.

68 Captain Alfred E. Bland, 'Letters to His Wife', 25 February 1916, IWM.

69 An unnamed Irish soldier, quoted in Michael MacDonagh, *The Irish on Somme* (London, 1917), p. 22.

70 For a more detailed discussion, also see Joanna Bourke, *Dismembering the Male: Men's Bodies, Britain, and the Great War* (London and Chicago, 1996).

71 Kenneth T. Henderson, *Khaki and Cassock* (Melbourne, 1919), p. 78.

72 For an interesting illustration of this, see Joseph Wambaugh, *The Choirboys* (New York, 1975), p. 58. Also see Jon Castelli, 'Society and the Vietnam Veteran', *Catholic World*, January 1971.

73 Jack Strahan, 'It's Just That Everyone Asks', in J. Topham (ed.), *Vietnam Literature Anthology*, revised and enlarged (Philadelphia, 1990), p. 20.

74 Unnamed female veteran, quoted in Jenny A. Schnaier, 'A Study of Women Vietnam Veterans and Their Mental Health Adjustment', in Charles R. Figley (ed.), *Trauma and Its Wake. Volume II: Traumatic Stress Theory, Research, and Intervention* (New York, 1986), p. 97.

75 'John', cited in Richard Moser, 'Talkin' the Vietnam Blues: Vietnam Oral History and Our Popular Memory of War', in D. Michael Shafer (ed.),

The Legacy of the Vietnam War in the American Imagination (Boston, 1990), p. 116.

76 Philip Caputo, *A Rumor of War* (London, 1977), p. xix.
77 Michael Herr, *Dispatches* (London, 1978), p. 169.

Epilogue

1 Niall Ferguson, *The Pity of War* (London, 1998), pp. 447 and 360.
2 Stuart Smyth, 'Back Home', in Jan Barry and W. D. Ehrhart (eds), *Demilitarized Zones. Veterans After Vietnam* (Perkasie, Pa., 1976), p. 172.
3 Unnamed Vietnam veteran quoted in Cindy Cook Williams, 'The Mental Foxhole: The Vietnam Veteran's Search for Meaning', *American Journal of Orthopsychiatry*, 53.1 (January 1983), p. 4.
4 Paul Sgroi, 'To Vietnam and Back', in Walter Capps (ed.), *The Vietnam Reader* (New York, 1991), p. 31.

BIBLIOGRAPHY

Only references cited in the endnotes have been included.

1 Australian War Memorial

'Allied Land Forces in S. W. Pacific Area. Training Ideas – No. 5', 1943, AWM 54, 943/1/24

Arnison, Mrs Barbara, 'Collection of Newspaper Cuttings', AWM PR89/6

Berryman, Lieutenant-General Sir F. H., 'Papers', AWM PR84/370

Casey, Edward (pseudonym J. W. Roworth), 'The Misfit Soldier', IWM 80/40/1

Crookston, Revd J., 'The Role of Chaplains', AWM 1DRL/0623

Davidson, Revd A. Irving, 'A Padre's Reminiscences', AWM 1DRL/0626

Ewen, Sergeant John Henry, 'Bougainville Campaign', 1944, AWM PR89/190

Gates, Private Peter, 'Letters to His Family', AWM PR83/218

Heming, Private Lionel (Leslie), 'When Down Came a Dragoon', AWM MSS 867

'History of Matters Affecting Army Women's Services', compiled by Assistant Adjutant-General (WS), August 1947 and 'History of the Australian Women's Army Service', 7 January 1947, AWM 54, 88/1/1[2]

Jason, Neville, 'Letters', AWM PR91/069

Jones, Geoffry R., 'A Short Memoir', AWM PR87/196

Jones, W. D., 'Precis on Jungle Fighting', 1942, AWM 54, 937/3/23

'Jungle Warfare Extracts', Second World War, AWM 54, 923/1/5

'J. W. N.', 'Mixed Batteries', AWM 54, 75/1/2

Kent, Lance, 'Autobiography', AWM MSS 1238

Laycock, Colonel, 'Commando Training in the Middle East', 1942, AWM 54, 422/7/8

Leslie, Major John, 'Diaries', AWM PR89/139

McCosker, Monsignor, 'Interview', AWM Keith Murdoch Sound Archives 599

Matheson, Lieutenant-Colonel, 'Notes on Jungle Combat', 1943, AWM 54, 642/2/9

Merrington, Chaplain Ernest N., 'With the Anzacs, 1914–1915', AWM 1DRL/496

Moore, Revd William Archibald, 'Experiences of a Chaplain in the A.I.F.', AWM 1DRL/0640

Nagle, William, 'Do You Remember When?', AWM PR89/148

Rogers, Revd Canon E. Jellicoe, 'Letter', AWM 1DRL/645

Rook, Canon William, 'Interview', AWM Keith Murdoch Sound Archive S571

Smith, Revd J., 'The Black Dragoon', AWM 1DRL/649

Treffry, Andrew, 'Letters to His Fiancé', AWM PR32

Willis, William K., 'Letters to Miss Luttrell', AWM 3DRL/6333

2 Imperial War Museum

Abraham, A. Jack, '1914–1918: Memoirs of a Non-Hero', c. 1973, IWM P191

Bacon, T. W., 'Letters', IWM P391

Barnett, Lieutenant-Colonel J. W., 'War Diary', IWM 90/37/1

Bate, Edwyn E. H., 'Memoirs – Vol. III. 1914–1918', IWM 90/37/1

Bere, Revd M. A., 'Papers', IWM 66/96/1

Bickerton, T. A., 'The Wartime Experiences of an Ordinary "Tommy" ', IWM 80/43/1

Bland, Captain Alfred E., 'Letters to His Wife', IWM 80/1/1

Bovingdon, Thomas Pincombe, 'Memoirs', IWM 81/34/1

Brennan, Private Anthony R. 'Diaries', IWM P262

Brett, G. A., 'Recollections', IWM 78/42/1

Brown, W. R. H., 'The Great War. Descriptive Diary', IWM 81/14/1

Bundy, Alfred E., 'War Diary', IWM P371

Casey, Edward, 'The Misfit Soldier', IWM 80/40/1

Chapman, Edward Frederick, 'Letters from France', IWM 92/3/1

Clarke, Nobby, 'Sniper on the Western Front', IWM 76/161/1

Clarke, William, 'Random Recollections of "14/18" ', IWM

Clements, H. T., 'Diaries', IWM 86/76/1

Colebrook, E. W., 'Letters Home', IWM 77/117/1

Colyer, Captain Wilfred Thomas, 'Memoirs', IWM 76/51/1

Cordy, John M., 'My Memories of the First World War', IWM86/30/1

Coventry, B. G., 'The Great War as I Sampled It', IWM 85/1/1

Cox, Cecil H., 'A Few Experiences of the First World War', IWM 88/11/1

Cumstie, Alex, 'Diary', IWM 94/28/1

Dance, H. S., 'Diary', IWM 84/58/1

Daniell, Major E. Henry E. , 'Letter to Mother', IWM 87/8/1

Davey, J., 'Letter to the Curator of the Imperial War Museum', IWM 85/28/1

Dawson, P. J., 'Some Incidents in the Development of Armaments for World War II', IWM 77/136/1

De Margry, F., 'From Saddle-Bag to Infantry Pack' and 'In Memoriam', IWM 82/22/1

Dennis, Gerald V., 'A Kitchener's Man's Bit (1916–1918)', IWM 78/58/1

Dible, Captain James Henry, 'Diary and Account', IWM Con Shelf

Doran, Private John, 'Letters', IWM Misc. 89 Item 1306

Evers, Revd Mervyn S., 'Memoirs', IWM 78/7/1

Foot, Richard C., 'Once a Gunner', IWM 86/57/1

Foukes, Edward, 'A V. C. Episode', IWM P190

Garry, K., 'Letter to Mother From France', IWM 77/95/1

Geddes, Colonel G. W., 'Papers', IWM 66/112/1

Godson, Captain E. A., 'Diaries', IWM P446

Hall, Percy Raymond, 'Recollections', IWM 87/55/1

Handley, John Starie, 'The Somme and the "Birdcage" 1916', IWM 92/36/1

Harker, R. [Bob] P., 'Letters', IWM Con Shelf

Hood, Revd C. I. O., 'Diary. Gallipoli 1915', IWM 90/7/1

Howeill, W. J. N., 'Papers', IWM F18.3

Hubbard, Private Arthur H., 'Letters Written May–November 1916', IWM Con Shelf

Hulse, Sir Edward, 'Letters', IWM 86/30/1

Hunt, Frederick, 'And Truly Serve', IWM 88/52/1

Jamieson, A. J., 'The Final Advance – September–November 1918', IWM 88/52/1

Johnston, Robert William Fairfield, 'The British Army Officer and the Great War' and 'Some Experiences in the Great War of 1914–1918', IWM 82/38/1

Kemp, Sid T., 'Remembrance. The 6th Royal West Kent Regiment 1914–1916', IWM 85/28/1

Kennedy, Captain M. D., 'Their Mercenary Calling', 1932, IWM P392

Kerr, William, 'Canadian Soldier', IWM 86/53/1

Kirby, W. R., 'The Battle of Cambrai, 1917', IWM 78/51/1

Kirkpatrick, Sir Ivone, 'Memoirs', IWM 79/50/1

Lockwood, Sydney W. D., 'First World War Memories 1914–1918', IWM 90/21/1 and 'No Man's Land', IWM

Luther, Rowland Myrddyn, 'The Poppies are Blood Red', IWM 87/8/1

McGregor, Private Peter, 'Letters', IWM 76/153/2

MacKay, R. L., 'Diary', IWM P374

McKeag, Captain H. T. A., 'Memoirs', IWM 86/2/1

McKerrow, Captain Charles K., 'Diaries and Letters', IWM 93/20/1

Markham, Sir Sydney Frank, 'Papers', IWM P46

Miller, Charles Cecil, 'A Letter from India to my Daughter in England', IWM 83/3/1

Morris, A. Ashurt, 'Diaries', IWM 68/3/1

Mudd, Jack W., 'Letters', IWM 82/3/1

Munro, Lieutenant Hugh A., 'Diary', IWM P374

Neville, W. P., 'Letters and Papers', IWM Con Shelf

Nightingale, Captain Guy Warneford, 'The 1915 Letters', IWM P216

Nixon, Clifford, 'A Touch of Memory', IWM Misc. 163 Item 2508

Ogden, C. K., 'Papers', IWM 81/28/1

Owen, Lieutenant Rowland H., 'Letters Home', IWM 90/37/1

Packham, F. M., 'Memoirs of an Old Contemptible 1912–1920', IWM P316

Palmer, E. C., 'It Happened Twice', IWM P370

Path, Captain Charles James Lodge, 'Memoir', IWM 66/304/1

Pease, Nicholas A., 'Retrospect', IWM 86/9/1

Plint, R. G., 'Memoirs', IWM 80/19/1

Pond, Major Bertie A., 'Memoirs', IWM 78/27/1

Pressey, William F., 'All for a Shilling a Day', IWM 77/84/1

Quinton, W. A., 'Memoirs', 1929, IWM 79/35/1

Reardon, Captain D. B., 'Diary', IWM 80/17/1

Ricketts, Victor G., 'Account of his Service', IWM 68/14/1

Roe, Brigadier W. Carden, 'Memoirs', IWM 77/165/1

Shaw, Norman, 'Papers', IWM 84/9/1

Sheard, Ernest, 'Manuscript Memoirs', IWM P285

Shipway, W. G., 'My Memories of the First World War', IWM 90/37/1

Smith, Lieutenant P. L., 'Papers', IWM 88/27/1

Spurrell, Hugh W., 'My Impressions of Life in the Great War', IWM 92/36/1

Stainton, Harold, 'A Personal Narrative of the War', IWM 78/11/1

Stephenson, J. W., 'With the Dukes in Flanders', IWM 78/36/1

Sturdy, Hiram, 'Illustrated Account of His Service on the Western Front With the Royal Regiment of Artillery', IWM Con Shelf

Taylor, Major O. P., 'Diaries of a Sometime Trench Mortar Man', IWM 92/3/1

Templer, Charles Gordon, 'Autobiography of an Old Soldier', IWM 86/30/1

Thomas, Revd Harold Augustine, 'A Parson', IWM Con Shelf

Turner, A. J., 'Zero Hour', IWM 81/21/1

Turner, Frank, 'Memo', IWM 83/23/1

Warren, Lieutenant Frank, 'Journals and Letters', IWM P443

Whitfeld, Brigadier G. H. P., 'Some Notes on the Doing [sic] of the 1st Btn. Royal Irish Rifles 1916–1918', IWM Con Shelf

Williams, Second Lieutenant Alfred Richard, 'Letters from the West Front', IWM 82/26/1

Williams, Miss D., 'Letters', IWM 85/4/1

Wollocombe, Lieutenant-Colonel T. S., 'Diary of the Great War', IWM 98/7/1

3 Newspapers, occasional periodicals, and parliamentary papers

The Advertiser
The Age
The Australian
British Medical Journal
British Parliamentary Papers
Capellanus
Catholic World
The Christian Century
Collier's
Current
The Daily Mail
The Daily Mirror
The Daily Telegraph Pictorial
The Globe
The Guardian
Hansard
Harper's Magazine
H. C. Parliamentary Debates
John Bull
The Lancet
Medical Journal of Australia
The Mill Hill Magazine
The Morning Adviser
The Nation
The Nation Review
The National Observer
The National Times
The New Republic
The New York Times

The New York Times Magazine
The New Yorker
Newsweek
The Observer
Society
The Sunday Pictorial
The Sunday Times
Time
The Times
Washington Monthly
The Washington Post
The Weekly Dispatch

4 Public Record Office

Cabinet Papers
Ministry of Pensions
War Office Papers

5 Other archives

Alexander, Charles Stewart, 'Letters to his Cousin', Auckland Institute and Museum Library MSS 92/70

Andrews, Captain D. W. J., 'Diary', CMAC RAMC 2021

'The Battle Hymn of Lieutenant Calley', Australian National Film and Sound Archive, 4ZZZZ-FM Archive Collection, 'My Lai Massacre', Cover 235340

Bentham, Private Walter, 'With the British Forces in France. Diary', CMAC RAMC 2010

'Brassard', 'The R. A. M. C.', *The Peeka Journal*, 15 (1916), CMAC RAMC 1428

Bratten, Private Frederick, 'Conscript RAMC', CMAC RAMC 1310

Burtchaell, Sir Charles, 'Papers and Reports', CMAC RAMC 44 Boc 66

Cantlie, Lieutenant-General Sir Neil, 'Papers', CMAC RAMC 465/10

Clarke, Firstbrooke and Grosvenor, 'Letters', in the Liddell Collection (Leeds)

Courtney, Henry Gother, 'Letters', Birmingham University Library Courtney Collection 8–10

Cousland, Lieutenant-Colonel Kenneth H., 'The Great War. A Former Gunner Looks Back', Liddell Hart Centre for Military Archives

Coverton, Captain Ralph H., 'Fifty Odd Years of Memories', Liddell Hart Centre for Military Archives

Cowell, Major-General Sir Ernest, 'Papers', CMAC RAMC 466/49

Darlington, Sir Henry C., 'Letters from Helles', Liddell Hart Centre for Military Archives

Downes, Alfred, 'Notes Taken at Southern Army School of Instruction, Brentwood. Course Commencing Oct. 16, 1916', Birmingham City Archives, MSS 547

Early, J. H., 'War Diary 1914–1918', Cambridgeshire County Record Office

Hodgson, John Slomm Riddell, 'Letters of a Young Soldier', Birmingham University Library Special Collection MSS 6/i/7

Houghton-Brown, Colonel Jack, 'Farmer-Soldier', Liddell Hart Centre for Military Archives

Idriess, Ion Llewellyn, 'Biographical Cuttings', Australian National Library

Morley, Private Frank E., 'War Diary', CMAC RAMC 964/1

Palmer, Major H. A., 'The Problem of the P & N Casualty – A Study of 12,000 Cases', 1944, CMAC RAMC 466/49

Red Devils. A Parachute Field Ambulance in Normandy, 1944, CMAC RAMC 695

Ridley, Captain E. D., 'Diary and Letters', Cambridge University Library Manuscripts Department, Acc. 7065

Scholes, Dorothy, 'Papers', Wigan Archives Service D/DZ.EHC

Scott, Dixon, 'Letters', Manchester City Council Local Studies Unit

Stevenson-Hamilton, Lieutenant-Colonel Vivian Edgar Olmar, 'Papers', Liddell Hart Centre for Military Archives

Vickers, Sir Geoffrey, 'Papers', Liddell Hart Centre for Military Archives

Williams, E. Gardiner, 'The Last Chapter', Liverpool Records Office Acc 2175

Wills, Dr H. W., 'Footnote to Medical History . . . General Aspects. Shell Shock', item 1, Liddle Collection

6 Published books and articles

A Book of British Heroes (London, 1914)

'A First Battalion Officer', 'Some Glimpses of the Battle of Givenchy', *The Braganza*, 1.1 (May 1916)

'A Flying Corps Pilot', *Death in the Air. The War Diary and Photographs of a Flying Corps Pilot* (London, 1933)

'An Officer', *Practical Bayonet-Fighting, with Service Rifle and Bayonet* (London, 1915)

'A Private', *The Private Life of a Private – Being Extracts from the Diary of a Soldier of Britain's New Army* (London, 1941)

Abercrombie, Clarence L., *The Military Chaplain* (Beverly Hills, 1977)

Abrams, Ray H., *Preachers Present Arms* (Philadelphia, 1933)

Adams, Bernard, *Nothing of Importance. A Record of Eight Months at the Front With a Welsh Battalion, October 1915 to June 1916* (Stevenage, 1988)

Adams, Carol J., 'Feminism, the Great War, and Modern Vegetarianism', in Helen M. Cooper, Adrienne Auslander Munich, and Susan Merrill Squier (eds), *Arms and the Woman. War, Gender, and Literary Representation* (Chapel Hill, 1989)

Addis, Elisabetta, 'Women and the Economic Consequences of Being a Soldier', in Addis and Valeria E. Russo (eds), *Women Soldiers. Images and Realities* (Basingstoke, 1994)

Adler, Bill (ed.), *Letters from Vietnam* (New York, 1967)

Adler, Morris H. and Phillip H. Gates, 'Veteran Status Complicating Psychotheraphy', *Journal of Nervous and Mental Disease*, 119 (January–June 1954)

Ahlstrom, Sydney E., *A Religious History of the American People* (New Haven, Connecticut, 1972)

Ahrenfeldt, Robert H., 'Military Psychiatry', in Sir Arthur Salisbury MacNalty and W. Franklin Mellor (eds), *Medical Services in the War. The Principal Medical Lessons of the Second World War Based on the Official Medical Histories of the United Kingdom, Canada, Australia, New Zealand and India* (London, 1968)

Ahrenfeldt, Robert H., *Psychiatry in the British Army in the Second World War* (London, 1958)

Alderson, E. A. H., *Pink and Scarlet. Or Hunting as a School for Soldiering* (London, 1913)

Alexander, Franz, 'Aggressiveness – Individual and Collective', in *The March of Medicine. The New York Academy of Medicine. Lectures to the Laity 1943* (New York, 1943)

Alexander, Franz, 'The Psychiatric Aspects of War and Peace', *The American Journal of Sociology*, xlvi.4 (January 1941)

Allanson-Winn, R. G. and C. Phillips-Wolley, *Broadsword and Singlestick* (London, 1890)

Allen, Revd E. L., *Pacifism as an Individual Duty* (London, 1946)

Allen, Frederick H., 'Homosexuality in Relation to the Problem of Human Difference', *The American Journal of Orthopsychiatry*, x (1940)

Allport, George, *The Nature of Prejudice*, first published in 1954 (Reading, Massachusetts, 1979)

Amalgamated Union of Building Trade Workers, *The Case of Sapper Hasties* (London, 1943)

Anderson, Charles, 'On Certain Conscious and Unconscious Homosexual

Responses to Warfare', *British Journal of Medical Psychology*, xx.2 (1945)

Andolsek, Commissioner L. J., 'Home and Hoping . . . The Vietnam Era Veteran', *Civil Service Journal*, 8.4 (April–June 1968)

Anisfield, Nancy, 'After the Apocalypse: Narrative Movement in Larry Heinemann's *Paco's Story*', in Owen W. Gilman, Jr., and Lorrie Smith (eds) *America Rediscovered. Critical Essays on Literature and Film of the Vietnam War* (New York, 1990)

'Anzac', *On the Anzac Trail. Being Extracts from the Diary of a New Zealand Sapper* (London, 1916)

Appy, Christian G., *Working-Class War. American Combat Soldiers and Vietnam* (Chapel Hill, 1993)

Archer, Dane, and Rosemary Gartner, *Violence and Crime in Cross-National Perspective* (New Haven, Connecticut, 1984)

Archer, Dane and Rosemary Gartner, 'Violent Acts and Violent Times: A Comparative Approach to Postwar Homicide Rates', *The American Sociological Review*, 41.6 (1976)

Archer, Captain E. S., *The Warrior. A Military Drama Illustrative of the Great War* (Durban, 1917)

Archer, John (ed.), *Male Violence* (London, 1994)

Arendt, Hannah, *Eichmann in Jerusalem: A Report on the Banality of Evil* (New York, 1965)

Armour, Major M. D. S., *Total War Training for Home Guard Officers and N.C.O.s* (London, 1942)

Armstrong, Lieutenant-Colonel N. A. D., *Fieldcraft, Sniping and Intelligence*, 5th edition (Aldershot, 1942)

Ashenhurst, Major J. T., 'Barbs, Bullets and Bayonets', *Canadian Army Journal*, 4.5 (October 1950)

Ashworth, Tony, *Trench Warfare 1914–1918. The Live and Let Live System* (London, 1980)

Augustine, Lieutenant-Colonel Robert W., 'The Relation of Physician and Physical Therapist', *U. S. Armed Forces Medical Journal*, 4.2 (1953)

Bacher, William A., (ed.), *The Treasury Star Parade* (New York, 1942)

Baden-Powell, R. S. S., *Sport in War* (London, 1900)

Baker, Mark, *Nam. The Vietnam War in the Words of the Men and Women Who Fought There* (London, 1982)

Bakhtin, Mikhail, *Rabelais and His World*, translated by H. Iswolsky (Bloomington, Indiana, 1985)

Balkind, Jon, 'A Critique of Military Sociology: Lessons from Vietnam', *Journal of Strategic Studies*, 1.3 (December 1978)

Ballard, Frank, *Mistakes of Pacifism or Why a Christian Can have Anything to Do with War* (London, 1915)

Ballard, John A. and Alliecia J. McDowell, 'Hate and Combat Behavior', *Armed Forces and Society*, 17.2 (winter 1991)

Baritz, Loren, *Backfire: A History of How American Culture Led Us Into Vietnam and Made Us Fight the Way We Did* (New York, 1985)

Barker, Ernest, *National Character and the Factors in its Formation* (London, 1927)

'Barnacle' [pseud. Revd Waldo Lovell Smith], *Talks to the Troops* (Toronto, 1944)

Barnes, Revd R. L., *A War-Time Chaplaincy* (London, 1939)

Barr, Norman I. and Leonard M. Zunin, 'Clarification of the Psychiatrist's Dilemma While in Military Service', *American Journal of Orthopsychiatry*, 41.4 (July 1971)

Barth, R. L., *A Soldier's Time. Vietnam War Poems* (Santa Barbara, 1987)

Bartlett, F. C., *Psychology and the Soldier* (Cambridge, 1927)

Basinger, Jeannie, *The World War II Combat Film. Anatomy of a Genre* (New York, 1986)

Bassett, George, *This Also Happened* (London, 1947)

Bassett, Revd R. H., 'The Chaplain with the West African Forces', *Journal of the Royal Army Chaplains' Department*, vii.49 (July 1950)

Bateman, Charles T., 'The Salvation Army and the War', *The Contemporary Review*, cix (January–June 1916)

Battery 'C' 11th. F.A., *From Arizona to the Huns* (Dijon, 1919)

Battye, Major C., 'Infantry Training', *Commonwealth Military Journal*, 4 (October 1913)

Baudelaire, Charles, *The Painter of Modern Life and Other Essays*, translated by Jonathan Mayne (London, 1964)

Bauman, Zygmunt, *Modernity and the Holocaust* (Oxford, 1989)

The Bayonet (Bayonet Fighting) (Aldershot, 1940)

'The Bayonet for Mounted Riflemen', *The Army Review*, v.1 (July 1913)

Beach, Commander Edward L., *Submarine!* (London, 1953)

Beard, Mary, *Women as a Force in History* (New York, 1962)

Bedborough, George, *Arms and the Clergy (1914–1918)* (London, 1934)

Beevor, Humphry, *Peace and Pacifism* (London, 1938)

Bell, G. K. A., *The Church and Humanity (1939–1946)* (London, 1946)

Bell, G. K. A., *Christianity and World Order* (Harmondsworth, 1940)

Bellah, J. W. and A. F. Clark, 'The Lunk Trainer', *Infantry Journal*, 52.3 (1943)

Benavidez, Master Sergeant Roy, with John R. Craig, *Medal of Honor. A*

Vietnam Warrior's Story (Washington, 1995)

Benedek, Therese, *Insight and Personality Adjustment. A Study of the Psychological Effects of War* (New York, 1946)

Berg, Charles, 'Clinical Notes on the Analysis of a War Neurosis', *British Journal of Medical Psychology*, xix.2 (1942)

Bergerud, Eric M., *Red Thunder, Tropic Lightning. The World of a Combat Division in Vietnam* (St Leonards, New South Wales, 1993)

Bergerud, Eric, *Touch with Fire. The Land War in the South Pacific* (New York, 1996)

Berkowitz, Leonard, 'The Frustration–Aggression Hypothesis Revisited', in Berkowitz (ed.), *Roots of Aggression* (New York, 1969)

Berkowitz, Leonard, 'Situational Influences on Aggression', in Jo Groebel and Robert A. Hinde (eds), *Aggression and War. Their Biological and Social Bases* (Cambridge, 1989)

Berlin, Irving N., 'Guilt as an Etiologic Factor in War Neuroses', *Journal of Nervous and Mental Disease*, 111 (January–June 1950)

Bernard, L. L., 'Are There Any Instincts?', *Journal of Abnormal and Social Psychology*, 14 (1920)

Bernard, L. L., *Instinct: A Study in Social Psychology* (New York, 1924)

Bernard, L. L., 'The Misuse of Instinct in the Social Sciences', *Psychological Review*, 28 (1921)

Best, Geoffrey, *Humanity in Warfare. The Modern History of the International Law of Armed Conflict* (London, 1980)

Best, Geoffrey, *War and Law Since 1945* (Oxford, 1995)

Bidwell, Shelford, *Modern Warfare. A Study of Men, Weapons and Theories* (London, 1973)

Bilton, Michael and Kevin Sims, *Four Hours in My Lai. A War Crime and its Aftermath* (London, 1992)

Bingham, Martin, 'Adjustment Problems of American Youth in Military Service', *Sociology and Social Research*, 31.1 (September–October 1946)

Bingham, Walter V., 'Psychological Services in the United States Army', *The Journal of Consulting Psychology*, 5.5 (September–October 1941)

Bird, Charles, 'From Home to the Charge: A Psychological Study of the Soldier', *The American Journal of Psychology*, xxviii.3 (July 1917)

Birmingham, George A., *A Padre in France* (London, 1918)

Bishop, Major William Avery, *Winged Warfare. Hunting Huns in the Air* (London, 1918)

Blackall, Captain C. W., *Songs from the Trenches* (London, 1915)

Blackam, Colonel Robert J., *Scalpel, Sword, and Stretcher. Forty Years of Work and Play* (London, 1931)

Blackburne, Revd Harry W., *This Also Happened On the Western Front. The Padre's Story* (London, 1932)

Bleckwenn, Colonel William J., 'Neuroses in the Combat Zone', *Annals of Internal Medicine*, 23.2 (August 1945)

Bleda, Paul R. and Robert H. Sulzen, 'The Effects of Simulated Infantry Combat Training on Motivation and Satisfaction', *Armed Forces and Society*, 6.2 (winter 1980)

Bobbin, Jack W., 'War to the Death', *The Boys' Journal*, iii.60 (7 November 1914)

Body, Captain O. G., 'Bush and Forest Fighting Against Modern Weapons', *The Army Quarterly*, viii (July 1924)

Bogacz, Ted, '"A Tyranny of Words": Language, Poetry, and Antimodernism in the First World War', *Journal of Modern History*, 58 (September 1986)

Bolitho, Hector, *Combat Report. The Story of a Fighter Pilot* (London, 1943)

Bolitho, Hector, 'Two in Twenty-Two Minutes', in *Slipstream. A Royal Air Force Anthology* (London, 1946)

Bomford, Nora, 'Drafts', in Catherine Reilly (ed.), *Scars Upon my Heart* (London, 1990)

Bonadeo, Alfredo, *Mark of the Beast. Death and Degradation in the Literature of the Great War* (Lexington, 1989)

Bond, Horace Mann, 'The Negro in the Armed Forces of the United States Prior to World War One', *Journal of Negro Education*, xii (1943)

Bone, J. 'The Spirit of the Bayonet', *The Incinerator*, 1.2 (June 1916)

A Book of British Heroes (London, 1914)

Borek, Colonel Ted B., 'Legal Services During the War', *Military Law Review*, 120 (spring 1988)

Bosworth, Edward Increase, *The Christian Witness in War* (New York, 1918)

Boullier, John A., *Jotting by a Gunner and Chaplain* (London, 1917)

Bourke, Joanna, *Dismembering the Male: Men's Bodies, Britain, and the Great War* (London and Chicago, 1996)

Bourke, Joanna, *The Misfit Soldier* (Cork, 1998)

Bourne, Peter G., 'Some Observations on the Psychosocial Phenomena Seen in Basic Training', *Psychiatry*, 30.2 (May 1967)

Bovet, Pierre, *The Fighting Instinct*, translated by J. Y. T. Greig (London, 1923)

Boyle, Richard, *The Flower of the Dragon. The Breakdown of the U.S. Army in Vietnam* (San Francisco, 1972)

Braceland, Commander Francis J. and Lieutenant-Commander Howard P. Rome, 'Problems of Naval Psychiatry', *War Medicine*, 6.4 (October 1944)

Breach, Revd G. C., 'So Fight I'. Thoughts Upon the Warfare in which Every Soul is Engaged and in which There Can Be No Neutrals (London, 1917)

Bredin, Brigadier A. E. C., A History of the Irish Soldier (Belfast, 1987)

Brereton, Captain Frederick Sadleir, With Rifle and Bayonet. A Story of the Boer War (London, 1900)

Breslau, Naomi and Glenn C. Davis, 'Posttraumatic Stress Disorder: The Etiologic Specificity of Wartime Stressors', American Journal of Psychiatry, 144.5 (May 1987)

Briscoe, Walter A., The Boy Hero of the Air. From Schoolboy to V.C. (London, 1921)

Briscoe, Walter A. and H. Russell Stannard, Captain Ball, V.C. The Career of Flight Commander Ball, V.C., D.S.O. (London, 1918)

Brittain, Vera, Testament of Youth, first published in 1930 (London, 1960)

Broadfoot, Barry, Six War Years 1939–1945. Memories of Canadians at Home and Abroad (Don Mills, Ontario, 1974)

Brockway, Fenner, Bermondsey Story. The Life of Alfred Salter (London, 1949)

Brody, Second Lieutenant John Bernard, 'Diary', in Brereton Greenhous (ed.), A Rattle of Pebbles: The First World War Diaries of Two Canadian Airmen (Ottawa, 1987)

Bromley, Marion, 'Feminism and Non-Violent Revolution', in Pam McAllister (ed.), Reweaving the Web of Life. Feminism and Nonviolence (Philadelphia, 1982)

Brooke, Revd Stopford A., A Discourse on War, first published in 1905 (London, 1916)

Broomhall, Marshall, 'Mine Own Vineyard'. Personal Religion and War (London, 1916)

Brosin, Henry W., 'Panic States and Their Treatment', American Journal of Psychiatry, 100 (1943–4)

Browning, Christopher R., Ordinary Men. Police Battalion 101 and the Final Solution in Poland (New York, 1992)

Brownrigg, Rear-Admiral Sir Douglas, Indiscretions of the Naval Censor (London, 1920)

Broyles, William, 'Why Men Love War', Esquire, November 1984

Brumwell, Revd P. Middleton, The Army Chaplain. The Royal Army Chaplains' Department. The Duties of Chaplains and Morale (London, 1943)

Buchan, John, Francis and Riversdale Grenfell (London, 1920)

Buckell, Revd A. C., The Greatest War. Six Addresses (London, 1915)

Bull, Paul B., Peace and War. Notes of Sermons and Addresses (London, 1917)

Bullough, Vern L. and Bonnie Bullough, Cross Dressing, Sex, and Gender (Philadelphia, 1993)

Burchard, Waldo W., 'Role Conflicts of Military Chaplains', *American Sociological Review*, 19.5 (October 1954)

Burchett, W. G., *Wingate's Phantom Army* (Bombay, 1944)

Burgess, Alan, *The Lovely Sergeant* (London, 1963)

Burk, James, 'Morris Janowitz and the Origins of Sociological Research on Armed Forces and Society', *Armed Forces and Society*, 19.2 (winter 1993)

Butler, Colonel A. G., *Official History of the Australian Army Medical Services 1914–1918. Volume II* (Canberra, 1940)

Butler, John Sibley, 'The Military as a Vehicle for Social Integration. The Afro-American Experience as Data', in Henry Dietz, Jerrold Elkin, and Maurice Roumani (eds), *Ethnicity, Integration and the Military* (Boulder, San Francisco, 1991)

Bychowski, Gustav, 'Personality Changes Characterising the Transition from Civilian to Military Life', *The Journal of Nervous and Mental Disease*, 100.3 (September 1944)

Cable, Boyd, *Air Men O' War* (London, 1918)

Caesar, Adrian, *Taking It Like a Man. Suffering, Sexuality, and the War Poets* (Manchester, 1993)

Caldicott, Helen, *Missile Envy. The Arms Race and Nuclear War* (New York, 1984)

Callan, John J., *With Guns and Wagons. A Day in the Life of an Artillery Chaplain* (London, 1918)

Calley, Lieutenant William L., *Body Count* (London, 1971)

Cameron, Revd Thomas, *The Happy Warrior* (London, 1939)

Campbell, D'Ann, 'Servicewomen of World War II', *Armed Forces and Society*, 16.2 (winter 1990)

Campbell, Brigadier I. R., 'Should We Study Military History?', *Australian Training Memorandum. Australian Military Forces*, 49 (June–July 1947)

Campbell, Revd R. J., *With Our Troops in France* (London, 1916)

Canadian War Records Office, *Thirty Canadian V.C.s 23rd April 1915 to 30th March 1918* (London, 1918)

Candler, Edmund, *The Year of Chivalry* (London, 1916)

Cannon, Walter C., *Bodily Changes in Pain, Hunger, Fear and Rage. An Account of Recent Researches into the Function of Emotional Excitement*, second edition, first published in 1915 (New York, 1929)

Caputo, Philip, *A Rumor of War* (London, 1977)

Carbery, Lieutenant-Colonel A. D., 'Some Medical Aspects of Recruiting for War', *Transactions of the Australasian Medical Congress (British Medical Association). Second Session: Dunedin, February 3 to 10, 1927* (1927)

Carey, Alex, *Australian Atrocities in Vietnam* (Sydney, 1968)

Carlton, Eric, *Massacres. An Historical Perspective* (Aldershot, 1994)

Carlyle, Thomas, *On Heroes, Hero Worship and the Heroic in History* (London, 1832)

Carr, Joyce, 'Just Like William', in Mavis Nicholson (ed.), *What Did You Do in the War, Mummy? Women in World War II* (London, 1995)

Carr, William, *A Time to Leave the Ploughshares. A Gunner Remembers 1917–1918* (London, 1985)

Carroll, John, *Token Soldiers* (Boronia, Victoria, 1983)

Carson, John, 'Army Alpha, Army Brass, and the Search for Army Intelligence', *ISIS*, 84.2 (June 1993)

Carson, Ralph Coburn, 'Recollections of the Front', in James Carson (compiler), *The Carsons of Monanton, Ballybay, Co. Monaghan, Ireland* (Lisburn, 1931)

Casserly, Major Gordon, *The Training of the Volunteers for War* (London, 1915)

Castle, M. J., 'The Role of Women in the Armed Forces', *The Australian Journal of Defence Studies*, 2.2 (November 1978)

Catchpool, T. Corder, *On Two Fronts* (London, 1918)

Cather, Willa, *One of Ours* (New York, 1922)

Catto, Alexander, *With the Scottish Troops in France* (Aberdeen, 1918)

Ceadel, Martin, 'Christian Pacifism in the Era of the Two World Wars', in W. J. Sheils (ed.), *The Church and War* (London, 1983)

Cederberg, Fred, *The Long Road Home. The Autobiography of a Canadian Soldier in Italy in World War Two* (Toronto, 1985)

Chaillou, A. J. M. A. and Leon MacAuliffe, *Morphologie medicale; étude des quatre types humains* (Paris, 1912)

Chamberlain, Captain W. R., 'Training the Functional Rifleman', *Canadian Army Journal*, 4.9 (February 1951)

Chandler, Andrew, 'The Church of England and the Obliteration Bombing of Germans in the Second World War', *English Historical Review* (October 1993)

Chapelle, Dickey, *What's a Woman Doing Here?* (New York, 1962)

Chapman, Abel, *On Safari. Big-Game Hunting in British East Africa With Studies in Bird Life* (London, 1908)

Chapman, Guy, *A Passionate Prodigality. Fragments of an Autobiography* (New York, 1966)

Chapman, Revd Reginald John, *With Our Troops in France* (London, 1916)

Child, Irvin L., 'Morale: A Bibliographical Review', *Psychological Bulletin*, 38 (1941)

Chisholm, Major-General G. B., 'Emotional Problems of Demobilization', in *Military Neuropsychiatry. Proceedings of the Association [for Research in Nervous and Mental Disease], December 15 and 16, 1944, New York* (Baltimore, 1946)

Chrisholm, Roderick, *Cover of Darkness* (London, 1953)

Christensen, Terry, *Reel Politics. American Political Movies from 'The Birth of a Nation' to 'Platoon'* (New York, 1987)

Clapman, H. S., *Mud and Khaki. Memories of an Incomplete Soldier* (London, 1930)

Clark, Robert A., 'Aggressiveness and Military Training', *American Journal of Sociology*, 51.5 (March 1946)

Clark, Ronald W., *The Rise of the Boffins* (London, 1962)

Clausen, Colonel Roy E. and Arlene K. Daniels, 'Role Conflicts and Their Ideological Resolution in Military Psychiatric Practice', *American Journal of Psychiatry*, 123.3 (September 1966)

Clifford, Joan, *Thank You, Padre. Memories of World War II* (London, 1989)

Clodfelter, Micheal [sic], 'Snipers', in J. Topham, *Vietnam Literature Anthology*, revised and enlarged (Philadelphia, 1990)

Cloutier, Marc G., 'Medical Care Behind Enemy Lines: A Historical Examination of Clandestine Hospitals', *Military Medicine*, 158.12 (December 1993)

Cobb, G. Belton, *Stand to Arms* (London, 1916)

Cobbold, W. N., *Poems on the War. March 21st to November 11th, 1918 and After the Armistice* (Cambridge, 1919)

Cohen, Lieutenant-Commander R. A. and Lieutenant J. G. Delano, 'Subacute Emotional Disturbances Induced by Combat', *War Medicine*, 7.5 (May 1945)

Colby, Elbridge, *The Profession of Arms* (New York, 1924)

Coleman, Major Jules V., 'The Group Factor in Military Psychiatry', *American Journal of Orthopsychiatry*, xvi (1946)

Collins, Captain D., 'Quick Kill', *Infantry. The Magazine of the Royal Australian Infantry Corps* (January 1969)

Collins, Canon L., *Faith Under Fire* (London, 1965)

Conde, Anne-Marie, '"The Ordeal of Adjustment": Australian Psychiatric Casualties of the Second World War', *War and Society*, 15.2 (October 1997)

Constable, Harold Strickland, *Something About Horses, Sport and War* (London, 1891)

Conway, Sir Martin, *The Crowd in Peace and War* (London, 1915)

Cooke, Alice M. P., *Irish Heroes in Red War* (Dublin, 1915)

Cooke, Miriam, *Women and the War Story* (Berkeley, 1996)

Coope, Robert, *Shall I Fight? An Essay on War, Peace, and the Individual* (London, 1935)

Cooper, Courtney Ryley, *Annie Oakley. Woman at Arms* (London, 1928)

Cooper, Helen, Adrienne Auslander Munich, and Susan Merrill Squier, Introduction, in their (eds), *Arms and the Woman. War, Gender, and Literary Representation* (Chapel Hill, 1989)

Copeland, Norman, *Psychology and the Soldier* (London, 1942)

Coppard, George, *With a Machine Gun to Cambrai* (London, 1980)

Cosby, Elisabeth, *Meditations of a Khaki Woman* (Amersham, Buckinghamshire, 1920)

'Courses in Psychology for the Students' Army Training Corps', *Psychological Bulletin*, 15 (1918)

Cox, Bryan, *Too Young to Die. The Story of a New Zealand Fighter Pilot in the Pacific War* (Ames, Iowa, 1989)

Crandell, William F., 'What did America Learn from the Winter Soldier Investigation?', *Vietnam General Journal. Nobody Gets off the Bus*, 5.14 (March 1994), on the Internet

Cravens, Hamilton, *The Triumph of Evolution. American Scientists and the Heredity–Environment Controversy 1900–1941* (Pennsylvania, 1978)

Crawford, Meredith P., 'Military Psychology and General Psychology', *American Psychologist*, 25 (1970)

Cronin, Cornelius A., 'Line of Departure. The Atrocity in Vietnam War Literature', in Philip K. Jason (ed.), *Fourteen Landing Zones. Approaches to Vietnam War Literature* (Iowa City, 1991)

Crook, Flight-Lieutenant D. M., *Spitfire Pilot* (London, 1942)

Crosby, Donald F., *Battlefield Chaplains. Catholic Priests in World War II* (Lawrence, Kansas, 1994)

Crozier, Brigadier-General F. P., *A Brass Hat in No Man's Land* (London, 1930)

Crozier, Brigadier-General F. P., *The Men I Killed* (London, 1937)

Cubis, Major R. M. C., 'An Academy of Military Art', *Australian Army Journal*, 162 (November 1962)

Culpin, Millais, *Psychoneuroses of War and Peace* (Cambridge, 1920)

Cuneo, Terence T., 'The Trail of the Iron Horse', in *Brave and Bold. Stories for Boys* (Oxford, 1943)

Curran, Surgeon Captain D., 'Functional Nervous States in Relation to Service in the Royal Navy', in Major-General Sir Henry Letheby Tidy (ed.), *Inter-Allied Conferences on War Medicine 1942–1945* (London, 1947)

Cushing, Harvey, *From a Surgeon's Journal 1915–1918* (London, 1936)

Cuthbert, Captain S. J., *'We Shall Fight in the Streets!' Guide to Street Fighting* (Aldershot, 1941)

Dain, Norman, 'Critics and Dissenters: Reflections on "Anti-Psychiatry" in the United States', *Journal of the History of the Behavioral Sciences*, 25 (January 1989)

Daly, Lieutenant-Colonel C. D., 'A Psychological Analysis of Military Morale', *The Army Quarterly*, xxxii (April 1936)

Daniels, Arlene K., 'The Captive Professional: Bureaucratic Limitations in the Practice of Military Psychiatry', *Journal of Health and Social Behavior*, 10.4 (December 1969)

Daniels, Arlene K., 'Military Psychiatry. The Emergence of a Subspecialty', in Eliot Freidson and Judith Lorber (eds), *Medical Men and Their Work* (Chicago, 1972)

Darrow, Clarence, *Crime: Its Cause and Treatment*, first published in 1922 (Montclair, 1972)

Davis, Newham Nathaniel, *Military Dialogues. On 'Active Service'* (London, 1900)

Dawkins, Colonel C. T., *Night Operations for Infantry* (Aldershot, 1916)

Dawson, Coningsby, *The Glory of the Trenches* (London, 1918)

Dawson, Captain Lionel, *Sport in War* (London, 1936)

Deane, Alan, 'The Battle of Mons', *The Boys' Journal*, iii.58 (24 October 1914)

Deardon, Harold, *Medicine and Duty. A War Diary* (London, 1928)

Defence and Security: What New Zealanders Want (Wellington, NZ, 1986)

De Man, Henry, *The Remaking of a Mind. A Soldier's Thoughts on War and Reconstruction* (London, 1920)

Demuth, Norman, *Harrying the Hun. A Handbook of Scouting, Stalking and Camouflage* (London, 1941)

Denman, Terence, *Ireland's Unknown Soldiers. The 16th (Irish) Division in the Great War, 1914–1918* (Dublin, 1992)

Dennis, D. J., *One Day at a Time. A Vietnam Diary* (St Lucia, 1992)

De Pauw, Linda Grant, 'Women in Combat. The Revolutionary War Experience', *Armed Forces and Society*, 7.2 (winter 1981)

Depew, Albert N., *Gunner Depew* (London, 1918)

DeRose, David J., 'A Dual Perspective: First-Person Narrative in Vietnam Film and Drama', in Owen W. Gilman, Jr., and Lorrie Smith (eds), *America Rediscovered. Critical Essays on Literature and Film of the Vietnam War* (New York, 1990)

Deutsch, Felix, 'Civilian War Neuroses and Their Treatment', *The Psychoanalytical Quarterly*, 13 (1944)

Devine, Major-General J. M., 'What is a Tough Soldier?', *Canadian Army Journal*, 3.6 (September 1949)

Diamond, Stanley, 'War and Disassociated Personality', in Morton Fried, Marvin Harris, and Robert Murphy (eds), *War. The Anthropology of Armed Conflict and Aggression* (New York, 1968)

Dicks, Shirley, *From Vietnam to Hell. Interviews with Victims of Post-Traumatic Stress Disorder* (Jefferson, North Carolina, 1990)

Dickson, Lovat, *Richard Hillary* (London, 1950)

Digby, Everard, *Tips for Padres. A Handbook for Chaplains* (London, 1917)

Dollard, John, Leonard W. Doob, Neal E. Miller, O. H. Mowrer, and Robert R. Sears, *Frustration and Aggression* (New Haven, Connecticut, 1939)

Donovan, Captain C. H. W., *With Wilson in Matabeleland or Sport and War in Zambesia* (London, 1894)

Dooley, Thomas P., *Irishmen or English Soldiers? The Times and World of a Southern Catholic Irish Man (1876–1916) Enlisted with the British Army During the First World War* (Liverpool, 1995)

Douglas, Keith, *The Complete Poems*, edited by Ted Hughes (Oxford, 1987)

Douie, Vera, *Daughters of Britain* (Oxford, 1949)

Dowd, Jerome, *The Negro in American Life* (New York, 1926)

Dower, John W., *War Without Mercy. Race and Power in the Pacific War* (London, 1986)

Dowley, Revd Powel M., 'The Conditions of a Just War', in Ashley Sampson (ed.), *This War and Christian Ethics* (Oxford, 1940)

Doyle, Lily, *Bound in Khaki* (London, 1916)

Drendel, Lou, . . . *And Kill Migs. Air to Air Combat in the Vietnam War* (Carrollton, 1974)

'The Drink that Failed', *The Yellow Band. The Journal of the 1st London, Sanitary Company, RAMC*, 1.5 (August–October 1916)

Driskell, James E. and Beckett Olmstead, 'Psychology and the Military. Recent Applications and Trends', *American Psychologist*, 44.1 (January 1989)

Drury, Revd William Edward, *Camp Follower. A Padre's Recollections of Nile, Somme and Tigris During the First World War* (Dublin, 1968)

Duffett, John, *Against the Crime of Silence. Proceedings of the Russell International War Crimes Tribunal* (New York, 1968)

Duffield, Sydney and Andrew G. Elliott, *Rough Stuff. For Home Guards and Members of H. M. Forces* (London, 1942)

Dugaw, Dianne, *Warrior Women and Popular Balladry 1650–1850* (Cambridge, 1989)

Dunbar, Flanders, 'Medical Aspects of Accidents and Mistakes in the Industrial Army and in the Armed Forces', *War Medicine*, 4.2 (August 1943)

Duncan, Donald, *A 'Green Beret' Blasts the War* (London, 1966)

Duncan, Donald, *The New Legions* (London, 1967)

Dundas, High, *Flying Start. A Fighter Pilot's War Years* (London, 1988)

Dunlan, Knight, 'The Cause and Prevention of War', *The Journal of Abnormal and Social Psychology*, 35.4 (October 1940)

Dunlap, Jack W., 'Psychologists and the Cold War', *The American Psychologist*, 10 (1955)

Dunlap, Commander Jack W., 'The Sensitive Adjustment of Men to Machines', *The Army Combat Forces Journal*, 5.3 (1954)

Dunn, Captain J. C., *The War the Infantry Knew 1914–1919*, first published in 1938 (London, 1987)

Durbin, E. F. M. and J. Bowlby, *Personal Aggressiveness and War* (London, 1939)

Eaker, Major-General H. H. and Colonel Ira C. Eaker, *Winged Warfare* (New York, 1941)

Eastman, James N., Walter Hanak, and Lawrence J. Paszek (eds), *Aces and Aerial Victories. The United States Air Force in Southeast Asia 1965–1973* (Washington, 1978)

Echols, Alice, 'New Feminism of Yin and Yang', in Ann Snitow et al. (eds), *Powers of Desire* (New York, 1983)

Edelman, Bernard (ed.), *Dear America. Letters Home from Vietnam* (New York, 1985)

Edelman, Isidore S., *Diseases of the Nervous System*, 8 (June 1947)

Edwards, Glen D., *Vietnam. The War Within* (Salisbury, South Australia, 1992)

Edwards, John Carver (ed.), 'Sergeant Jones Goes to War. Extracts from a U.S. Artilleryman's Diary, 1918', *The Army Quarterly and Defence Journal*, 104.1 (October 1973)

Egendorf, Arthur, et al., *Legacies of Vietnam* (New York, 1981)

Egendorf, Arthur, 'Vietnam Veteran Rap Groups and Themes of Postwar Life', *Journal of Social Issues*, 31.4 (1975)

Ehrhart, William Daniel, 'Full Moon', in Larry Rottmann, Jan Barry, and Basil T. Paquet (eds), *Winning Hearts and Minds. War Poems by Vietnam Veterans* (New York, 1972)

Ehrhart, William Daniel, 'Guerrilla War', in Ehrhart (ed.), *Unaccustomed Mercy. Soldier-Poets of the Vietnam War* (Lubbock, Texas, 1989)

Ehrhart, William D., 'Why I Did It', *The Virginia Quarterly Review*, 56.1 (winter 1980)

Eisenhart, R. Wayne, 'You Can't Hack it Little Girl: A Discussion of the Covert Psychological Agenda of Modern Combat Training', *Journal of*

Social Issues, 31.4 (1975)

Eley, Beverley, *Ion Idriess* (Sydney, 1995)

Elliott, Andrew, *Shoot to Kill. A Book Which May Save your Life* (London, 1941)

Elliot, Andrew G., 'J. B.', and 'Scientist', *The Home Guard Encyclopedia* (London, 1942)

Ellis, Joseph and Robert Moore, *School for Soldiers. West Point and the Profession of Arms* (New York, 1974)

Elshtain, Jean Bethke, *Women and War* (Chicago, 1995)

Eltinge, LeRoy, *Psychology of War* (London, 1918)

Elton, Oliver, *C. E. Montague. A Memoir* (London, 1929)

Elwood-Akers, Virginia, *Women War Correspondents in the Vietnam War, 1961–1975* (Metuchen, New Jersey, 1988)

Emerson, Gloria, 'Hey, Lady, What are You Doing Here?', *McCalls*, 98 (August 1971)

Empey, Arthur Guy, *First Call. Guide Posts to Berlin* (New York, 1918)

English, Horace B., 'How Psychology can Facilitate Military Training – A Concrete Example', *The Journal of Applied Psychology*, xxvi (1942)

English, Raymond, *The Pursuit of Purpose. An Essay on Social Morale* (London, 1947)

Enloe, Cynthia H., *Ethnic Soldiers. State Security in Divided Societies* (Harmondsworth, 1980)

Erith Urban District Council, *The Effect of War Films on Child Opinion. Report of an Investigation. Erith, 1935* (Erith, 1935)

Ervine, Thomas Alexander, 'Undiminished Memories', in Michael Hall (ed.), *Sacrifice on the Somme* (Newtownabbey, 1993)

Everett, Arthur, Kathryn Johnson, and Harry F. Rosenthal, *Calley* (New York, 1971)

Ewell, Lieutenant-General Julian J. and Major-General Ira A. Hunt, *Sharpening the Combat Edge: The Use of Analysis to Reinforce Military Judgment* (Washington, DC, 1974)

'Ex-Private X', *War is War* (London, 1930)

Eyles, Allen, *John Wayne and the Movies* (London, 1976)

Fahey, James J., *Pacific War Diary 1942–1945* (Boston, 1963)

Fairbairn, W. R. D., *From Instinct to Self: Selected Papers of W. R. D. Fairbairn. Volume II: Applications and Early Contributions* (Northvale, New Jersey, 1994)

Fairbairn, W. R. D., 'Is Aggression an Irreducible Factor?', *British Journal of Medical Psychology*, xviii.2 (1940)

Falk, Richard A., Gabriel Kolko, and Robert Jay Lifton (eds), *Crimes of War*

(New York, 1971)

Fallon, Captain David, *The Big Fight (Gallipoli to the Somme)* (London, 1918)

Faris, Ellsworth, 'Are Instincts Data or Hypotheses?', *American Journal of Sociology*, 27 (1921–2)

Fawcett, Millicent Garrett, 'Women's Work in War Time', *The Contemporary Review*, cvi (June–December 1914)

'F. B. M.' [Frank Bertrand Merryweather], *The Defiance of Death. Being Some Thoughts on the Death of a Brave Soldier* (London, 1918)

Feilding, Rowland *War Letters to a Wife. France and Flanders, 1915–1919* (London, 1929)

Feld, M. D., 'Arms and the Woman. Some General Considerations', *Armed Forces and Society. An Interdisciplinary Journal*, 4.4 (summer, 1978)

Ferguson, Niall, *The Pity of War* (London, 1998)

Field, G. C., *Pacifism and Conscientious Objection* (Cambridge, 1945)

Findlay, Revd J. L., *Fighting Padre. (Memoirs of an Army Chaplain)* (London, 1941)

Finnemore, John, *A Boy Scout with the Russians* (London, 1915)

Finnemore, John, *Foray and Fight. Being the Story of the Remarkable Adventures of an Englishman and an American in Macedonia* (London, 1906)

Finnemore, John, *Two Boys in War-Time* (London, 1900 and 1928)

Fisher, James J., *The Immortal Deeds of Our Irish Regiments in Flanders and the Dardanelles* (Dublin, 1916)

FitzGerald, Frances, *Fire in the Lake. The Vietnamese and the Americans in Vietnam* (Boston, 1972)

Fitzpatrick, David, 'Militarism in Ireland, 1900–1922', in Thomas Bartlett and Keith Jeffery (eds), *A Military History of Ireland* (Cambridge, 1996)

'F. Ll. H.', 'The Chaplain's Duty in Battle', *The Chaplains' Magazine. Middle East*, 1.2 (Easter 1943)

Fleming, Michael and Roger Manwell, *Images of Madness. The Portrayal of Insanity in the Feature Film* (London, 1985)

Floherty, John J., *The Courage and the Glory* (Philadelphia, 1942)

Forbes, Wing-Commander Athol and Squadron-Leader Hubert Allen, *The Fighter Boys* (London, 1942)

Ford, Daniel, *Incident at Muc Wa* (London, 1967)

Ford, Major E. S., 'Principles and Problems of Maintenance of Fighter-Bomber Pilots', *War Medicine*, 8.1 (July 1945)

Forsythe, Colonel George I. and Lieutenant-Colonel Harold H. Dunwood, 'Solidarity and the Mass Army', *The Army Combat Forces Journal*, 5.9 (April 1955)

Fosdick, Harry Emerson, *The Challenge of the Present Crisis* (New York, 1917)

Foss, William and Cecil Gerahty, *The Spanish Arena* (London, 1938)

44 Indian Airborne Division. Training Pamphlet No. 1. The Medical Services in the Division (np, 1944)

Fraser, George MacDonald, *The Hollywood History of the World* (London, 1988)

Fraser, George MacDonald, *Quartered Safe Out Here. A Recollection of the War in Burma* (London, 1992)

Freeman, Derek, 'Human Aggression in Anthropological Perspective', in J. D. Carthy and F. J. Ebling (eds), *The Natural History of Aggression* (London, 1964)

French, Gerald, *The Life of Field Marshal Sir John French* (London, 1931)

Friedman, Leon (ed.), *The Law of War. A Documentary History*, vols. 1 and 2 (New York, 1972)

Fuller, J. F. C., 'The Foundations of the Science of War', *The Army Quarterly*, I (October 1920–January 1921)

Fuller, Captain J., *Hints on Training Territorial Infantry. From Recruit to Trained Soldier* (London, 1913)

Fuller, Major-General J. F. C., *Lectures on F. S. R. II* (London, 1931)

Fuller, Captain J. F. C., *Training Soldiers for War* (London, 1914)

Fussell, Paul, *Wartime: Understanding and Behaviour in the Second World War* (Oxford, 1989)

Futterman, Samuel and Eugene Pumpion-Mindlin, 'Traumatic War Neuroses Five Years Later', *American Journal of Psychiatry*, 108 (December 1951)

Gabriel, Richard A., *The Painful Field: The Psychiatric Dimension of Modern War* (New York, 1988)

Gabriel, Richard A., *To Serve With Honor. A Treatise on Military Ethics and the Way of the Soldier* (Westport, Connecticut, 1982)

Gallagher, Nancy W., 'The Gender Gap in Popular Attitudes Toward the Use of Force', in Ruth H. Howes and Michael R. Stevenson (eds), *Women and the Use of Military Force* (Boulder, 1993)

Gallup, George H., *The Gallup Poll*, vol. 1 (New York, 1972)

Garber, Major W. E., 'Every Rifleman Must be an Aggressive Fighter', *Canadian Army Journal*, 6.6 (January 1953)

Gardner, Brian, *The Terrible Rain. The War Poets 1939–1945* (London, 1966)

Garnegie, Margaret and Frank Shields, *In Search of Breaker Morant. Balladist and Bushveldt Carbineer* (Armadale, 1979)

Garner, Major H. H., 'Psychiatric Casualties in Combat', *War Medicine*, 8.5 (1945)

Garrard, Robert L., 'Combat Guilt Reactions', *North Carolina Medical Journal*, 10.9 (September 1949)

Garrett, Stephen A., *Ethics and Airpower in World War II. The British Bombing of German Cities* (New York, 1993)

Garton, Stephen, *The Cost of War. Australians Return* (Melbourne, 1996)

Geare, Revd William Duncan, *Letters of an Army Chaplain* (London, 1918)

Geiger, J., 'Hidden Professional Roles: The Physician as Reactionary, Reformer, Revolutionary', *Social Policy*, 1 (March–April 1974)

George, David Lloyd, *The Great Crusade: Extracts from Speeches Delivered During the War* (London, 1918)

Gershen, Martin, *Destroy or Die: The True Story of My Lai* (New York, 1971)

Gerzon, Mark, 'The Soldier', in Reese Williams (ed.), *Unwinding the Vietnam War. From War into Peace* (Seattle, 1987)

Gethyn-Jones, Canon J. E., *A Territorial Army Chaplain in Peace and War. A Country Cleric in Khaki 1938–1961* (East Wittering, West Surrey, 1988)

'G. G. A.', 'The Bayonet for Mounted Riflemen', *The Army Review*, v.1 (July 1913)

Gibb, Lieutenant-Colonel J. W., *Training in the Army* (London, nd)

Gibbs, Philip, *Realities of War* (London, 1920)

Gibbs, Sir Philip, *The War Dispatches* (Isle of Man, 1964)

Gilchrist, Donald, *Castle Commando* (Edinburgh, 1960)

Gillespie, R. D., *Psychological Effects of War on Citizen and Soldier* (London, 1942)

Ginzberg, Eli, *Breakdown and Recovery* (New York, 1950)

Ginzberg, Eli, *Patterns of Performance* (New York, 1959)

Ginzberg, Eli, John L. Herma, and Sol W. Ginsburg, *Psychiatry and Military Manpower Policy. A Reappraisal of the Experience in World War II* (New York, 1953)

Gioglio, Gerald R., *Days of Decision. An Oral History of Conscientious Objectors in the Military During the Vietnam War* (Trenton, New Jersey, 1989)

Givens, R. Dale and Martin A. Nettleship (eds), *Discussions on War and Human Aggression* (Paris, 1976)

Glass, Albert J., Introduction, in Peter G. Bourne (ed.), *The Psychology and Physiology of Stress with Reference to Special Studies of the Vietnam War* (New York, 1969)

Glass, Albert J., 'Preventative Psychiatry in the Combat Zone', *U.S. Armed Forces Medical Journal*, iv.1 (1953)

Glass, Colonel Albert J. and Lieutenant-Colonel Calvin S. Drayer, 'Italian Campaign (1 March 1944–2 May 1945), Psychiatry Established at Division Level', in Lieutenant-General Hal B. Jennings (ed.), *Neuropsychiatry in World War II. Volume II. Overseas Theatres* (Washington, DC, 1973)

Glass, Albert J., Francis J. Ryan, Ardie Lubin, C. V. Ramana, and Anthony C. Tucker, 'Psychiatric Predictions and Military Effectiveness. Part II', *U.S.*

Armed Forces Medical Journal, vii.11 (November 1956)

Gleason, Arthur and Helen Hayes Gleason, *Golden Lads* (New York, 1916)

Glover, Edward, *The Psychology of Fear and Courage* (Harmondsworth, 1940)

Glover, Edward, *War, Sadism and Pacifism. Further Essays on Group Psychology and War* (London, 1947)

Glover, Edward, *War, Sadism and Pacifism. Three Essays* (London, 1933)

Godson, E. A., *The Great War 1914–1918. Incidents, Experiences, Impressions and Comments of a Junior Officer* (Hertford, nd)

Goldhagen, Daniel Jonah, *Hitler's Willing Executioners. Ordinary Germans and the Holocaust* (London, 1996)

Goldman, George S., 'The Psychiatrist's Job in War and Peace', *Psychiatry*, 9.3 (August 1946)

Goldman, Nancy Loring and Richard Stites, 'Great Britain and the World Wars', in Goldman (ed.), *Female Soldiers – Combatants or Noncombatants? Historical and Contemporary Perspectives* (Westport, Connecticut, 1982)

Goldstein, Joseph, Burke Marshall, and Jack Schwartz (eds), *The Peers Commission Report* (New York, 1976)

Gordon, Huntly, *The Unreturning Army. A Field-Gunner in Flanders, 1917–18* (London, 1967)

Gore, Charles, *The War and the Church and Other Addresses* (London, 1914)

Gould, Nat, *Lost and Won. A Tale of Sport and War* (London, 1916)

Graham, Stephen, *The Challenge of the Dead* (London, 1921)

Graham, Stephen, *A Private in the Guards* (London, 1919)

Grant, Amy Gordon, *Letters from Armageddon. A Collection Made During the World War* (Boston, 1930)

Graves, Donald E., '"Naked Truths for the Asking". Twentieth-Century Military Historians and the Battlefield Narrative', in David A. Chartes, Marc Milner, and J. Brent Wilson (eds), *Military History and the Military Profession* (Westport, Connecticut, 1992)

Graves, Robert, *Goodbye to All That*, first published in 1929 (London, 1977)

Gray, Revd Arthur Herbert, *The Captain* (London, 1921)

Gray, J. Glenn, *The Warriors: Reflections on Men in Battle* (New York, 1959)

Great Advance. Tales from the Somme Battlefield (London, 1916)

Green, Lieutenant-Colonel A., 'Revolution in the Military Profession', *Australian Army Journal*, 61 (June 1954)

Green, L. C. *Essays on the Modern Law of War* (New York, 1985)

Greener, W. W., *Sharpshooting for Sport and War* (London, 1900)

Greenhaw, Wayne, *The Making of a Hero. The Story of Lieut. William Calley* (Louisville, Kentucky, 1971)

Greenson, Ralph R., *On Loving, Hating and Living Well* (Madison, 1992)

Greenwell, Graham H., *An Infant in Arms. War Letters of a Company Officer 1914–1918*, first published in 1935 (London, 1972)

Greiber, Major Marvin F., 'Narcosynthesis in the Treatment of the Noncombatant Psychiatric Casualty Overseas', *War Medicine*, 8.2 (August 1945)

Grey, C. G., *British Fighter Planes* (London, 1941)

Gribble, Francis, *Women in War* (London, 1916)

Griffiths, Margaret W., *Hazel in Uniform* (London, 1945)

Grinker, Roy R. and John P. Spiegel, *Men Under Stress* (London, 1945)

Guest, John, *Broken Images. A Journal* (London, 1949)

Gullett, Henry ('Jo'), *Not as a Duty Only. An Infantryman's War* (Melbourne, 1976)

Gurney, Ivor, *War Letters*, selected by R. K. R. Thornton (Ashington, Northumberland, 1983)

Gwynn, Denis, *The Life of John Redmond* (London, 1932)

Gwynn, Denis, *Redmond's Last Years* (London, 1919)

Gwynn, Captain Stephen L., 'Irish Regiments', in Felix Lavery (ed.), *Great Irishmen in War and Politics* (London, 1920)

Hackworth, Colonel David H. and Julie Sharman, *About Face* (Sydney, 1989)

Halberstadt, Hans, *Green Berets. Unconventional Warriors* (London, 1988)

Haldane, J. B. S., *Callinicus. A Defence of Chemical Warfare* (London, 1925)

Hall, Donald P., 'Stress, Suicide, and Military Service During Operation Uphold Democracy', *Military Medicine*, 161.3 (March 1996)

Hall, G. Stanley, *Morale. The Supreme Standard of Life and Conduct* (New York, 1920)

Hall, G. Stanley, 'Morale in War and After', *The Psychological Bulletin*, 15 (1918)

Hall, G. Stanley, 'Practical Applications of Psychology as Developed by the War', *The Pedagogical Seminary. A Quarterly*, xxvi (1919)

Hall, James Norman, *Kitchener's Mob. The Adventures of an American in the British Army* (London, 1916)

Hall, Norris F., 'Science in War', in Hall, Zechariah Chafee, Jr., and Manley O. Hudson (eds), *The Next War* (Cambridge, 1925)

Hall, Robert A., *Fighters from the Fringe: Aborigines and Torres Strait Islanders Recall the Second World War* (Canberra, 1995)

Hammer, Richard, *The Court-Martial of Lt. Calley* (New York, 1971)

Hampson, Norman, 'Corvette', in Brian Gardner, *The Terrible Rain. The War Poets 1939–1945* (London, 1966)

Hankey, Donald, *A Student in Arms. Second Series* (London, 1917)

Hanna, Henry, *The Pals at Suvla Bay. Being the Record of 'D' Company of the 7th Royal Dublin Fusiliers* (Dublin, 1916)

Hansen, J. T., A. Susan Owen, and Michael Patrick Madden, *Parallels. The Soldiers' Knowledge and the Oral History of Contemporary Warfare* (New York, 1992)

Hardin, A. N., *The American Bayonet 1776–1964* (Philadelphia, 1964)

Hardy, Revd E. J., *The British Soldier. His Courage and Humour* (London, 1915)

Hargreaves, Reginald, *Women-at-Arms. Their Famous Exploits Throughout the Ages* (London, 1930)

Haritos-Fatouros, Mika, 'The Official Torturer: A Learning Model for Obedience to the Authority of Violence', *Journal of Applied Social Psychology*, 18.13 (1988)

Hartley, Jenny (ed.), *Hearts Undefeated. Women's Writing of the Second World War* (London, 1995)

Harvey, H. E., *Battle-Line Narratives 1915–1918* (London, 1928)

Hastings, D. W., D. G. Wright, and B. C. Glueck, *Psychiatric Experiences of the Eighth Air Force* (New York, 1944)

Heath, Clark W., W. L. Woods, L. Brouha, C. C. Seltzer, and A. V. Bock, 'Personnel Selection: A Short Method for Selection of Combat Soldiers', *Annals of Internal Medicine*, 19 (1943)

Heath, Clark W., *What People Are. A Study of Normal Young Men* (Cambridge, Massachusetts, 1946)

Hecker, Major Arthur O., First Lieutenant Marvin R. Plessett, and First Lieutenant Philip C. Grana, 'Psychiatry Problems in Military Service During the Training Period', *The American Journal of Psychiatry*, 99 (1942–43)

Heinemann, Larry, *Paco's Story* (London, 1987)

Heinl, Robert D., 'My Lai in Perspective: The Court-Martial of William L. Calley', *Armed Forces Journal*, 21 December 1970

Helmer, John, *Bringing the War Home. The American Soldier in Vietnam and After* (New York, 1974)

Hemingway, Kenneth, *Wings Over Burma* (London, 1944)

Henderson, Mrs Charles, 'The Women's Voluntary Reserve', leaflet (np)

Henderson, Kenneth T., *Khaki and Cassock* (Melbourne, 1919)

Hering, Captain Eugene R., 'The Medical Officer's Responsibility to Command', *U. S. Armed Forces Medical Journal*, 4.2 (1953)

Herr, Michael, *Dispatches* (London, 1978)

Hersey, John, *Into the Valley. A Skirmish of the Marines* (London, 1943)

Hersh, Seymour M., *Cover-Up. The Army's Secret Investigation of the*

Massacre at My Lai 4 (New York, 1972)

Herzog, Tobey C., *Vietnam War Stories. Innocence Lost* (London, 1992)

Hesketh-Prichard, Major H., 'The Hunter in Modern War', *Land and Water*, 3 (April 1919)

Hesketh-Prichard, H., *Sniping in France* (London, 1920)

Hesketh-Prichard, Major H., 'Some Memories of Sniping and Observation', *The Cornhill Magazine*, new series, xlvii (July–December 1919)

Hessin, Arthur L., 'Neuropsychiatry in Airborne Divisions', in Lieutenant-General Hal B. Jennings (ed.), *Neuropsychiatry in World War II. Volume II. Overseas Theatres* (Washington, DC, 1973)

'H. G.', *R. A. F. Occasions* (London, 1941)

Hickey, Revd R. M., *The Scarlet Dawn* (Campbellton, N.B., 1949)

Higgins, Marguerite, *War in Korea* (New York, 1951)

Hilberg, Rail, *The Destruction of the European Jews* (London, 1961)

Hill, Peggy, 'Wartime Bride', in Mavis Nicholson (ed.), *What Did You Do in the War, Mummy? Women in World War II* (London, 1995)

Hillary, Richard, *The Last Enemy* (London, 1942)

Hinsley, Cardinal, *The Bond of Peace and Other War-Time Addresses* (London, 1941)

Hitchcock, Captain F. C., *'Stand To'. A Diary of the Trenches 1915–1918* (London, 1937)

Hocking, William Ernest, *Morale and Its Enemies* (New Haven, Connecticut, 1918)

Hogan, Commander B. W., 'Psychiatric Observations of Senior Medical Officer on Board Aircraft Carrier U.S.S. *Wasp* During Action in Combat Areas, at Time of Torpedoing, and Survivor's Reactions', *The American Journal of Psychiatry*, 100 (1943–4)

Holbrook, Stewart H., *None More Courageous: American War Heroes of Today* (New York, 1942)

Holdsworth, W. W., 'Impressions of a Hospital Chaplain', *The Contemporary Review*, cix (May 1916)

Holland, Henry Scott, *So As By Fire. Notes on the War. Second Series* (London, 1916)

Hollingshaw, August B., 'Adjustment to Military Life', *American Journal of Sociology*, li.5 (March 1946)

Holm, Thomas M., 'American Indian Veterans and the Vietnam War', in Walter Capps (ed.), *The Vietnam Reader* (New York, 1991)

Holm, Tom, 'Forgotten Warriors: American Indian Servicemen in Vietnam', *The Vietnam Generation*, 1.2 (spring 1989)

Holman, Gordon, *Commando Attack* (London, 1942)

Holmes, Richard, *Firing Line* (London, 1985)

Holmes, Richard L., *On War and Morality* (Princeton, 1989)

Honeywell, Roy J., *Chaplains of the United States Army* (Washington, DC, 1958)

Hoover, A. J., *God, Germans, and Britain in the Great War. A Study in Clerical Nationalism* (New York, 1989)

Hopkins, Pryns Charles, *The Psychology of Social Movements. A Psycho-Analytic View of Society* (London, 1938)

Horan, James D. and Gerald Frank, *Out in the Boondocks. Marines in Action in the Pacific. 21 U. S. Marines Tell Their Story* (New York, 1943)

Horowitz, Mardi J. amd George F. Solomon, 'A Prediction of Delayed Stress Syndromes in Vietnam Veterans', *Journal of Social Issues*, 31.4 (1975)

Horton, Revd Robert Forman, *The War. Its Cause and Its Conduct* (London, 1914)

Housman, Laurence (ed.), *War Letters of Fallen Englishmen* (London, 1930)

Hovland, Carl I., Arthur A. Lumsdaine, and Fred D. Sheffield, *Experiments on Mass Communication. Volume III* (Princeton, 1949)

Howard, Ann, *You'll be Sorry!* (Sydney, 1990)

Howe, Adrian, 'Anzac Mythology and the Feminist Challenge', in Joy Damousi and Marilyn Lake (eds), *Gender and War. Australians at War in the Twentieth Century* (Cambridge, 1995)

Hughes, Robert, *Through the Waters. A Gunnery Officer in H.M.S. Scylla 1942–43* (London, 1956)

Hughes, Spencer Leigh, *Things That Don't Count* (London, 1916)

Hume, David, 'Of National Character', in Hume, *Essays: Moral, Political and Literary* (Oxford, 1963)

Hunt, George P., *Coral Comes High* (New York, 1946)

Hunt, William A., 'The Relative Incidence of Psychoneurosis Among Negroes', *Journal of Consulting Psychology*, xi (1947)

Huntington, Samuel P., *The Soldier and the State* (New York, 1967)

Huston, Nancy, 'Tales of War and Tears of Women', *Women's Studies International Forum*, 5 (1982)

Idriess, Ion Llewellyn, *The Desert Column. Leaves from the Diary of an Australian Trooper in Gallipoli, Sinai, and Palestine* (Sydney, 1932)

Idriess, Ion Llewellyn, *Guerrilla Tactics* (Sydney, 1942)

Idriess, Ion Llewellyn, *Lurking Death* (Sydney, 1942)

Idriess, Ion Llewellyn, *Must Australia Fight?* (Sydney 1939)

Idriess, Ion Llewellyn, *Sniping. With an Episode from the Author's Experiences During the War of 1914–18* (Sydney, 1942)

Iglitzin, Lynne B., 'War, Sex, Sports, and Masculinity', in L. L. Farrar, Jr., (ed.), *War. A Historical, Political, and Social Study* (Santa Barbara, 1978)

44 Indian Airborne Division. Training Pamphlet No. 1. The Medical Services in the Division (np, 1944)

Indochina Curriculum Group (eds), *Front Lines. Soldiers' Writings from Vietnam* (Cambridge, Massachusetts, 1975)

Ingram, Right Revd Arthur F. Winnington, *The Church in Time of War* (London, 1915)

Ingram, Revd Arthur F. Winnington, *A Message for the Supreme Moment Delivered to the Assembled Clergy of the London Dioceses by the Bishop of London, November 23rd, 1915, at St Martin's-in-the-Fields* (London, 1915)

Intercom. A Journal for Australian Defence Force Chaplains

Irwin, Anthony S., *Infantry Officer. A Personal Record* (London, 1943)

James, Lieutenant-Colonel A. A., 'Emotional Adjustment and Morale in War', *Canadian Army Journal*, 2.5-8 (August–September 1948 and October–November 1948)

James, William, *'The Moral Equivalent of War' and Other Essays and Selections from 'Some Problems of Philosophy'*, edited by John K. Roth (New York, 1971)

James, William, *Principles of Psychology*, vol. 2 (London, 1910)

Janowitz, Morris, 'Consequences of Social Science Research on the U.S. Military', *Armed Forces and Society*, 8.4 (summer 1982)

Jeffery, George Rutherford, 'Some Points of Interest in Connection with the Psychoneurosis of War', *The Journal of Mental Science*, lxvi.273 (April 1920)

Jennings, Lieutenant-General Hal B., *Neuropsychiatry in World War II. Volume II. Overseas Theatres* (Washington, DC, 1973)

Johns, Captain W. E., 'Worrals in the Wilds', *Girls' Own Paper* (November 1945)

Johnston, Francis J., 'Solidarity – Sociologist Comments', *The Army Combat Forces Journal* 5.10 (May 1955)

Johnston, George H., *The Toughest Fighting in the World* (New York, 1944)

Johnston, Stanley, *The Grim Reapers* (London, 1945)

Johnstone, Tom and James Hagerty, *The Cross on the Sword. Catholic Chaplains in the Forces* (London, 1996)

Jones, Bruce E., *War Without Windows. A True Account of a Young Army Officer Trapped in an Intelligence Cover-Up in Vietnam* (New York, 1987)

Jones, Revd Dennis, *The Diary of a Padre at Suvla Bay* (London, 1916)

Jones, E. Griffith, *The Challenge of Christianity to a World at War* (London, 1915)

Jones, Ernest, 'Psychology and War Conditions', *The Psychoanalytic*

Quarterly, 14 (1945)

Jones, Ernest, 'War and Individual Psychology', *The Sociological Review*, viii (1915)

Jones, Ernest, 'War Shock and Freud's Theory of the Neurosis', in S. Ferenczi, Karl Abraham, Ernest Simmel, and Jones (eds), *Psycho-Analysis and the War Neuroses* (London, 1921)

Jones, Major Franklin Del, 'Experiences of a Division Psychiatrist in Vietnam', *Military Medicine*, 132.12 (December 1967)

Jones, James, *The Thin Red Line* (New York, 1962)

Jones, Ken D. and Arthur F. McClure, *Hollywood at War. The American Motion Picture and World War II* (South Brunswick, 1973)

Jourdain, Lieutenant-Colonel H. F. N. , *Ranging Memories* (Oxford, 1934)

'The Jungle Belongs to Us', *Australian Army Journal*, 85 (May 1956)

'Junius Redivivus', *The Holy War. Diabolus Extremes: Generosity and Avarice. A Satire* (London, 1915)

Juskalian, Lieutenant-Colonel George, 'Why Didn't They Shoot More?', *Army Combat Forces Journal*, 5.2 (September 1954)

Kardiner, A., 'Forensic Issues in the Neuroses of War', *The American Journal of Psychiatry*, 99 (1942–43)

Karsten, Peter, 'Irish Soldiers in the British Army, 1792–1922: Suborned or Subordinate?', *Journal of Social History*, 17 (1983)

Karsten, Peter, *Law, Soldiers and Combat* (Westport, Connecticut, 1978)

Karsten, Peter, '"Professional" and "Citizen" Officers: A Comparison of Service Academy and ROTC Officer Candidates', in Charles C. Moskos (ed.), *Public Opinion and the Military Establishment* (Beverly Hills, 1971)

Kaufman, M. Ralph and Lindsay E. Beaton, 'South Pacific Area', in Lieutenant-General Hal B. Jennings (ed.), *Neuropsychiatry in World War II. Volume II. Overseas Theatres* (Washington, DC, 1973)

Keast, William R., 'The Training of Enlistment Replacements', in Robert R. Palmer, Bell I.Wiley, and Keast (eds), *United States Army in World War II. The Army Ground Forces. The Procurement and Training of Ground Combat Troops* (Washington, DC, 1948)

Keating, Joseph, 'Tyneside Irish Brigade', in Felix Lavery (ed.), *Great Irishmen in War and Politics* (London, 1920)

Keeling, Frederic Hillersdon, *Keeling Letters and Recollections* (London, 1918)

Kellett, Anthony, 'Combat Motivation', in Gregory Belenky (ed.), *Contemporary Studies in Combat Psychiatry* (New York, 1987)

Kelman, Herbert C. and Lee H. Lawrence, 'Assignment of Responsibility in the Case of Lt. Calley: Preliminary Report on a National Survey', *Journal of Social Issues*, 28.1 (1972)

Kemp, Peter, *Mine Were of Trouble* (London, 1957)

Kennedy, Revd E. J., *With the Immortal Seventh Division*, 2nd edition (London, 1916)

Kennedy, Revd G. A. Studdert, *More Rough Rhymes of a Padre* (London, 1920)

Kennedy, Revd G. A. Studdert, *Rough Talks by a Padre. Delivered to Officers and Men of the B. E. F.* (London, 1918)

Kerr, S. Parnell, *What the Irish Regiments have Done* (London, 1916)

Kertzer, Morris N., *With an H on my Dog Tag* (New York, 1947)

Kettle, Tom M., *The Ways of War* (London, 1917)

Kiernan, R. H., *Captain Albert Ball* (London, 1933)

Kiggell, Lieutenant-General L. E., *Training of Divisions for Offensive Action* (London, 1916)

Killingray, David, ' "The 'Rod of Empire' ": The Debate Over Corporal Punishment in the British African Colonial Forces, 1888–1946', *Journal of African History*, 35 (1994)

King, William, ' "Our Men in Vietnam": Black Media as a Source of the Afro-American Experience in Southeast Asia', *The Vietnam Generation*, 1.2 (spring 1989)

Kinnard, Douglas, 'Vietnam Reconsidered: An Attitudinal Study of U. S. Army General Officers', *The Public Opinion Quarterly*, xxxix.4 (1975–6)

Kirshner, L., 'Counter-Transference Issues in the Treatment of the Military Dissenter', *American Journal of Orthopsychiatry*, 43 (1973)

Kitchin, Dennis, *War in Aquarius. Memoir of an American Infantryman in Action Along the Cambodian Border During the Vietnam War* (Jefferson, North Carolina, 1994)

Knox, Collie, *Heroes All* (London, 1941)

Koestler, Arthur, 'The Birth of a Myth. In Memory of Richard Hillary', *Horizon*, vii.40 (April 1943)

Koestler, Arthur, *The Yogi and the Commissar and Other Essays* (London, 1945)

Kohn, Richard, 'Commentary', in Horst Boog (ed.), *The Conduct of the Air War in the Second World War. An International Comparison* (Oxford, 1992)

Kolko, Gabriel, 'War Crimes and the American Conscience', in Jay W. Baird (ed.), *From Nuremberg to My Lai* (Lexington, Massachusetts, 1972)

Kolko, Gabriel, 'War Crimes and the Nature of the Vietnam War', *Journal of Contemporary Asia*, 1.1 (autumn 1970)

Kovic, Ron, *Born on the Fourth of July*, first published in 1976 (Aylesbury, 1990)

Krammer, Arnold, 'Japanese Prisoners of War in America', *Pacific Historical*

Review, 52 (1983)

Kren, George M., 'The Holocaust: Moral Theory and Immoral Acts', in Alen Rosenberg and Gerald E. Myers (eds), *Echoes from the Holocaust. Philosophical Reflections on a Dark Time* (Philadelphia, 1988)

Kulka, Richard A. et al., *Trauma and the Vietnam War Generation. Report of Findings from the National Veterans Readjustment Study* (New York, 1990)

Kyngdon, Lieutenant-Colonel C. W. T., 'The AMF Gold Medal Prize Essay, 1948–49', *Australian Army Journal*, 10 (December–January 1949–50)

Laffin, John, *Women in Battle* (London, 1967)

Lampe, David, *The Last Ditch* (London, 1968)

Lanchbery, Edward, *Against the Sun. The Story of Wing Commander Roland Beaumont, D.S.O., O.B.E., D.F.C. Pilot of the Canberra and the P.1* (London, 1955)

Landecker, Werner S., 'Sociological Research and the Defense Program', *Sociology and Social Research*, xxvi.2 (November–December 1941)

Lane, Mark, *Conversations with Americans* (New York, 1970)

Lang, Daniel, *Casualties of War* (New York, 1969)

Langtry, Lieutenant-Colonel J. O., 'Man-the-Weapon: Neglected Aspects of Leader Training', *Australian Army Journal*, 202 (March 1966)

Langtry, Major J. O., 'Tactical Implications of the Human Factors in Warfare', *Australian Army Journal*, 107 (April 1958)

Lanier, Henry Wysham, *The Book of Bravery. Being True Stories in an Ascending Scale of Courage* (London, 1918)

Large, Lofty, *One Man's War in Korea* (Wellingborough, Northamptonshire, 1988)

Laski, Professor H. J., *The Germans – Are They Human? A Reply to Sir Robert Vansittart* (London, 1941)

Lasswell, Harold D., *Propaganda Technique in the World War* (London, 1927)

Laufer, Robert S., M. S. Gallops, and Ellen Frey-Wouters, 'War Stress and Trauma: The Vietnam Veteran Experience', *Journal of Health and Social Behavior*, 25 (1984)

Laurie, Clayton D., 'The Ultimate Dilemma of Psychological Warfare in the Pacific: Enemies Who Don't Surrender, and GIs Who Don't Take Prisoners', *War and Society*, 14.1 (May 1996)

Law, Francis, *A Man at Arms. Memoirs of Two World Wars* (London, 1983)

Lawrence, Hal, *A Bloody War. One Man's Memories of the Canadian Navy, 1939–1945* (Toronto, 1979)

Lawson, Jacqueline E., '"She's a Pretty Woman . . . For a Gook". The Misogyny of the Vietnam War', in Philip K. Jason (ed.), *Fourteen Landing Zones. Approaches to Vietnam War Literature* (Iowa City, 1991)

Lazell, Dr Edward W., 'Psychology of War and Schizophrenia', *The Psychoanalytic Review*, vii (1920)

Leask, G. A., *V. C. Heroes of the War* (London, 1917)

Leather, Lieutenant E. Hartley, *Combat Without Weapons* (Aldershot, 1942)

Le Bon, Gustave, *Psychologie des foules* (1985)

Le Queux, William, *German Atrocities. A Record of Shameless Deeds* (London, 1914)

Lee, Ulysses Grant, *The Employment of Negro Troops* (Washington, DC, 1966)

Legge, Major-General S. F., 'Soldier, Scientist or Socialite', *Australian Army Journal*, 65 (October 1954)

Leggett, Captain J. K., 'The Human Factor in Warfare', *Australian Army Journal*, 183 (August 1964)

Lehrack, Otto J., *No Shining Armour. The Marines at War in Vietnam. An Oral History* (Lawrence, Kansas, 1992)

Lester, David, 'The Association Between Involvement in War and Rates of Suicide and Homicide', *The Journal of Social Psychology*, 131.6 (1991)

Lester, David, 'War and Personal Violence', in G. Ausenda (ed.), *Effects of War on Society* (San Marino, 1992)

Letters from the Front. Being a Record of the Part Played by Officers of the [Canadian] Bank [of Commerce] in the Great War 1914–1919, vol. 1 (Toronto, 1920)

Leventman, Seymour and Paul Comacho, 'The "Gook" Syndrome: The Vietnam War as a Racial Encounter', in Charles R. Figley and Seymour Leventman (eds), *Strangers at Home. Vietnam Veterans Since the War* (New York, 1990)

Levy, Charles J., 'A. R. V. N. as Faggots: Inverted Warfare in Vietnam', *Trans-Action* (October 1971)

Levy, Howard and David Miller, *Going to Jail. The Political Prisoner* (New York, 1970)

'Lewis Gunner', *Tactical Handling of Lewis Gun with Notes on Instruction, Etc.* (London, 1918)

Lewis, Cecil, *Sagittarius Rising* (London, 1936)

Lewy, Guenter, *America in Vietnam* (New York, 1978)

Lewy, Guenter, *Peace and Revolution. The Moral Crisis of American Pacifism* (Grand Rapids, Michigan, 1988)

Lewy, Guenter, 'Superior Orders, Nuclear Warfare, and the Dictates of Conscience', in Richard A. Wasserstrom (ed.) *War and Morality* (Belmont, California, 1970)

Lidstone, Squadron Leader R. A., *Bloody Bayonets. A Complete Guide to Bayonet Fighting* (Aldershot, 1942)

Lifton, Robert Jay, 'Existential Evil', in Nevitt Sanford and Craig Comstock (eds), *Sanctions for Evil* (San Francisco, 1971)

Lifton, Robert Jay, *Home from the War. Vietnam Veterans: Neither Victims Nor Executioners* (London, 1974)

Lifton, Robert Jay, 'The Postwar World', *Journal of Social Issues*, 31.4 (1975)

Lindeman, Lieutenant-Colonel K. E., 'The Psychology of Fear', *Canadian Army Journal*, xiv.5 (January 1960)

Lindsell, Major W. G., 'War Exhaustion', *Journal of the Royal Artillery*, liii (1926–7)

Line, Colonel William, 'Morale and Leadership', *Canadian Army Journal*, 6.1 (April 1952)

Linklater, Eric, *The Art of Adventure* (London, 1947)

Linwood, Captain R. J., 'The Sniper – Part I', *Australian Infantry*, xx.2 (December 1980)

Little, Roger W., 'Buddy Relations and Combat Performance', in Morris Janowitz (ed.), *The New Military. Changing Patterns of Organization* (New York, 1964)

Livingstone, Dr Gordon S., 'Healing in Vietnam', in Richard A. Falk, Gabriel Kolko, and Robert Jay Lifton (eds), *Crimes of War* (New York, 1971)

Lizotte, Alan J. and David J. Bordua, 'Military Socialization, Childhood Socialization, and Current Situation: Veterans' Firearms Ownership', *Journal of Political and Military Socialisation*, 8.2 (fall 1980)

Llorens, David, 'Why Negroes Re-enlist', *Ebony*, 23 August 1968

Lodge, Sir Oliver J., *Raymond or Life and Death with Examples of the Evidence for Survival of Memory and Affection After Death*, 10th edition (London, 1918)

Logan, Captain H. Meredith, 'Military Training To-Day', *The Army Quarterly*, vi (April 1923)

Lomperis, Timothy J., *'Reading the Wind'. The Literature of the Vietnam War. An Interpretive Critique* (Durham, North Carolina, 1987)

Louttit, C. M., 'Psychology During the War and Afterwards', *Journal of Consulting Psychology*, viii.1 (January–February 1944)

Ludwig, Major Alfred O., 'Malingering in Combat Soldiers', *The Bulletin of the U. S. Army Medical Department*, ix, supplemental number (November 1949)

Ludwig, Alfred O., 'Neuroses Occurring in Soldiers After Prolonged Combat Exposure', *Bulletin of the Menninger Clinic*, 11.1 (January 1947)

Lyall, Gavin (ed.), *The War in the Air, 1939–1945. An Anthology of Personal Experience* (London, 1968)

Lynn, Escott, *In Khaki for the King. A Tale of the Great War* (London, 1915)

McAfee, John, 'Open Season', in J. Topham (ed.), *Vietnam Literature Anthology*, revised and enlarged (Philadelphia, 1990)

McBride, Herbert W., *A Rifleman Went to War* (Marines, North Carolina, 1935)

McCamley, Megan Llewellyn, 'Absent Without Leave', in *'Off Duty'. An Anthology. Verse and Short Stories by Men and Women of the Forces – Members of the War Service Women's Club (English Speaking Union) at the Stanley Palace, Chester* (Chester, 1945)

McCann, Sean, *The Fighting Irish* (London, 1972)

McCarthy, John, 'Aircrew and "Lack of Moral Fibre" in the Second World War', *War and Society*, 2.2 (September 1984)

McCarthy, Mary, *Medina* (New York, 1972)

McCudden, James Byford, *Flying Fury*, first published in 1918 (London, 1930)

McCulloch, Joseph, *We Have Our Orders* (London, 1944)

MacCurdy, John T., *The Psychology of War* (London, 1917)

MacCurdy, John T., *The Structure of Morale* (Cambridge, 1943)

MacCurdy, John T., *War Neuroses* (Cambridge, 1918)

MacDonagh, Michael, *The Irish at the Front* (London, 1916)

MacDonagh, Michael, *The Irish on the Somme* (London, 1917)

McDonald, Edward C., 'Social Adjustment to Militarism', *Sociology and Social Research*, 29.6 (July–August 1945)

MacDonald, Colonel Sir John, 'The Knife in Trench Warfare', *Journal of the Royal United Service Institute*, lxii (1917)

MacDonald, Walter, 'Interview with a Guy Named Fawkes, U.S. Army', in William D. Ehrhart (ed.), *Unaccustomed Mercy. Soldier-Poets of the Vietnam War* (Lubbock, Texas, 1989)

McDougall, Murdoch C., *Swiftly They Struck. The Story of No. 4 Commando* (London, 1954)

McDougall, William, *An Introduction to Social Psychology* (London, 1908)

McDougall, William, *An Introduction to Social Psychology*, 9th edition (London, 1915)

McEniry, Howard, *A Marine Dive-Bomber Pilot at Guadalcanal* (Tuscaloosa, Alabama, 1987)

McKay, Gary, *In Good Company. One Man's War in Vietnam* (Sydney, 1987)

McKay, Gary, *Vietnam Fragments. An Oral History of Australians at War* (St Leonards, New South Wales, 1992)

MacKenna, Robert William, *Through a Tent Door*, first published 1919 (London, 1930)

MacKenna, Robert William, *Verses* (Edinburgh, 1897)

MacKenzie, De Witt, *Men Without Guns* (Philadelphia, 1945)

MacKenzie, J. S., *Arrows of Desire. Essays on Our National Character and Outlook* (London, 1920)

MacKenzie, S. P., 'Morale and the Cause: The Campaign to Shape the Outlook of Soldiers in the British Expeditionary Force, 1914–1918', *Canadian Journal of History*, 25 (1990)

McKernan, Michael, *Australian Churches at War. Attitudes and Activities of the Major Churches 1914–1918* (Sydney, 1980)

McKernan, Michael, *Padre. Australian Chaplains in Gallipoli and France* (Sydney, 1986)

McLaglen, Captain Leopold, *Bayonet Fighting for War* (London, 1916)

McLuskey, J. Fraser, *Parachute Padre* (London, 1951)

McNeel, B. H., 'War Psychiatry in Retrospect', *The American Journal of Psychiatry*, 102 (1945–6)

McNeill, Ian, 'Australian Army Advisers. Perceptions of Enemies and Allies', in Kenneth Maddock and Barry Wright (eds), *War. Australia and Vietnam* (Sydney, 1987)

MacPhail, Sir Andrew, *Official History of the Canadian Forces in the Great War 1914–19. The Medical Services* (Ottawa, 1925)

Macklin, Major-General W. H. S., 'Military Law', *Canadian Army Journal*, viii.1 (January 1954)

Macksey, Kenneth and William Woodhouse (eds), *The Penguin Encyclopedia of Modern Warfare* (London, 1991)

Macky, N. L., 'Weapon and Target', *The Army Quarterly*, xxxiii (January 1937)

Maddock, Kenneth, 'Going over the Limit? – The Question of Australian Atrocities', in Maddock (ed.), *Memories of Vietnam* (Sydney, 1991)

Mahan, Jack L. and George A. Clum, 'Longitudinal Prediction of Marine Combat Effectiveness', *The Journal of Social Psychology*, 83 (1971)

Mahedy, William P., *Out of the Night. The Spiritual Journey of Vietnam Veterans* (New York, 1986)

Malo, Jean-Jacques and Tony Williams (eds), *Vietnam War Films* (Jefferson, North Carolina, 1994)

Major, Revd H. D. A., 'Sentimentalists and Casuists', *Modern Churchman* (August 1917)

Manchester, William, *Goodbye Darkness. A Memoir of the Pacific War* (Boston, 1980)

Mann, Emily, *Still Life* (New York, 1982)

Mann, Leon, 'Attitudes Towards My Lai and Obedience to Orders: An Australian Survey', *Australian Journal of Psychology*, 25.1 (1973)

Mannock, Major Edward 'Mick', *The Personal Diary* (London, 1966)

Mantell, David Mark and Robert Panzarella, 'Obedience and Responsibility', *British Journal of Social and Clinical Psychology*, 15 (1976)

Manual of Military Law (1914 and 1944)

Manzen, Captain P. P., 'Philosophy, Psychology and the Army', *Australian Army Journal*, 188 (January 1965)

Marin, Peter, 'Living with Moral Pain', *Psychology Today*, 15.11 (November 1981)

'Mark VII', *A Subaltern on the Somme in 1916* (London, 1927)

Marks, Private Richard E., *The Letters of Pfc. Richard E. Marks, USMC* (Philadelphia, 1967)

Marlowe, N., 'The Mood of Ireland', *British Review*, xi.1 (July 1915)

Marquis, Donald G., 'The Mobilization of Psychologists for War Service', *Psychological Bulletin*, 41 (1944)

Marren, Colonel John J., 'Psychiatric Problems in Troops in Korea During and Following Combat', *U. S. Armed Forces Medical Journal*, vii.5 (May 1956)

Marrin, Albert, *The Last Crusade. The Church of England in the First World War* (Durham, North Carolina, 1974)

Marsden, George M., *Fundamentalism and American Culture. The Shaping of Twentieth Century Evangelicalism 1870–1925* (New York, 1980)

Marshall, Kathryn, *In the Combat Zone. An Oral History of American Women in Vietnam 1966–1975* (Boston, 1987)

Marshall, Colonel S. L. A., *Men Against Fire. The Problem of Battle Command in Future War* (New York, 1947)

Maskin, Meyer, *Psychiatry*, 9 (May 1946)

'Maxims for the Leader', *42nd East Lancashire Division, Handbook* (Aldershot, 1918)

Maxwell, W. N., *A Psychological Retrospect of the Great War* (London, 1923)

May, Mark A., *A Social Psychology of War and Peace* (New Haven, Connecticut, 1943)

Mead, Margaret, 'Alternatives to War', in Morton Fried, Marvin Harris, and Robert Murphy (eds), *War. The Anthropology of Armed Conflict and Aggression* (New York, 1968)

Mead, Margaret, 'Warfare is Only an Invention – Not a Biological Necessity', in Leon Bramson and George W. Goethals (eds), *War: Studies from Psychology, Sociology, Anthropology* (New York, 1964)

Mee, Patrick, *Marine Gunner. Twenty-Two Years in the Royal Marine Artillery* (London, 1935)

Meier, Norman C., *Military Psychology* (New York, 1943)

Mellor, Major Tony, *Machine-Gunner* (London, 1944)

Mendelsohn, Everett, Merritt Roe Smith, and Peter Weingart (eds), *Science, Technology, and the Military*, 2 vols. (Dordrecht, 1988)

Mercer, John, *Mike Target* (Lewes, Sussex, 1990)

Merewether, Lieutenant-Colonel J. W. B. and Sir Frederick Smith, *The Indian Corps in France* (London, 1917)

Merritt, William E., *Where the Rivers Ran Backwards* (Athens, 1989)

Meynell, Viola, *Julian Grenfell* (London, 1918)

Middlebrook, Martin, *The Battle of Hamburg. Allied Bomber Forces Against a German City in 1943* (London, 1980)

Milgram, Stanley, 'Behaviour Study of Obedience', in A. Etzioni and W. Wenglinsky (eds), *War and Its Prevention* (New York, 1970)

Milgram, Stanley, *Obedience to Authority: An Experimental View* (New York, 1974)

Milgram, Stanley, 'Some Conditions of Obedience and Disobedience to Authority', *Human Relations*, 18.1 (February 1965)

The Military Chaplaincy. A Report to the President by the [U.S.A.] President's Committee on Religion and Welfare in the Armed Forces (Washington, DC, 1 October 1950)

Miller, Major Charles W., 'Delayed Combat Reactions in Air Force Personnel', *War Medicine*, 8.4 (October 1945)

Miller, John, *Guadalcanal: The First Offensive* (Washington, DC, 1949)

Miller, Merle, Introduction, in Don Congdon (ed.), *Combat: War in the Pacific* (London, 1958)

Miller, Neal E., 'The Frustration-Aggression Hypothesis', *Psychological Review*, 48 (1941)

Miller, Neal E. and Richard Bugelski, 'Minor Studies in Aggression: II. The Influence of Frustrations Imposed by the In-Group on Attitudes Expressed Toward Out-Groups', *The Journal of Psychology*, 25 (1948)

Millington, Major M. S. F., 'The Women Who Served the Guns', in *On Target. The Great Story of Ack-Ack Command. Official Souvenir* (London, 1955)

Mitchell, Alan W., 'Sergeant-Pilot James Allen Ward, V.C.', in Derek Tangye (ed.), *Went the Day Well* (London, 1942)

Mitchell, Mairin, *Storm over Spain* (London, 1937)

Mitgang, Herbert (ed.), *Civilians Under Arms. The American Soldier – Civil War to Korea as He Revealed Himself in His Own Words in 'The Stars and Stripes', Army Newspaper* (Cleveland, Ohio, 1959)

Mithers, Carol Lynn, 'Missing in Action: Women Warriors in Vietnam', *Cultural Critique*, 3 (spring 1986)

Monard, Lieutenant S. H., 'Fuel and Ashes', *The Incinerator*, 1.2 (June 1916)

Money-Kyrle, R. E., 'The Development of War. A Psychological Approach', *British Journal of Medical Psychology*, xvi.3 (1937)

Montagu, M. F. Ashley (ed.), *Man and Aggression* (New York, 1968)

Montague, C. E., *Disenchantment* (London, 1922)

Montgomery of Alamein, *Morale in Battle: Analysis* (np, 30 April 1946)

Montgomery of Alamein, 'Raw Material', *Australian Army Journal*, 3 (October–November 1948)

Moore, Clyde B., 'Some Psychological Aspects of War', *The Pedagogical Seminary. A Quarterly*, xxiii (1916)

Moore, J. Howard, *Savage Survivals* (London, 1916)

Moran, Lord, *The Anatomy of Courage* (London, 1945)

Morgan, Lloyd, *Habit and Instinct* (London, 1896)

Morrow, John H., 'Knights of the Sky. The Rise of Military Aviation', in Frans Coetzee and Marilyn Shevin-Coetzee (eds), *Authority, Identity and the Social History of the Great War* (Providence, 1995)

Moser, Richard, 'Talkin' the Vietnam Blues: Vietnam Oral History and Our Popular Memory of War', in D. Michael Shafer (ed.), *The Legacy of the Vietnam War in the American Imagination* (Boston, 1990)

Moskos, Charles C., *The American Enlisted Man* (New York, 1970)

Moskos, Charles, 'Army Women', *The Atlantic Monthly* (August 1990), and on the Internet

Mosley, Nicholas, *Julian Grenfell. His Life and the Times of His Death 1888–1915* (London, 1976)

Mosse, George L., *Fallen Soldiers. Reshaping the Memory of the World Wars* (New York, 1990)

Mosse, George, 'The Knights of the Sky and the Myth of the War Experience', in Robert A. Hinde and Helen E. Watson (eds), *War: A Cruel Necessity? The Bases of Institutional Violence* (London, 1995)

Motley, Mary Penick (ed.), *The Invisible Soldier. The Experience of the Black Soldier, World War II* (Detroit, 1975)

Mott, F. W., *War Neuroses and Shell Shock* (London, 1919)

Mouldin, Bill, *Up Front* (New York, 1945)

Moynihan, Michael (ed.), *Greater Love. Letters Home 1914–1918* (London, 1980)

Mullen, Thomas J., '"Mick" Mannock. The Forgotten Ace', *The Irish Sword*, x.39 (winter 1971)

Murphy, Audie, *To Hell and Back* (London, 1956)

Murray, Professor George Gilbert Aime, *Herd Instinct: For Good and Evil* (London, 1940)

Murrow, Edward R., *This is London* (London, 1941)

Myers, Charles S., *Shell Shock in France 1914–1918. Based on a War Diary* (Cambridge, 1940)

Myrer, Anton, *Once an Eagle* (New York, 1968)

Nash, Mary, '"Milicianas" and Homefront Heroines: Images of Women in Revolutionary Spain (1936–1939)', *History of European Ideas*, 11 (1989)

Needles, William, 'The Regression of Psychiatry in the Army', *Psychiatry*, 9.3 (August 1946)

Nelson, Scott H. and E. Fuller Torrey, 'The Religious Functions of Psychiatry', *American Journal of Orthopsychiatry*, 43.3 (April 1973)

Neville, Captain J. E. H., *The War Letters of a Light Infantryman* (London, 1930)

Newman, Richard A., 'Combat Fatigue: A Review to the Korean Conflict', *Military Medicine*, 129.10 (October 1964)

Niles, John J., *Singing Soldiers* (New York, 1927)

Nolan, Keith William, *Sappers in the Wire. The Life and Death of Firebase Mary Ann* (College Station, 1975)

Norden, Eric, *American Atrocities in Vietnam* (Sydney, 1966)

Oakley, John, *A Gentleman in Khaki. A Story of the South African War* (London, 1900)

O'Brien, Tim, *The Things They Carried* (New York, 1990)

'The O. C.', 'Birth of a Battery', *Tackle 'Em Low*, 1.1 (December 1939)

O'Callahan, Father Joseph T., *I Was Chaplain on the 'Franklin'* (New York, 1961)

O'Connell, Robert L., *Of Arms and Men. A History of War, Weapons, and Aggression* (New York, 1989)

'An Officer', *Practical Bayonet-Fighting, with Service Rifle and Bayonet* (London, 1915)

O'Leary, Shawn, *Spikenard and Bayonet. Verse from the Front Line* (Melbourne, 1941)

Olley, Randy and Herbert H. Kraus, 'Variables Which May Influence the Decision to Fire in Combat', *The Journal of Social Psychology*, 92 (1974)

Oplinger, Jon, *Quang Tri Cadence. Memoir of a Rifle Platoon Leader in the Mountains of Vietnam* (Jefferson, North Carolina, 1993)

Oppenheim, Lassa, *International Law. A Treatise*, vol. II (1906)

Opton, Edward M., 'It Never Happened and Besides They Deserved It', in Nevitt Sanford and Craig Comstock (eds), *Sanctions for Evil* (San Francisco, 1971)

O'Rahilly, Professor Alfred, *Father William Doyle S. J. A Spiritual Study* (London, 1925)

Orr, Captain G. M., 'Some Moral Factors in War', *The Commonwealth Military Journal*, 1 (May 1911)

Orr, Philip, *The Road to the Somme. Men of the Ulster Division Tell Their Story* (Belfast, 1987)

Ottley, Roi, *'New World A-Comin": Inside Black America* (Boston, 1943)

Ouditt, Sharon, *Fighting Forces, Writing Women. Identity and Ideology in the First World War* (London, 1994)

'Outis', 'Has Recruiting in Ireland Been Satisfactory?', *The United Service Magazine*, new series, li (April–September 1915)

Owen, Lieutenant-Colonel David Lloyd, *The Desert My Dwelling Place* (London, 1957)

Owen, Harold and John Bell (eds), *Wilfred Owen. Collected Letters* (London, 1961)

Owen, Wilfred, 'Apologia Pro Poemate Meo', in Jon Stallworthy (ed.), *Wilfred Owen: The Complete Poems and Fragments*, vol. 1 (London, 1983)

Palmer, General Bruce, *The 25-Year War. America's Military Role in Vietnam* (New York, 1984)

Parks, David, *GI Diary* (New York, 1968)

Parks, Major William H., 'Command Responsibility for War Crimes', *Military Law Review*, 62 (1973)

Parks, Major W. Hays, 'Crimes in Hostilities. Part I', *Marine Corps Gazette* (August 1976)

Parrish, Randall, *Shea of the Irish Brigade. A Soldier's Story* (London, 1914)

Paterson, T. T., *Morale in War and Work. An Experiment in the Management of Men* (London, 1955)

Pear, T. H., *Are There Human Instincts?* (Manchester, 1943)

Pearn, O. P. Napier, 'Psychoses in the Expeditionary Force', *The Journal of Mental Science*, lxv (April 1919)

Peat, Harold R., *The Inexcusable Lie* (New York, 1923)

Peat, Harold R., *Private Peat* (Indianapolis, 1917)

Peberdy, G. R., 'Moustaches', *Journal of Mental Science*, 107.446 (January 1961)

Pedersen, Chet, 'Wastelands', in J. Topham (ed.), *Vietnam Literature Anthology*, revised and enlarged (Philadelphia, 1990)

Peers, Lieutenant-General William R., *The My Lai Inquiry* (New York, 1979)

Pen Pictures of the War by Men at the Front. Vol. I. The Campaign in Natal to the Battle of Colenso (London, 1900)

Perry, W. J., *War and Civilisation* (Manchester, 1918)

Peterson, Donald B., 'Discussion', *The American Journal of Psychiatry*, 123.7 (January 1967)

Phillpotts, Eden, *The Human Boy and the War* (London, 1916)

Pilisuk, Marc, 'The Legacy of the Vietnam Veteran', *Journal of Social Issues*, 31.4 (1975)

Pinkham, Henry William, *Was Christ a Pacifist?* (Melrose, Massachusetts, 1917)

Pitman, Patricia, 'Private Hatred', in Mavis Nicholson (ed.), *What Did You Do in the War, Mummy? Women in World War II* (London, 1995)

Plater, Charles, *A Primer of Peace and War. The Principles of International Morality*, edited for the Catholic Social Guild (London, 1915)

Playne, Caroline, *Society at War 1914–1916* (London, 1931)

Plowman, Max, *War and the Creative Impulse* (London, 1919)

Polner, Murray, *No Victory Parades. The Return of the Vietnam Veteran* (London, 1971)

Porter, William C., 'Military Psychiatry and the Selective Services', *War Medicine*, 1.3 (May 1941)

Pratt, Major Dallas and Abraham Neustadter, 'Combat Record of Psychoneurotic Patients', *The Bulletin of the U.S. Army Medical Department*, vii.9 (September 1947)

'Preventative Psychiatry', *The Bulletin of the U.S. Army Medical Department*, vi.5 (November 1946)

Price, Second Lieutenant Harold Warnica, 'Diary', in Brereton Greenhous (ed.), *A Rattle of Pebbles: The First World War Diaries of Two Canadian Airmen* (Ottawa, 1987)

Priday, H. G. L., *The War from Coconut Square. The Story of the Defence of the Island Bases of the South Pacific* (Wellington, New Zealand, 1945)

Priestley, J. B., *British Women Go to War* (London, 1943)

Prugh, Major-General George S., *Law at War: Vietnam 1964–1973* (Washington, DC, 1975)

Pruitt, Colonel Francis W., 'Doctor–Patient Relationships in the Army', *U.S. Armed Forces Medical Journal*, v.2 (February 1954)

Psychology for the Armed Forces (Washington, DC, 1945)

Psychology for the Fighting Man. Prepared for the Fighting Man Himself by a Committee of the National Research Council with the Collaboration of Science Service as a Contribution to the War Effort, second edition (Washington, DC, 1944)

Purcell, William, *Woodbine Willie. An Anglican Incident. Being Some Account of the Life and Times of Geoffrey Anketell Studdert Kennedy, Poet, Prophet, Seeker After Truth, 1883–1929* (London, 1962)

Pyle, Ernie, *Here is Your War* (New York, 1943)

Pym, Revd T. W. and Revd Geoffrey Gordon, *Papers from Picardy by Two Chaplains* (London, 1917)

Quester, George H., 'The Problems', in Nancy Loring Goldman (ed.), *Female Soldiers – Combatants or Noncombatants? Historical and Contemporary Perspectives* (Westport, Connecticut, 1982)

Quinn, Major Kathryn, 'Women and the Military Profession', in Hugh Smith (ed.), *The Military Profession in Australia* (Canberra, 1988)

Radine, Lawrence B., *The Taming of the Troops. Social Control in the United States Army* (Westport, Connecticut, 1977)

Ralya, Lynn L., 'Some Surprising Beliefs Concerning Human Nature Among Pre-Medical Psychology Students', *The British Journal of Educational Psychology*, xv.ii (June 1945)

Rashan, Cadet Ian, 'The Spirit of the Bayonet', *The Blimp* (Cambridge, 1917)

Raven, Charles E., *Musings and Memories* (London, 1931)

Raws, Lieutenant-Colonel W. L., 'Discipline and *Moral*', *The Australian Military Journal*, v (April 1914)

Raws, Lieutenant-Colonel W. L., 'Home Training', *The Australian Military Journal*, 7 (January 1916)

Raymond, Ernest, *Tell England. A Study in a Generation* (London, 1922)

Read, James Morgan, *Atrocity Propaganda 1914–1919* (New Haven, Connecticut, 1941)

Reckill, Maurice B., *God's Opportunity. A Guide to Clear Thought and Discussion About the War* (London, c.1940)

Redmond, Major William, *Trench Pictures from France* (London, 1917)

Rees, John Rawlings, *The Shaping of Psychiatry by War* (London, 1945)

Reid, Brian Holden, *J. F. C. Fuller: Military Thinker* (London, 1987)

Reid, Professor H. M. B., *Theology After the War* (Glasgow, 1916)

Rendulic, Dr (General) Lothar, 'The Change in Tactics Through Atomic Weapons', *Canadian Army Journal*, xi.1 (January 1957)

Reyburn, Wallace, *Glorious Chapter. The Canadians at Dieppe* (London, 1943)

Richardson, Lewis F., 'The Persistence of National Hatred and the Changeability of Objects', *British Journal of Medical Psychology*, xxii.3 (1949)

Richey, Paul, *Fighter Pilot. A Personal Record of the Campaign in France* (London, 1944)

Ridgway, General Matthew B., 'Man – The Vital Weapon', *Australian Army Journal*, 79 (December 1955)

Rieber, Robert W. (ed.), *The Psychology of War and Peace. The Image of the Enemy* (New York, 1991)

Ringnaldo, Don, *Fighting and Writing the Vietnam War* (Jackson, 1994)

Ripley, Lieutenant-Colonel Herbert S. and Major Stewart Wolf, 'Mental Illness Among Negro Troops Overseas', *American Journal of Psychiatry*,

103 (1946–7)

Rivaz, Flight Lieutenant Richard Charles, *Tail Gunner* (London, 1943)

Robertson, Squadron Leader A. K., 'Theirs Not to Reason Why? An Essay in Military Ethics', *Defence Force Journal*, 1 (November–December 1976)

Robinson, Captain Ernest H., *Rifle Training for War. A Textbook for Home Guards* (London, 1940)

Robinson, Captain Ernest H., *Rifle and Carton. Some Notes on Target Shooting with the .22 Calibre Rifle, as a Sport and in Preparation for War* (London, 1914)

Robinson, Vee, *On Target* (Wakefield, West Yorkshire, 1991)

Rogerson, C. H., 'Letter to Dr. D. Ewen Cameron, 18 July 1940', *The American Journal of Psychiatry*, 97.2 (1941)

Rogerson, Sidney, *Twelve Days*, first published in 1930 (Norwich, 1988)

Rome, Lieutenant-Commander Howard P., 'Motion Pictures as a Medium of Education', *Mental Hygiene*, 30 (January 1946)

Roscoe, Revd J. E., *The Ethics of War, Spying and Compulsory Training* (London, nd)

Rose, D. E., 'Psychology and the Armed Forces', *Australian Journal of Psychology*, 10.1 (June 1958)

Rosenberg, Elizabeth, 'A Clinical Contribution to the Psychopathology of the War Neuroses', *The International Journal of Psycho-Analysis*, xxiv (1943)

Ross, Jane, 'Veterans in Australia: The Search for Integration', in Jeff Doyle and Jeffery Grey (eds), *Australia R&R. Representations and Reinterpretations of Australia's War in Vietnam* (Chevy Chase, MD, 1991)

Rottmann, Larry, Jan Barry, and Basil T. Paquet (eds), *Winning Hearts and Minds. War Poems by Vietnam Veterans* (New York, 1972)

Roussy, Dr. G. and J. Lhermitte, *The Psychoneuroses of War* (London, 1917)

Rowe, Captain W. C., 'Ethics of Surrender', *Army Training Memorandum. Australian Military Force*, 44 (October 1946)

Rundell, J. R., R. J. Ursano, H. C. Holloway, and D. R. Jones, 'Combat Stress Disorders and the U. S. Air Force', *Military Medicine*, 155 (1990)

Sach, Wing Commander R. H., 'The Soldier's Dilemma', *Defence Force Journal*, 12 (September–October 1978)

Sagan, Eli, *Cannibalism. Human Aggression and Cultural Form* (New York, 1970)

Salmon, Thomas W., *The Care and Treatment of Mental Diseases and War Neuroses ('Shell Shock') in the British Army* (New York, 1917) (and in *Mental Hygiene*, i.4 (October 1917))

Salt, Henry S., *The Creed of Kinship* (London, 1935)

Samelson, Franz, 'World War I Intelligence Testing and the Development of

Psychology', *Journal of the History of the Behavior Sciences*, 133 (1977)

Sandes, Flora, *The Autobiography of a Woman Soldier. A Brief Record of Adventure with the Serbian Army, 1916–1919* (London, 1927)

Sandes, Flora, *An English Woman-Sergeant in the Serbian Army* (London, 1916)

Santoli, Al, *Everything We Had. An Oral History of the Vietnam War by Thirty-Three American Soldiers Who Fought It* (New York, 1981)

Sassoon, Siegfried, *Diaries 1915–1918*, edited by Rupert Hart-Davis (London, 1983)

Saul, Commander Leon J., Commander Howard Rome, and Edwin Leuser, 'Desensitization of Combat Fatigue Patients', *The American Journal of Psychiatry*, 102 (1945–6)

Saunders, Hilary St George, *The Green Beret. The Story of the Commandos 1940–1945* (London, 1949)

Schloerb, Rolland W., 'An Uneasy Conscience About Killing', in Dewitte Holland (ed.), *Sermons in American History. Selected Issues in the American Pulpit 1630–1967* (Nashville, 1971)

Schnaier, Jenny A., 'A Study of Women Vietnam Veterans and Their Mental Health Adjustment', in Charles R. Figley (ed.), *Trauma and Its Wake. Volume II: Traumatic Stress Theory, Research, and Intervention* (New York, 1986)

Schroeder, Eric James, 'Two Interviews: Talks with Tim O'Brien and Robert Stone', *Modern Fiction Studies*, 30.1 (spring 1984)

Scott, Mrs Dawson (ed.), *The Official Handbook of the Women's Defence Relief Corps*, third (enlarged) edition (Southall, Middlesex, 1915)

Scott, J. P., 'Biology and Human Aggression', *American Journal of Orthopsychiatry*, 40.4 (July 1970)

Scott, J. P., 'Biology and the Control of Violence', *International Journal of Group Tensions*, 3.3-4 (1973)

Scott, J. P., 'Biological Basis of Human Warfare: An Interdisciplinary Problem', in Muzafer Sherif and Carolyn W. Sherif (eds), *Interdisciplinary Relationships in the Social Sciences* (Chicago, 1969)

Scott, Thomas D., 'Tactical Training for Ground Combat Forces', *Armed Forces and Society*, 6.2 (winter 1980)

Scott, Wilbur J., *The Politics of Readjustment. Vietnam Veterans Since the War* (New York, 1993)

Seago, Dorothy W., 'Stereotypes: Before Pearl Harbour and After', *The Journal of Psychology*, 23 (1947)

Segal, David R., Nora Scott Kinzer, and John C. Woelfel, 'The Concept of Citizenship and Attitudes Toward Women in Combat', *Sex Roles*, 3.5 (1977)

Segal, Robert A. (ed.), *In Quest of the Hero* (Princeton, 1990)

Seidenberg, Willa and William Short (eds), *A Matter of Conscience. GI Resistance During the Vietnam War* (Andover, Massachusetts, 1991)

Sellers, W. E., *From Aldershot to Pretoria. A Story of Christian Work Among Our Troops in South Africa* (London, 1900)

Serraillier, Ian, 'The New Learning', in Brian Gardner, *The Terrible Rain. The War Poets 1939–1945* (London, 1966)

Servern, Mark, *The Gambordier. Giving Some Account of the Heavy and Siege Artillery in France 1914–1918* (London, 1930)

Service, Robert W., *The Rhymes of a Red-Cross Man* (London, 1916)

Seton-Karr, Henry, *The Call to Arms 1900–1901* (London, 1902)

Sewall, May Wright, *Women, World War and Permanent Peace* (San Francisco, 1915)

Sgroi, Paul, 'To Vietnam and Back', in Walter Capps (ed.), *The Vietnam Reader* (New York, 1991)

Shafer, D. Michael, 'The Vietnam Combat Experience: The Human Legacy', in Shafer (ed.), *The Legacy. The Vietnam War in the American Imagination* (Boston, 1990)

Shatan, Chaim F., 'The Grief of Soldiers: Vietnam Combat Veterans' Self-Help Movement', *The American Journal of Orthopsychiatry*, 43.4 (July 1973)

Shawcross, Edith, 'The Hand that Rocks the Cradle', in Keith Briant and Lyall Wilkes (eds), *Would I Fight?* (Oxford, 1938)

Shay, Jonathan, *Achilles in Vietnam. Combat Trauma and the Undoing of Character* (New York, 1994)

Sheppard, H. R. L., *We Say 'No'. The Plain Man's Guide to Pacifism* (London, 1935)

Shimeld, J. A., *Hints on Military Instruction. A Concise Presentation of Valuable Instructional Knowledge for the Military Officer, Warrant Officer and N. C. O.* (Sydney, 1941)

Shirley, Lieutenant-Colonel W., *Morale. The Most Important Factor in War* (London, 1916)

Shute, Carmel, 'Heroines and Heroes: Sexual Mythology in Australia 1914–1918', in Joy Damousi and Marilyn Lake (eds), *Gender and War. Australians at War in the Twentieth Century* (Cambridge, 1995)

Silber, Earle, 'Adjustment to the Army. The Soldiers' Identification with the Group', *U.S. Armed Forces Medical Journal*, v.9 (September 1954)

Sillman, Leonard R., 'Morale', *War Medicine*, 3.5 (May 1943)

Sillman, Leonard R., 'A Psychiatric Contribution to the Problem of Morale', *Journal of Nervous Mental Disorder*, 97 (1943)

Silverman, Maurice, 'Causes of Neurotic Breakdown in British Service Personnel Stationed in the Far East in Peacetime', *The Journal of Mental Science*, xcvi.403 (April 1950)

Simon, William J., 'My Country', in Larry Rottmann, Jan Barry, and Basil T. Paquet (eds), *Winning Hearts and Minds. War Poems by Vietnam Veterans* (New York, 1972)

Simpson, J. G., 'The Witness of the Church in Time of War', in G. K. A. Bell (ed.), *The War and the Kingdom of God* (London, 1915)

Singh, Saint Nihal, *India's Fighters: Their Mettle, History and Services to Britain* (London, 1914)

Singh, Saint Nihal, *India's Fighting Troops* (London, 1914)

Sinker, Revd John, *The War, Its Deeds and Lessons. Addresses Delivered in Lytham Parish Church* (London, 1916)

Sisson, Colin P., *Wounded Warriors. The True Story of a Soldier in the Vietnam War and of the Emotional Wounds Inflicted* (Auckland, 1993)

Skennerton, Ian, *The British Sniper. British and Commonwealth Sniping and Equipments 1915–1983* (London, 1983)

Skinner, Revd Leslie, *'The Man Who Worked on Sundays'. Personal War Diary* (Stoneleigh, Surrey, nd)

Sledge, Eugene B. 'Sledgehammer', *With the Old Breed at Peleliu and Okinawa* (New York, 1990)

Slim, Field-Marshal Sir William, *Courage and Other Broadcasts* (London, 1957)

Smelser, Neil J., 'Some Determinants of Destructive Behavior', in Nevitt Sanford and Craig Comstock (eds), *Sanctions for Evil* (San Francisco, 1971)

Smigel, Erwin O., 'The Place of Sociology in the Army', *Sociology and Social Research*, xxvi.6 (July–August 1942)

Smith, G. Elliott and T. H. Pear, *Shell Shock and its Lessons* (Manchester, 1919)

Smith, Hugh, 'Conscientious Objection to Particular Wars: Australia's Experience During the Vietnam War, 1965–1972', *War and Society*, 8.1 (May 1990)

Smith, Malvern Van Wyk, *Drummer Hodge. The Poetry of the Anglo–Boer War, 1899–1902* (Oxford, 1978)

Smithies, Edward (compiler), *Men in the Air. Men and Women Who Built, Serviced, and Flew War Planes Remember the Second World War* (London, 1990)

Smuts, Robert W., 'The Negro Community and the Development of Negro Potential', *The Journal of Negro Education*, xxvi.4 (fall 1957)

Smyth, Stuart, 'Back Home', in Jan Barry and W. D. Ehrhart (eds), *Demilitarized Zones. Veterans After Vietnam* (Perkasie, Pa., 1976)

Smyth, Brigadier The Rt. Hon. Sir John, *In This Sign Conquer. The Story of the Army Chaplains* (London, 1968)

Sniping, Scouting and Patrolling (Aldershot, nd)

Sobel, Major Raymond, 'Anxiety-Depressive Reactions After Prolonged Combat Experience – The "Old Sergeant" Syndrome', *The Bulletin of the U.S. Army Medical Department*, ix, supplemental number (November 1949)

Soderbergh, Peter A., *Women Marines in the Korean War Era* (Westport, Connecticut, 1994)

Solano, Captain E. J., *The Pacifist Lie: A Book for Sailors and Soldiers* (London, 1918)

Sommers, Cecil, *Temporary Heroes* (London, 1917)

Southard, E. E., *Shell-Shock and Other Neuro-Psychiatric Problems Presented in Five Hundred and Eighty Nine Case Histories from the War Literature, 1914–1918* (Boston, 1919)

Spaight, James Molony, *Air Power and War Rights* (London, 1924)

Sparks, Blair W. and Oliver K. Niess, 'Psychiatric Screening of Combat Pilots', *U.S. Armed Forces Medical Journal*, vii.6 (June 1956)

Sparrow, Lieutenant-Colonel J. H. A. (compiler), *Morale* (London, 1949)

Spender, Harold, 'Ireland and the War', *Contemporary Review*, cx (November 1916)

Spiegel, Captain Herbert X., 'Psychiatric Observations in the Tunisian Campaign', *American Journal of Orthopsychiatry*, xiv (1944)

Spiegel, Captain Herbert X., 'Psychiatry in an Infantry Battalion in North Africa', in Lieutenant-General Hal B. Jennings (ed.), *Neuropsychiatry in World War II. Volume II. Overseas Theatres* (Washington, DC, 1973)

Spiller, Roger J., 'S. L. A. Marshall and the Ratio of Fire', *R.U.S.I. Journal*, 133 (winter 1988)

Springer, Claudia, 'Military Propaganda: Defense Department Films from World War II and Vietnam', *Cultural Critique*, 3 (spring 1986)

Sprott, Thomas Henry, *Christianity and War. Five Addresses Delivered to the Student Christian Movement Conference. Te Awamutu. January, 1916* (Wellington, 1916)

Spurling, Cuthberg, 'The Secret of the English Character', *The Contemporary Review*, cx (November 1916)

Spurr, Frederick C., *Some Chaplains in Khaki, An Account of the Work of Chaplains of the United Navy and Army Board* (London, 1916)

Stanishich, Major-General Milija, 'Command Initiative', *Canadian Army*

Journal, xiv.2 (spring 1960)

Stanley, Sandra Carson and Mady Wechsler Segal, 'Military Women in NATO: An Update', *Armed Forces and Society*, 14.4 (summer 1980)

Stapleton, Stephen, 'The Relations Between the Trenches', *The Contemporary Review*, cxi (January–June 1917)

Staub, Ervin, *The Roots of Evil. The Origins of Genocide and Other Group Violence* (Cambridge, 1989)

Staub, Ervin, 'The Psychology and Culture of Torture and Torturers', in Peter Suedfeld (ed.), *Psychology and Torture* (New York, 1990)

Stein, M. L., *Under Fire: The Story of America's War Correspondents* (New York, 1969)

Stein, Captain Martin, 'Neurosis and Group Motivation', *The Bulletin of the U. S. Army Medical Department*, vii.3 (March 1947)

Stekel, W., *Sadism and Masochism: The Psychology of Hatred and Cruelty*, vol. 1 (New York, 1953)

Stevens, Rutherford B., 'Racial Aspects of Emotional Problems of Negro Soldiers', *The American Journal of Psychiatry*, 103 (1946–7)

Stevenson, Clare and Honor Darling (eds), *The W.A.A.F. Book* (Sydney 1984)

Stewart, George and Henry B. Knight, *The Practice of Friendship* (New York, 1918)

Stokes, Anson Phelps, 'American Race Relations in War Time', *The Journal of Negro Education*, xiv (1945)

Stone, Gerald L., *War Without Heroes* (Melbourne, 1966)

Stouffer, Samuel A., et al., *The American Soldier: Adjustment During the Army Life. Volume I* (Princeton, 1949)

Stouffer, Samuel A., et al., *The American Soldier: Combat and Its Aftermath. Volume II* (Princeton, 1949)

Strachey, Alix, *The Unconscious Motives of War. A Psycho-Analytic Contribution* (London, 1957)

Strahan, Jack, 'The Medic', in J. Topham (ed.), *Vietnam Literature Anthology*, revised and enlarged (Philadelphia, 1990)

Strahan, Jack, 'It's Just that Everybody Asks', in J. Topham (ed.), *Vietnam Literature Anthology*, revised and enlarged (Philadelphia, 1990)

Strahan, Jack, Peter Hollenbeck, and R. L. Barth, *Vietnam Literature Anthology: A Balanced Perspective* (Philadelphia, 1985)

Strange, Commander Robert E. and Captain Dudley E. Brown, 'Home from the War: A Study of Psychiatric Problems in Viet Nam Returnees', *American Journal of Psychiatry*, 127.4 (October 1970)

Strayer, Richard and Lewis Ellenhorn, 'Vietnam Veterans: A Study Exploring Adjustment Patterns and Attitudes', *Journal of Social Issues*, 31.4 (1975)

Strecker, Edward A. and Kenneth E. Appel, *Psychiatry in Modern Warfare* (New York, 1945)

Streimer, Jeffrey and Christopher Tennant, 'Psychiatric Aspects of the Vietnam War. The Effect on Combatants', in Kenneth Maddock and Barry Wright (eds), *War. Australia and Vietnam* (Sydney, 1987)

Sturdevant, Charles O., 'Residuals of Combat Induced Anxiety', *The American Journal of Psychiatry*, 103 (1946–7)

Summerskill, Edith, *A Woman's World* (London, 1967)

Sutherland, Edwin H., *On Analyzing Crime*, first published in 1943 (Chicago, 1973)

Swanwick, Helen Mana Lucy, *I Have Been Young* (London, 1935)

Swanwick, Helen Mana Lucy, *Women and War* (London, 1915)

Swiers, George, '"Demented Vets" and Other Myths. The Moral Obligation of Veterans', in Harrison E. Salisbury (ed.), *Vietnam Reconsidered. Lessons from a War* (New York, 1985)

Tatar, Maria, *Lustmord. Sexual Murder in Weimar Germany* (Princeton, 1995)

Tavris, C., 'The Mismeasure of Women', in J. D. Goodchilds (ed.), *Psychological Perspectives on Human Diversity in America* (Washington, DC, 1991)

Taylor, Flight Lieutenant A. F., 'Flight to Rostock', in Squadron Leader R. Raymond and Squadron Leader David Langdon (eds), *Slipstream. A Royal Air Force Anthology* (London, 1946)

'T. B. H.', 'The Complaint of the RAMC', *The Rifle Splint. A Weekly Return of the 3rd Sub-Division Training Centre, Royal Army Medical Corps*, 12 February 1915

Teicher, Joseph D., '"Combat Fatigue" or Death Anxiety Neurosis', *The Journal of Nervous and Mental Disease*, 117 (January–June 1953)

Temple, William, *A Conditional Justification of War* (London, 1940)

Temple, William, *Thoughts in War-Time* (London, 1940)

'Temporary Chaplain', *The Padre* (London, 1916)

Tennant, Eleonora, *Spanish Journey. Personal Experiences of the Civil War* (London, 1936)

Terkel, Studs, *'The Good War'. An Oral History of World War Two* (London, 1985)

Terman, Lewis M., 'The Mental Test as a Psychological Method', *Psychological Review*, 31 (1924)

Terry, Wallace, *Bloods. An Oral History of the Vietnam War by Black Veterans* (New York, 1984)

'T. F. N.', *A Few Helpful Hints on Drill by the Adjutant, 1st B. N. V. R.* (Norwich, 1918)

Theweleit, Klaus, *Male Fantasies*, 2 vols. (Minneapolis, 1987)

Thomas, Edward, *Collected Poems* (London, 1920)

Thompson, C. J. S., *Ladies or Gentlemen? Women Who Posed as Men and Men Who Impersonated Women* (New York, 1993)

Thompson, Kenrick S., Alfred C. Clarke, and Simon Dinitz, 'Reactions of My-Lai: A Visual-Verbal Comparison', *Sociology and Social Research*, 58.2 (January 1974)

Thompson, R. W., *Men Under Fire* (London, 1946)

Thorburn, A. Douglas, *Amateur Gunners. The Adventures of an Amateur Soldier in France, Salonica and Palestine in The Royal Field Artillery* (Liverpool, 1933)

Thornton, Lieutenant-Colonel G. E. and Major H. de L. Walters, *Aids to Weapon Training. Some Ideas on Improvisation* (Aldershot, 1941)

Tiede, Tim, *Calley: Soldier or Killer?* (New York, 1971)

Tiplady, Thomas, *The Kitten in the Crater and Other Fragments from the Front* (London, 1917)

Tomedi, Rudy, *No Bugles, No Drums. An Oral History of the Korean War* (New York, 1993)

Treadwell, Mattie E., *United States Army in World War II. Special Studies. The Women's Army Corps* (Washington, DC, 1954)

Tregaskis, Richard, *Guadalcanal Diary* (New York, 1943)

Treves, Frederick, *The Tale of a Field Hospital* (London, 1901)

Trosman, Harry and I. Hyman Weiland, 'Application of Psychodynamic Principles to Psychotherapy in Military Service', *U.S. Armed Forces Medical Journal*, viii.9 (September 1957)

Trotter, Wilfred, *Instincts of the Herd in Peace and War* (London, 1916)

Tureen, Lieutenant-Colonel Louis L. and Major Martin Stein, 'The Base Section Psychiatric Hospital', *The Bulletin of the U.S. Medical Department*, ix, supplemental number (November 1949)

Turquet, P. M., 'Aggression in Nature and Society (II)', *British Journal of Medical Psychology*, xxii.3 (1949)

Tweedy, Henry Hallam, 'The Ministry and the War', in E. Hershey Sneath (ed.), *Religion and the War* (New Haven, Connecticut, 1918)

Twine, C. J., *Bayonet Battle Training. A Realistic and Practical Series of Exercises on the Use of the Training Stick and Dummy* (Aldershot, 1942)

Tytler, Lieutenant-Colonel Neil Fraser, *Field Guns in France* (London, 1922)

Uhl, Michael, *Vietnam. A Soldier's View* (Wellington, New Zealand, 1971)

Ungerson, Colonel B., *Personnel Selection* (London, 1953)

United Telecasters Sydney, 'Interview Typescript for the Television Programme "Australians at War"', nd

Universal Military Training – No! (np, 1919)

U.S. Army, *Principles of Quick Kill. Training Text 23-71-1* (Fort Benning, Georgia, May 1967)

The U.S. Army Field Manual, 'The Law of Land Warfare' (1956)

U.S. Department of War. Army Service Forces. Special Service Division, *Soldier Shows, Staging Area and Transport Entertainment Guide Comprising Blackouts, Sketches, Quizzes, Parodies, and Games* (Washington, DC, 1944)

U.S. Headquarters Army Air Force Training Division, *An Appraisal of Wartime Training of Individual Specialists in Army Air Forces* (Washington, DC, 1946)

U.S. Manual for Court Martials (1969)

Valentine, C. W., *The Human Factor in the Army. Some Applications of Psychology to Training, Selection, Morale and Discipline* (Aldershot, 1943)

Valentine, C. W., *Principles of Army Instruction With Special Reference to Elementary Weapon Training* (Aldershot, 1942)

Van Devanter, Lynda and Joan A. Furey (eds), *Visions of War, Dreams of Peace. Writings of Women in the Vietnam War* (New York, 1991)

Vann, Gerald, *Morality and War* (London, 1939)

Vansittart, Sir Robert, *Black Record: Germans Past and Present* (London, 1941)

Vaughan, Right Revd John S., 'Thoughts on the Present War', *Irish Ecclesiastical Record*, 5th series, iv (July–December 1914)

Vaughan, Dame Helen Gwynne, *Service with the Army* (London, 1941)

Verkamp, Bernard J., *The Moral Treatment of Returning Warriors in Early Medieval and Modern Times* (Scranton, 1993)

Vidich, Arthur J. and Maurice R. Stein, 'The Dissolved Identity in Military Life', in Stein, Vidich, and David Manning White (eds), *Identity and Anxiety. Survival of the Person in Mass Society* (Glencoe, Illinois, 1960)

Vietcong Atrocities and Sabotage in South Vietnam (np, 1966)

'Vietnam Veterans Against the War', *The Winter Soldier Investigation. An Inquiry into American War Crimes* (Boston, 1972)

Villies-Stuart, Captain J. P., 'The Bayonet', *The Army Review*, 1.2 (October 1911)

Von Clausewitz, Karl, *On War*, edited and translated by Michael Howard and Peter Paret (London, 1993)

Wace, Henry, *The Christian Sanction of War. An Address at the Service of Intercession for the King's Naval and Military Forces, Held in Canterbury Cathedral: August 9th. 1914* (London, 1914)

Wagner, Philip S., 'Psychiatric Activities During the Normandy Offensive, June 20–August 20, 1944', *Psychiatry*, 9.4 (November 1946)

Walker, A. T., *Field Craft for the Home Guard* (Glasgow, 1940)

Walker, Adrian, *A Barren Place. National Servicemen in Korea, 1950–1954* (London, 1994)

Walker, A. S., *Australians in the War of 1939–1945. Series 5: Medical. Vol. I: Clinical Problems of the War* (Canberra, 1952)

Walker, A. S., *Medical Services of the R.A.N. and R.A.A.F. With a Section on Women in the Army Medical Services* (Canberra, 1961)

Walker, Joe, *Parsons and War and Other Essays in War Time* (Bradford, 1917)

Walker, Keith, *A Piece of My Heart. The Stories of 26 American Women who Served in Vietnam* (Novato, CA, 1985)

Walker, Rowland, *Commando Captain* (London, 1942)

Wallace, Anthony F. C., 'Psychological Preparations for War', in Morton Fried, Marvin Harris, and Robert Murphy (eds), *War. The Anthropology of Armed Conflict and Aggression* (New York, 1968)

Wallace, Edgar, *Heroes All. Gallant Deeds of the War* (London, 1914)

Wallis, Revd Canon John, *With God's Blessing and a Green Beret. A Pilgrimage* (Poole, Dorset, 1994)

Walzer, Michael, *Just and Unjust Wars* (New York, 1977)

Wambaugh, Joseph, *The Choirboys* (New York, 1975)

Wanty, Major-General E., 'The Offensive', *Canadian Army Review*, xii.4 (October 1958)

War Office, *Army Medical Department Bulletin No. 36* (London, June 1944)

War Office, *Home Guard Training 1952* (London, 1952)

War Office, *Infantry Training. Volume I. Infantry Platoon Weapons Pamphlet No. 10. Sniping 1951* (London, 1951)

War Office, *Infantry Training. Vol. II* (London, 1926)

War Office, *Instructions for the Training of Platoons for Offensive Action 1917* (London, 1917)

War Office, *Notes for Infantry Officers on Trench Warfare* (London, 1917)

War Office, *Notes for Young Officers* (London, 1917)

War Office, *Notes from the Front. Part III* (London, 1915)

War Office, *Platoon Training* (War Office, 1919)

War Office, *Psychiatric Disorders in Battle, 1951* (London, 1951)

War Office, *Sniping* (London, 1951)

War Office, *Statistics of the British Empire During the Great War 1914–1920* (London, 1922)

War Office, *The Training and Employment of Bombers* (London, 1916)

War Office General Staff, *Bayonet Training, 1916* (London, 1916)

Waring, George L., *Chaplain's Duties and How Best to Accomplish His Work* (Washington, DC, 1912)

Watkins, Owen Spencer, *With French in France and Flanders* (London, 1915)

Watson, Bruce Allen, *When Soldiers Quit. Studies in Military Disintegration* (Westport, Connecticut, 1997)

Watson, John Broadus, *Behaviour: An Introduction to Comparative Psychology* (New York, 1914), *Behaviourism* (New York, 1925)

Watson, Peter, *War on the Mind. The Military Uses and Abuses of Psychology* (London, 1978)

Waugh, Evelyn, *The Diaries of Evelyn Waugh* (London, 1976)

Wavell, Sir A. P., *Allenby* (London, 1940)

Wavell, Field Marshal Earl, *The Good Soldier* (London, 1948)

Wavell, General Sir Archibald, *Speaking Generally. Broadcasts, Orders and Addresses in Time of War (1939–43)* (London, 1946)

Wayne, Hilary, *Two Odd Soldiers* (London, 1946)

Webb, Clement, *In Time of War. Addresses Upon Several Occasions* (Oxford, 1918)

Wecter, Dixon, *When Johnny Comes Marching Home* (Cambridge, Massachusetts, 1944)

Weeden, Craig, 'My Lie', in Jan Barry and W. D. Ehrhart (eds), *Demilitarized Zones. Veterans After Vietnam* (Perkasie, Pa., 1976)

Weinberg, S. Kirson, 'The Combat Neuroses', *The American Journal of Sociology*, li.5 (March 1946)

Weingartner, James J., 'Massacre at Biscari: Patton and an American War Crime', *The Historian*, lii.1 (November 1989)

Weingartner, James J., 'Trophies of War: U.S. Trophies and the Mutilation of Japanese War Dead, 1941–1945', *Pacific Historical Review*, lxi.1 (February 1992)

Weinstein, Major Edwin, 'The Fifth U.S. Army Neuropsychiatric Centre – "601st" ', in Lieutenant-General Hal B. Jennings (ed.), *Neuropsychiatry in World War II. Volume II. Overseas Theatres* (Washington, DC, 1973)

Weinstein, Major Edwin A. and Lieutenant-Colonel Calvin S. Drayer, 'A Dynamic Approach to the Problem of Combat-Induced Anxiety', *The Bulletin of the U.S. Army Medical Department*, ix, supplemental number (November 1949)

Weinstein, Major Edwin A. and Captain Martin H. Stein, 'Psychogenic Disorders of the Upper Gastrointestinal Tract in Combat Personnel', *War Medicine*, 8.5 (November–December 1945)

Westmoreland, W. C., 'Mental Health – An Aspect of Command', *Military Medicine* (March 1963)

Westmoreland, W. C., *A Soldier Reports* (New York, 1976)

Whateley, Dame Leslie, *As Thoughts Survive* (London, 1949)

White, Major C. B. B., 'The Study of War', *The Commonwealth Military Journal*, 1 (July 1911)

White, Right Revd Gilbert, 'The Doctrine of Non-Resistance', *The Commonwealth Military Journal*, 4 (October 1913)

White, William A., *Thoughts of a Psychiatrist on the War and After* (New York, 1919)

Whitehouse, A. G. J., *Hell in the Heavens. The Adventures of an Aerial Gunner in the Royal Flying Corps* (London, 1938)

Wilde, Oscar, *The Ballad of Reading Gaol* (London, 1896)

Wiley, Bell I., 'The Building and Training of Infantry Divisions', in Robert R. Palmer, Wiley, and William R. Keast (eds), *United States Army in World War II. The Army Ground Forces. The Procurement and Training of Ground Combat Troops* (Washington, DC, 1948)

Wilkins, Leslie T., *The Social Survey. Prediction of the Demand for Campaign Stars and Medals* (London, 1949)

Wilkinson, Alan, *The Church of England and the First World War* (London, 1978)

Wilkinson, Alan, *Dissent or Conform? War, Peace, and the English Churches 1900–1945* (London, 1986)

Wilkinson-Lathan, R. J., *British Military Bayonets from 1700 to 1945* (London, 1967)

Wilkinson, Spenser, *First Lessons in War* (London, 1914)

Willenson, Kim, *The Bad War. An Oral History of the Vietnam War* (New York, 1987)

Williams, A. Hyatt, 'A Psychiatric Study of Indian Soldiers in the Arakan', *British Journal of Medical Psychology*, xxviii.3 (1950)

Williams, Charles H., *Sidelights on Negro Soldiers* (Boston, 1923)

Williams, Cindy Cook, 'The Mental Foxhole: The Vietnam Veteran's Search for Meaning', *American Journal of Orthopsychiatry*, 53.1 (January 1983)

Williams, Robin M., 'Some Observations on Sociological Research in Government During World War II', *American Sociological Review*, 11.5 (October 1946)

Williams, Wade, *Infantry Attack* (Sydney, 1955)

Williamson, Robert, 'Women Mobilise', leaflet (np nd)

Wilmer, Harry A., 'The Healing Nightmare: A Study of the War Dreams of Vietnam Combat Veterans', in Reese Williams (ed.), *Unwinding the Vietnam War. From War into Peace* (Seattle, 1987)

Wilson, Barbara M. (ed.), *Ontario and the First World War 1914–1918. A Collection of Documents* (Toronto, 1977)

Wilson, Desmond P. and Jessie Horack, 'Military Experience as a Determinant of Veterans' Attitudes', in *Studies Prepared for the President's Commission on an All-Volunteer Armed Force* (Washington, DC, 1970)

Wilson, Revd J. M., 'Christ's Sanction as Well as Condemnation of War', *Hibbert Journal*, xiii.4 (July 1915)

Wilson, N. A. B., 'Application of Psychology in the Defence Departments', in C. A. Mace and P. E. Vernon (eds), *Current Trends in British Psychology* (London, 1953)

Wilson, Ruth Danenhower, *Jim Crow Joins Up. A Study of Negroes in the Armed Forces of the United States* (New York, 1945)

Winnifrith, Douglas P., *The Church in the Fighting Line* (London, 1915)

With the First Canadian Contingent (Toronto, 1915)

Winton, Edw., 'The Clergy and Military Service', *The Contemporary Review*, cix (January–June 1916)

Wintringham, Tom, *The Home Guard Can Fight. A Summary of Lectures Given at the Osterley Park Training School for Home Guard* (London, 1941)

Wittkower, Eric and J. P. Spillane, 'A Survey of the Literature of Neuroses in War', in Emanuel Miller (ed.), *The Neuroses in War* (London, 1940)

Wolff, Tobias, *In Pharaoh's Army. Memories of a Lost War* (London, 1994)

Wolfle, Dael, 'Military Training and the Useful Parts of Learning Theory', *Journal of Consulting Psychology*, x (1946)

Wood, Lieutenant-Colonel H. F., 'The Case of the Fainting Soldier', *Canadian Army Journal*, xii.4 (October 1958)

Woolf, Virginia, *A Room of One's Own*, first published in 1929 (Harmondsworth, 1945)

Woolf, Virginia, *Three Guineas* (London, 1938)

Wright, Ronald Selby, *The Padre Presents. Discussions About Life in the Forces* (Edinburgh, 1944)

'XYZ', *A General's Letters to his Son on Minor Tactics* (London, 1918)

Yehuda, Rachel, Steven M. Southwick, and Earl L. Giller, 'Exposure to Atrocities and Severity of Chronic Posttraumatic Stress Disorder in Vietnam Combat Veterans', *American Journal of Psychiatry*, 149.3 (March 1992)

Young, Lieutenant-Colonel James, *With the 52nd (Lowland) Division in Three Continents* (Edinburgh, 1920)

Yorkes, R. M. and C. S. Yoakum, *Army Mental Tests* (New York, 1920)

Zahn, Gordon C., *Chaplains in the R.A.F. A Study in Role Tension* (Manchester, 1969)

Zeligs, Lieutenant-Commander Meyer A., 'War Neurosis. Psychiatric

Experiences and Management on a Pacific Island', *War Medicine*, 6.3 (September 1944)

Zilboorg, Gregory, 'Fear of Death', *Psychoanalytic Quarterly*, 12 (1943)

Zimbardo, Philip, Craig Haney, and Curtis Banks, 'Interpersonal Dynamics in a Simulated Prison', *International Journal of Criminology and Penology*, I (1973)

INDEX

INDEX